Multi-UAV Networks

Multi-UAV Networks

Editors

Zhihong Liu
Shihao Yan
Yirui Cong
Kehao Wang

Basel • Beijing • Wuhan • Barcelona • Belgrade • Novi Sad • Cluj • Manchester

Editors

Zhihong Liu
College of Intelligence and
Technology, National
University of Defense
Technology
Changsha, China

Shihao Yan
School of Science, Edith
Cowan University
Perth, Australia

Yirui Cong
College of Intelligence and
Technology, National
University of Defense
Technology
Changsha, China

Kehao Wang
School of Information
Engineering, Wuhan
University of Technology
Wuhan, China

Editorial Office
MDPI
St. Alban-Anlage 66
4052 Basel, Switzerland

This is a reprint of articles from the Special Issue published online in the open access journal *Drones* (ISSN 2504-446X) (available at: https://www.mdpi.com/journal/drones/special_issues/32K8N7DLNM).

For citation purposes, cite each article independently as indicated on the article page online and as indicated below:

Lastname, A.A.; Lastname, B.B. Article Title. *Journal Name* **Year**, *Volume Number*, Page Range.

ISBN 978-3-03928-605-8 (Hbk)
ISBN 978-3-03928-606-5 (PDF)
doi.org/10.3390/books978-3-03928-606-5

Contents

About the Editors

Zhihong Liu

Zhihong Liu received a Ph.D. degree in computer science from the National University of Defense Technology (NUDT), in 2016. He was a visiting student with the David R. Cheriton School of Computer Science, University of Waterloo, Waterloo, ON, Canada, from 2013 to 2015. He is currently an Associate Professor with the College of Intelligence Science and Technology, NUDT. He has authored or co-authored more than 50 publications in peer-reviewed journals and international conferences. His research interests include learning-based robotic control, UAV swarming, and reinforcement learning. He is a regular reviewer for several prominent journals and conferences.

Shihao Yan

Shihao Yan received a Ph.D. degree in Electrical Engineering from the University of New South Wales (UNSW), Sydney, Australia, in 2015. He received a B.S. in Communication Engineering and an M.S. in Communication and Information Systems from Shandong University, Jinan, China, in 2009 and 2012, respectively. He was a Postdoctoral Research Fellow in the Australian National University, a University Research Fellow in Macquarie University, and a Senior Research Associate in the School of Electrical Engineering and Telecommunications, UNSW, Sydeny, Australia. He is currently a Senior Lecturer in the School of Science, Edith Cowan University (ECU), Perth, Australia. He is also the Theme Lead of Emerging Technologies for Cybersecurity in the Security Research Institute (SRI) at ECU. He was a Technical Co-Chair and Panel Member of a number of IEEE conferences and workshops, including the IEEE GlobeCOM 2018 Workshop on Trusted Communications with Physical Layer Security and IEEE VTC 2017 Spring Workshop on Positioning Solutions for Cooperative ITS. He was also awarded the Endeavour Research Fellowship by the Department of Education, Australia. His current research interests are in the areas of signal processing for wireless communication security and privacy, including covert communications, covert sensing, location spoofing detection, physical layer security, IRS-aided wireless communications, and UAV-aided communications.

Yirui Cong

Yirui Cong is an associate professor with the National University of Defense Technology, Changsha, China. He received a Ph.D. degree from the Australian National University in 2018. His research interests include distributed control and filtering theory, multi-UAV cooperative localization, set-membership filtering theory and applications, and networked control under communication constraints. He has published over 20 papers in top journals such as IEEE Transactions on Automatic Control and IEEE Transactions on Wireless Communications.

Kehao Wang

Kehao Wang is a professor with Wuhan University of Technology, Wuhan, China. He received a Ph.D degree from the Department of Computer Science, the University of Paris-Sud XI, Orsay, France, in 2012. From February 2013 to August 2013, he was a postdoc with the HongKong Polytechnic University. From December 2015 to December 2017, he was a visiting scholar in the Laboratory for Information and Decision Systems, Massachusetts Institute of Technology, Cambridge, MA. USA. His research interests include stochastic optimization, operation research, scheduling, wireless network communications, and embedded operating systems. He has published over 80 papers in top journals such as *IEEE Transactions on Signal Processing* and *IEEE Transactions on Communications*.

Article

Object Detection in Drone Video with Temporal Attention Gated Recurrent Unit Based on Transformer

Zihao Zhou [1], Xianguo Yu [2,*] and Xiangcheng Chen [3]

[1] School of Automation, Wuhan University of Technology, Wuhan 430070, China; stars_zihao@whut.edu.cn

[2] College of Intelligence Science and Technology, National University of Defense Technology, Changsha 410073, China

[3] School of Artificial Intelligence, Anhui University, Hefei 230039, China; chenxgcg@ustc.edu

* Correspondence: yuxianguo11@nudt.edu.cn; Tel.: +86-1887-4883-361

Abstract: Unmanned aerial vehicle (UAV) based object detection plays a pivotal role in civil and military fields. Unfortunately, the problem is more challenging than general visual object detection due to the significant appearance deterioration in images captured by drones. Considering that video contains more abundant visual features and motion information, a better idea for UAV based image object detection is to enhance target appearance in reference frame by aggregating the features in neighboring frames. However, simple feature aggregation methods will frequently introduce the interference of background into targets. To solve this problem, we proposed a more effective module, termed Temporal Attention Gated Recurrent Unit (TA-GRU), to extract effective temporal information based on recurrent neural networks and transformers. TA-GRU works as an add-on module to bring existing static object detectors to high performance video object detectors, with negligible extra computational cost. To validate the efficacy of our module, we selected YOLOv7 as baseline and carried out comprehensive experiments on the VisDrone2019-VID dataset. Our TA-GRU empowered YOLOv7 to not only boost the detection accuracy by 5.86% in the mean average precision (mAP) on the challenging VisDrone dataset, but also to reach a running speed of 24 frames per second (fps).

Keywords: drone video object detection; deformable transformer; recurrent neural network; feature aggregation

Citation: Zhou, Z.; Yu, X.; Chen, X. Object Detection in Drone Video with Temporal Attention Gated Recurrent Unit Based on Transformer. *Drones* **2023**, *7*, 466. https://doi.org/10.3390/drones7070466

Academic Editor: Dimitrios Zarpalas

Received: 30 May 2023
Revised: 4 July 2023
Accepted: 10 July 2023
Published: 12 July 2023

1. Introduction

Recently, computer vision researchers are becoming more and more interested in the field of video object detection (VOD). Images obtained by moving platforms often suffer from appearance deterioration due to motion blur, partial occlusion, and rare poses, especially by unmanned aerial vehicles (UAV). These issues have hindered advanced image-based object detectors from reaching a higher standard for challenging real-world scenarios such as drone vision.

Previous VOD methods [1–4] attempted to leverage the rich temporal and motion context in videos. Some methods [4,5] utilize the motion information extracted by an extra optical flow net to guide feature fusion. However, it is difficult to obtain accurate flow features for videos. In contrast, some methods [6,7] attempt to exploit the video context by long short-term memory networks (LSTM). LSTM combines temporal features from different video frames with a forgetting gate and an update gate. Nevertheless, when it comes to UAV images, LSTM-based VOD methods are proven to introduce a significant amount of noise into targets due to the rapid changes in appearance and the small size of objects in drone footage. Other approaches [1,8] leverage deformable convolution to estimate object motion and utilize the displacement to align features in multiple frames. Recently, the transformer model has been utilized to learn video context features for object detection [3], and this method obtains state-of-the-art results on the ImageNet-VID dataset.

Deformable Detection Transformer (Deformable DETR) is employed in this method to boost object detection performance on drone videos.

Our philosophy is to acquire rich, high accuracy while maintaining a lower computation burden. We made some tune-ups according to previous works on video object detection. We chose Gated Recurrent Unit (GRU), Deformable Detection Transformer (Deformable DETR), DeformAlign module, and temporal attention and fusion module to compose our temporal processing module. The proposed method captures temporal context from known video frames to enhance target features in the current frame. To achieve this, we employ the GRU to fuse features of different images. Notably, we depart from the commonly used Convolutional Gated Recurrent Unit (Conv-GRU) and instead use a Deformable Detection Transformer (Deformable DETR) in place of convolution. This modification enables our network to better focus on areas relevant to the targets being detected. Additionally, drone videos often suffer from significant degradation in visual quality; we employ deformable convolution to effectively learn the deviations between the features of reference frame and temporal features. This process enables us to accurately align the temporal features with the reference frame features by taking into account the offsets. Notably, the frames that are most relevant to a given reference frame are probably its immediate neighbors. To reflect this, neighboring frames are always assigned with higher weights than frames far from the reference frame in the proposed temporal attention module. We then use a weighted fusion method to combine the aligned temporal features with the reference frame features, resulting in a set of fused features that are subsequently fed into the detection network to generate detection results for the reference frame.

The main contributions are summarized as follows:

- We proposed an effective and efficient TA-GRU module to model the temporal context between videos. It is very effective to handle appearance deterioration in drone vision, and it can easily be used to promote existing static image object detectors to effective video object detection methods.
- We proposed a new state-of-the-art video object detection method, which not only achieves top performance on the VisDrone2019-VID dataset, but also runs in real-time.
- Compared to previous works, we integrated recurrent neural networks, transformer layers, and feature alignment and fusion modules to create a more effective module for handling temporal features in drone videos.

2. Related Work

Image-based Object Detection: Image-based detectors can be categorized broadly into two groups: two-stage detectors and one-stage detectors. Two-stage detectors first generate region proposals and then refine and classify them. Some representative methods in this category include R-CNN [9], SSD [10], RetinaNet [11], Fast RCNN [12], and Faster R-CNN [13]. While two-stage detectors tend to be more accurate, they are also slower. On the other hand, one-stage detectors are usually faster but less accurate, as they directly predict the region proposals based on the feature map. Relevant research in the field of object detection includes various iterations of the YOLO series, such as YOLOv5 [5], YOLOX [14], and YOLOv7 [6]. In our work, we utilized YOLOv7 as the base detector and extended its capabilities for video object detection.

Video Object Detection: Compared to image object detection, video object detection provides more comprehensive information about targets, including motion and richer appearance details. In recent years, researchers have tried to utilize neighboring frame features to enhance reference frame features. However, the presence of varying offsets in each frame poses a significant challenge to effectively utilizing these features. Previous studies attempted to address this issue by aligning the neighboring frame features with the reference frame features. Alternatively, some methods choose to overlook the offsets in each frame and instead use specialized modules to extract temporal information from videos.

Feature Aggregation: To address the issue of significant degradation in the visual quality of drone videos, various previous methods focus on feature aggregation. This technique involves enhancing the reference features by combining the features of adjacent frames. For instance, FGFA [4] and THP [15] utilize the optical flow produced by FlowNet [14] to model motion relations and align various frames. Alternatively, the optical-flow-based framework [5] categorizes video images based on the background, acquires the optical flow of the input sequence using FlowNet [14], and eventually aggregates the optical flow to model motion relations. Nevertheless, flow-warping-based techniques have some drawbacks. Firstly, drone videos frequently comprise numerous small objects, which make it challenging to accurately extract optical flow. Secondly, it is important to note that obtaining optical flow demands a considerable amount of computational resources, which can make real-time detection a challenging task. In contrast, some other approaches employ deformable convolution to compute the offsets in different frames. This method allows for the adaptive adjustment of convolutional kernel parameters to obtain corresponding offsets. For instance, STSN [8] utilizes stacked 6-layer deformable convolutional layers to gradually aggregate the temporal contexts. TCE-Net [1] takes into account that the contribution of neighboring frames to the reference frame may differ. To align frames, it uses a single deformable convolutional layer and a temporal attention module, which assigns weights to frames based on their respective contributions. However, the task of drone video object detection presents significant challenges, and relying solely on a single deformable convolutional layer can make it difficult to accurately compute the offset between neighboring frames and the reference frame. To avoid introducing excessive computation, simply increasing the number of deformable convolutional layers is not the ideal solution. Our approach, however, is to utilize the GRU module in our TA-GRU to transfer temporal features and incorporate a temporal context enhanced aggregation module to obtain the fusion features that are then fed to the detection network. This method allows us to avoid the need for aligning every neighboring frame with a reference frame and instead adopt a frame-by-frame alignment strategy, which not only reduces computation but also enhances alignment accuracy.

Some recent studies have utilized recurrent neural networks, such as long short-term memory networks (LSTM), to propagate temporal features that contain previous video features. STMN [6] and Association LSTM [7] attempt to model object association between different frames by applying LSTM or its variants. However, the object association modeled by these methods is often imprecise, particularly in drone videos. On the other hand, Conv-GRU utilizes convolution to replace linear calculation, which introduces significant challenges to the GRU module originally used for calculating sequences. TPN [16] adopts a unique method of object tracking which differs from general video object detections. The proposed approach involves linking multiple frames of the same object to generate a segment of tube, which is then fed into an ED-LSTM network to capture temporal context. However, this method introduces significant background noise that can compromise the accuracy of the results. To address this issue, recent research has explored the use of transformers for video object detection. TransVOD [3] demonstrated that incorporating self-attention and cross-attention modules can improve the model's focus on the target regions. Building on this work, our TA-GRU method aggregates temporal features and applies deformable attention instead of convolution to enhance performance. We elaborate on the details of TA-GRU in Section 3.

3. Proposed Method

To enable both high accuracy and high efficiency for UAV based image object detection, we proposed a new, highly effective video object detection framework termed TA-GRU YOLOv7. Particularly, we designed four effective modules including, the Temporal Attention Gated Recurrent Unit (TA-GRU) to enhance attention to target features in the current frame and improve the accuracy of motion information extraction between frames; the Temporal Deformable Transformer Layer (TDTL) to reduce additional computational overhead and strengthen the target features; a new deformable alignment module (DeformAlign) to extract motion information and align features from two frames; as well as a temporal attention based fusion module (TA-Fusion) to integrate useful information from temporal features into the current frame feature.

3.1. Overview of TA-GRU YOLOv7

YOLOv7 is one of the most popular image object detectors, at present, due to its great balance between speed and accuracy. Compared with the previous YOLO structures, the backbone of YOLOv7 has a more intensive hop connection structure, which makes it possible to extract richer and more diverse features from the input image. At the same time, it uses an innovative downsampling structure that could reduce the number of parameters while maintaining high accuracy, making it highly efficient and effective. It uses max pooling and features with a step size of 2×2 for parallel extraction and compression. As a mature and representative image object detector, YOLOv7 has already reached a performance bottleneck. Therefore, we selected it as the baseline for our study on how to effectively enhance the performance of current single frame image detection algorithms in the context of video object detection problems.

The architecture of TA-GRU YOLOv7 is illustrated in Figure 1, which takes multiple frames of a video clip as inputs and generates detection results for each frame as outputs. In order to make sure that our TA-GRU module can be handily applied to various single frame image detectors, we retained all structures of YOLOv7 and only added our TA-GRU module at the neck of YOLOv7, which serves to confirm the efficacy of our module. Our TA-GRU module contains four main components: Temporal Attention Gated Recurrent Unit (TA-GRU) to propagate temporal features, Temporal Deformable Transformer Layer (TDTL) to increase the attention on target regions, DeformAlign to model object motion and align the features from frame-to-frame, and temporal attention and temporal fusion module (TA-Fusion) to aggregate features from videos.

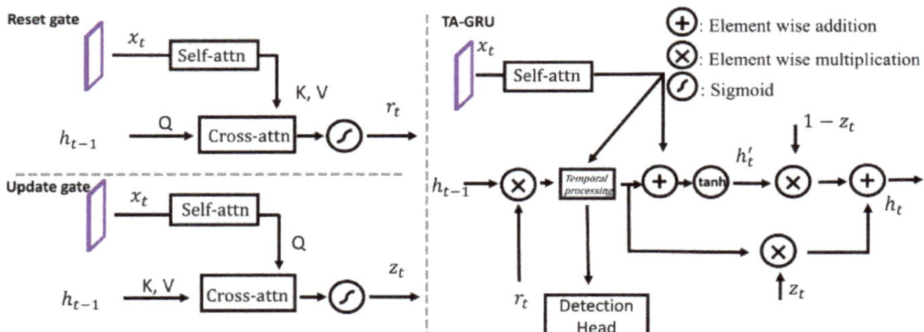

Figure 1. Architecture of TA-GRU. In TA-GRU, input features x_t interact with temporal features h_{t-1} through a temporal processing module (temporal alignment and fusion) to obtain enhanced features to feed to the detection head. Additionally, temporal features h_{t-1} will be updated by update gate features z_t, where $self\_attn$ is self-deformable transformer layer, $cross\_attn$ is cross-deformable transformer layer.

Analysis of Model Complexity. Here, we analyzed the model complexity of our proposed modules to the existing object detectors. These methods have two main computational loads: 1. feature extraction network from the backbone $C_{backbone}$; 2. detection head C_{head}. Therefore, the total computational complexity is $O(C_{backbone} + C_{head})$.

In our proposed models, we introduced a simple but effective module $C_{temporal}$ to extract the temporal information in drone videos. Therefore, during the training process, the computational complexity of our model is defined as $O\left(C_{backbone} + C_{head} + C_{temporal}\right)$. We only increased the computational overhead required for the temporal processing module. Adding our module only increases the parameters of the model from 37,245,102 parameters to 46,451,727 parameters. However, it surpassed the baseline in terms of detection accuracy by a significant margin.

3.2. Model Design

Convolutional Gated Recurrent Unit (Conv-GRU). Gated Recurrent Unit (GRU) neural network is a recurrent neural network, a variant of LSTM. Based on LSTM neural network, the cell structure is optimized to reduce parameters and accelerate training speed. The overall Convolutional Gated Recurrent Unit (Convolutional GRU) architecture is shown in Figure 2. There, x_t is the feature extracted by backbone, which uses convolution to compute the update gate and reset gate to renew the temporal feature h_{t-1}. However, this temporal feature contains a lot of background from previous frames due to the complexity of drone images. To solve this problem, we aimed to incorporate additional modules into the Conv-GRU architecture to perform deeper temporal feature processing and improve its attention towards the target region. Additionally, inspired by TransVOD [3], we explored the effectiveness of deformable transformer layers in drone video object detection tasks. Based on our experiments, incorporating deformable transformer layers prior to subsequent temporal feature processing enables our network to effectively concentrate on the target area and, to a certain extent, mitigates the impact of background information on the temporal feature processing stage.

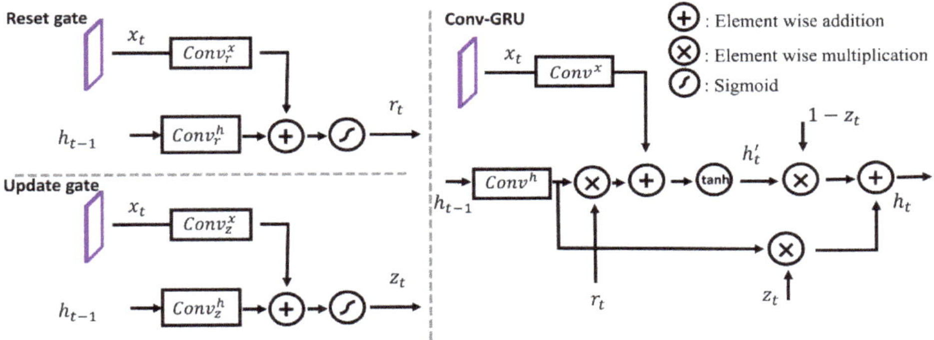

Figure 2. Architecture of Conv-GRU. It is constituted by the reset gate, update gate, and main body of Conv-GRU.

The calculation formula is shown in Equation (1):

$$\begin{aligned}
z_t &= \sigma(W_{xz} * x_t + W_{hz} * h_{t-1}) \\
r_t &= \sigma(W_{xr} * x_t + W_{hr} * h_{t-1}) \\
h'_t &= tanh(W_x * x_t + r_t \circ (W_h * h_{t-1})) \\
h_t &= (1 - z_t) \circ h'_t + z_t \circ h_{t-1}
\end{aligned} \tag{1}$$

where σ is mean Sigmoid activation function, tanh is tanh activation function, $\circ$ is element-wise multiplication, $*$ is convolution, x_t is input features extracted by backbone, and W_{xz}, W_{hz}, W_{xr}, W_{hr}, W_x, W_h are the 2D convolutional kernels whose parameters are optimized end-to-end.

Temporal Attention Gated Recurrent Unit (TA-GRU). Different from the original Conv-GRU, we modified it to make it extend to drone video object detections. The overall Temporal Attention Gated Recurrent Unit (TA-GRU) architecture is shown in Figure 1. We used it to propagate temporal features to more effectively retain temporal information; we chose deformable transformer layer to replace the original convolutional layer and added the temporary processing module (temporal alignment and fusion) to aggregate input and temporal features. The deformable transformer layer can enable the model to focus more effectively on target areas, and it is better at handling temporal inputs than traditional convolutional layers, resulting in improved performance compared to traditional convolutional layers. In TA-GRU module, the temporal features are propagated frame-by-frame between inputs to improve each frame appearance features. The final outputs will be batch inputs. The specific formula is shown in Equation (2):

$$
\begin{aligned}
Z_t &= \sigma(cross\_attn(W_{hz}, self\_attn(W_{xz}, x_t), H_{t-1})) \\
R_t &= \sigma(cross\_attn(W_{hr}, self\_attn(W_{xr}, x_t), H_{t-1})) \\
H'_t &= tanh((self\_attn(W_x, x_t) + Tem\_prc(R_t \circ H_{t-1})) \\
H_t &= (1 - Z_t) \circ H'_t + Z_t \circ H'_t
\end{aligned}
\tag{2}
$$

where σ is mean Sigmoid activation function, tanh is tanh activation function, $\circ$ is element-wise multiplication, $self\_attn$ is self-deformable attention, $cross\_attn$ is cross-deformable attention, Tem_prc is the mean after temporal processing on temporal features, x_t is input features extracted by backbone, and W_{xz}, W_{hz}, W_{xr}, W_{hr}, W_x are the weight matrix of deformable attention whose parameters are optimized end-to-end.

Temporal Deformable Transformer Layer (TDTL). To our knowledge, there are a lot of tiny objects in drone videos, which will introduce much background. Previous work [17] has addressed this issue by adding a transformer layer at the neck of the model to enhance the features extracted from the backbone. However, the general transformer layer [18] will introduce much computation overhead. The viewpoint in DETR [19] suggests that the more relevant area to the target area is often its nearby area. Furthermore, a video object detector was built using a deformable transformer within TransVOD [3] and attained satisfactory detection outcomes. Therefore, we utilized a deformable transformer layer to build our Temporal Deformable Transformer Layer (TDTL). This module will make the model pay more attention on target areas to improve the features. As shown in Figure 3, the deformable transformer layer only assigns a small, fixed number of keys for each query. Given an input feature map $x \in R^{C \times H \times W}$, let i index a 2D reference point p_i. The deformable attention feature is calculated by Equation (3):

$$
DeformAttn(x, p_i) = \sum_{n=1}^{N} W_n \left[\sum_{k=1}^{K} A_{nik} W'_n x(p_i + \Delta p_{nik}) \right]
\tag{3}
$$

where n indexes the attention head, k indexes the sampled keys, Δp_{nik} and A_{nik} denote the sampling offset and attention weight of the kth sampling point in the nth attention head, respectively, and the scalar attention weight A_{nik} lies in range [0, 1], normalized by $\sum_{k=1}^{K} A_{nik} = 1$.

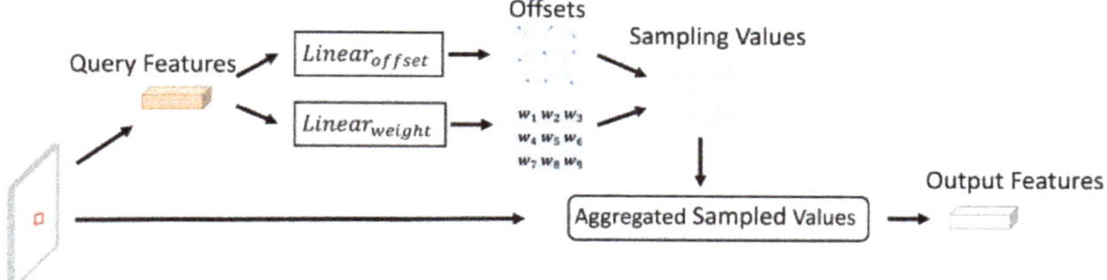

Figure 3. Architecture of Temporal Deformable Transformer Layer (TDTL). To reduce huge computation overhead on a typical transformer layer, the deformable transformer layer only attends to a small set of key sampling points around the reference.

We chose self-deformable attention to improve the attention on target areas of the input features, then used cross-deformable attention to complete the interaction with temporal features. By implementing this approach, our model becomes more adept at emphasizing the features of the current frame during the update of temporal features while also giving due consideration to the previous temporal features when determining which information should be preserved. This enhanced flexibility enables our network to focus more precisely on the specific areas of interest.

DeformAlign. We noticed that same object features are usually not spatially aligned across frames due to video motion. Without proper feature alignment before aggregation, the object detector may generate numerous false recognitions and imprecise localizations. Therefore, recent works [1,8] have utilized deformable convolution [20] to compute offset caused by movement between different frames to align different frame features. The architecture of the DeformAlign module is shown in Figure 4. Different from the deformable convolution module, we needed model motion in different frames so we used an extra convolution layer to simply fuse different frame features. Then, we used two different convolutions to compute the offsets and corresponding weights by choosing the fused features as inputs and utilized the offsets and weights to align neighboring frame features to the reference frame features. Given the prevalence of small targets in drone imagery, where the inter-frame motion of these targets may not be substantial, we found that a single layer of deformable convolution was sufficient to effectively capture their motion information.

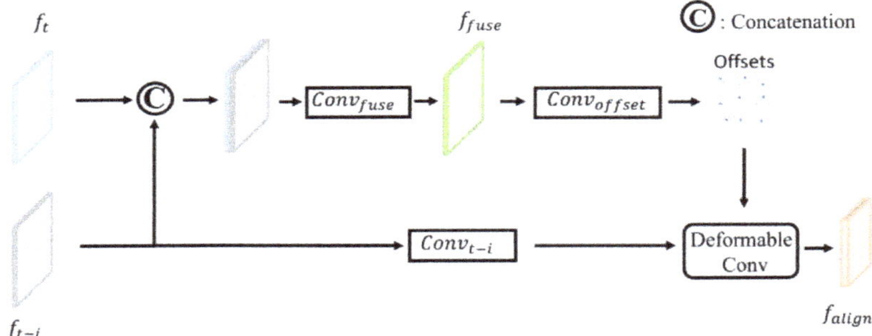

Figure 4. Architecture of DeformAlign. We used an extra convolution layer to simply fuse different features connected by channel. We used bilinear interpolation in a deformable convolution module to align the features of neighboring frames to the reference frame.

Two-dimensional convolution samples were positioned on a uniformly spaced grid R, and we used weight W to sum the sampling values. For example, when we use a convolution that $kernel\_size = 3 \times 3, stride = 1$ to compute the pixel at position p_0, we can obtain the corresponding new value on feature map y by following Equation (4):

$$y(p_0) = \sum_{i=1}^{N} W_{p_i} \cdot x(p_0 + \Delta p_i) \tag{4}$$

where $N = kernel - size$, $\Delta p_i = \{(-1, -1), (-1, 0) \cdots (1, 1)\}$, W_{p_i} is the corresponding weight at $p_0 + \Delta p_i$.

Deformable convolution introduces two additional convolutional layers to adaptively calculate offset Δp_n and weight Δw_n. We can compute the aligned pixel at p_0 by following Equation (5):

$$y_{align}(p_0) = \sum_{i=1}^{N} W_{p_i} \cdot x(p_0 + \Delta p_i + \Delta p_n) \cdot \Delta w_n \tag{5}$$

It uses bilinear interpolation to achieve the process of $p_0 + \Delta p_i + \Delta p_n$.

Temporal Attention and Temporal Fusion Module (TA-Fusion). TCE-Net [1] notices that there are different contributions to reference frame features in different frame features. The goal of temporal attention is to compute frame similarity in an embedding space to focus on 'when' it is important given neighboring frames. Intuitively, at location p, if the aligned features f_{align} are close to reference features f_t, they should be assigned higher weights. Here, dot product similarity metric is used to measure the similarity. Additionally, temporal fusion is proposed to aggregate features from neighboring frames to model temporal context. We used a $1 \times 1 \times C$ convolutional network to fuse the aligned temporal features with the features of the current frame. During the training process, the parameters of the fusion network were adaptively updated, enhancing the efficiency of feature fusion in our model and improving overall performance.

The weights of temporal attention map are estimated by Equation (6):

$$M_t(p) = \sigma(f_{align}(p) \cdot f_t(p)) \tag{6}$$

where σ is Sigmoid activation function. The architecture of the temporal attention and temporal fusion module is shown in Figure 5.

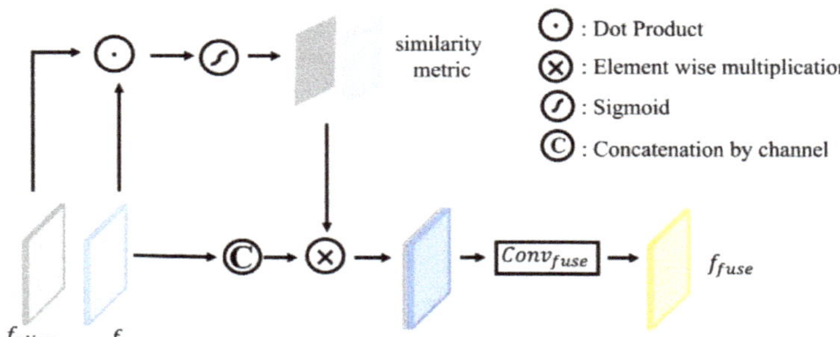

Figure 5. Architecture of Temporal Attention and Temporal Fusion module. We used dot product to measure the similarity of f_{align} and f_t. Then, we used the similarity metric to assign different weights for each frame. Finally, we chose a $1 \times 1 \times C$ convolution layer to aggregate features.

As shown in Figure 5, the temporal attention maps have the same spatial size with f_t and are then multiplied in a pixel-wise manner to the original aligned features f_{align}.

4. Experiments

4.1. Training Dataset and DETAILS

Training Dataset. We trained and tested on the VisDrone2019-VID dataset [21], which includes 288 video clips taken by the UAV platform at different angles and heights. All videos are fully annotated with object bounding box, object category, and tracking IDs. There are 10 object categories ('pedestrian', 'people', 'bicycle', 'car', 'van', 'truck', 'tricycle', 'awning-tricycle', 'bus', 'motor') consisting of 261,908 images, 24,201 for training images, 2846 for validate images, and 6635 for test images. Unlike other general video object detection datasets, there are a lot of tiny objects and severe appearance deterioration in it. Thus, we needed a video object detection method that could aggregate extensive tiny object features to solve the appearance deterioration. Mean average precision (mAP) (average of all 10 IoU thresholds, ranging from [0.5:0.95]) and AP50 were used as the evaluation metric.

Implementation Details. Our modules rely on one NVIDIA RTX3090 GPU for both training and testing. Additionally, our experiments show that the diversity of the video clips in VisDrone2019-VID is significantly lower when compared to ImageNet-VID. Hence, it was necessary for us to perform additional data processing on VisDrone2019-VID. Referring to the method in TCE-Net [1], we chose a temporal stride predictor that took the differences between features t and features k to select which frames to aggregate. This predictor takes the differences between features t and features k, i.e., $(f_t - f_k)$, as input and predicts the deviation score between frame t and frame k. The deviation score is formally defined as the motion intersection-over-union (IoU). If IoU < 0.5, the temporal stride is set to 1. If 0.5 < IoU < 0.7, the temporal stride is set to 2. Furthermore, if IoU > 0.7, the temporal stride is set to 4. Inspired by FGFA [4], we firstly used VisDrone2019-DET to pretrain our model by setting $batch\_size = 1$. We then used the pretrained model weights as the resume model to continue training on VisDrone2019-VID. Because the VisDrone2019-VID training set is a bit small, we only trained the model on VisDrone2019-VID trainset for 70 epochs, and the first 2 epochs were used for warm-up. We used an SGD optimizer for training and 5×10^{-4} as the initial learning rate with the cosine learning rate schedule. The learning rate of the last epoch decays to 0.01 of the initial learning rates. Considering the small objects in the drone image, we assigned the size of the image to 1280 pixels. The important parameters of the training process were set, as shown in Table 1.

Table 1. Training parameter setting table.

Parameters	Setup
Epochs	70
Batch Size	4
Image Size	1280×1280
Initial Learning Rate	2×10^{-4}
Final Learning Rate	2×10^{-6}
Momentum	0.937
Weight-Decay	5×10^{-4}
Image Scale	0.6
Image Flip Left-Right	0.5
Mosaic	0
Image Translation	0.2
Image Rotation	0.2
Image Perspective	2×10^{-5}

Data Analysis. Based on our past experience, it is crucial to analyze the dataset thoroughly before designing and training a model in order to construct an effective one. Upon reviewing the VisDrone2019-VID dataset, we observed the presence of numerous small objects, as well as some appearance deterioration such as part occlusion, motion blur, and video defocus. Therefore, there is an urgent need to develop a simple yet effective VOD framework that can be fully end-to-end.

In Figure 6, there are numerous objects smaller than 4 pixels. While these objects aided in training our temporal aggregated module, they should not be included in the computation of the model loss function. Typical methods for handling the ignore regions in the VisDron2019-VID dataset involve replacing them with gray squares. However, our experiments show that this approach can result in a loss of image information, particularly in UAV images, which is not conducive to training the temporal aggregated module. To prevent our model from detecting ignore regions and to retain useful training information, we chose to map the predicted bounding box back to the original images and set the intersection-over-union (IoU) to 0.7 about ground truth bounding box and ignore regions, thus excluding the ignore regions from loss calculation. This method has proven to be more effective than simply replacing the ignored regions with gray squares, resulting in a 0.72% increase in mean average precision (mAP).

Figure 6. Illustration of the 'ignore region' in the VisDrone2019-VID dataset.

4.2. Comparisons to State-of-the-Art

Table 2 shows the comparison of TA-GRU YOLOv7 with other state-of-the-art methods. With the use of temporal post-processing techniques, D&T adopts a well-designed tubelet rescore technique, while others use Seq-NMS. The results demonstrate that our TA-GRU module is effective when compared to image-based object detectors such as YOLOv7.

Table 2. Object detection results on VisDrone2019-VID.

Methods	mAP (%)	AP50 (%)	Aggregate Frames	FPS
TA-GRU YOLOv7	24.57	48.79	2	24
YOLOv7 [22]	18.71	40.26	-	45
TA-GRU YOLOX	19.41	40.59	2	-
YOLOX [23]	16.86	35.62	-	-
TA-GRU YOLOv7-tiny	0.165	0.296	2	-
YOLOv7-tiny	0.103	0.212	-	-
FGFA [4]	18.33	39.71	21	4
STSN [8]	18.52	39.87	27	-
D&T [24]	17.04	35.37	-	-
FPN [25]	16.72	39.12	-	-
CornerNet [26]	16.49	35.79	-	-
CenterNet [27]	15.75	34.53	-	-
CFE-SSDv2 [28]	21.57	44.75	-	21

Table 2 shows that TA-GRU YOLOv7 achieves a higher mean average precision (mAP) than YOLOv7, with an improvement of 5.86% mAP. Moreover, the computational overhead introduced by our method is small, which provides strong evidence for its effectiveness. Compared with FGFA (18.33% mAP), TA-GRU YOLOv7 obtains 24.57% mAP, outperforming it by 6.24%. Furthermore, TA-GRU YOLOv7 only aggregates a temporal feature and reference frame feature, while FGFA is 21. Additionally, with a deformable convolution detector and temporal post-processing, STSN obtains 18.52% mAP. However, TA-GRU YOLOv7 obtains 24.57% mAP, which is about 6.05% higher than it. The detection effect of some scenes is shown in Figure 7.

Figure 7. Examples of the detection effect.

Table 3, presented below, illustrates the detection outcomes of our model across various categories in the VisDrone2019-VID dataset. Our model has achieved outstanding detection performance across the vast majority of categories.

Table 3. The detection of our network on various categories on VisDrone2019-VID dataset.

Classification	mAP (%)	AP50 (%)	P	R
all	24.57	48.79	0.577	0.513
pedestrian	29.1	68.1	0.658	0.668
people	18.5	49.7	0.56	0.579
bicycle	30.1	65.0	0.572	0.663
car	40.1	63.6	0.737	0.636
van	26.0	43.7	0.725	0.422
truck	32.1	56.1	0.636	0.578
tricycle	16.7	37.8	0.537	0.428
awning-tricycle	12.9	25.8	0.496	0.261
bus	25.4	33.4	0.368	0.345
motor	14.3	42.3	0.481	0.547

4.3. Ablation Study and Analysis

In order to evaluate the effectiveness of our proposed methods, we conducted a series of experiments to analyze the impact of key components. This section provides a detailed analysis of our findings, including experimental results and insights into how each component contributes to the overall success of our approach.

Ablation for TA-GRU YOLOv7

Table 4 compares our TA-GRU YOLOv7 with the single-frame baseline and its variants.

Table 4. Accuracy and runtime of different methods on VisDrone2019-VID validation. The runtime contains data processing, which is measured on one NVIDIA RTX3090 GPU.

Methods	(a)	(b)	(c)	(d)	(e)	(f)
Conv-GRU?		√	√	√		
TA-GRU?					√	√
DeformAlign?			√	√	√	√
Temporal Attention and Temporal Fusion module?				√	√	√
end-to-end training?	√	√	√	√	√	
mAP (%)	18.71	16.55	20.03	23.96	24.57	23.82
AP50 (%)	40.26	37.29	41.68	47.61	48.79	47.23

Method (a) is the single-frame baseline. It has a mAP of 18.71% using YOLOv7. It outperforms the video detector, FGFA, by 0.38%. This indicates that our baseline is competitive and serves as a valid reference for evaluation.

Method (b) is a naive feature aggregation approach and a degenerated variant of TA-GRU YOLOv7, which uses Conv-GRU to aggregate temporal features. The variant is also trained end-to-end in the same way as TA-GRU YOLOv7. The mAP decreases to 16.55%, 2.16% shy of baseline (a). This indicates that using traditional feature fusion networks to directly aggregate complex drone video features can potentially introduce background interference.

Method (c) adds the DeformAlign module into (b) to align neighboring frame features to the reference frame features. It obtains a mAP of 20.03%, 1.32% higher than that of (a) and 3.48% higher than that of (b). This result suggests that when features are aligned to the same spatial position, it enhances the fusion of effective features in the fusion network. However, introducing noise remains inevitable.

Method (d) adds the temporal attention and temporal fusion module to (c). It increases the mAP score from 20.03% to 23.96%. Figure 8 shows that images with distinct appearance features are assigned varying weights depending on how similar they are to the features of the reference frame. This also effectively eliminates the impact of noise information from adjacent frames on the features of the reference frame.

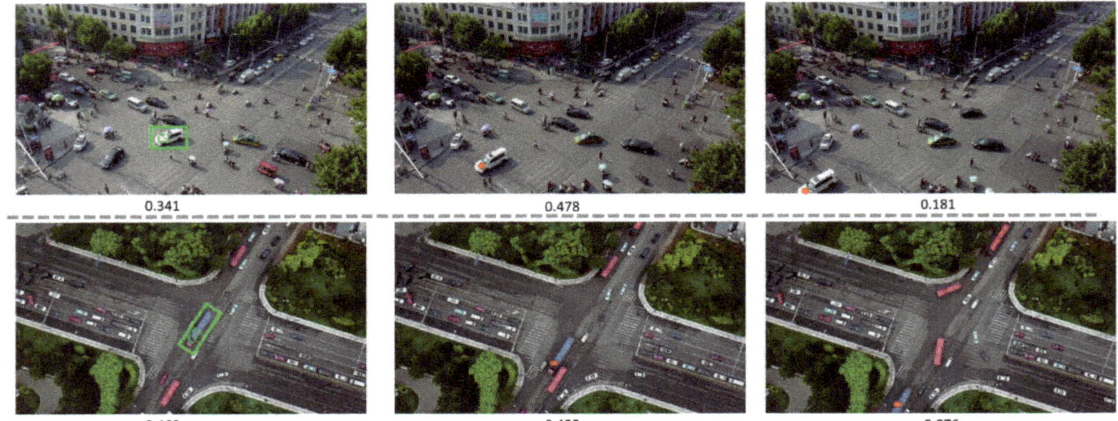

Figure 8. Images with distinct appearance features are assigned varying weights. The weight is determined by both the distance and similarity to the reference frame.

Method (e) is the proposed TA-GRU YOLOv7 method, which uses deformable attention to replace the convolution layer in (d). It increases the mAP score from 23.96% to 24.57%. It suggests that the deformable attention make model pays more attention to target areas to effectively promote the information from nearby frames in feature aggregation. The proposed TA-GRU YOLOv7 method improves the overall mAP score by 5.86% compared to the single-frame baseline (a).

Method (f) is a degenerated version of (e) without using end-to-end training. It takes the feature and the detection sub-networks from the single-frame baseline (a). During training, these modules are fixed and only the embedding temporal extracted module is learnt. It is clearly worse than (e). This indicates the importance of end-to-end training in TA-GRU YOLOv7.

5. Conclusions

This work presents an accurate, simple yet effective VOD framework in a fully end-to-end manner. Because our approach focuses on improving feature quality, it would be complementary to existing single frameworks for better accuracy in video frames. Our primary contribution is the integration of recurrent neural networks, transformer layers, and feature alignment and fusion modules. Ablation experiments show the effectiveness of our modules. Together, the proposed model not only achieves a 24.57% mAP score on VisDorne2019-VID, but also reaches a running speed of 24 frames per second (fps). However, more annotation data and precise motion estimation may be beneficial for improvements. Indeed, our module currently lacks proficiency in handling long-term motion information, and the degradation of appearance characteristics in various objects within UAV images presents a challenge for our module's ability to effectively learn temporal information. Addressing this issue is a key objective for our next stage of development.

Author Contributions: Conceptualization, Z.Z.; methodology, Z.Z.; software, Z.Z.; validation, Z.Z.; formal analysis, Z.Z. and X.Y.; investigation, Z.Z. and X.Y.; resources, X.Y.; data curation, Z.Z. and X.Y.; writing—original draft preparation, Z.Z.; writing—review and editing, Z.Z. and X.Y.; visualization, Z.Z.; supervision, X.C.; project administration, X.Y.; funding acquisition, X.Y. All authors have read and agreed to the published version of the manuscript.

Funding: This work was supported by the National Natural Science Foundation of China under Grant 61973309 and the Natural Science Foundation of Hunan Province under Grant 2021JJ20054.

Data Availability Statement: The data presented in this study are openly available in The Vision Meets Drone Object Detection in Video Challenge Results (VisDrone-VID2019) at https://github. com/VisDrone/VisDrone-Dataset, accessed on 29 May 2023.

Conflicts of Interest: The authors declare no conflict of interest. The funders had no role in the design of the study; in the collection, analyses, or interpretation of data; in the writing of the manuscript, or in the decision to publish the results.

References

1. He, F.; Gao, N.; Li, Q.; Du, S.; Zhao, X.; Huang, K. TCE-Net. In Proceedings of the AAAI Conference on Artificial Intelligence, Hilton, NY, USA, 7–12 February 2020; pp. 10941–10948. [CrossRef]
2. Shi, Y.; Wang, N.; Guo, X. YOLOV: Making Still Image Object Detectors Great at Video Object Detection. In Proceedings of the AAAI Conference on Artificial Intelligence, Washington, DC, USA, 7–14 February 2023; Volume 37, pp. 2254–2262.
3. Zhou, Q.; Li, X.; He, L.; Yang, Y.; Cheng, G.; Tong, Y.; Ma, L.; Tao, D. TransVOD: End-to-End Video Object Detection with Spatial-Temporal Transformers. *IEEE Trans. Pattern Anal. Mach. Intell.* **2022**, *45*, 7853–7869. [CrossRef] [PubMed]
4. Zhu, X.; Wang, Y.; Dai, J.; Yuan, L.; Wei, Y. Flow-Guided Feature Aggregation for Video Object Detection. In Proceedings of the 2017 IEEE International Conference on Computer Vision (ICCV), Venice, Italy, 22–29 October 2017. [CrossRef]
5. Fan, L.; Zhang, T.; Du, W. Optical-Flow-Based Framework to Boost Video Object Detection Performance with Object Enhancement. *Expert Syst. Appl.* **2020**, *170*, 114544. [CrossRef]
6. Xiao, F.; Lee, Y.J. Video Object Detection with an Aligned Spatial-Temporal Memory. In Proceedings of the Computer Vision—ECCV 2018, Munich, Germany, 8–14 September 2018; Lecture Notes in Computer Science. Springer: Cham, Switzerland, 2018; pp. 494–510. [CrossRef]
7. Lu, Y.; Lu, C.; Tang, C.-K. Online Video Object Detection Using Association LSTM. In Proceedings of the 2017 IEEE International Conference on Computer Vision (ICCV), Venice, Italy, 22–29 October 2017. [CrossRef]
8. Bertasius, G.; Torresani, L.; Shi, J. Object detection in video with spatiotemporal sampling networks. In Proceedings of the European Conference on Computer Vision (ECCV), Munich, Germany, 8–14 September 2018; pp. 331–346.
9. Girshick, R.; Donahue, J.; Darrell, T.; Malik, J. *Rich Feature Hierarchies for Accurate Object Detection and Semantic Segmentation*; Cornell University: Ithaca, NY, USA, 2013.
10. Liu, W.; Anguelov, D.; Erhan, D.; Szegedy, C.; Reed, S.; Fu, C.Y.; Berg, A.C. SSD: Single Shot MultiBox Detector. In Proceedings of the Computer Vision—ECCV 2016, Amsterdam, The Netherlands, 11–14 October 2016; Lecture Notes in Computer Science. Springer: Berlin/Heidelberg, Germany, 2016; pp. 21–37. [CrossRef]
11. Lin, T.; Goyal, P.; Girshick, R.; He, K.; Dollár, P. Focal Loss for Dense Object Detection. In Proceedings of the 2017 IEEE International Conference on Computer Vision (ICCV), Venice, Italy, 22–29 October 2017. [CrossRef]
12. Girshick, R. Fast R-CNN. *arXiv* **2015**, arXiv:1504.08083.
13. Ren, S.; He, K.; Girshick, R.; Sun, J. Faster R-CNN: Towards Real-Time Object Detection with Region Proposal Networks. *IEEE Trans. Pattern Anal. Mach. Intell.* **2016**, *39*, 1137–1149. [CrossRef] [PubMed]
14. Dosovitskiy, A.; Fischer, P.; Ilg, E.; Hausser, P.; Hazirbas, C.; Golkov, V.; Van Der Smagt, P.; Cremers, D.; Brox, T. FlowNet: Learning Optical Flow with Convolutional Networks. *arXiv* **2015**, arXiv:1504.06852.
15. Zhu, X.; Dai, J.; Yuan, L.; Wei, Y. Towards high performance video object detection. In Proceedings of the IEEE Conference on Computer Vision and Pattern Recognition, Salt Lake City, UT, USA, 18–23 June 2018; pp. 7210–7218.
16. Kang, K.; Li, H.; Xiao, T.; Ouyang, W.; Yan, J.; Liu, X.; Wang, X. Object Detection in Videos with Tubelet Proposal Networks. In Proceedings of the 2017 IEEE Conference on Computer Vision and Pattern Recognition (CVPR), Honolulu, HI, USA, 21–26 July 2017. [CrossRef]
17. Zhu, X.; Lyu, S.; Wang, X.; Zhao, Q. TPH-YOLOv5: Improved YOLOv5 Based on Transformer Prediction Head for Object Detection on Drone-Captured Scenarios. In Proceedings of the 2021 IEEE/CVF International Conference on Computer Vision Workshops (ICCVW), Montreal, BC, Canada, 11–17 October 2021. [CrossRef]
18. Liu, Z.; Lin, Y.; Cao, Y.; Hu, H.; Wei, Y.; Zhang, Z.; Lin, S.; Guo, B. Swin Transformer: Hierarchical Vision Transformer Using Shifted Windows. *arXiv* **2021**, arXiv:2103.14030.
19. Zhu, X.; Su, W.; Lu, L.; Li, B.; Wang, X.; Dai, J. Deformable DETR: Deformable Transformers for End-to-End Object Detection. *arXiv* **2020**, arXiv:2010.04159.
20. Dai, J.; Qi, H.; Xiong, Y.; Li, Y.; Zhang, G.; Hu, H.; Wei, Y. Deformable Convolutional Networks. In Proceedings of the 2017 IEEE International Conference on Computer Vision (ICCV), Venice, Italy, 22–29 October 2017. [CrossRef]
21. Zhu, P.; Du, D.; Wen, L.; Bian, X.; Ling, H.; Hu, Q.; Peng, T.; Zheng, J.; Wang, X.; Zhang, Y.; et al. VisDrone-VID2019: The Vision Meets Drone Object Detection in Video Challenge Results. In Proceedings of the 2019 IEEE/CVF International Conference on Computer Vision Workshop (ICCVW), Seoul, Republic of Korea, 27–28 October 2019. [CrossRef]
22. Wang, C.Y.; Bochkovskiy, A.; Liao, H.Y.M. YOLOv7: Trainable Bag-of-Freebies Sets New State-of-the-Art for Real-Time Object Detectors. *arXiv* **2022**, arXiv:2207.02696.
23. Ge, Z.; Liu, S.; Wang, F.; Li, Z.; Sun, J. YOLOX: Exceeding YOLO Series in 2021. *arXiv* **2021**, arXiv:2107.08430.

24. Feichtenhofer, C.; Pinz, A.; Zisserman, A. Detect to Track and Track to Detect. In Proceedings of the International Conference on Computer Vision, Venice, Italy, 22–29 October 2017; ICCV: Paris, France, 2017; pp. 3057–3065.
25. Lin, T.Y.; Dollár, P.; Girshick, R.; He, K.; Hariharan, B.; Belongie, S. Feature Pyramid Networks for Object Detection. In Proceedings of the 2017 IEEE Conference on Computer Vision and Pattern Recognition (CVPR), Honolulu, HI, USA, 21–26 July 2017; CVPR: New Orleans, LA, USA, 2017; pp. 936–944.
26. Law, H.; Deng, J. Cornernet: Detecting Objects as Paired Keypoints. In Proceedings of the 15th European Conference on Computer Vision, Munich, German, 8–14 September 2018; ECCV: Aurora, CO, USA, 2018; pp. 765–781.
27. Zhou, X.; Wang, D.; Krähenbühl, P. Objects as points. *arXiv* **2019**, arXiv:1904.07850.
28. Zhao, Q.; Sheng, T.; Wang, Y.; Ni, F.; Cai, L. Cfenet: An accurate and efficient single-shot object detector for autonomous driving. *arXiv* **2018**, arXiv:1806.09790.

Article

Intelligent Mining Road Object Detection Based on Multiscale Feature Fusion in Multi-UAV Networks

Xinkai Xu [1,2], Shuaihe Zhao [3,4], Cheng Xu [2,*], Zhuang Wang [2], Ying Zheng [1], Xu Qian [1] and Hong Bao [2]

1   School of Mechanical Electronic & Information Engineering, China University of Mining & Technology-Beijing, Beijing 100083, China
2   Beijing Key Laboratory of Information Service Engineering, Beijing Union University, Beijing 100101, China
3   The School of Automation, Beijing Institute of Technology, Beijing 100081, China
4   Aerospace Shenzhou Aerial Vehicle Ltd., Tianjin 300301, China
*   Correspondence: xucheng@buu.edu.cn

Abstract: In complex mining environments, driverless mining trucks are required to cooperate with multiple intelligent systems. They must perform obstacle avoidance based on factors such as the site road width, obstacle type, vehicle body movement state, and ground concavity-convexity. Targeting the open-pit mining area, this paper proposes an intelligent mining road object detection (IMOD) model developed using a 5G-multi-UAV and a deep learning approach. The IMOD model employs data sensors to monitor surface data in real time within a multisystem collaborative 5G network. The model transmits data to various intelligent systems and edge devices in real time, and the unmanned mining card constructs the driving area on the fly. The IMOD model utilizes a convolutional neural network to identify obstacles in front of driverless mining trucks in real time, optimizing multisystem collaborative control and driverless mining truck scheduling based on obstacle data. Multiple systems cooperate to maneuver around obstacles, including avoiding static obstacles, such as standing and lying dummies, empty oil drums, and vehicles; continuously avoiding multiple obstacles; and avoiding dynamic obstacles such as walking people and moving vehicles. For this study, we independently collected and constructed an obstacle image dataset specific to the mining area, and experimental tests and analyses reveal that the IMOD model maintains a smooth route and stable vehicle movement attitude, ensuring the safety of driverless mining trucks as well as of personnel and equipment in the mining area. The ablation and robustness experiments demonstrate that the IMOD model outperforms the unmodified YOLOv5 model, with an average improvement of approximately 9.4% across multiple performance measures. Additionally, compared with other algorithms, this model shows significant performance improvements.

Keywords: multisystem collaboration; 5G-multi-UAV systems; multiscale feature fusion; pyramid model

Citation: Xu, X.; Zhao, S.; Xu, C.; Wang, Z.; Zheng, Y.; Qian, X.; Bao, H. Intelligent Mining Road Object Detection Based on Multiscale Feature Fusion in Multi-UAV Networks. *Drones* **2023**, *7*, 250. https://doi.org/10.3390/drones7040250

Academic Editors: Zhihong Liu, Shihao Yan, Yirui Cong and Kehao Wang

Received: 28 February 2023
Revised: 31 March 2023
Accepted: 3 April 2023
Published: 5 April 2023

1. Introduction

Smart mines integrate modern information, control technology, and mining technology to achieve the goals of efficiency, safety, and environmental protection. The use of automatic driving applications in mining trucks reduces manual demand in key production links at open-pit mines while promoting efficient collaboration. This development is supported by 5G and other networking technologies that enable all-around communication between vehicles, roads, and management platforms, with the ultimate goal of optimizing mine operations [1]. The four scenarios for 5G automatic driving in mine operations are loading, transportation, unloading, and operational support. To effectively realize these processes while improving accuracy and efficiency, there are requirements of remote control driving systems for mining trucks as well as cooperation among various construction machinery through the integration of cutting-edge multisystems aimed at resolving the travel path planning issues specifically associated with mining truck movements.

Internationally, there is continuing growth in the demand for industrial-grade drones, which are widely used in numerous industries. With the high growth in demand for logistics and transport expected in the coming years, the drone transport market holds great promise but also faces risks and challenges. Unmanned aerial systems (UASs) are versatile, advanced high-tech equipment with extensive scientific, social, and strategic applications that have the potential to trigger transformative industrial and societal change. UAS equipment is widely used in a variety of scenarios, such as pesticide spraying, courier transport delivery, video filming, and aerial inspection and monitoring, resulting in greater convenience in terms of human labor.

Worldwide, the development of civil drones is still in its early stages, but countries around the globe recognize the potential of drones in both military and civilian applications, leading to the issuance of numerous policies and regulatory documents [2]. In the realm of intelligent mining, advancements have been made from truck intelligence to collaborative intelligence through the integration of vehicle-road-cloud superagents [3]. This innovative approach allows for autonomous networked intelligent integration, unlocking limitless possibilities for networked systems to overcome traditional constraints on time and space interaction within open-pit mining areas.

Currently, 5G-multi-UAV applications remain at a nascent stage within open-pit mining areas, including auxiliary driving and early warning services, among others. The next phase will involve more complex applications serving not only L4 automatic driving but also manned driving for different levels of automatic driving scenarios [4].

In Figure 1, the 5G network is shown as an integrated vehicle-road-cloud scenario. This paper proposes the IMOD model, which combines 5G-multi-UAV and deep learning methods for open-pit mining environments. The IMOD model uses data sensors to monitor surface data in real time and transmit them to driverless mining trucks. The unmanned mine truck utilizes a deep learning algorithm to identify obstacles in real time and build a driving area that avoids static, dynamic, and multiple continuous obstacles. The IMOD model ensures the safety of driverless vehicles, personnel, and equipment by ensuring the smooth running of vehicles while avoiding obstacles.

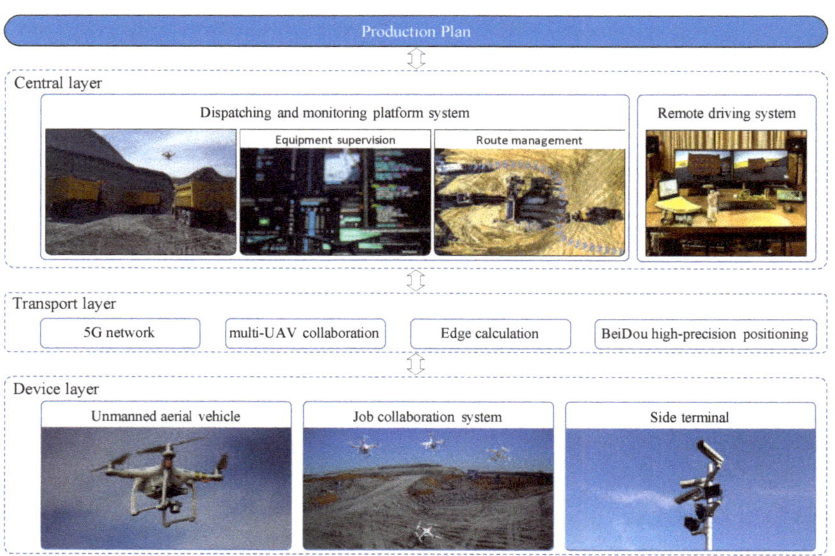

Figure 1. A 5G-multi-UAV vehicle-road-cloud integration scene in an open-air intelligent mining area.

The contributions of this paper include: (1) proposing an IMOD automatic driving model based on 5G-multi-UAV for enhancing safety in mining areas; (2) constructing an obstacle image dataset through field collection and manual marking; and (3) improving multiscale obstacle detection capabilities through cross-modal data fusion.

The second section presents related work, followed by a presentation of the IMOD autopilot model based on 5G-multi-UAV in Section 3. In Section 4, we provide experimental analysis results, followed by our conclusions in Section 5.

2. Related Work

2.1. Multisystem Collaboration Scenarios and Applications in Open-Pit Mines

Automated driving in open-pit mines continues to adhere to the standard production workflow of drilling, blasting, mining, transportation, and discharging [5]. Considering the process of mining transport operations and platooning, automatic driving scenarios in mines can be classified into three categories: loading, transporting, and unloading. Additionally, there are maintenance support scenarios, such as refueling and water replenishment, that facilitate the aforementioned operational processes. To implement intelligent networked automatic driving applications effectively, the realization of remote control driving capabilities for mining trucks is required. Moreover, there is a demand for seamless operation coordination between these trucks and other construction machinery as well as accurate planning of their travel path to ensure safe autonomous operation.

In this scenario, an unmanned mining truck travels to a loading point, where it receives payloads from excavators, shovels, and other equipment. The entire process involves communication between the mining trucks, equipment, and cloud platforms during entry, loading, and transportation to the designated destination along with providing updates such as on position/speed/direction/acceleration. Loading equipment also sends positional and directional information for efficient operations.

During abnormal conditions, triggering an emergency brake feature that initiates remote control mode provides increased safety measures by triggering alarms alerting excavators about any danger. Cloud-based systems provide troubleshooting assistance, resolving issues experienced during material transport.

For autonomous driving in mining trucks, a cloud platform is utilized for planning paths while integrating environmental information. The vehicle interacts with other vehicles/equipment/cloud platforms, ensuring safe driving via functions such as forward collision warning and over-the-horizon perception while being capable of emergency braking followed by remote takeover, if required.

The unloading process requires communication/cooperation among various pieces of equipment (bulldozers/loaders/cloud platforms). In using planned tasks/paths, assistance is provided by the surrounding environment perception resulting in real-time status/information exchange between the mining truck/unloading equipment, leading to efficient cooperation. Again, it should be capable of emergency braking followed by a remote takeover in case of any abnormal situations.

Finally, refueling/water replenishment/maintenance tasks require organizing maintenance or overhaul tasks by the cloud platform when necessary, with support task execution planning via coordination between the truck/platform detecting faults or insufficient oil/water using planned paths periodically broadcasting real-time status/task information with availability for remote takeover in case of abnormalities.

2.2. 5G-Based Multi-UAV Collaboration Technology in Mining Areas

Using complex networks such as 5G/4G/MEC and V2X, along with cloud computing, big data, and artificial intelligence technologies, we can achieve ubiquitous network connectivity between vehicles, roads, people, and cloud service platforms. This enables environmental awareness and integrated computing while allowing for decision control across end-users, road management systems, and cloud architectures. These advancements provide safer, more efficient, and intelligent solutions for automatic vehicle driving as well as traffic optimization services that promote green practices [6].

The integration of 5G and multi-UAV facilitates business operations by primarily focusing on three major points: network convergence to connect vehicles and roads to the 5G network, data fusion through MEC for processing interactive or roadside sensing data, and business integration by syncing roadside sensing results with the cloud while supporting on-demand, multicast, and roadside broadcast for both 5G and multi-UAV. This collaboration between vehicles, roads, and clouds is a crucial part of the overall business process.

To enhance infrastructure services as well as network connections, this method proposes utilizing human vehicle roads as service objects while incorporating edge clouds for digital infrastructure support. The mining area multisystem cooperation approach based on 5G-multi-UAV provides redundant information service across multiple channels along with high-speed slicing interconnection and collaborative perception for the vehicle-road-cloud integration; this involves end-to-end communication protection paired with high-precision positioning services, resulting in a unified service system including one network platform that caters to various terminals in different scenarios.

Platform: Establishing a 5G vehicle-road collaborative service platform to accelerate scene innovation through technological advancements and open standard interfaces.

Map: Integrating high-precision maps with applications to improve driving safety, offering unified basic coordinates and supporting environmental perception assistance, path planning, and decision-making with real-time updates on the edge map information.

Network: Creating a 5G-multi-UAV wireless scene library for exploring optimal access experience for various scenarios. Intelligent network connection hierarchical services cater to the differentiated needs of different IoV businesses based on demand awareness [7].

Roadside: Building multisource fusion sensing platforms across all-weather self-developed vehicle-roadside-cloud systems facilitating low delay/high-efficiency transmission of roadside sensor data from vehicles, enabling supercomputing fusion processing at MEC-edge cloud nodes via collaborative communication computing methodologies.

Application: Utilizing cloud-side collaborations via 5G + edge cloud technology providing early vehicle-road collaboration warning services to realize multi-UAV collaboration defined under IoV's scene service logo defining unicast/multicast broadcast services.

The implementation of drones as a flying platform poses difficulties due to their unique operating characteristics. Communication and safety risks are also high. However, various overhead angles for aerial photography can be achieved through drone usage. In the field of security, drones have been employed in dispatching security communications, commanding assisted patrols, and performing line planning in cases involving electricity. Drones are also widely used in consulting, resource exploration, urban planning, and other areas.

When it comes to secure communication with drones, it is important to pay attention to physical layer security technology, as conventional techniques do not guarantee secure transmission against brute-force cracking attempts [8]. Physical layer security techniques that can be adopted for UAVs include beamforming, artificial noise (AN), power allocation, and cooperative interference, among others.

2.3. Obstacle Detection for Unmanned Mine Trucks

With the remarkable advancements in GPU performance, many scholars worldwide have recently utilized deep learning methods for target detection. The design principle of convolutional neural networks is to simulate synapses of brain neurons by establishing different characteristic neuron links. With regard to target detection, these networks focus on developing different levels of receptive fields and high-dimensional feature information through multiscale feature extraction and classification via a multilevel cascade design. Region-CNN is the first algorithm that successfully applied deep learning to target detection [9]. It uses CNNs to extract features from region proposals and then performs SVM classification and bbox regression, resulting in greater accuracy than traditional methods. Its successor, fast R-CNN, utilizes shared convolution features for improved detection efficiency while also reducing computational complexity with high accuracy [10]. Another study [11] proposed an engineering vehicle detection algorithm based on faster R-CNN, which adjusts the position of ROI pooling layers and adds a convolution layer in the feature classification part, thereby enhancing model accuracy. Furthermore, ref. [12] introduces several differently sized RPNs into traditional faster R-CNN structures, allowing for larger vehicle detection, whilst [13], building upon faster R-CNN, improved object feature extraction by combining multilayered feedforwarding alongside the output of each context layer, enriching robustness against smaller or occluded targets that may confound other models. In the literature [14], there is a suggestion for improving domain adaptive fast R-CNN algorithms by refining their respective region proposal network (RPN) configuration; multiscale training helps mine difficult samples during secondary training, leading toward expanded small-target capability-albeit at a considerable computational expense.

Given the slow execution speed of the R-CNN algorithm, the one-stage detection method pioneered in [15] creatively integrated candidate frame extraction and feature classification, developing several versions [16]. The you only look once (YOLO) target detection algorithm enables higher accuracy and faster detection. Another example [17] improved YOLOv4 by adjusting the size of the detection layer for smaller objects and replacing the backbone with CSPLocknet-19, effectively achieving a good average accuracy (mAP) and frames per second (FPS) on low-cost edge hardware. In [18], an improved vehicle detection method using the YOLOv5 network under different traffic scenarios was proposed, utilizing flip-mosaic to enhance the perception of small targets, thereby increasing the accuracy of detection while reducing false positives. By adding an SSH module after YOLOv7's pafpn structure to merge context information, the small object detection ability was improved. In recent years, the one-stage target detection algorithm has gained popularity across various fields due to its strong generalization performance and fast processing.

The open-pit mining environment is complex and dynamic, necessitating the use of a target detection algorithm based on convolutional neural networks. The selected algorithm must satisfy the real-time and high-precision requirements for obstacle detection by unmanned mining trucks. After analyzing the existing algorithms, we selected the YOLOv5 network based on its suitability for detecting obstacles within an open-pit mining area [19]. Obstacle detection using only two-dimensional image methods often produces inaccurate distance information; multisensor fusion that combines stereo vision with laser radar can provide more accurate results. Currently, YOLOv5 and YOLOX algorithms demonstrate favorable obstacle detection performance. Notably, YOLOV5-s and YOLOX-s are lightweight models recommended for mobile deployment devices but still require improvement in detecting occluded and small-scale targets.

Automatic driving solutions in the field of mining primarily comprise three modules: a central control system, automatic driving trucks, and other engineering vehicle coordination kits. In cloud-based remote monitoring, the scheduling platform serves as the central control system, whereas automatic driving trucks operate with perception abilities to make intelligent decisions automatically, thereby reducing operational costs while increasing transportation efficiency through improved safety measures between vehicles such as excavator coordination kits or other vehicle support tools. With the integration of these solutions into practices such as travel path planning during onsite applications or collaborative management activities located within various sections throughout the mine site, significant improvements have been made toward safer operations for lower overall cost.

3. 5G-Multi-UAV-Based IMOD Autonomous Driving Model

The complex and variable background information present on unstructured roads within open-pit mining areas, coupled with the varying size and characteristics of obstacles as well as dependence on natural lighting conditions, resulting in frequent changes to road information, pose a challenge for obstacle detection [20]. Although such roads are typically simple pavements, frequently being traversed by heavy-duty trucks increases the likelihood of pavement damage and deformation.

The one-stage target detection algorithm YOLOv5 utilizes mosaic image enhancement and adaptive anchor frame calculation at its input. Its backbone network integrates focus and CSP structures, whereas the neck module uses FPN structures to enhance semantic information across different scales. The PAN structure fosters location awareness across these scales, thereby improving multiscale target detection performance. The lightweight YOLOv5-s and YOLOX-s target detection algorithms have strong performance in detecting targets across multiple scales and are well-suited for deployment on resource-constrained devices.

Based on the 5G-multi-UAV architecture, Figure 2 displays the IMOD collaborative system for a mining scene. Using the YOLOv5-s network structure as a basic framework, this paper presents an IMOD obstacle target detection model that adapts feature fusion to address challenges related to high-density occlusion of targets, low detection accuracy, and miss rate in detecting small-scale targets within open-pit mining areas. The specific improvements are summarized as follows:

(1) To cope with the negative impact of adjacent scale feature fusion on models, we propose utilizing a feature fusion factor and improving the calculation method. By increasing effective samples post-fusion, this approach improves learning abilities toward small and medium-sized scale targets.

(2) To enhance the detection accuracy of smaller targets in open-pit mining areas, reinforcing shallow feature layer information extraction via added shallow detection layers is crucial.

(3) Adaptively selecting appropriate receptive field features during model training can help tackle insufficient feature information extraction in scenes containing vehicles and pedestrians with significant scaling changes. Therefore, an adaptive receptive field fusion module based on the concept of an RFB [21] network structure is proposed.

(4) For efficiently detecting dense small-scale targets with high occlusion, we introduce StrongFocalLoss as a loss function while incorporating the CA attention mechanism to alter model focus toward relevant features, resulting in improved algorithmic accuracy.

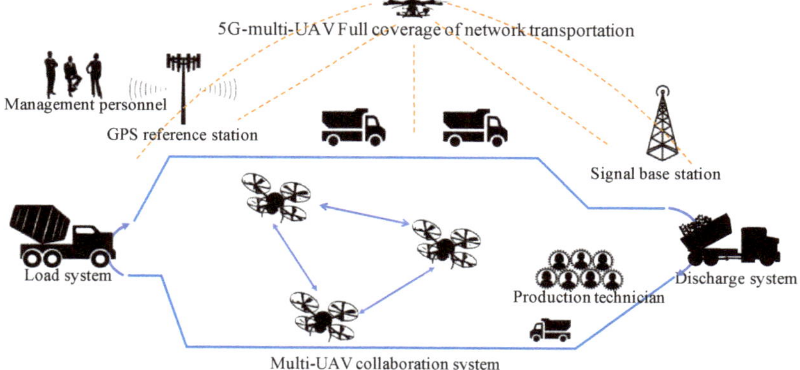

Figure 2. The IMOD mining scene collaborative system architecture based on 5G-multi-UAV.

3.1. Effective Fusion of Adjacent Scale Features

In small-scale target detection tasks involving occluding vehicles and pedestrians, the difficulty lies in extracting feature information due to their limited scale. Shallow networks have limited capacity to learn such information, whereas deep networks fail to provide sufficient support for shallow networks, thus impacting the successful detection of small targets. The neck network of the YOLOv5-s algorithm utilizes the bidirectional feature pyramids and horizontal connection structures of PAFPN for different scale feature fusion. However, some scales exhibit large response values across adjacent feature maps, leading to the identification of only one layer during network learning based on rough response value estimation, resulting in poor detection accuracy and convergence effectiveness.

The challenge in multisystem collaborative target detection stems from variations in image scale, sparse distribution of targets, a high number of targets, and small target size. Balancing the computational demand for processing high-resolution UAV images with limited computing power presents additional difficulties. To address these problems, the IMOD model uses three layers of feature maps that differ in size to detect objects at different scales. The model employs YOLOv5 as its base feature network and leverages inter-layer connections to extract more semantically informative features that facilitate effective object recognition while minimizing interference information. Selective channel expansion used by the IMOD does not excessively impact the model's parameter size and avoids unnecessary training operations, thereby maximizing detection accuracy while ensuring algorithmic speediness.

The effectiveness of the same sample can vary in characteristic maps of different scales. Deep and shallow layers contribute differently to target information at various scales, and their impact on other layers has both advantages and disadvantages. To alleviate the negative effects of feature fusion, it is necessary to adjust the participation rate of deep features in shallow feature learning by filtering out invalid samples during adjacent layer transmission. This ensures more effective samples are available for learning on deep feature maps, which improves detection performance for targets of different sizes. An improved adjacent scale feature fusion strategy is proposed here to address these challenges. The expression for the FPN's feature fusion process is as follows:

$$P_i = f_{conv}(f_{lateral}(C_i) + a_i^{i+1} * f_{upsample}(P_{i+1})). \tag{1}$$

Here, C_i and P_{i+1} signify the feature map of layer i prior to and after feature fusion, respectively. The term $f_{lateral}$ represents the one-in-FPN horizontal connection by convolution operation, whereas $f_{upsample}$ denotes an operation that increases the resolution twofold. In addition, f_{conv} indicates a convolution operation for processing features, whereas a_i^{i+1} signifies the factor for fusing features that should be multiplied when transferring layer $i + 1$ feature maps into those of layer i.

This study derived its proposed feature fusion factors through statistical analysis with calculated corresponding target numbers for each layer using this formula:

$$a_i^{i+1} = N_{P_{i+1}} / N_{P_i} \tag{2}$$

The proposed method in [22] utilizes an attention module for calculating the fusion factor, incorporating the BAM attention mechanism from [23] and enhancing the efficiency of feature fusion between adjacent scales, as illustrated in Figure 3. The feature fusion factor is computed to alleviate the negative effects during the feature fusion process, with its formula expressed as follows:

$$a_i^{i+1} = M_s(C_{i'}, P_{i+1}) \cdot M_c(C_{i'}, P_{i+1}) \tag{3}$$

Here, $C_{i'}$ represents the feature map obtained by C_i after a 1×1 convolution operation, whereas $P_{i+1'}$ denotes a feature map that was likewise processed through a twofold upsampling operation from P_{i+1}. Both M_s and M_c denote spatial and channel attention modules used within the adjacent scale feature high-efficiency fusion (AFHF) module, as depicted in Figure 3.

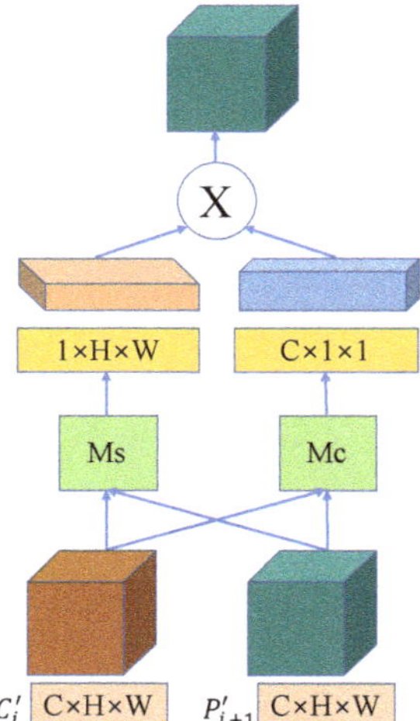

Figure 3. Adjacent scale feature high-efficiency fusion (AFHF).

The spatial attention module plays a crucial role in analyzing the differences between adjacent feature maps at different layers of transmission and filtering out invalid samples passed from deep to shallow layers. The M_s module is represented by the following formula:

$$M_s(C_i', P_{i+1}') = \sigma(f_5(Softmax'(f_1 f_3 f_3 f_1(C_i')) - Softmax'(f_1 f_3 f_3 f_1(P_{i+1}')))) \tag{4}$$

Here, σ represents the sigmoid activation function, and f_1, f_3, and f_5 specify convolution operations with varying kernel sizes and shared parameter information. Additionally, $Softmax'$ refers to an operation involving multiplication across the row and column dimensions of feature maps after a softmax operation has been applied. Overall, these techniques aid in accurately identifying invalid sample data within drone imagery datasets at various depths of analysis.

Each feature map channel contains a significant amount of information. The channel attention module (CAM) focuses on the meaningful content within the feature map, and, in conjunction with the spatial attention module (SAM), it can more effectively process features at the channel level to reduce meaningless channels for improved performance. The formula for the MC module is:

$$M_c(C_i', P_{i+1}') = \sigma(MLP(GAP(C_i'); GAP(P_{i+1}'))) \tag{5}$$

Here, "GAP" represents global average pooling operation, while MLP represents a multilayer perceptron with a hidden layer composed of fully connected layers along with ReLU activation that reduces channel dimensions to $1/r$ times their original size before expanding them back out again. The specific experiment uses r = 16 in this instance, whereas sigmoid is used as an activation function.

3.2. Multiscale Wide Field-of-View Adaptive Fusion Module

In open-pit mining environments, heavily obscured targets such as vehicles and pedestrians present challenges due to the scale of the involved objects. In such scenarios, contextual information can be utilized to improve recognition performance. The YOLOv5-s algorithm leverages SPPF spatial pyramid pooling modules to increase the perceptual field while segregating critical contextual features from multiple sources for fusion purposes. However, this approach may hinder feature extraction accuracy during target detection by interfering with the extraction of significant key features, resulting in inadequate information capture.

Therefore, this paper proposes using an adapted RFB-s network structure, which effectively increases the receptive field area for adaptive fusion whilst addressing shortcomings experienced with previous approaches. Figure 4 illustrates the improved methodology employed in our study.

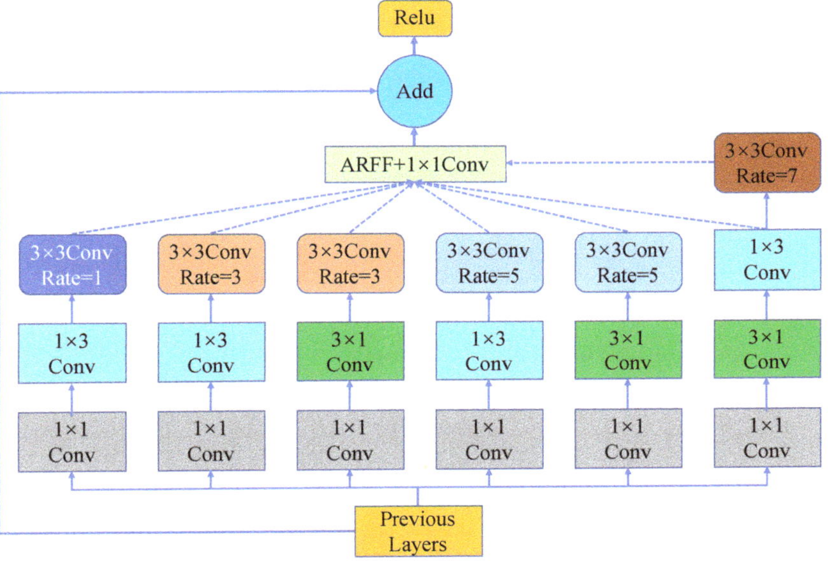

Figure 4. SRFB-s module of IMOD.

The proposed RFB-s module employs several techniques to optimize the structural design. First, the input feature map is subjected to a 1×1 convolution operation, which reduces both channel count and computational load. Asymmetric convolutional layers are then used to further reduce parameter size before applying 3×3 dilated convolutions that expand the feature perceptual field across three rates (1, 3, and 5). Each such branch undergoes stitching, followed by another round of 1×1 convolution fusion operations that yield a final output fulfilling the fusion requirements of each stage. Critically, our approach incorporates shortcut regularization [24], which not only accelerates training but also reconciles issues around exploding or vanishing gradient flow via the merging of multiscale perceptual features with their original counterparts.

To further enhance the receptive field, the improved module for object detection in drone imagery, referred to as SRFB-s (strong receptive field block), adopts the overall structure of multi-branch null convolution. It replaces 3×3 convolutions with more efficient 1×3 and 3×1 asymmetric convolutions to reduce parameter count and computational effort. The module also includes additional perceptual field branches to provide a wider range of features, including contextual information. Additionally, it utilizes the ASFF network [25] to adaptively fuse feature maps and selects optimal fusion methods based on scale targets during training to prevent irrelevant background noise degradation while enhancing detection capabilities under occlusion, large-scale changes, etc.

3.3. Attention Mechanism and Loss Function Optimization

In object detection in open-pit mining scenes, challenges concerning considerable scale variations together with occlusions are encountered. During feature extraction, the model integrates extensive amounts of invalid feature information stemming from background clutter as well as undetected targets, which negatively affect valid target information extraction. To mitigate such concerns, researchers worldwide utilize attention mechanisms that highlight crucial features while ignoring irrelevant data, thus improving overall model performance. The coordinate attention (CA) module, introduced in [26], filters out invalid details, instead emphasizing relevant ones by incorporating novel encoding methods along two spatial directions, integrating coordinate information into generated attention maps for lightweight networks.

The role of the loss function is essential to enhance object detection and localization in open-pit mining scene models. The loss function comprises three critical components: localization loss, confidence loss, and classification loss. The formula is described as follows:

$$Loss = Localization_{Loss} + Confidence_{Loss} + Classification_{Loss} \tag{6}$$

The localization loss function shows varying effectiveness depending on different detection scenarios. To enhance open-pit scenario detection of occluded targets, small-scale targets, and medium-sized targets, this model employs StrongFocalLoss for calculating target frame localization loss; it applies the FocalLoss function for confidence loss in contrast to the conventionally employed cross-entropy loss function BCEclsloss. The experiment uses the StrongFocalLoss loss function instead of cross-entropy for classification loss, resulting in more effective object recognition.

$$SFL(\sigma) = -|y - \sigma|^{\alpha}((1 - y)\lg(1 - \sigma) + y\lg(1 - \sigma)) \tag{7}$$

where σ represents the prediction, and $|y - \sigma|^{\alpha}(\alpha \geq 0)$, with scaling factor α, is a quality label ranging from 0 to 1 that denotes the absolute value of distance. This hyperparameter controls the downscaling rate, which can be set for optimal performance; in the literature, examples such as the recent study of Yuan et al. [27] suggest an experimentally specific variety of this parameter, where $\alpha = 2$.

In open-pit mining scenarios, targets are often occluded or densely distributed at small scales, causing overlaps between candidate frames and leading to reduced classification accuracy within model loss functions subject to non-maximal suppression post-processing analyses. These deficiencies may be compensated for by introducing SFL into models that uncover obscured targets or those appearing on dense small-scale settings found in open-pit mining scenes more accurately relative to earlier work.

3.4. Improved Multiscale Obstacle Object Detection Model

The YOLOv5-s algorithm detects large, medium, and small targets using three scales of 20×20, 40×40, and 80×80. Due to the presence of a considerable number of small-scale and densely occluded targets in open-pit scenes, target features mostly exist at shallow network layers. To improve the accuracy of detecting smaller objects, we add a 160×160 shallow detection layer. However, as model complexity increases with this change, we reduced the deep feature layer to 20×20. This reduction allows for obtaining enough information required for larger-scale object detection, such as trucks or forklifts, leading to significant decreases in accuracy overall.

Comparing models in terms of improvements based on multiple metrics indicates that adding an extra shallow layer substantially improves model performance without inferring complexities unacceptable for practical deployment while still retaining the crucial 20×20 deep detection layer necessary for the precise identification of large-scale objects such as trucks or sprinklers. Figure 5 shows our improved IMOD model architecture, which includes both the original scales and one added scale per the recommendations.

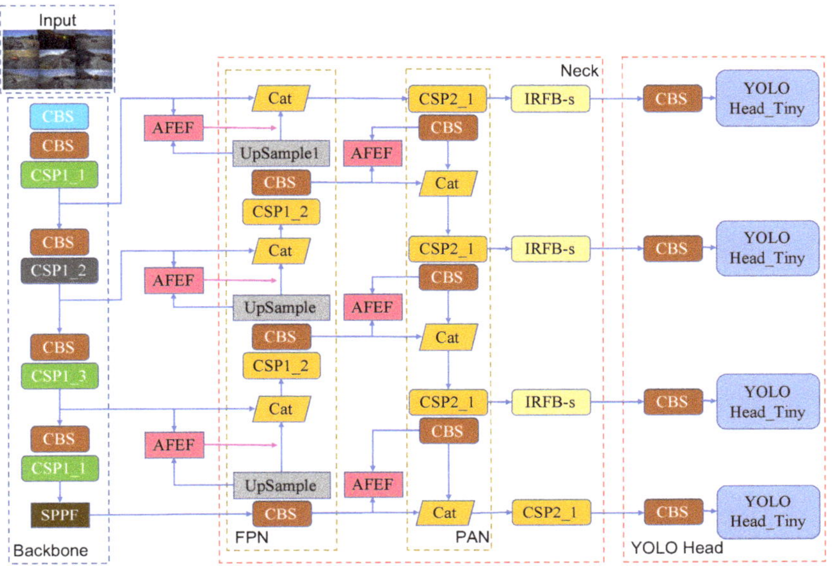

Figure 5. The structure of the IMOD model.

4. Experimental Analysis

4.1. Network Model Ablation Study

To fully examine the efficacy of the three proposed improvement strategies in this paper, their impact on the performance of YOLOv5-s was investigated by conducting ablation and robustness experiments. Evaluation metrics including parameter count, weight size, computation volume (GFLOPS), mean average precision (mAP), and single-frame detection time (FPS) were evaluated. The mAP was computed at an intersection over a union threshold of 0.5 using the following formula:

$$P = TP/(TP/FP) \tag{8}$$

$$R = TP/(TP + FP) \tag{9}$$

$$mAP = \frac{1}{C}\sum_{i=1}^{c} AP_i = \frac{1}{C}\int_0^1 P(R)dR \tag{10}$$

where P is the accuracy rate, TP is true positive cases, FP is false positive cases, and FN is false negative cases.

Figure 6 presents the training loss curves of localization, classification, and confidence losses, indicating that the improved model converges faster than before. Furthermore, Figure 7 reports an enhanced mAP achieved by the improved model.

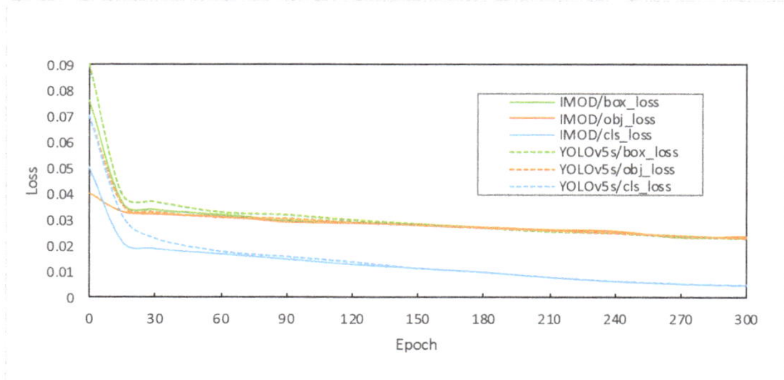

Figure 6. Comparison of training loss curves.

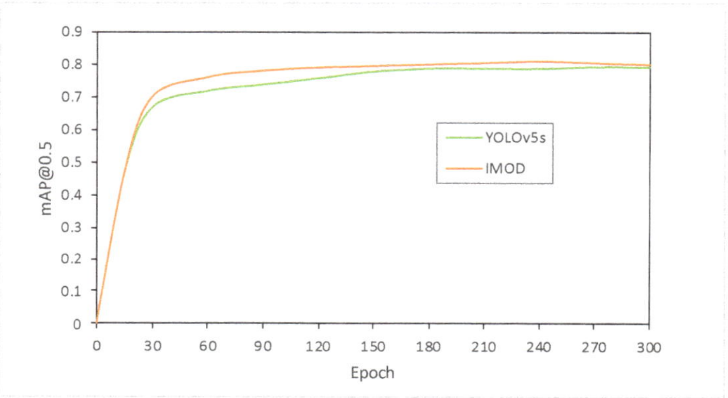

Figure 7. Validation set mAP comparison map.

4.2. Robustness Experiment

The team utilizes intelligent driving data centers to acquire on-road datasets for large-scale real-world scenarios, one of which is the BUUISE-MO dataset shown in Figure 8, a mine obstacle image dataset.

Figure 8. The BUUISE-MO mine obstacle image dataset.

The BUUISE-MO dataset has a picture resolution of 1920 × 1080, a training set of 7220, and a test set of 2500, for a total of 9720 images, as shown in Figure 8. The dataset contains 15 categories including truck, forklift, car, excavator, person, signboard, and others, with 6041 (large), 9230 (medium), and 12,043 (small) labels. This dataset is appropriate for the task of detecting small objects.

4.3. Comparative Experiment

4.3.1. Network Model Ablation Experiment

To verify the feasibility of the improved strategy and to explore its effect on model performance, different improvement methods were integrated into the original network structure. Ablation experiments were performed on the BUUISE-MO dataset to evaluate all aspects of the index, such as changes in model structure, parameter count, and computational complexity due to these improvements. Table 1 reports specific ablation experiment results, where ✔ denotes integration with our proposed method. The proposed strategy displays varying levels of improvement across both datasets without impacting real-time detection capabilities. Additionally, it facilitates higher detection accuracy for target objects within open-pit mining areas while validating feasibility through the successful completion of the ablation study.

Table 1. Experimental results of the ablation study on the IMOD model using the BUUISE-MO dataset.

CA	SRFB-s	AFHF	SFL	4-Scale	mAP@0.5	FPS
	✔				0.495	144
		✔			0.494	146
✔					0.481	155
			✔		0.478	154
				✔	0.492	152
	✔	✔			0.514	123
✔	✔	✔			0.515	122
✔	✔	✔	✔		0.528	120
✔	✔	✔	✔	✔	0.543	120

In this article, the IMOD model is proposed, which employs 5G-based multi-UAV and deep learning methods to enhance unmanned mining vehicle behavior via multisystem coordination in an open-pit mining environment. An autonomous mine obstacle image dataset is constructed and experimentally analyzed to address difficulties with recognizing small-scale targets amidst complex road scenes in mining areas. The IMOD model effectively enhances safety for driverless vehicles and personnel/equipment within mining zones. Future work should focus on improving the accuracy of small target recognition under abnormal illumination conditions and addressing error correction resulting from data desynchronization caused by multisystem coordinated power outages.

The BUUISE-MO dataset was chosen for quantitative and qualitative analysis to demonstrate the degree of improvement achieved by the improved algorithm on various targets. Table 2 displays the average accuracy of different enhanced algorithms for each target.

Upon comparison of Tables 1 and 2, it was determined that the inferior performance of the model in open-pit mining environments can be attributed to an increase in occluded targets and smaller targets. Although the SRFB-S module and AFHF module have been demonstrated to improve the detection accuracy for trucks, signboards, excavators, persons, forklifts, and cars within this scene, real-time detection is sacrificed as a result. The addition of both CA attentional modules and SFL loss functions resulted in improvements of 0.3% and 0.2%, respectively, without compromising other features. Implementing four detection branches presents the potential for significantly enhancing small target detection accuracy at the cost of increased model complexity.

Table 2. Average accuracy of improved models for various types of targets.

Algorithm	Truck	Signboard	Excavator	Person	Car	Forklift
YOLOv5-s	0.732	0.556	0.537	0.556	0.413	0.365
v5s-SRFB-s	0.754	0.582	0.563	0.587	0.406	0.398
v5s-AFHF	0.756	0.576	0.564	0.573	0.414	0.386
v5s-CA	0.726	0.562	0.541	0.555	0.417	0.373
v5s-SFL	0.744	0.546	0.542	0.546	0.414	0.313
v5s-4-scale	0.757	0.574	0.565	0.573	0.41	0.384
IMOD	0.815	0.592	0.612	0.596	0.489	0.396

4.3.2. Robustness Experiment

To validate the robustness and generalization of our improved model in complex road scenarios with significant occlusion and shadow factors, we utilized the BUUISE-MO dataset [28] for experimental verification. After undergoing thorough training involving 320 iterations, our model achieved stability while producing results during validation that are detailed in Table 3.

Table 3. Experimental results showing improved model robustness on the BUUISE-MO dataset.

Algorithm	mAP@0.5	FPS	Parameter/M	GFLOPs
YOLOv5-s	0.466	155	7.08	16.2
YOLOv5-m	0.512	129	20.85	47
IMOD	0.535	120	11.19	20.4

To further validate whether the improved algorithm can be generalized to basic object detection scenarios, we removed small objects in the BDD100K dataset that have a significant impact on detection and evaluated our approach using both the BDD100K dataset (with six classes) and VOC dataset. After training for 320 epochs, the model showed stable performance, as demonstrated in Figures 9 and 10.

The data analysis presented in the above table indicates that, although the algorithm can be applied to basic object detection scenarios, its performance does not exhibit significant improvements. However, when compared with the YOLOv5-s algorithm, the enhanced approach demonstrates improved detection accuracy in complex road-scene datasets. This finding provides evidence that the enhanced model is both robust and generalizable.

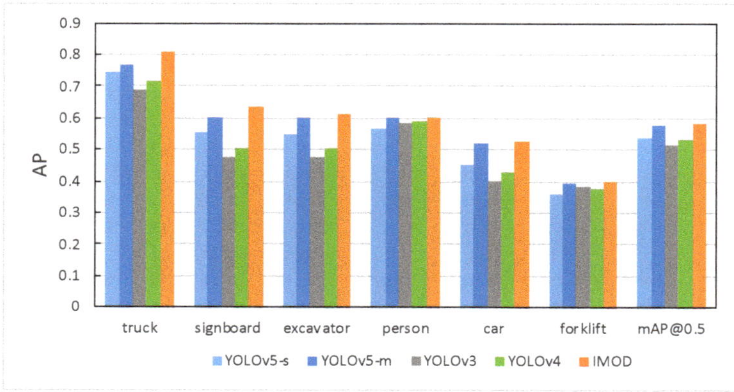

Figure 9. Comparison of detection results of different models for major targets.

Figure 10. Comparison of core parameters computing performance.

4.3.3. Comparative Experiment

To demonstrate the efficacy of the IMOD algorithm, we compared it with four state-of-the-art algorithms, namely faster R-CNN, YOLOv4, YOLOv5-s, and YOLOv5-m, using the BDD100K dataset [29] as a benchmark. Detection accuracy and speed were employed as evaluation metrics. The experimental results are presented in Table 4 for comparison purposes.

Table 4. Performance comparison of different algorithms on the BDD100K dataset.

Algorithm	Backbone	mAP@0.5	FPS (V100)
YOLOv4	CSPDarknet53	0.452	65
IMOD	Darknet53	0.318	12
YOLOv5-s	CSPDarknet53	0.467	105
YOLOv5-m	CSPDarknet53	0.511	89
IMOD	CSPDarknet53	0.534	80

According to the results in Table 4, obtained from different algorithmic models applied on the BDD100K dataset, the enhanced YOLOv5 algorithm outperforms other prevalent models based on both detection accuracy and speed metrics. In contrast to the YOLOv5-s and YOLOv5-m algorithms, with swift performance but lower detection accuracy, the enhanced YOLOv5 algorithm manifests superiority over real-time model operation conditions while delivering prominent overall performance, resulting in improved object recognition in intricate road scenarios. This has practical significance, demonstrating its beneficial applicability scope.

5. Conclusions

We proposed an object detection algorithm for complex road scenes in open-pit mining environments, aiming to address the problems of low detection accuracy, false detection, and missed detection of road occlusion targets and small-scale targets. Our algorithm is based on adaptive feature fusion using the YOLOv5-s algorithm as a starting point. We introduce a feature fusion factor to reduce negative impacts caused by adjacent scale fusion strategies, increase effective samples after feature fusion, and improve learning ability for small- to medium-sized targets. Additionally, we propose an improved receptive field module that extracts more target feature information from shallow feature layers. Finally, we introduce a CA attention mechanism and StrongFocalLoss loss function to improve model accuracy for dense occlusion targets and small-scale targets.

We autonomously collect and construct a mine obstacle image dataset to facilitate experimental testing of our approach. Our results show that our approach effectively addresses issues of blockage and small-scale target recognition in complex road scenarios found in mining areas, with use cases demonstrating the IMOD model increases the safety of unmanned vehicles while preserving equipment fidelity required at industrial scales.

Future work will involve improving recognition accuracy under abnormal illumination conditions as well as correcting errors due to data synchrony caused by multisystem shutdowns during network operations. Developing lightweight architectures toward facilitating deployment on mobile terminals while simultaneously enhancing overall model accuracy is also essential.

Author Contributions: Conceptualization and methodology, X.X. and C.X.; software and validation, C.X. and Z.W.; formal analysis, Y.Z.; investigation, S.Z.; writing—original draft preparation, X.X.; writing—review and editing, C.X., X.X. and Z.W.; project administration, H.B. and X.Q. All authors have read and agreed to the published version of the manuscript.

Funding: This work is supported in part by a key project of the National Nature Science Foundation of China (Grant No. 61932012), in part by the National Natural Science Foundation of China (Grant No. 62102033), and in part by Support for high-level Innovative Teams of Beijing Municipal Institutions (Grant No. BPHR20220121).

Data Availability Statement: Not applicable.

Conflicts of Interest: The authors declare no conflict of interest.

References

1. Gao, Y.; Ai, Y.; Tian, B.; Chen, L.; Wang, J.; Cao, D.; Wang, F.Y. Parallel end-to-end autonomous mining: An IoT-oriented approach. *IEEE Internet Things J.* **2019**, *7*, 1011–1023. [CrossRef]
2. Ko, Y.; Kim, J.; Duguma, D.G.; Astillo, P.V.; You, I.; Pau, G. Drone Secure Communication Protocol for Future Sensitive Applications in Military Zone Number: 6. *Sensors* **2021**, *21*, 2057. [CrossRef] [PubMed]
3. Xu, C.; Wu, H.; Liu, H.; Gu, W.; Li, Y.; Cao, D. Blockchain-oriented privacy protection of sensitive data in the internet of vehicles. *IEEE Trans. Intell. Veh.* **2022**, *8*, 1057–1067. [CrossRef]
4. Chen, S.; Hu, J.; Shi, Y.; Zhao, L.; Li, W. A vision of C-V2X: Technologies, field testing, and challenges with chinese development. *IEEE Internet Things J.* **2020**, *7*, 3872–3881. [CrossRef]
5. Zhang, X.; Guo, A.; Ai, Y.; Tian, B.; Chen, L. Real-time scheduling of autonomous mining trucks via flow allocation-accelerated tabu search. *IEEE Trans. Intell. Veh.* **2022**, *7*, 466–479. [CrossRef]
6. Ma, N.; Li, D.; He, W.; Deng, Y.; Li, J.; Gao, Y.; Bao, H.; Zhang, H.; Xu, X.; Liu, Y.; et al. Future vehicles: Interactive wheeled robots. *Sci. China Inf. Sci.* **2021**, *64*, 1–3. [CrossRef]
7. Pan, Z.; Zhang, C.; Xia, Y.; Xiong, H.; Shao, X. An Improved Artificial Potential Field Method for Path Planning and Formation Control of the Multi-UAV Systems. *IEEE Trans. Circuits Syst. II Express Briefs* **2022**, *69*, 1129–1133. [CrossRef]
8. Krichen, M.; Adoni, W.Y.H.; Mihoub, A.; Alzahrani, M.Y.; Nahhal, T. Security Challenges for Drone Communications: Possible Threats, Attacks and Countermeasures. In Proceedings of the 2022 2nd International Conference of Smart Systems and Emerging Technologies (SMARTTECH), Riyadh, Saudi Arabia, 9–11 May 2022; pp. 184–189.
9. Girshick, R.; Donahue, J. Trevor DARRELL a Jitendra MALIK. Rich feature hierarchies for accurate object detection and semantic segmentation. In Proceedings of the IEEE Computer Society Conference on Computer Vision and Pattern Recognition, Columbus, OH, USA, 23–28 June 2014; pp. 580–587.
10. Ren, S.; He, K.; Girshick, R.; Sun, J. Faster r-cnn: Towards real-time object detection with region proposal networks. *Adv. Neural Inf. Process. Syst.* **2015**, *28*, 1497. [CrossRef] [PubMed]
11. Xiang, X.; Lv, N.; Guo, X.; Wang, S.; El Saddik, A. Engineering vehicles detection based on modified faster R-CNN for power grid surveillance. *Sensors* **2018**, *18*, 2258. [CrossRef] [PubMed]
12. Ghosh, R. On-road vehicle detection in varying weather conditions using faster R-CNN with several region proposal networks. *Multimed. Tools Appl.* **2021**, *80*, 25985–25999. [CrossRef]
13. Luo, J.Q.; Fang, H.S.; Shao, F.M.; Zhong, Y.; Hua, X. Multi-scale traffic vehicle detection based on faster R-CNN with NAS optimization and feature enrichment. *Def. Technol.* **2021**, *17*, 1542–1554. [CrossRef]
14. Yin, G.; Yu, M.; Wang, M.; Hu, Y.; Zhang, Y. Research on highway vehicle detection based on faster R-CNN and domain adaptation. *Appl. Intell.* **2022**, *52*, 3483–3498. [CrossRef]
15. Redmon, J.; Divvala, S.; Girshick, R.; Farhadi, A. You only look once: Unified, real-time object detection. In Proceedings of the IEEE Conference on Computer Vision and Pattern Recognition, Las Vegas, NV, USA, 27–30 June 2016; pp. 779–788.
16. Qiu, Z.; Bai, H.; Chen, T. Special Vehicle Detection from UAV Perspective via YOLO-GNS Based Deep Learning Network. *Drones* **2023**, *7*, 117. [CrossRef]
17. Koay, H.V.; Chuah, J.H.; Chow, C.O.; Chang, Y.L.; Yong, K.K. YOLO-RTUAV: Towards real-time vehicle detection through aerial images with low-cost edge devices. *Remote Sens.* **2021**, *13*, 4196. [CrossRef]

18. Zhang, Y.; Guo, Z.; Wu, J.; Tian, Y.; Tang, H.; Guo, X. Real-Time Vehicle Detection Based on Improved YOLO v5. *Sustainability* **2022**, *14*, 12274. [CrossRef]
19. Ultralytics. YOLOv5. Available online: https://github.com/ultralytics/yolov5 (accessed on 3 February 2023).
20. Lu, X.; Ai, Y.; Tian, B. Real-time mine road boundary detection and tracking for autonomous truck. *Sensors* **2020**, *20*, 1121. [CrossRef]
21. Liu, S.; Huang, D. Receptive field block net for accurate and fast object detection. In Proceedings of the European Conference on Computer Vision (ECCV), Munich, Germany, 8–14 September 2018; pp. 404–419.
22. Wu, H.; Xu, C.; Liu, H. S-MAT: Semantic-Driven Masked Attention Transformer for Multi-Label Aerial Image Classification. *Sensors* **2022**, *22*, 5433. [CrossRef]
23. Park, J.; Woo, S.; Lee, J.Y.; Kweon, I.S. Bam: Bottleneck attention module. *arXiv* **2018**, arXiv:1807.06514.
24. Geirhos, R.; Jacobsen, J.; Michaelis, C.; Zemel, R.; Brendel, W.; Bethge, M.; Wichmann, F. Shortcut Learning in Deep Neural Networks. *Nat. Mach. Intell.* **2020**, *2*, 665–673. [CrossRef]
25. Cheng, X.; Yu, J. RetinaNet with difference channel attention and adaptively spatial feature fusion for steel surface defect detection. *IEEE Trans. Instrum. Meas.* **2020**, *70*, 1–11. [CrossRef]
26. Hou, Q.; Zhou, D.; Feng, J. Coordinate attention for efficient mobile network design. In Proceedings of the IEEE/CVF Conference on Computer Vision and Pattern Recognition, Nashville, TN, USA, 20–25 June 2021; pp. 13708–13717.
27. Yuan, J.; Wang, Z.; Xu, C.; Li, H.; Dai, S.; Liu, H. Multi-vehicle group-aware data protection model based on differential privacy for autonomous sensor networks. *IET Circuits Devices Syst.* **2023**, *17*, 1–13. [CrossRef]
28. Li, M.; Zhang, H.; Xu, C.; Yan, C.; Liu, H.; Li, X. MFVC: Urban Traffic Scene Video Caption Based on Multimodal Fusion. *Electronics* **2022**, *11*, 2999. [CrossRef]
29. Yu, F.; Chen, H.; Wang, X.; Xian, W.; Chen, Y.; Liu, F.; Madhavan, V.; Darrell, T. BDD100K: A Diverse Driving Dataset for Heterogeneous Multitask Learning. In Proceedings of the IEEE/CVF Conference on Computer Vision and Pattern Recognition, Seattle, WA, USA, 13–19 June 2020; pp. 2636–2645.

Article

Decentralized Multi-UAV Cooperative Exploration Using Dynamic Centroid-Based Area Partition

Jianjun Gui [1,†], Tianyou Yu [1,†], Baosong Deng [1,*], Xiaozhou Zhu [1] and Wen Yao [1,2]

[1] Defense Innovation Institute, Chinese Academy of Military Science, Beijing 100071, China
[2] Intelligent Game and Decision Laboratory, Beijing 100071, China
[*] Correspondence: dbs@nudt.edu.cn
[†] These authors contributed equally to this work.

Abstract: Efficient exploration is a critical issue in swarm UAVs with substantial research interest due to its applications in search and rescue missions. In this study, we propose a cooperative exploration approach that uses multiple unmanned aerial vehicles (UAVs). Our approach allows UAVs to explore separate areas dynamically, resulting in increased efficiency and decreased redundancy. We use a novel dynamic centroid-based method to partition the 3D working area for each UAV, with each UAV generating new targets in its partitioned area only using the onboard computational resource. To ensure the cooperation and exploration of the unknown, we use a next-best-view (NBV) method based on rapidly-exploring random tree (RRT), which generates a tree in the partitioned area until a threshold is reached. We compare this approach with three classical methods using Gazebo simulation, including a Voronoi-based area partition method, a coordination method for reducing scanning repetition between UAVs, and a greedy method that works according to its exploration planner without any interaction. We also conduct practical experiments to verify the effectiveness of our proposed method.

Keywords: path planning; collaborative exploration; area partition; swarm UAVs

Citation: Gui, J.; Yu, T.; Deng, B.; Zhu, X.; Yao, W. Decentralized Multi-UAV Cooperative Exploration Using Dynamic Centroid-Based Area Partition. *Drones* **2023**, *7*, 337. https://doi.org/10.3390/drones7060337

Academic Editor: Carlos Tavares Calafate

Received: 5 April 2023
Revised: 10 May 2023
Accepted: 20 May 2023
Published: 23 May 2023

1. Introduction

Using swarm unmanned aerial vehicles (UAVs) to execute collaborative missions in local areas has become an important direction of development in unmanned system applications in recent years. Small UAVs are characterized by their ease of deployment and agility [1], but the problem of collaborative control has always been challenging, especially in search and rescue missions [2,3] where the environment structure may change and the UAV swarm may not have prior knowledge of the task area. This requires the ability of autonomy to be able to avoid obstacles and accurately reach the task area, as well as the ability of intelligence to build environment maps and even identify specific targets [4–6].

In this study, we consider using a UAV swarm for exploring unknown areas and dispatching multiple UAVs to collaboratively map specific areas, in order to achieve rapid area reconnaissance. Therefore, we hope that the system can run efficiently and handle the problem of perception overlap or motion interference between UAVs [7–10]. Each participating individual can maintain a relative balance in task partitioning appropriately and dynamically. At the same time, the system needs to have a certain degree of robustness so that the task can be completed even if some individuals in the swarm are damaged [11].

Based on the above considerations, we propose a decentralized collaborative exploration method in this paper. The key of our method is a novel dynamic centroid-based partition algorithm, allowing the work area of each UAV to be dynamically adjusted as the mission progresses. The framework structure of the entire system is shown in Figure 1. Each UAV independently runs localization, mapping, partitioning, and planning, with only a small amount of pose and weight parameters shared among the network. After the initial partitioning is generated, to ensure independence in their work, each individual UAV prioritizes finding unexplored target directions in their own area. Within the partitioning calculation, based on the receding horizon next-best-view (NBV) exploration planner [12], our proposed method generates rapidly-exploring random tree (RRT) [13] with a constrained range, and the task can be quantitatively represented. By composing the information of the UAV swarm's pose and task weight, we adopt the concept of centroid, whose variation results from changes in area partitioning, ensuring flexibility in mutual work. In the mapping module, Octomap [14] is used to provide a quantitative expression as free, occupied, and unknown for exploration.

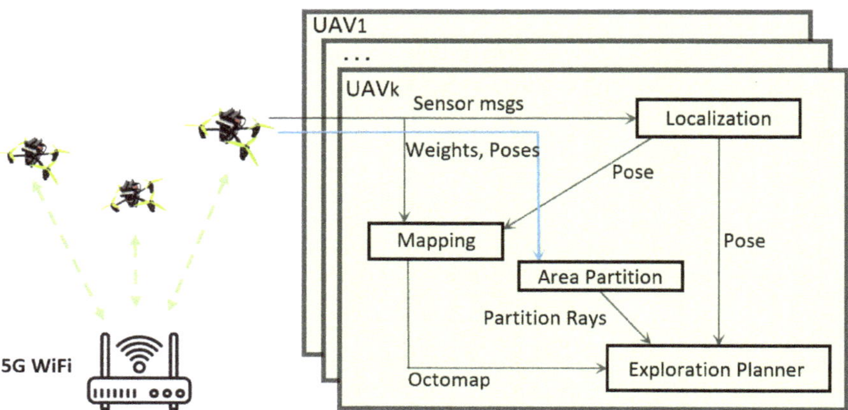

Figure 1. System framework. The modules of localization, mapping, partition, and planning are run independently in each UAV. All of the working UAVs connect to a 5G WiFi for information exchange. Poses and platform weights are shared to dynamically adjust the partition areas. The details of partition and weight calculation are discussed in Section 4.

The main contributions of this paper are as follows:

- We propose a dynamic centroid-based area partition method that takes into account both the positions and current tasks of each UAV during the exploration process. This approach ensures that the UAVs that detect fewer tasks are allocated more mission area in the next step, maximizing the efficiency of the exploration process.
- The NBV exploration planner has been improved by making modifications so that it only samples within a dynamically partitioned area. Additionally, we have innovatively formulated a weight to describe task quantity for the purpose of more effective partitioning.
- The performance of the proposed method is validated by comparing with three classic multi-UAV cooperative exploration methods in both indoor and outdoor simulation environments. The practical experiment is also conducted to verify its feasibility.

The remainder of this paper is organized as follows. In Section 2, the works related to swarm UAVs' cooperative exploration and mapping are reviewed. Section 4 is devoted to describing the details of our method, and in Section 5, comparative experiments in simulation environments are given. The results of practical experiments are discussed in Section 6. Finally, Section 7 concludes this paper and points out potential future improvements.

2. Related Work

In recent years, significant progress has been made within the academic community regarding collaborative exploration by multiple UAVs in unknown environments [15]. In light of our research focus, this paper will examine three areas related to collaborative exploration: methods for multi-UAV coordination, multi-UAV mapping, and exploration in unknown environments.

2.1. Multi-UAV Coordination for Exploration

Using multiple UAVs to increase exploration efficiency is a common practice, and related issues have been extensively studied [16]. One classic method involves maximizing overall utility while minimizing the potential overlap of measurements among UAVs [8]. This idea has been employed in many works such as [10,11,17]. However, as the number of UAVs increases, uncertainty [18] and redundant scanning between them become more prevalent, especially in larger environments where the sensor range is relatively small compared to the scale of the environment.

In conventional multi-agent allocation problems, a TSP-greedy allocation (TSGA) planner with ideal centralized architecture and communication assumptions is utilized to optimize global utility [19]. This approach considers the whole global task, which may be time-consuming for collecting tasks in the center. Alternatively, a dynamic Voronoi partition has been utilized in [7,10] to assign different target locations to individual UAVs, guaranteeing the separateness between them. However, this area partition-based method may not always be optimal as it does not consider the exploring process of each UAV, resulting in less efficient task allocation.

Therefore, in this paper, a dynamic centroid-based area partition is proposed, which considers the exploration process of each UAV for more reasonable task allocation. When a UAV has an insufficient number of candidates, it will be assigned a larger partitioned area to explore. The partition is processed dynamically to adapt to changing situations.

2.2. Multi-UAV Mapping for Exploration

To perform target selection and quantitative calculation in planning, it is necessary to have a map that depicts the environment and further exploration areas. Two representative volumetric mapping methods used in UAV exploration are truncated signed distance field (TSDF) [20] and occupancy [14]. When employing multi-UAV mapping methods, the key issue is often the map merging [21]. Previous works such as [8,17] involve each UAV maintaining its local map and correcting odometry errors while exploring. They then transmit their local maps with uncertain information to a central work station who can combine local maps into a global one for further optimization. In [22], sensor messages are shared among UAVs, and Gaussian mixture models (GMMs) are adopted to assist the exploration planner of each UAV. In [11], two maps are utilized: a low-resolution map for navigation and a high-resolution map for reconstruction. In order to achieve efficient coordination in a decentralized method, it is crucial to share the global map message among UAVs as quickly as possible. This is one of the central issues that we address in this paper.

2.3. Exploration in Unknown Environments

While fully functional UAVs possess autonomous sensing and computing capabilities, the exploration planner enables them to independently perform tasks in unknown environments. Existing works fall into two categories when executing under unknown: frontier-based methods [23–25] and sampling-based methods [11,13,26,27]. With given frontier clusters [28] or sampled viewpoints [27], an information-theoretic measure is optimized to calculate information gain, resulting in reduced map uncertainty. The frontier-based method explicitly computes the boundary between the known and unknown areas and assigns UAVs to frontiers iteratively, but the frontier selection process can be time-consuming as it traverses all surface voxels in a large environment [23]. Some methods reject unsuitable frontiers during selection [1] to ease the computational burden. On the other hand, the sampling-based method randomly selects viewpoints in free areas, such as the rapidly-exploring random tree (RRT) [13] and probabilistic roadmap planner (PRM) [3], which deliver speed and probabilistic completeness. However, these two methods could converge locally.

The two mentioned categories were widely used in the exploration planning of a single UAV. However, for multi-UAV exploration, a coordination module is required to prevent collisions and redundancies. The NBV method [12] is commonly utilized in such scenarios. This method iteratively selects viewpoints in free space to refresh candidates' paths, ensuring a consistent update rate. Our proposed method follows this approach by integrating the strengths of the sampling-based method. This enables frequent recollection of viewpoints to avoid collisions and facilitate flexible collaboration between UAVs.

3. Problem Statement

The task of multi-UAV exploration in an unknown environment performs the process of exploring and mapping iteratively. A 3D workspace $\mathcal{WS}$ of known size is given before the task for establishing the concerned area; all UAVs will explore the workspace. Exploration processing by a UAV team contains N identical UAV with four degrees of freedom, as the 3D position $[x, y, z]^T \in \mathbb{R}^3$ and the yaw angle $\psi \in \mathcal{S}^1$. The UAV state can be described as $\mathbf{x} = [x, y, z, \psi]^T$. In each platform, a depth camera is equipped to collect the environment information with a certain field of view.

The environment is reconstructed by an Octomap $\mathcal{M}$, and the occupancy probability of each gird $m \in \mathcal{M}$ is initialized as $P(m) = 0.5$. The posterior occupancy probability $P(m \mid \mathbf{x}_{1:t}, z_{1:t})$ is updated by the depth measurement $z_{1:t}$ and the UAV state $\mathbf{x}_{1:t}$ from initial time to current time t. The grids in the map will be gradually scanned by the sensor and identified as either free grids $\mathcal{M}_f = \{m \mid P(m \mid \mathbf{x}_{1:t}, z_{1:t}) < P_{free}, m \in \mathcal{M}\}$ or occupied grids $\mathcal{M}_o = \{m \mid P(m \mid \mathbf{x}_{1:t}, z_{1:t}) > P_{occ}, m \in \mathcal{M}\}$. P_f and P_o are given thresholds. Given a map $\mathcal{M}$ at time t, the receding horizon exploration planner decides an optimal path $\mathcal{T}^*$ in every period. To seek the $\mathcal{T}^*$ for the UAV so that it gathers measurements that reduce unknown space and maintain coordination, a cost function is formulated to measure the value of the candidate path, considering the uncertainty of the map $\mathcal{M}$, the UAV team information RT, the location of waypoints in path $\mathcal{T}$, and the time cost of the path $c(\mathcal{T})$.

$$\mathcal{T}^* = \arg \max_{\mathcal{T}} f(\mathcal{M}, RT, \mathcal{T}, c(\mathcal{T})). \tag{1}$$

UAVs visit unknown spaces independently according to the outputs of the exploration planner. We assume that the UAVs are equipped with an accurate localization system. From the initial state, the UAVs are deployed and set with a connected network and a known relative position as in a practical task application.

4. Method

This paper describes the implementation of a decentralized structure, as illustrated in Figure 1. Each platform performs RRT-based planning, area partitioning, and mapping independently. The core parts of our proposed modules are as follows.

4.1. Dynamic Centroid-Based Area Partition

The area for each available UAV is partitioned using a dynamic centroid-based method, and each UAV is only responsible for its own refined area. Assuming that all of the UAVs work with the same efficiency using the same exploring planner on the same platform, a supposed idea is allocating UAVs equal assignments to prevent any UAV falls idle prematurely. However, it is not easy to count the tasks of all individual UAVs and re-allocate them in a decentralized structure.

To quantify the assignments of each UAV, we introduce the number of effective candidate nodes per unit space. The candidate nodes are generated by an RRT-based planner described in Section 4.3. In the area partition method, the candidate viewpoints are randomly sampled in the whole given workspace $\mathcal{WS}$, and nodes are generated when viewpoints are placed in a given partitioned area $\mathcal{PA}$. When effective candidate nodes N_e, which can lead to environmental observation and the space volume $V_{\mathcal{PA}}^{E}$, of the executable area are known, the weight w can be calculated as:

$$w = N_e / V_{\mathcal{PA}}^{E}. \tag{2}$$

The executable space volume $V_{\mathcal{PA}}^{E}$ in the $\mathcal{PA}$ can be approximated as Equation (3). The $V_{\mathcal{PA}}^{E}$ represents the volume of a space, which is free, and the RRT-based planner in Section 4.3 can build a tree there.

$$V_{\mathcal{PA}}^{E} \approx V_{WS} / \frac{C_{loop}}{N_T} = V_{WS} \times N_T / C_{loop}. \tag{3}$$

The number of node sampling loops C_{loop} is counted to calculate the size of the executable exploration area. With the executable space bigger, the C_{loop} is smaller because it is more likely to generate a candidate in $\mathcal{PA}$ successfully. An average number of samples required to generate a node can be formulated by N_T / C_{loop}.

The proposed dynamic centroid-based area partition calculates a virtual centroid $X_c = (x_c, y_c)$ adjusted from the information of platform weights w_i for UAV_i and its positions $X_i = (x, y)$ for the 2D projection from the poses, expressed in Equation (4). UAVs explore 3D space and map a 3D Octomap, while the two-dimensional coordinates of UAVs for partitioning the 3D area are considered to be the simplest form of calculation.

$$X_c = \frac{\sum_{i=1}^{N} w_i \cdot X_i}{\sum_{i=1}^{N} w_i}. \tag{4}$$

As illustrated in Figure 2 and depicted in Equations (5) and (6), the partition ray Prs for the area partition are generated according to the included angle between the j th and $(j+1)$ th UAV when starting from the virtual centroid. θ_{Pr_j} is the angle from x axis to Pr_j in a counterclockwise direction.

$$Pr_j(T) = X_c + (\cos\theta_{Pr_j}, \sin\theta_{Pr_j})T, \ T \geqslant 0, \tag{5}$$

$$\theta_{Pr_{j1}} / \theta_{Pr_{j2}} \propto w_{j+1} / w_j. \tag{6}$$

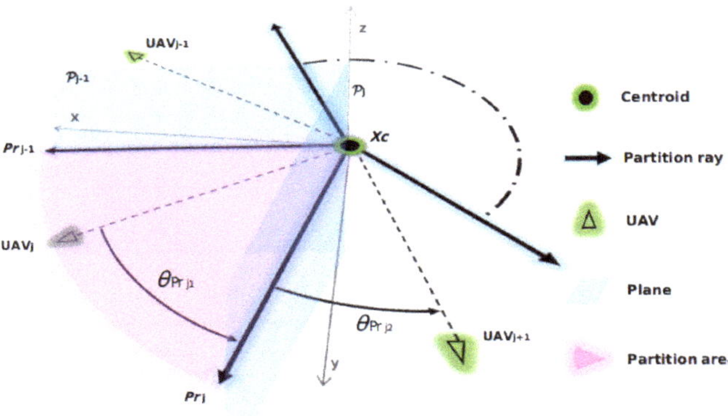

Figure 2. The partitioning process of UAV_j. Each UAV computes a virtual centroid, and a 3D self-responsible area is partitioned by the nearest two planes $\mathcal{P}$, defined by partition rays Pr and the axis of the centroid.

For UAV_j, its exploring space $\mathcal{PA}_j$ is bounded by two vertical planes $\mathcal{P}_j$ and $\mathcal{P}_{j-1}$, which are defined by partition rays and the axis of the centroid, see Equation (7). As illustrated in Figure 2, the pink area in the sector space marks the two-dimensional plane projection of the exploration space allocated ahead. The partition area is formulated by Equation (8). After the coordinates of the three-dimensional point under the x and y axes are converted to the polar coordinate system, the angle should be between $\theta_{Pr_{j-1}}$ and θ_{Pr_j}.

$$\mathcal{P}_j = \left\{ p \mid \overrightarrow{\left(\cos\left(\theta_{Pr_j} + \frac{\pi}{2}\right), \sin\left(\theta_{Pr_j} + \frac{\pi}{2}\right), 0\right)} \cdot \vec{p} = \mathbf{0} \right\},$$ (7)

$$\mathcal{PA}_j = \{(x, y, z) \mid x = \rho\cos\theta, y = \rho\sin\theta, z \in R, \rho \in R^+, \quad \theta_{Pr_{j-1}} < \theta < \theta_{Pr_j}\}.$$ (8)

The core computing of the area partition is carried out on every UAV, and the platform weights and positions are shared between UAVs in real-time. With the change of them, every platform can adjust the working $\mathcal{PA}$ dynamically. The UAV with more detected tasks in the same volume space will attain more weight, the virtual centroid and the dividing planes will tend to be closer to it, and its $\mathcal{PA}$ in the next iteration will shrink. This tactic ensures the area assignment is in a relatively balanced condition, thus making the whole exploration more efficient.

4.2. Distributed Ray-Cast Mapping

A numerically environmental expression is necessary for cooperative exploring. In a unified space, a shared environment could be divided into the unknown, occupied, or free parts. Due to its simple and fast searching character, the OctoMap is adopted in our method. Within it, the unknown, occupied, or free space can be represented by the cubic volume (voxel) in an octree.

In our decentralized setup, all recent sensor outputs, including poses, are shared among the UAVs, as depicted in Figure 1. The use of 5G WiFi ensures high-quality information sharing. We assume that the initial relative position is known, and a perfect localization node runs on each UAV to provide accurate real-time position data. By using the known initial relative position, the poses can be transformed into a unified coordinate system. Consequently, each UAV maintains its own global map with a unified coordinate system to enable cooperative planning, as illustrated in Figure 3. Collecting and aligning information within the team as much as possible promotes collaboration. In this mode, each UAV can consider the most comprehensive global information available.

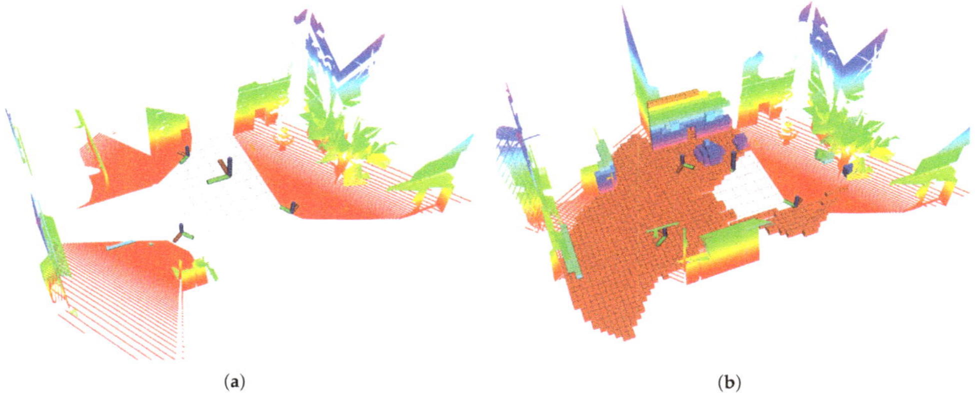

(a)   (b)

Figure 3. Distributed ray-cast mapping. As (**a**) shows, three small axes represent the UAV, and the big axis represents the origin of a unified coordinate system. The depth measurement is visualized by point clouds of various colours, where blue denotes objects that are far away and red indicates those that are in close proximity. As the UAV wander, (**b**) shows the map expressed by Octomap simultaneously.

4.3. Decentralized Cooperative Exploration Planner

The approach presented in this paper constitutes an enhancement of a receding horizon NBV planner [12]. The planner is based on a sampling strategy that leverages RRT to generate plausible target points and feasible trajectories in free space. Each UAV runs the same planner, leveraging the same hardware and algorithms. The planning process is formalized in Algorithm 1. The RRT algorithm iteratively generates random points throughout the workspace, stopping when a point is placed within the $\mathcal{PA}$ of the corresponding UAV.

Algorithm 1 Decentralized Cooperative Exploration Planner

1: **for Exploration** is not over **do**
2:     Obtain the latest weights and poses of all UAVs
3:     Do the area partition
4:     Initialize the RRT T
5:     Initialize the $N_e = 0$, $C_{loop} = 0$.
6:     **for** $N_T < N_{init}$ *or* $G(n_{best}) = 0$ **do**
7:       **if** $N_T < N_{threshold}$ **then**
8:         **while** True **do**
9:           $C_{loop} = C_{loop} + 1$.
10:           generate candidate $C \leftarrow X \sim Uniform(\mathcal{WS})$
11:           **if** C is in $\mathcal{PA}$ **then**
12:             generate node $n \leftarrow$ **Connect**(C, T)
13:             **if** $G(n)! = 0$ **then**
14:               $N_e = N_e + 1$
15:             **end if**
16:             **break**
17:           **end if**
18:       **else**
19:         generate candidate $C \leftarrow X \sim Uniform(\mathcal{WS})$
20:         generate node $n \leftarrow$ **Connect**(C, T)
21:       **end if**
22:       **if** $N_T > N_{MAX}$ **then**
23:         **Exploration** is over and break
24:       **end if**
25:       Update n_{best} and *BestBranch*
26:     **end for**
27:     update weight $w = N_e \times C_{loop} / N_T$
28:     Share the latest weight in the team
29:     Execute the planned path
30: **end for**

Throughout the process, the planner generates a tree T consisting of nodes in the free space of an OctoMap $\mathcal{M}$. Each node n corresponds to a potential viewpoint, with its state denoted by $\xi = (x, y, z, \psi)^T$, reflecting position and yaw. The tree is constrained to remain in the collision-free space to guarantee safe planning. For UAVj, the best node n^j_{best} is selected based on Equation (9), which considers the feasibility of the path. The function $G(n)$ reflects the gain of the parent node, and a novel information value (with related parameters) can be expressed as $g(\mathcal{M}, n, \mathcal{P}_j, \mathcal{P}_{j-1})$ in Equation (10). This optimization aims to minimize Equation (1) considering $\mathcal{M}$ for information measurements and path planning, nodes corresponding to the path, and $\mathcal{P}_j, \mathcal{P}_{j-1}$ aggregating team information to enable coordination.

$$n^j_{best} = \arg max_{n \in T_j} G(n), \tag{9}$$

$$G(n) = G(n_{parent}) + g(\mathcal{M}, n, \mathcal{P}_j, \mathcal{P}_{j-1}), \tag{10}$$

and

$$g(\mathcal{M}, n, \mathcal{P}_j, \mathcal{P}_{j-1}) = \sum_{v \in FOV(\xi) \cap \mathcal{M}} V(v) \times e^{-\lambda c(\sigma)} \prod f. \tag{11}$$

In this context, the function $V(v)$ takes a value of 1 if the voxel v is unexplored, and 0 otherwise. Our objective is to explore the unknown space, and the visible voxels within the field of view of the sensor are accumulated to form a basic visibility score. The cost of a path from the initial node n_0 to n is determined by the RRT algorithm and denoted by $c(\sigma)$. To enable collaboration among the UAVs, the product function $\prod f$ is utilized, as shown in Figure 4. Each plane $\mathcal{P}$ exerts a repulsive factor of $f = 1 + 1/d$, where d is the distance between the candidate node and the plane $\mathcal{P}$. The term $e^{-\lambda c(\sigma)}$ causes the visibility score to decrease smoothly as the path cost increases, while $e^{-\prod 1+1/d}$ drives the UAVs away from the $\mathcal{P}_j$ and $\mathcal{P}_{j-1}$.

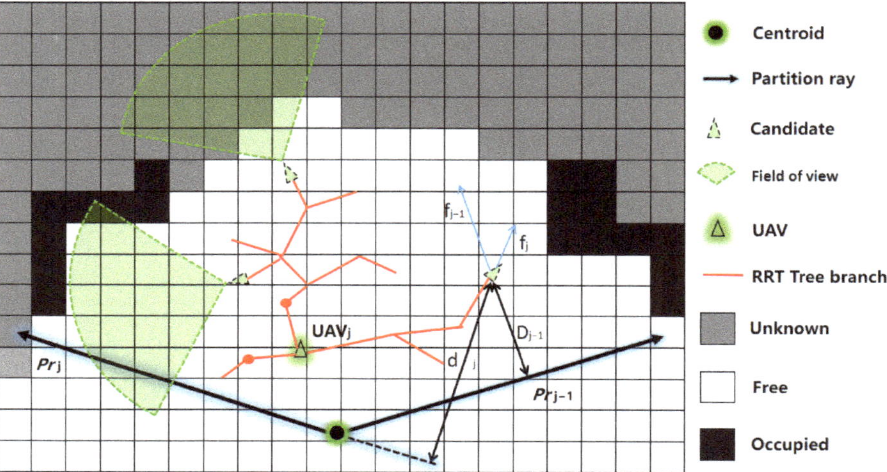

Figure 4. The planning process of UAV$_j$. The d_j and d_{j-1} are calculated to formulate the factor f_j and f_{j-1}.

The algorithm initializes the value of n_0 in tree T as the current state of the UAV during the first initialization. Otherwise, the initial tree would be set as the previous best branch. The *BestBranch* is defined as the branch from n_0 to n_{best}. To ensure that sufficient environmental information is obtained while generating nodes, the tree T must have a minimum of N_{init} nodes, and loops continue until a valuable solution is obtained with $G(n_{best}) = 0$. To prevent the UAV from idling due to unreasonable area partitioning, the partition constraint is disabled when the number of tree nodes N_T exceeds a certain threshold $N_{threshold}$. Once $N_T > N_{MAX}$, the exploration is deemed to be completed. In each iteration, the first segment of the *BestBranch* is considered the planned path. The weight w can be updated using Equation (12) according to Equation (2). Since $\mathcal{WS}$ is assumed to be given and $V_{\mathcal{WS}}$ is constant and known, it is omitted.

$$w = N_e / V_{PA}^E \approx N_e \times Sampleloop / (V_{\mathcal{WS}} \times N_T). \tag{12}$$

The algorithm presented in this paper is intended to be executed on each UAV independently. The planner and the area partition calculation are interdependent, and a smaller partition area can make it more difficult for the planner to generate an effective trajectory, resulting in a smaller weight. In turn, a smaller weight can lead to a larger partition being provided to the UAV during the partition area calculation.

5. Evaluation in Simulation

To demonstrate the superiority of the proposed method, we compared it with three representative algorithms. The first was the greedy method for group application, which did not employ cooperative settings (referred to as "greedy" in this context). The second [11] was the classic method that discounted the information gain based on the repeating area (referred to as "coordination" in this context). In this method, the gain was reduced based on the area within the sensor range of the current best nodes of other UAVs. The third method [7] used dynamic Voronoi partitions to assign different target locations to individual UAVs, minimizing duplicated exploration areas (referred to as "Voronoi" in this context). All of the three algorithms were decentralized and used RRT to generate candidate points.

Simulation experiments were conducted using Gazebo. All of the methods were tested with the same virtual UAVs and environmental settings. Each UAV was equipped with a depth camera that had a field of view $[60, 90]°$ in the vertical and horizontal directions. For the indoor scenario, the camera was mounted with a downward pitch angle of $15°$. For the outdoor scenario, it was mounted with a pitch angle of $35°$. For all of the simulation experiments, the maximum velocity was set as $\dot{\psi}_{\max} = 0.5$ rad/s and $v_{\max} = 0.25$ m/s, while the size of the collision detection box was assumed to be $0.5 \times 0.5 \times 0.3$ m^3.

5.1. Indoor Scenario

For the indoor scenario, a regular single-story space with limited furniture and weighing structures but no other obstacles was used with dimensions of $20 \times 12 \times 3$ m^3, as shown in Figure 5a. The proposed method was tested 20 times on teams consisting of 2, 3, 4, and 5 UAVs with the same initial position, where the relative distances between UAVs were less than 50 cm. Each UAV was equipped with a depth camera, and the maximum length of RRT was set to 1 m, while the maximum sensor range was $d_{\max}^{sensor} = 4$ m. The partition constraint became invalid when iterative tree nodes reached a threshold, $N_T > N_{threshold} = 30$, and exploration also stopped when $N_T > N_{MAX} = 150$. The minimum node number of each RRT iteration was $N_{init} = 15$, and the space within the maximum planner range of $d_{\max}^{planner} = 2.5$ m was used to calculate the gain, with the value of λ set to 0.5.

To evaluate efficiency, the exploration completion time was measured using a team of 5 UAVs as shown in Figure 6a. The exploration completion degree was plotted over time to reflect the exploration process, as depicted in Figure 6b. The results showed that with an increase in the number of UAVs, the exploration time decreased for all of the methods, albeit at a lower rate of decline for larger teams. For two UAVs, the proposed and Voronoi methods show similar performance with mean values of 185.4 s and 191.8 s, respectively. For a team of 5 UAVs, our proposed method outperformed other methods with lower variance and mean values of 86.2 s, while the voronoi, coordination, and greedy methods took 98 s, 125.5 s, and 185.4 s, respectively. The proposed method ensured efficient exploration with fewer scan repetitions by spacing out UAVs, which led to less backtracking and smaller variance. On the other hand, more space would be allocated to the UAVs that detect fewer tasks in our method, thus resulting in overall efficiency.

In addition, Figure 6b showed a turning point during exploration, where the speed of exploration began to differ at about 50 s. The proposed method experienced the exploration bottleneck later than the voronoi, coordination, and greedy methods, respectively. The greedy method failed to cooperate, leading to multiple UAVs exploring the same area, and scan repetitions. Although the coordination method cooperated through a designed gain, it could not avoid scan repetitions. The Voronoi-based partition method restricted individual UAVs to their partitioned areas without considering detection efficiency for unknown spaces, and this could cause UAVs to fall idle when such areas became fewer.

Figure 5. Simulation environments and their exploration results. (**a**) A $20 \times 12 \times 3$ m^3 indoor scenario, a regular single-story space; (**b**) a $40 \times 40 \times 9$ m^3 outdoor scenario, a typical urban community. The colored points on the floor represent the initial positions of UAVs, with red indicating 2 UAVs, orange for 3 UAVs, yellow for 4 UAVs, and green for 5 UAVs. (**c**,**d**) are the exploration results for both scenarios. (The following figures in this paper have the same meaning, where the spatial structure and depth information are depicted using grids of different colours.)

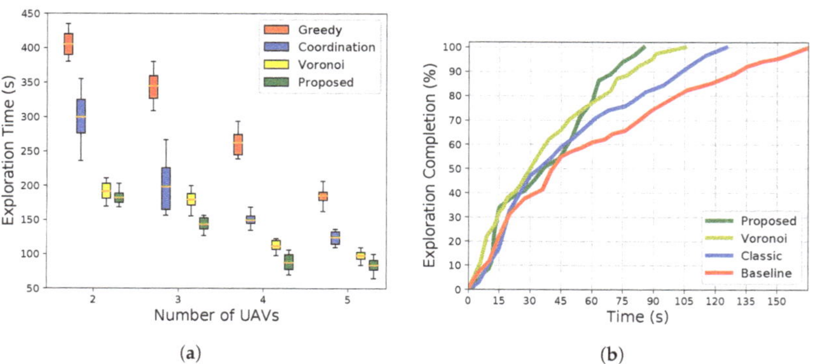

Figure 6. Numerical analysis of indoor simulation. (**a**) shows the exploration time of a team with 2, 3, 4, and 5 UAVs. (**b**) shows the team of 5 UAVs in one trial. Three other algorithms are compared.

5.2. Large Outdoor Scenario

The second scenario involves an outdoor space measuring $40 \times 40 \times 9 \text{ m}^3$, which is a typical urban community with multiple buildings and complex spatial structures, including spatial dead ends, as shown in Figure 5b. Since the outdoor environment is more extensive, the maximum length of the RRT planner is set to 3 m based on empirical experiments. During the trials, many depth value frames exceeding 7 m were observed, and the trial settings were adjusted to $d_{max}^{sensor} = 8.5 \text{ m}$ with a mounting pitch of $35°$ to obtain more ground measurements, where $d_{max}^{planner} = 3.5 \text{ m}$ with $N_{init} = 30$, $N_{threshold} = 80$, and $N_{MAX} = 250$.

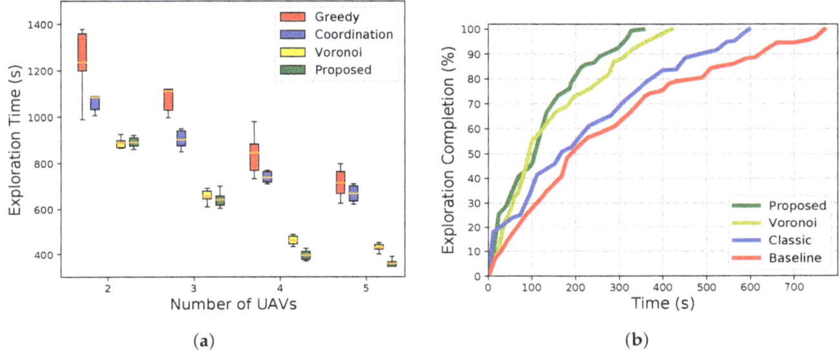

Figure 7. Numerical analysis of outdoor simulation. (**a**) shows the exploration time of the team with 2, 3, 4, and 5 UAVs. (**b**) shows the team of 5 UAVs in one trial. Three other algorithms are compared.

For each planner, 20 trials were conducted for teams consisting of 2, 3, 4, and 5 UAVs, starting from the same initial position, where the relative distances between UAVs were less than 100 cm. Similar to the indoor scenario, the results of the algorithms (indicated in Figure 7) show that the proposed method outperforms the voronoi, coordination, and greedy methods, with a mean completion time of 376.7 s for a team of 5 UAVs. In contrast, the voronoi, coordination, and greedy methods gave mean completion times of 428.8 s, 657.8 s, and 771.5 s, respectively. The experiment also shows that as the scenario size increases, the greedy method, without cooperation, becomes increasingly random and unintelligent. The variance of the results is significantly larger in the outdoor scenario.

In both scenarios, the exploration completion rate exhibits a decreasing trend. This trend can be attributed to the fact that at the start of the exploration, there were many unknown areas, and the UAVs can find enough task points with fewer sampling times. Regardless of the efficiency of the planning algorithm, the UAVs could detect unknown spaces that were not pre-included in the planning, which could lead to efficient exploration in the beginning. As the environment has been continuously explored, the unknown area has decreased, and the time required to calculate the targets has become longer. Especially with the use of the receding horizon method, the UAVs often re-visited those optimal targets due to being confined to local optima, which would slow down the rate of exploration at the later stage.

6. Evaluation in Practical Experiments

To further validate the proposed method, practical indoor experiments were conducted with three self-assembly UAVs equipped with depth cameras flying in a room with obstacles, as shown in Figure 8b. A $10 \times 8 \times 3 \text{ m}^3$ bounding box was used to constrain the space for exploration. Due to the UAV structure, the cameras could only be mounted with a downward pitch angle of $5°$ on the front side, and the UAVs' precise location was ensured through the use of VICON, a motion capture system, for safe piloting. Parameter values for the practical experiments were set based on the simulation experiments conducted for the indoor scenario. Although limitations such as hardware restrictions, network bandwidth, and flight trajectory control were not the primary focus of this paper, multiple trials were carried out to ensure the proposed method's usability. Table 1 summarizes some of the critical parameters employed in the practical experiments.

(a)   (b)

Figure 8. The practical experiments. (**a**) shows the initial status of three UAVs; they are placed on the same side of a room. (**b**) shows three UAVs are performing exploration in one trial; a $10 \times 8 \times 3$ m^3 virtual boundary is set to bound the exploring space.

Table 1. Important parameters in practical experiments.

Parameter	Value	Parameter	Value
v_{max}	0.1 m/s	$\dot{\psi}_{max}$	0.5 rad/s
$d_{max}^{planner}$	1.5 m	d_{max}^{sensor}	2.5 m
λ	0.5	MaxTreeLength	0.8 m
N_{init}	10	N_{MAX}	100
$N_{threshold}$	30	Mounting pitch	5°

The proposed algorithm was tested in a practical experiment involving a team of three UAVs, as shown in Figure 8. The UAVs were initially positioned closely together on the same side of the exploration area, which is typical for real deployments. The exploration process was repeated 20 times, with a maximum exploration time of 232.2 s, a minimum of 194.6 s, and an average of 209.4 s. The decentralized nature of the planner ensures that the UAVs can perform their tasks robustly, with interruptions to one UAV having no impact on the work of others, as demonstrated in Figure 9. The effectiveness and usability of the proposed method in a practical scenario are demonstrated by the exploration maps at six different sampling times, as shown in Figure 10. The virtual centroid in three colors dynamically changes during the exploration process, and the working area of each UAV is partitioned reasonably and iteratively. The UAV denoted as the yellow on the left moves gradually to the lower-left area after completing its task in the upper-left corner and collaborates with the UAV in the lower-right section to adjust the task areas. In the final map of Figure 10, the gap areas on the ground were detected, which were affected by the range of the depth camera. We can assume that using more robust sensors such as 3D LiDAR could alleviate this phenomenon, but such an approach requires greater consideration of the comprehensiveness of the experimental system and its applicability to different settings, which needs to be further considered in future research.

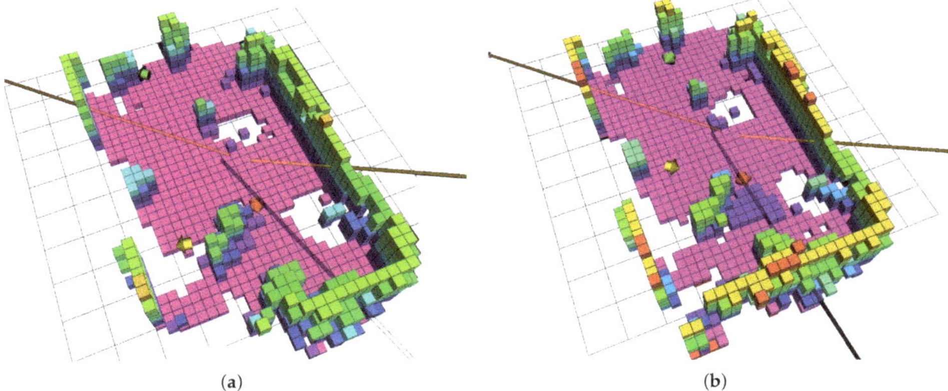

(a)   (b)

Figure 9. Robust coordination case. (**a**) shows UAV represented by the red arrow has stopped exploring due to insufficient power at an early stage; (**b**) shows other UAVs continue to finish the task. The yellow one helps the red to explore the bottom right corner of this environment.

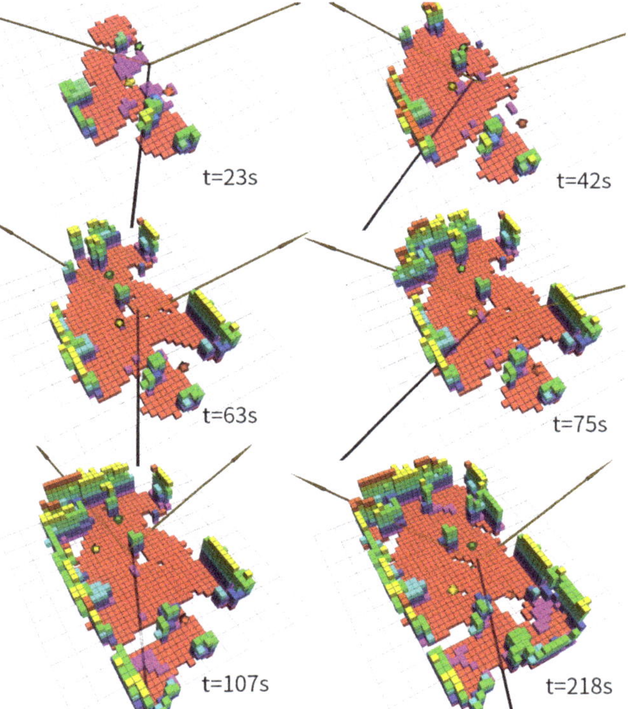

Figure 10. The mapping process of one practical experiment. The sampling times display the complete process of three UAVs collaborating on exploration, with each UAV's designated area being continuously updated. The UAV represented by the yellow icon on the left gradually moves towards the lower region, collaborating with the other UAVs to adjust the exploration area. In the final map, gaps on the ground were influenced by the depth camera's perception range. It is assumed that using more powerful sensors such as 3D LiDAR [29] may mitigate this phenomenon, but this approach necessitates further consideration of the experimental system's applicability.

7. Conclusions and Future Work

To improve the efficiency and reliability of cooperative exploration in unknown environments, a dynamic centroid-based area partition method has been implemented in this study, which takes into account both position and task information throughout the process. This method assigns more space to UAVs that detect fewer tasks, which ensures that each UAV undertakes the exploration mission evenly. Additionally, an NBV exploration method has been improved by sampling candidates in their partitioned area and formulating the weight, which also quantifies the task. The proposed method has been compared to three other methods in two simulation environments and found to be faster than traditional methods. Practical experiments have also demonstrated the effectiveness of this approach. However, the randomness of the sampling process introduces variability, which could lead to local minima. This sampling uncertainty, resulting in an unreasonable partition, prevents us from providing a stable and precise exploration process description in different trials with the same setting. To address this issue, future research will investigate the use of frontier-based methods that provide an accurate number of tasks, allowing for more rational partitioning results. Additionally, alternative sensing methods or equipment could be considered to enhance the perception of swarm UAVs.

Author Contributions: Conceptualization, J.G. and B.D.; methodology, J.G. and T.Y.; software, T.Y.; validation, T.Y., J.G. and X.Z.; formal analysis, T.Y. and J.G.; investigation, J.G.; resources, X.Z. and W.Y.; data curation, T.Y. and J.G.; writing—original draft preparation, T.Y.; writing—review and editing, J.G.; visualization, J.G.; supervision, B.D and W.Y.; project administration, B.D.; funding acquisition, B.D. All authors have read and agreed to the published version of the manuscript.

Funding: This work was supported by the National Natural Science Foundation of China (No. 61902423).

Data Availability Statement: Data available on request from the authors.

Conflicts of Interest: The authors declare no conflict of interest.

References

1. Cieslewski, T.; Kaufmann, E.; Scaramuzza, D. Rapid exploration with multi-rotors: A frontier selection method for high speed flight. In Proceedings of the 2017 IEEE/RSJ International Conference on Intelligent Robots and Systems (IROS), Vancouver, BC, Canada, 24–28 September 2017; pp. 2135–2142. [CrossRef]
2. Wang, C.; Chi, W.; Sun, Y.; Meng, M.Q.H. Autonomous Robotic Exploration by Incremental Road Map Construction. *IEEE Trans. Autom. Sci. Eng.* **2019**, *16*, 1720–1731. [CrossRef]
3. Xu, Z.; Deng, D.; Shimada, K. Autonomous UAV Exploration of Dynamic Environments Via Incremental Sampling and Probabilistic Roadmap. *IEEE Robot. Autom. Lett.* **2021**, *6*, 2729–2736. [CrossRef]
4. Jung, S. Bridge Inspection Using Unmanned Aerial Vehicle Based on HG-SLAM: Hierarchical Graph-Based SLAM. *Remote Sens.* **2020**, *12*, 3022. [CrossRef]
5. Wang, J.; Wu, Y.X.; Chen, Y.Q.; Ju, S. Multi-UAVs collaborative tracking of moving target with maximized visibility in urban environment. *J. Frankl. Inst.* **2022**, *359*, 5512–5532. [CrossRef]
6. Pan, T.; Gui, J.; Dong, H.; Deng, B.; Zhao, B. Vision-Based Moving-Target Geolocation Using Dual Unmanned Aerial Vehicles. *Remote Sens.* **2023**, *15*, 389. [CrossRef]
7. Hu, J.; Niu, H.; Carrasco, J.; Lennox, B.; Arvin, F. Voronoi-Based Multi-Robot Autonomous Exploration in Unknown Environments via Deep Reinforcement Learning. *IEEE Trans. Veh. Technol.* **2020**, *69*, 14413–14423. [CrossRef]
8. Simmons, R.; Apfelbaum, D.; Burgard, W.; Fox, D.; Moors, M.; Thrun, S.; Younes, H. *Coordination for Multi-Robot Exploration and Mapping*; AAAI Press: Palo Alto, CA, USA 2020; pp. 852–858.
9. Yu, J.; Tong, J.; Xu, Y.; Xu, Z.; Dong, H.; Yang, T.; Wang, Y. SMMR-Explore: SubMap-based Multi-Robot Exploration System with Multi-robot Multi-target Potential Field Exploration Method. In Proceedings of the 2021 IEEE International Conference on Robotics and Automation (ICRA), Xi'an, China, 30 May–5 June 2021; pp. 8779–8785. [CrossRef]
10. Masaba, K.; Li, A.Q. GVGExp: Communication-Constrained Multi-Robot Exploration System based on Generalized Voronoi Graphs. In Proceedings of the 2021 International Symposium on Multi-Robot and Multi-Agent Systems (MRS), Cambridge, UK, 4–5 November 2021; pp. 146–154. [CrossRef]
11. Mannucci, A.; Nardi, S.; Pallottino, L. Autonomous 3D Exploration of Large Areas: A Cooperative Frontier-Based Approach. In Proceedings of the Modelling and Simulation for Autonomous Systems, Rome, Italy, 24–26 October 2017; Mazal, J., Ed.; Springer International Publishing: Cham, Switzerland, 2018; pp. 18–39. [CrossRef]

12. Bircher, A.; Kamel, M.; Alexis, K.; Oleynikova, H.; Siegwart, R. Receding Horizon "Next-Best-View" Planner for 3D Exploration. In Proceedings of the 2016 IEEE International Conference on Robotics and Automation (ICRA), Stockholm, Sweden, 16–21 May 2016; pp. 1462–1468. [CrossRef]
13. Lindqvist, B.; Agha-Mohammadi, A.A.; Nikolakopoulos, G. Exploration-RRT: A multi-objective Path Planning and Exploration Framework for Unknown and Unstructured Environments. In Proceedings of the 2021 IEEE/RSJ International Conference on Intelligent Robots and Systems (IROS), Prague, Czech Republic, 27 September–1 October 2021; pp. 3429–3435. [CrossRef]
14. Hornung, A.; Wurm, K.M.; Bennewitz, M.; Stachniss, C.; Burgard, W. OctoMap: An efficient probabilistic 3D mapping framework based on octrees. *Auton. Robot.* **2013**, *34*, 189–206. [CrossRef]
15. Yu, T.; Deng, B.; Gui, J.; Zhu, X.; Yao, W. Efficient Informative Path Planning via Normalized Utility in Unknown Environments Exploration. *Sensors* **2022**, *22*, 8429. [CrossRef] [PubMed]
16. Ju, S.; Wang, J.; Dou, L. MPC-Based Cooperative Enclosing for Nonholonomic Mobile Agents Under Input Constraint and Unknown Disturbance. *IEEE Trans. Cybern.* **2023**, *53*, 845–858. [CrossRef]
17. Fox, D.; Jonathan, K.O.; Konolige, K.; Limketkai, B.; Schulz, D.; Stewart, B. Distributed Multirobot Exploration and Mapping. *Proc. IEEE* **2006**, *94*, 1325–1339. [CrossRef]
18. Tang, Y.; Chen, Y.; Zhou, D. Measuring uncertainty in the negation evidence for multi-source information fusion. *Entropy* **2022**, *24*, 1596. [CrossRef] [PubMed]
19. Hardouin, G.; Moras, J.; Morbidi, F.; Marzat, J.; Mouaddib, E.M. Next-Best-View planning for surface reconstruction of large-scale 3D environments with multiple UAVs. In Proceedings of the 2020 IEEE/RSJ International Conference on Intelligent Robots and Systems (IROS), Las Vegas, NV, USA, 24 October 2020–24 January 2021; pp. 1567–1574. [CrossRef]
20. Oleynikova, H.; Taylor, Z.; Fehr, M.; Siegwart, R.; Nieto, J. Voxblox: Incremental 3D Euclidean Signed Distance Fields for on-board MAV planning. In Proceedings of the 2017 IEEE/RSJ International Conference on Intelligent Robots and Systems (IROS), Vancouver, BC, Canada, 24–28 September 2017; pp. 1366–1373. [CrossRef]
21. Li, H.; Tsukada, M.; Nashashibi, F.; Parent, M. Multivehicle Cooperative Local Mapping: A Methodology Based on Occupancy Grid Map Merging. *IEEE Trans. Intell. Transp. Syst.* **2014**, *15*, 2089–2100. [CrossRef]
22. Corah, M.; O'Meadhra, C.; Goel, K.; Michael, N. Communication-Efficient Planning and Mapping for Multi-Robot Exploration in Large Environments. *IEEE Robot. Autom. Lett.* **2019**, *4*, 1715–1721. [CrossRef]
23. Schmid, L.; Reijgwart, V.; Ott, L.; Nieto, J.; Siegwart, R.; Cadena, C. A Unified Approach for Autonomous Volumetric Exploration of Large Scale Environments Under Severe Odometry Drift. *IEEE Robot. Autom. Lett.* **2021**, *6*, 4504–4511. [CrossRef]
24. Zhou, B.; Zhang, Y.; Chen, X.; Shen, S. FUEL: Fast UAV Exploration Using Incremental Frontier Structure and Hierarchical Planning. *IEEE Robot. Autom. Lett.* **2021**, *6*, 779–786. [CrossRef]
25. Lee, E.M.; Choi, J.; Lim, H.; Myung, H. REAL: Rapid Exploration with Active Loop-Closing toward Large-Scale 3D Mapping using UAVs. In Proceedings of the 2021 IEEE/RSJ International Conference on Intelligent Robots and Systems (IROS), Prague, Czech Republic, 27 September–1 October 2021; pp. 4194–4198. [CrossRef]
26. Zhu, H.; Cao, C.; Xia, Y.; Scherer, S.; Zhang, J.; Wang, W. DSVP: Dual-Stage Viewpoint Planner for Rapid Exploration by Dynamic Expansion. In Proceedings of the 2021 IEEE/RSJ International Conference on Intelligent Robots and Systems (IROS), Prague, Czech Republic, 27 September–1 October 2021; pp. 7623–7630. [CrossRef]
27. Schmid, L.; Pantic, M.; Khanna, R.; Ott, L.; Siegwart, R.; Nieto, J. An Efficient Sampling-Based Method for Online Informative Path Planning in Unknown Environments. *IEEE Robot. Autom. Lett.* **2020**, *5*, 1500–1507. [CrossRef]
28. Charrow, B.; Kahn, G.; Patil, S.; Liu, S.; Goldberg, K.; Abbeel, P.; Michael, N.; Kumar, V. Information-Theoretic Planning with Trajectory Optimization for Dense 3D Mapping. In Proceedings of the Robotics: Science and Systems, Rome, Italy, 13–17 July 2015; Volume 11, pp. 3–12. [CrossRef]
29. Li, S.; Tian, B.; Zhu, X.; Gui, J.; Yao, W.; Li, G. InTEn-LOAM: Intensity and Temporal Enhanced LiDAR Odometry and Mapping. *Remote Sens.* **2022**, *15*, 242. [CrossRef]

Article

BCDAIoD: An Efficient Blockchain-Based Cross-Domain Authentication Scheme for Internet of Drones

Gongzhe Qiao [1], Yi Zhuang [1,*], Tong Ye [1] and Yuan Qiao [2]

[1] College of Computer Science and Technology, Nanjing University of Aeronautics and Astronautics, Nanjing 211100, China; qgz@nuaa.edu.cn (G.Q.); yetong@nuaa.edu.cn (T.Y.)
[2] Harbin Electric Power Bureau, STATE GRID Corporation of China, Harbin 150050, China; qiaoyuan_qy@163.com
* Correspondence: zy16@nuaa.edu.cn

Abstract: During long-distance flight, unmanned aerial vehicles (UAVs) need to perform cross-domain authentication to prove their identity and receive information from the ground control station (GCS). However, the GCS needs to verify all drones arriving at the area it is responsible for, which leads to the GCS being unable to complete authentication in time when facing cross-domain requests from a large number of drones. Additionally, due to potential threats from attackers, drones and GCSs are likely to be deceived. To improve the efficiency and security of cross-domain authentication, we propose an efficient blockchain-based cross-domain authentication scheme for the Internet of Drones (BCDAIoD). By using a consortium chain with a multi-chain architecture, the proposed method can query and update different types of data efficiently. By mutual authentication before cross-domain authentication, drones can compose drone groups to lighten the authentication workload of domain management nodes. BCDAIoD uses the notification mechanism between domains to enable path planning for drones in advance, which can further improve the efficiency of cross-domain authentication. The performance of BCDAIoD was evaluated through experiments. The results show that the cross-domain authentication time cost and computational overhead of BCDAIoD are significantly lower those of than existing methods when the number of drones is large.

Keywords: blockchain; Internet of Drone; cross-domain authentication; security protocol; multi-UAVs; group policy

Citation: Qiao, G.; Zhuang, Y.; Ye, T.; Qiao, Y. BCDAIoD: An Efficient Blockchain-Based Cross-Domain Authentication Scheme for Internet of Drones. *Drones* **2023**, *7*, 302. https://doi.org/10.3390/drones7050302

Academic Editors: Zhihong Liu, Shihao Yan, Yirui Cong and Kehao Wang

Received: 3 April 2023
Revised: 28 April 2023
Accepted: 3 May 2023
Published: 4 May 2023

1. Introduction

The increasing demand for unmanned aerial vehicles (UAVs) and Internet of Drones (IoD) in civil and military applications has been noticed in the last few decades [1]. IoD usually consists of a number of drones, ground control stations (GCS), and a communication network for data exchange. Drones can obtain information through sensors and exchange information through the network. They can also communicate with GCS and other drones for proper navigation [2]. The available space of the air traffic network (ATN) is much larger than that of the ground traffic network (GTN). Reasonable use of the ATN can reduce the burden of the GTN. Furthermore, using ATN can also avoid the congestion of the GTN. Therefore, many companies try to use drones as air transport tools for cargo transportation, so as to promote logistics efficiency [3].

During long-distance flights, drones are likely to enter other IoD domains where they need to pass identity authentication and obtain necessary information such as flight routes to continue their tasks. In the IoD, wireless communication between drones is easy to eavesdrop on, resulting in information leakage [4–7]. Furthermore, attackers can conduct replay attacks or identity forgery to interrupt the IoD [8–10]. In addition, once drone transportation forms a complete industry, the number of drones may be quite large. The resource overhead of complex authentication mechanisms, such as computation and storage, is a huge challenge for identity authentication servers [11–13]. At the same time, it also increases the cross-domain waiting time for drones to obtain necessary information.

Centralized authentication techniques [14–16] are widely adopted in traditional information systems. However, in an IoD environment, if conventional centralized authentication approaches are applied, the workload of the authentication service center increases exponentially as the system scales up [17]. Therefore, traditional authentication technology is not suitable for the IoD environment. At present, there are also some studies using public key infrastructure (PKI), trusted third party (TTP), or blockchain technology to verify the identity of the drone and the authenticity of the task [18,19]. The existing methods are shown in Figure 1a. However, the existing studies rarely consider the efficiency of cross-domain authentication in the case of a large number of drones. Additionally, when the number of drones is large, the time cost of cross-domain authentication is also very high.

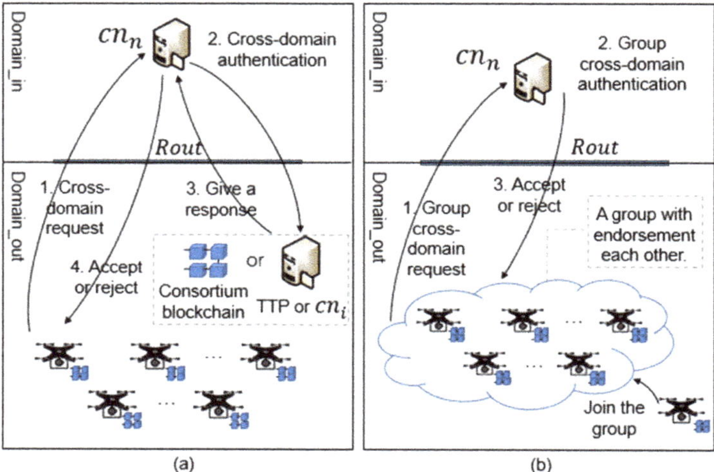

Figure 1. Comparison of existing methods and this paper. (**a**) Ideas of existing methods. (**b**) The idea of the method proposed in this paper.

When facing a large number of drones, existing drone cross-domain authentication methods lead to a surge in communication and computational costs for the IoD, increase the waiting time of the drones, and further reduce the overall operational efficiency of the IoD. At the same time, the existing methods cannot guarantee the anonymity of drones when performing the task, and attackers can obtain the connection between the sender and receiver through the identity and task information of the drone, which may leak the user's privacy. Therefore, existing methods cannot meet the cross-domain authentication requirements for large-scale drone cargo transportation scenarios. Therefore, we propose a novel cross-domain authentication scheme for the IoD based on blockchain technology.

Blockchain has the advantages of decentralization, openness, being tamper-proof, traceability, and anonymity [20]. The decentralization is consistent with the distributed network structure of the IoD. Furthermore, the openness can well satisfy the requirements for drones to join and leave the network freely [21]. The advantages of being tamper-proof and traceability ensure the reliability and non-repudiation of data in the network, and the advantage of anonymity protects the data by addressing with address instead of addressing with identity. However, there are three potential challenges to using blockchain for authentication. (1) In the IoD, there are various types of data and the scale of data is large, meaning the single-chain structure may lead to query inefficiency and throughput bottleneck. (2) In the authentication process, a large number of drones may cause a high authentication delay time and increase the waiting time of drones. (3) The authentication server needs to authenticate the UAV identity information, verify the transaction in the blockchain, and upload the task information. A large number of drones to be authenticated may cause server interruptions.

To address the above challenges, we propose an efficient cross-domain authentication scheme based on blockchain. As shown in Figure 1b, the main idea is that drones should authenticate each other before cross-domain authentication to form a mutually endorsed group. To ensure that the identities and tasks of drones in each group are real and trusted, we designed an authentication mechanism based on domain signature, encryption, and domain private chain. In this way, drones with the same out-of-domain range (Rout) compose a drone group. Considering the continuity of drone tasks in the process of cross-domain tasks, we designed a notification mechanism between domains combined with the concepts of token, permission, and authority. The main contributions of this paper are as follows.

(1) We propose an efficient blockchain-based cross-domain authentication scheme for the Internet of Drones (BCDAIoD). By using a consortium chain with a multi-chain architecture, the proposed method can query and update different types of data efficiently, which can also facilitate the domain management node to manage and control the drones. Additionally, we describe the mission model of the drones.

(2) In order to improve the efficiency of the cross-domain authentication of drones, we designed an establishment method of drone groups and a group cross-domain authentication method based on blockchain, encryption, and challenge–response game. By mutual authentication before cross-domain authentication, drones can compose drone groups to lighten the authentication workload of domain management nodes and improve the efficiency of cross-domain authentication.

(3) We propose a notification mechanism between domains that can enable the management node of the next domain to know the task information of the drones in advance. The management node of the next domain can plan space resources reasonably and plan the flight path for drones in advance, which can also ensure the continuity of the tasks of drones.

The remainder of this article is organized as follows. The literature is reviewed in Section 2. The framework of BCDAIoD, the consortium blockchain architecture, and the mission model of the drones are presented in Section 3. In Section 4, we first propose the single-drone cross-domain authentication method. Then, we propose the establishment mechanism of drone groups, the drone group cross-domain authentication method, and the notification mechanism between domains. Section 5 describes the simulation experiments and shows the experimental evaluation results of the BCDAIoD method. In Section 6, we analyze the security of BCDAIoD. Finally, Section 7 concludes this paper.

2. Related Works

(1) IoD management scheme

IoD has received widespread attention due to its potential application prospects. Therefore, there are studies targeting the management scheme of IoD. In order to facilitate the information acquisition of drones and users in IoD, Al-Hilo et al. [22] proposed a collaborative and management framework between UAVs and roadside units. Arafeh et al. [23] proposed a blockchain-based UAV management method that can verify the authenticity of information in IoD networks. By using blockchain and trust policies, García-Magariño et al. [24] proposed a UAV management approach which can also maintain security in IoD by corroborating information about events from different sources. However, the above UAV management framework mainly focuses on data security and neglects the management efficiency when facing a large number of drones. Additionally, the blockchain-based methods have not been able to reasonably partition the data storage on the chain, leading to low query efficiency and data isolation.

(2)   Cross-domain authentication scheme based on cryptography

During the task execution, drones need to verify their identity through cross-domain authentication. Meanwhile, in order to address potential threats, researchers have proposed some cross-domain authentication schemes based on cryptography. To protect drones against various types of possible attacks, Wazid et al. [25] proposed a remote authentication and key management scheme. Srinivas et al. [26] designed a three-factor authentication scheme which relies on elliptic-curve cryptography (ECC). Tanveer et al. [27] leveraged ECC along with symmetric encryption and hash function, and proposed a robust authentication mechanism for IoD. For the sensitive environment in which an attacker might trap data from the open network channel, Jan et al. [28] proposed a verifiably secure ECC-based authentication scheme for IoD. Additionally, Rajamanickam et al. [29] proposed an ECC-based authentication protocol for insider attack protection in IoD scenarios that can protect the IoD from several known attacks, especially insider attacks. Ever et al. [30] proposed a secure authentication framework using elliptic-curve crypto-systems to ensure data confidentiality. However, the above methods guarantee the security of IoD by performing multiple process parameter calculations on the device, registration center, and control center, which increase the computational burden.

(3)   Cross-domain authentication scheme based on blockchain

Considering the interoperability and complexity characteristics of IoD systems, many existing research studies have applied blockchain technology in the cross-domain authentication area of IoT systems. To solve the single point of failure problem, Feng et al. [31] employed blockchain and multiple signatures based on threshold sharing to build a cross-domain authentication framework. To avoid reliance on a trusted third party, Shen et al. [32] introduced a consortium blockchain to construct trust among different domains and present an efficient blockchain-assisted secure device authentication mechanism. By using a hierarchy of local and global smart contracts, Gauhar et al. [33] proposed a decentralized blockchain-based authentication mechanism which uses a proof-of-authenticity mechanism to find and retrieve device hashes stored in local blockchain. Zhang et al. [34] proposed a thoroughly cross-domain authentication scheme based on blockchain, allowing participants from different domains with different settings to complete the authentication. However, the above methods neglect the anonymity of drones when performing the task, which may leak the user's privacy. Additionally, the methods do not take into account the communication and computational costs of IoD when facing a large number of drones.

3. Overview of BCDAIoD

3.1. The BCDAIoD Framework

To improve the efficiency of data query, BCDAIoD uses a multi-chain architecture to reasonably partition the data storage on the consortium blockchain. Additionally, domain private chains are used to ensure the anonymity of drones. The framework of the proposed BCDAIoD scheme is shown in Figure 2. The BCDAIoD framework includes four layers: an application layer, service support layer, data storage layer, and network layer.

In the network layer, a P2P network is used for communication between consortium nodes (CNs) and UAVs. Additionally, CNs can transmit information through the P2P network, such as mission information. Each CN manages a certain domain and maintains a private chain. The UAVs can also communicate with each other through the P2P network.

In the data storage layer, the UAV device information, task information, address information of CNs, smart contracts, and user registration information are stored in the consortium blockchain (CBC) in a specified format. At the same time, the CNs store the details of the above information in local databases. The architecture of the consortium blockchain is described in Section 3.2. For identity authentication and UAV grouping, the CNs also maintain a private chain to store the device ID on PBC (Pid) and the out-of-domain range ($Rout$) information of UAVs.

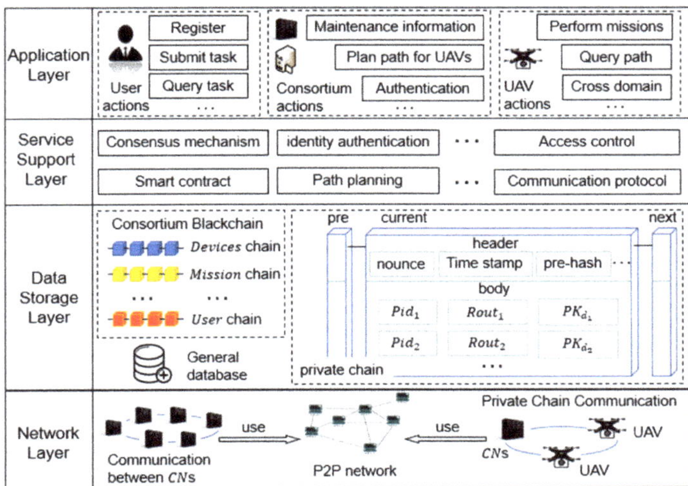

Figure 2. The BCDAIoD framework.

The service support layer provides support for users, CNs, and UAVs to interact with the data storage layer and the network layer. It mainly includes the consensus mechanism, smart contract, identity authentication mechanism, access control strategy, path planning algorithm, and communication protocol. The consensus mechanism can ensure that the information of each block is consistent. The smart contract can conduct trusted transactions in the form of commitment without a trusted third party. The identity authentication mechanism and access control strategy can ensure that UAVs enter the correct domains and obtain their own path information. By using the path planning algorithm, the CNs obtain the next domain IDs and calculate the paths in their domains for UAVs. The communication protocol supports the communication among users, UAVs, and CNs.

By using the functions of the application layer, users can submit registration applications to the CNs, release tasks, and query the status of tasks. The CNs can manage the information of the UAV devices, tasks, and paths, as well as perform identity authentication and path planning. The UAVs can query task information, view path information, and submit cross-domain authentication requests to perform tasks. To facilitate the introduction of subsequent methods, Table 1 lists the key symbols and their definitions.

Table 1. The symbol definitions.

Symbol	Definition	Symbol	Definition
CBC	Consortium blockchain	PBC	Private blockchain
CN	Consortium node	$CNid$	ID of consortium node
Pid_i	Device ID of i on PBC	Did_i	Device ID of i on CBC
Mid	Task ID	$GList$	Group member list
$Rout$	Out-of-domain range	$Tout$	Expected time out of domain
d_i	Drone marked with i	PK_i	Public key of i
SK_i	Private key of i	$Enc(*)$	Elliptic curve encryption
PBH_i	Block number of the PBC block that includes the task information of i	ks	Session key
$Token$	Cross-domain token	$H(*)$	Hash function

Table 1. *Cont.*

Symbol	Definition	Symbol	Definition
T_{stamp}	Time stamp of the *Token*	M_{d_j}	ID verification information
ack	Challenge	rsp_i	Response of i

3.2. Architecture of Consortium Blockchain

The scale of data to be processed by the CNs is very large and there are various types of data. All the data being stored in a single chain leads to low query efficiency, poor data isolation, and difficulty to expand. Therefore, we designed the multi-chain architecture of the CBC to improve the efficiency of the CNs. As shown in Figure 3, the multi-chain architecture of the CBC mainly includes the *Mission* chain, *User* chain, *Address* chain, *Devices* chain, and *Contract* chain. The CNs are also responsible for the private blockchain (PBC) in their domain. Similar to cellular mobile communication networks, the proposed framework achieves service coverage in large-scale open areas by combining CNs. Additionally, there are cross-domain channels between the domains, so as to facilitate the formation of groups and the cross-domain authentication of UAVs with the same $Rout$.

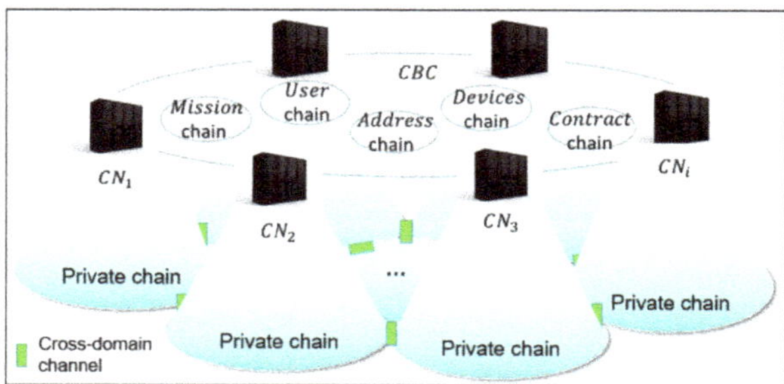

Figure 3. The multi-chain architecture of CBC.

The *Mission* chain mainly stores task information, and the data structure can be expressed as Equation (1), where Mid is the current task ID; Did is the drone's ID on the CBC that can be provided to or queried by the CNs; $CNid_n$ is the ID of the CN that is responsible for the next domain; $CNid_d$ is the ID of the CN that manages the destination domain; $CNid_c$ is the CN ID of the current domain; $Rout$ is the expected range out of the current domain; and $Tout$ represents the current expected time out of the domain.

$$Mission = (Mid, Did, CNid_n, CNid_d, CNid_c, Rout, Tout) \tag{1}$$

The *User* chain mainly stores the necessary information of the registered users. The stored information can be expressed as Equation (2), where Uid is the user ID for the consortium authentication, $CNid_r$ is the CN ID of the registration place, and Fu is the current user account balance. The other registration details of the user are directly stored in the local database of the registration place's CN.

$$User = (Uid, CNid_r, Fu) \tag{2}$$

The *Address* chain mainly stores the domain range information of the CN, which can be expressed as Equation (3), where $CNid$ represents the ID of a CN; PK_{cn} represents the public key of the CN; and RG_{cn} represents the domain range of the CN.

$$Address = (CNid, PK_{cn}, RG_{cn}) \tag{3}$$

The *Devices* chain mainly stores the necessary information of the UAV registered in the *CBC*, which can be expressed as Equation (4), where PK_d is the public key of a UAV; *Mod* is the module of the UAV; and *Pol* is the execution strategy of the UAV. The registration details of other UAVs can be directly stored in the local databases of the registration place's *CN*.

$$Devices = (Did, PK_d, CNid_r, Mod, Pol) \tag{4}$$

The *Contract* chain mainly stores the contract information, which can be expressed as Equation (5), where *Cid* is the current contract ID, *V* is the current contract version, and *Cont* is the current contract content.

$$Contract = (Cid, V, Cont) \tag{5}$$

The *PBC* chain mainly stores the necessary information for the intradomain authentication of the UAVs, which can be expressed as Equation (6), where *Pid* represents the temporary ID of a UAV in a domain.

$$PBC = (Pid, Rout, PK_d) \tag{6}$$

3.3. Mission Model of Drone

We designed a scenario of a drone delivery in an Internet of Drones environment, as shown in Figure 4. Each domain *CN* in the consortium blockchain can be regarded as a server with strong computing power and storage capacity, which is responsible for drone management and scheduling in a certain location area. The drone delivery process mainly includes three phases: registration, task release, and task execution.

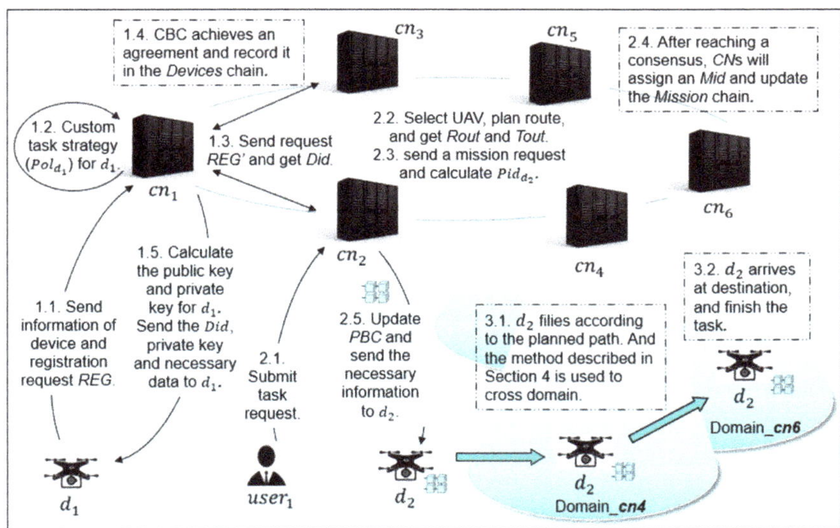

Figure 4. Scenario of drone delivery.

3.3.1. Registration Mechanism

In the proposed framework, both drones and users need to register in the *CBC*. The drone registration process includes the following steps. The user registration process is similar to the drone registration process, which is not described in detail here.

Step 1. Drone d_j submits a registration request $REG = \left(Rrequest, mac_{d_j}, inf_{d_j} \right)$ to cn_i, where *Rrequest* denotes registration request, mac_{d_j} denotes the hardware ID of d_j, and inf_{d_j}

denotes the detailed information of d_j, such as drone model, maximum sailing distance, and payload.

Step 2. After receiving the registration request, cn_i determines the acceptable task type of d_j (Pol_{d_j}), such as selecting tasks with reasonable distance according to power consumption.

Step 3. Then, cn_i sends the registration request $REG' = \left(Rrequest,\ mac_{d_j}, CNid_i, Pol_{d_j} \right)$ to the other nodes in the CBC.

Step 4. After the members of the CBC reach a consensus, the CNs assign a device ID in the CBC (Did_{d_j}) to the d_j and update the *Devices* chain. Furthermore, cn_i calculates the public key (PK_{d_j}) and private key (SK_{d_j}) for d_j. Then, cn_i sends the SK_{d_j} to d_j.

Step 5. cn_i sends the Did_{d_j} and its public key PK_{cn_i} to d_j. Additionally, cn_i calculates $Enc_{SK_{cn_i}} \left(Did_{d_j} \right)$ and sends it to d_j. Then, cn_i stores the registration time, model, and other detailed information of d_j in the local database.

3.3.2. Task Release Mechanism

When a user submits a delivery task request, the CN in charge of the current domain selects a suitable drone to perform the task. The process includes the following steps.

Step 1. When a user submits a delivery task request to the cn_i, cn_i queries the available drone from the local database, selects an appropriate drone d_k according to the acceptable task type, and plans a delivery path for d_k. Additionally, cn_i calculates $Rout$ and $Tout$.

Step 2. At the same time, cn_i queries the *Address* chain to find the ID of the destination domain ($CNid_d$) and the next domain ($CNid_n$) that d_k will pass through.

Step 3. Then, cn_i sends a mission request *mission* $= (Did, CNid_n,\ CNid_d,\ CNid_c, Rout, Tout)$ to the other CNs in the CBC and calculates the ID on the PBC (Pid_{d_k}) for d_k. Then, cn_i uses SK_{cn_i} to generate $Enc_{SK_{cn_i}} \left(Pid_{d_k} \right)$.

Step 4. After the members of the CBC reach a consensus, the CNs assign the mission ID (Mid) and update the *Mission* chain with *mission'* $= (Mid,\ Did,\ CNid_n,\ CNid_d, CNid_c, Rout, Tout)$.

Step 5. Then, cn_i updates the PBC, and sends the necessary information to the drone d_k. The information is shown in Table 2. Then, d_k starts the task. Correspondingly, d_k updates its own PBC during the execution of the task.

Table 2. The symbol definitions.

Symbol	Definition
Flight path	The flight path in the current domain.
Pid_{d_k}	Device ID of d_k on the PBC.
SK_{d_k}	Private key of d_k.
PK_{cn_i}	Public key of cn_i for intradomain authentication between drones.
PK_{cn_n}	Public key of cn_n (the CN of the next domain) for cross-domain authentication.
$Enc_{SK_{cn_i}} \left(Did_{d_k} \right)$	The identification for cross-domain authentication.
$Enc_{SK_{cn_i}} \left(Pid_{d_k} \right)$	The identification for intradomain authentication.
PBC	The private blockchain of the current domain.
PBH_{d_k}	The block number of the PBC block that includes the task information of d_k.

Step 6. Finally, the cn_i stores the detailed information in the local database and removes the drone d_k from the available drones of the local database.

Furthermore, we make the following assumptions:

(1) We consider that each CN has the public keys of the other CNs, and the private key of each CN is not leaked; (2) The private key of the drone carrying out the task is not

leaked; (3) Attackers cannot deduce the private key from the public key, or it takes too much time.

3.3.3. Task Execution Method

After the task starts, the drone d_k flies to the destination according to the planned path. If d_k needs to pass through other domains, the method described in Section 4 is used for cross-domain authentication. After successful cross-domain authentication, d_k enters the next domain and receives the necessary information from the CN, in a similar method to Step 5 in the task release phase. In this way, d_k flies to the destination domain.

When d_k reaches the destination domain and completes the task, the CN in charge of the destination domain (cn_d) publishes a task completion confirmation, $mission_F = (Mid, Did, 0, CNid_d, CNid_d, 0, 0)$, in the $Mission$ chain to prove that the task has been completed.

Finally, the cn_d is responsible for recycling the drone d_k. By querying the Did of d_k and other information in the $Devices$ chain, cn_d stores the necessary information of d_k in the local database and updates the available device database.

4. Cross-Domain Authentication of Drones

Drones may fly to other domains during the execution of tasks. Therefore, we designed a UAV cross-domain authentication method combining public key infrastructure and blockchain technology. In this section, we first propose a single-drone cross-domain authentication method. Then, considering that the cross-domain authentication of a large number of UAVs may cause a long waiting time for UAVs, we propose an establishment mechanism of UAV groups and a drone group cross-domain authentication method. Additionally, we propose a notification mechanism between domains to let the CNs prepare for UAV cross-domain and path planning in advance.

4.1. Single-Drone Cross-Domain Authentication Method

The single-drone cross-domain authentication method is shown in Figure 5. This method uses public key infrastructure and a challenge–response mechanism to ensure the authenticity of the UAV's identity, and ensures the authenticity of the UAV's task by querying the $Mission$ chain of the CBC. At the same time, a session key can also be generated during this process.

When drone d_j wants to fly to the next domain ($Domain_n$), d_j firstly send its cross-domain request (CRM_{d_j}) to the CN in charge of $Domain_n$ (cn_n). The CRM_{d_j} can be expressed as Equation (7), where $Crequest$ represents the cross-domain request and $Enc_{SK_{cn_i}}\left(Did_{d_j}\right)$ represents the Did of d_j encrypted with the private key of the current domain CN.

$$CRM_{d_j} = \left(Crequest, Enc_{SK_{cn_i}}\left(Did_{d_j}\right)\right) \tag{7}$$

After receiving the CRM_{d_j} from d_j, cn_n uses the PK_{cn_i} to decrypt the $Enc_{SK_{cni}}\left(Did_{d_j}\right)$ to obtain the Did_{d_j}. Then, cn_n checks whether there is the incomplete $Mission$ chain transaction $trans^*$ of the corresponding Did_{d_j}. The $trans^*$ can be generated through the notification mechanism (detailed in Section 4.4). Then, cn_n searches the $Devices$ chain to obtain the public key (PK_{dj}) of d_j according to the Did_{d_j}. If $trans^*_{d_j}$ exists, cn_n sends a random number x with PK_{d_j} encryption to d_j as a challenge ack. Then, d_j decrypts ack with SK_{d_j} to obtain x, and sends back $x + 1$ and a random number y with PK_{cn_n} encryption as a response, $rsp_j = Enc_{PK_{cn_n}}(x + 1 \| y)$. Then, cn_n checks rsp_j to determine whether d_j has the declared identity and obtains y. If d_j passes the verification, cn_n sends back a response, $rsp_{cn} = Enc_{PK_{d_j}}(y + 1)$, as a confirmation message. Then, the session key between d_j and cn_n can be generated by $ks = H(x \| y)$.

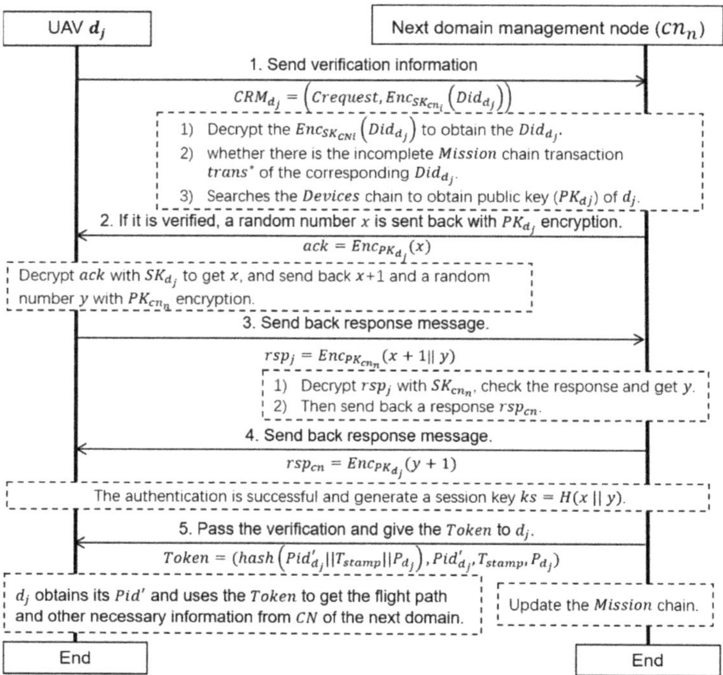

Figure 5. The single-drone cross-domain authentication method.

In this way, cn_n can determine the identity and task information of d_j. Then, cn_n sends the $Token$ to d_j. The $Token$ of d_j can be expressed as Equation (8), where, Pid'_{d_j} represents the device ID of d_j in the $Domain_n$, T_{stamp} represents the time stamp of the $Token$, P_{d_j} represents the permission that d_j has for obtaining the necessary information, and $hash\left(Pid'_{d_j}||T_{stamp}||P_{d_j}\right)$ represents the hash value of the combination of Pid'_{d_j}, T_{stamp}, and P_{d_j}.

$$Token = \left(hash\left(Pid'_{d_j}||T_{stamp}||P_{d_j}\right), Pid'_{d_j}, T_{stamp}, P_{d_j}\right) \tag{8}$$

Finally, d_j obtains its Pid'_{d_j} and uses the $Token$ to obtain the flight path and other necessary information from the CN of the next domain. At the same time, cn_n updates the *Mission* chain by using the method proposed in Section 4.4.

In the UAV cargo transportation scenario, there are many UAVs flying to the same next domain. Therefore, we propose a method of drone group cross-domain authentication to improve the efficiency of UAV cross-domain authentication. The idea of this method is that UAVs compose a group through mutual authentication before the cross-domain authentication. In this way, the proposed method can lighten the authentication workload of the CN and improve the speed of the UAV cross-domain authentication. The method is mainly divided into two stages: (1) the formation of a UAV group (in Section 4.2), and (2) the cross-domain authentication of a UAV group (in Section 4.3).

4.2. Establishment Mechanism of UAV Groups

The establishment mechanism of UAV groups mainly includes two parts: (1) the method of building a new drone group and (2) the method of joining a drone group. In the process of building or joining a drone group, drones need to use verification strategies to verify each other. The verification strategy is shown in Figure 6.

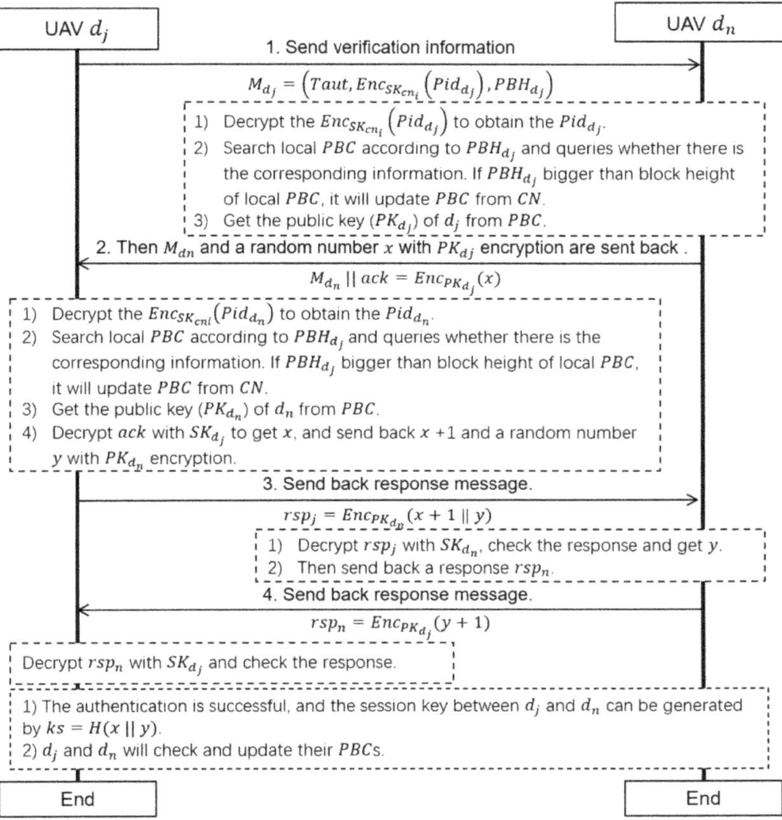

Figure 6. Two-way authentication strategy between drones.

When drone d_j and drone d_n start the verification, d_j firstly sends its verification information M_{d_j} to d_n. M_{d_j} can be expressed as Equation (9), where, $Taut$ represents the two-way authentication request, $Enc_{SK_{cn_i}}\left(Pid_{d_j}\right)$ represents the Pid of d_j encrypted with the private key of the current domain CN, and PBH_{d_j} represents the PBC block height of the block that includes the task information of d_j.

$$M_{d_j} = \left(Taut, Enc_{SK_{cn_i}}\left(Pid_{d_j}\right), PBH_{d_j}\right) \tag{9}$$

After receiving the verification information from d_j, d_n uses the PK_{cn_i} to decrypt the $Enc_{SK_{cn_i}}\left(Pid_{d_j}\right)$ to obtain the Pid_{d_j}. Then, d_n searches the local PBC according to PBH_{d_j} and queries whether the corresponding information is there. If PBH_{d_j} is bigger than the block height of the local PBC, it updates the PBC from the CN. After that, d_n obtains the public key (PK_{d_j}) of d_j from the PBC. Then, a random number x is encrypted by PK_{d_j} as a challenge ack, and d_n sends its M_{d_n} and ack to d_j.

After receiving the M_{d_n} and ack, d_j decrypts the $Enc_{SK_{cni}}(Pid_{d_n})$ to obtain the Pid_{d_n}. Additionally, d_j searches the local PBC according to PBH_{d_j} and queries whether the corresponding information is there. If PBH_{d_j} is bigger than the block height of the local PBC, it updates the PBC from the CN. Then, d_j obtains the public key (PK_{d_n}) of d_n from the PBC. Additionally, d_j decrypts the ack with SK_{d_j} to obtain x, and sends back $x + 1$ and a random number y with PK_{d_n} encryption as a response, $rsp_j = Enc_{PK_{d_n}}(x+1||y)$. According to the notification mechanism between domains described in Section 4.4, local $PBCs$ saved by drones store task information for a period of time in the future, and the $PBCs$ can be updated by drones after mutual authentication. Therefore, in theory, drones do not need or rarely need to update blocks through the CN, and they only need to update blocks through the CN at most once during a two-way authentication period.

Next, d_n decrypts rsp_j with SK_{d_n} and checks whether $x+1$ is received within a certain period of time to determine whether d_j has the declared identity. Then, d_n obtains y and sends back a response, $rsp_n = Enc_{PK_{d_j}}(y+1)$, to d_j. After receiving the rsp_n, d_j decrypts the rsp_n with SK_{d_j} and checks the response. After successful identity authentication, the session key between d_j and d_n can be generated by $ks = H(x \,||\, y)$. Then, d_j and d_n can communicate with each other and update their $PBCs$.

By using the proposed verification strategy, drones can confirm each other's identity, generate session keys, and update the $PBCs$. In the process of moving, drones try to join a group or build a new one, as shown in Figure 7.

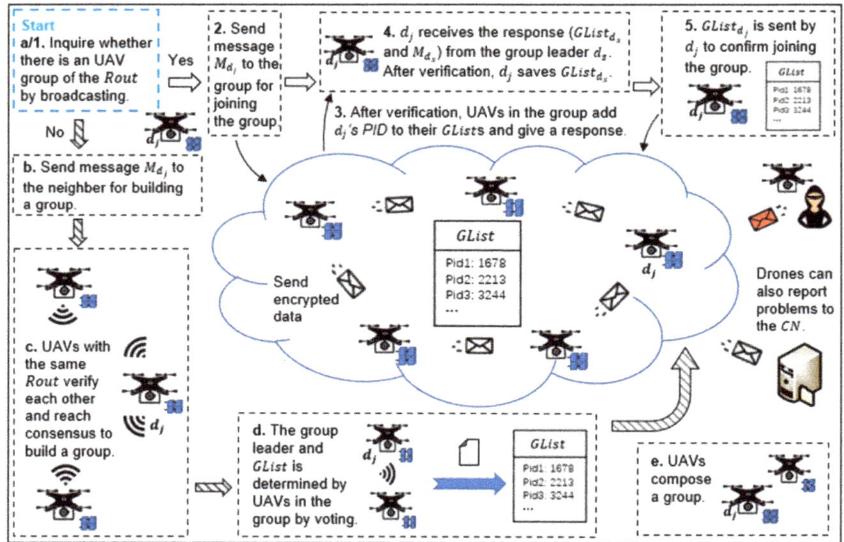

Figure 7. The establishment method of UAV groups.

(1) Method of building a new drone group

During flight in the current domain, a drone d_j broadcasts to inquire whether there is a drone group with the same *Rout*. If there is no drone group, d_j tries to build a new group and continues to broadcast to inquire whether there are drones with the same *Rout*. If d_j finds other drones with the same *Rout*, the drones and d_j send each other verification information, verify each other (as shown in Figure 6), and then reach consensus to build a group. Usually, the number of initial group members is small (about 2–4 drones). In order to stabilize the drone group, the initial members need to authenticate each other and build communication links with a session key.

Each drone group selects a group leader d_l by voting. For cross-domain authentication, d_l generates a member list, $GList$, of the drone group. The $GList$ can be expressed as Equation (10), where Pid_{d_i} represents the Pid of d_i. Then, the drones compose a drone group successfully.

$$GList = \{ Pid_{d_i} \mid \forall \, d_i \text{ in the group} \} \tag{10}$$

(2) Method of joining a drone group

If d_j finds a group after broadcasting, it sends verification information M_{d_j} to the group leader (d_l) to join the group. Then, d_l randomly chooses one drone to verify the d_j together. After that, d_l determines whether d_j can join the group according to the authentication result. If d_j passes the verification, d_l adds the *Pid* and public key of d_j to the $GList_{d_l}$ maintained by itself, and send its $GList_{d_l}$ to d_j. Then, d_j saves the $GList_{d_l}$ and broadcasts its own $GList_{d_j}$ as a confirmation of joining the group. Additionally, the other drones in the group update their $GList$.

At the same time, in order to prevent the loss of drones, the drones in the group send inquiry and response signals regularly. Additionally, drones with forged identities cannot build or join a group because they cannot send the correct response.

4.3. Drone Group Cross-Domain Authentication Method

The drone group cross-domain authentication method proposed in this paper is shown in Figure 8. When a drone group is ready for cross-domain authentication, the group leader of the group (d_l) sends a group cross-domain request (GCM_{d_l}) to the CN of the next domain (cn_n). The GCM_{d_l} sent by d_l can be expressed as Equation (11), where $GCrequest$ represents the group cross-domain request, $Enc_{SK_{cn_i}}(Did_{d_l})$ represents the Did of d_l encrypted with the private key of the current domain CN (cn_i), and $GList_{d_l}$ is the group member list of d_l.

$$GCM_{d_l} = \left(GCrequest,\ Enc_{SK_{cn_i}}(Did_{d_l}), GList_{d_l} \right) \tag{11}$$

After receiving the GCM_{d_l} from d_l, cn_n verifies the group leader through the single-drone cross-domain authentication method proposed in Section 4.1. Then, cn_n obtains the $GList_{d_l}$. If d_l passes validation, cn_n sends d_l a *Token*. After receiving the *Token*, d_l sends a group cross-domain signal to the drone group. Then, the other drones in the group send group cross-domain requests to cn_n. The group cross-domain request sent by d_j (GCM_{d_j}) can be expressed as Equation (12), where $Enc_{PK_{cn_n}}\left(Pid_{d_j}, x_j \right)$ represents the device ID on the PBC and a random number x_j encrypted with the public key of cn_n. Additionally, $Enc_{PK_{cn_n}}\left(Pid_{d_j}, x_j \right)$ is generated by d_j after d_j joins or builds a group.

$$GCM_{d_j} = \left(GCrequest, Enc_{PK_{cn_n}}\left(Pid_{d_j}, x_j \right), Enc_{SK_{cn_i}}\left(Did_{d_j} \right) \right) \tag{12}$$

After receiving the GCM_{d_j}, cn_n decrypts the $Enc_{SK_{cn_i}}\left(Did_{d_j} \right)$ to obtain the Did of d_j. Additionally, cn_n checks whether the incomplete *Mission* chain transaction $trans^*$ of the corresponding Did_{d_j} is there. Then, cn_n searches the *Devices* chain to obtain the public key of d_j. Additionally, cn_n decrypts the $Enc_{PK_{cn_n}}\left(Pid_{d_j}, x_j \right)$ to obtain the Pid and x_j of d_j. If Pid_{d_j} is in the $GList_{d_l}$, cn_n generates the *Token* for d_j according to the equivalence between Pid and Did. Next, cn_n sends a response, $rsp_{cn}^j = Enc_{PK_{d_j}}(y_j)$, to d_j. After decrypting rsp_{cn}^j and obtaining y_j, d_j can generate a session key by $ks_j = H(x_j \,\|\, y_j)$.

In this way, cn_n verifies the drones and distributes the *Tokens* to the drones in the group. Finally, the drones in the group obtain their *Pids* and use their *Tokens* to obtain the flight path and other necessary information from cn_n. At the same time, cn_n updates the *Mission* chain by using the method proposed in Section 4.4.

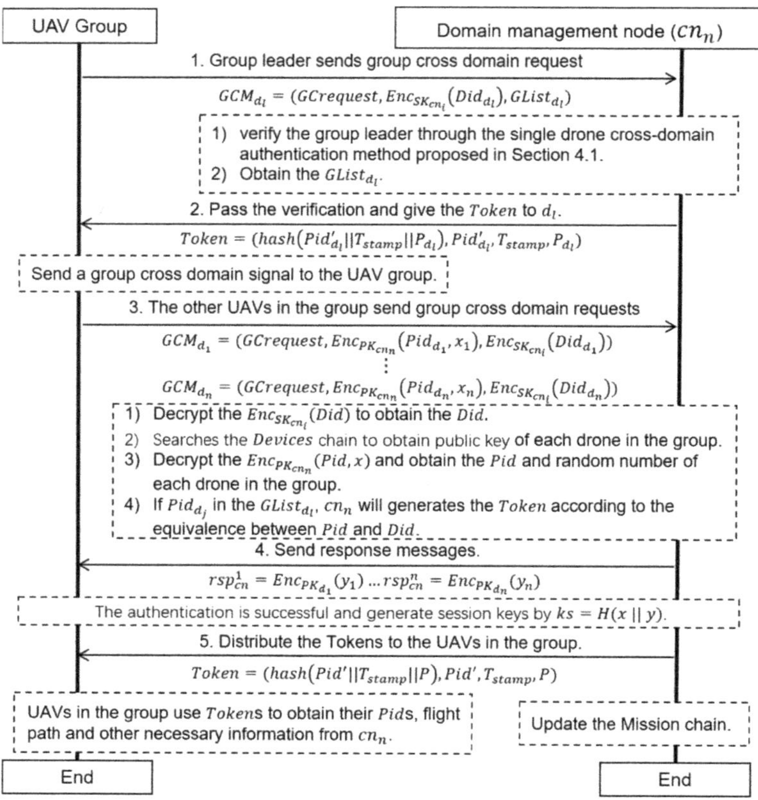

Figure 8. Drone group cross-domain authentication strategy.

4.4. Notification Mechanism between Domains

When a drone enters a new domain, the CN of the domain needs to publish a transaction to the *Mission* chain for uploading and updating the task information of the drone. After reaching a consensus, a CN packages a certain number of transactions and generates a new block on the *Mission* chain. The CNs of the other domains query the *Mission* chain at regular intervals and collect the task information of drones flying to their own domains. In this way, the CNs can plan the path for the drones in advance according to the $Rout$, the destination, and other information of the task. The specific method is as follows.

In the domain $Domain_i$, the CN of $Domain_i$ (cn_i) uses Algorithm 1 to find the transactions of the next domain ($CNid_n$), which is $Domain_i$, at regular intervals. Firstly, the algorithm obtains the latest block height of the *Mission* chain. Then, it searches blocks that have not been queried to obtain the transactions that include the latest drone task information. By comparing the $CNid_n$ in the transaction ($trans.CNid_n$) and the $CNid$ of cn_i ($cn_i.CNid$), the algorithm determines whether the next domain in the transaction is the current domain, and then saves the task information. Then, cn_i obtains the list of transactions ($List_{trans}$) and the latest block height (NH_{bc}).

Algorithm 1. Task information query algorithm.

Input: Mission, LH_{bc} // Mission chain and the block height of the last query.
Output: $List_{trans}$, NH_{bc} // List of transactions and the latest block height.
1. Initialize variable NH_{bc} // Initialize the latest block height.
2. Initialize variable $List_{trans}$ // Initialize the list of transactions.
3. NH_{bc} = getHeight(*Mission*) // Obtain the latest block height of the *Mission* chain.
4. for *block* in range(LH_{bc},NH_{bc}) // Search blocks that have not been queried.
5.   *trans* = read(*block*)   // Read the transactions on the blocks.
6.   if(*trans*.$CNid_n$ = = cn_i.*CNid*)   // Determine whether the next domain in the *trans* is the current domain.
7. $List_{trans}$.add(*trans*)   // Save the task information.
8.   else continue
9.   end if
10. end for
11. return $List_{trans}$, NH_{bc}

When it has obtained the $List_{trans}$ and the relevant task information, cn_i calls on Algorithm 2 to preprocess the tasks. For each transaction in the $List_{trans}$, the algorithm receives $Mid, Did, CNid_d, Rout$, and $Tout$ from the transaction. Then, it calls on the path planning algorithm to plan a flight path for the drone and obtain the CN ID of the next domain ($CNid_n'$), the range out of the current domain ($Rout'$), and the current expected time cost out of the domain (TC^*). For drone cross-domain authentication, the algorithm reads the *Address* chain and obtains the public key (PK_{cn_n}) of $CNid_n'$. Then, it reads the *Devices* chain and obtain the PK_d of the drone. Additionally, it generates the device ID in this domain (Pid') and the permission (P) for the drone. Then, the algorithm submits the transaction ($Pid', Rout', PK_d$) to the PBC. In this way, the PBC of the drone currently flying to the next domain has the information about the drones performing tasks in that domain for a period of time in the future. Additionally, it generates an incomplete *Mission* chain transaction, $trans^* = \left(Mid, Did, CNid_n', CNid_d, CNid_c', Rout', TC^* \right)$, and the TC^* in $trans^*$ is updated when the drone arrives. At the same time, cn_i packages the PBC transactions and generates a new block on the PBC at certain intervals, or the number of transactions meets the requirement.

Algorithm 2. Task preprocessing algorithm.

Input: $List_{trans}$
Output: $trans^*, Pid', P, PK_{cnn}$
1. Initialize variable $Rout', TC^*$ // Initialization range and expected time cost out of the current domain.
2. Initialize variable $CNid_n'$ // Initialization variable *CNid* of the next domain.
3. Initialize variable $CNid_c' = cn_i$.*CNid*   // Initialization variable *CNid* of the current domain.
4. for each transaction in $List_{trans}$
5.   Get $Mid, Did, CNid_d, Rout, Tout$ from the transaction.
6.   Use path planning algorism to plan flight path for the drone and get $CNid_n', Rout_c'$, and TC^*.
7.   Read *Address* chain and get the PK_{cnn} of $CNid_n'$.
8.   Read *Devices* chain and get the PK_d of the drone.
9.   Create Pid' and P // Generate device ID in this domain and the permission for the drone.
10. submit($Pid', Rout', PK_d$) ->PBC // Submit the PBC transaction to the PBC.
11. $trans^* = \left(Mid, Did, CNid_n', CNid_d, CNid_c', Rout', TC^* \right)$ // Generate the *Mission* chain transaction.
12. end for
13. cn_i packages PBC transactions and generates a new block on the PBC at certain intervals or the number of transactions meets the requirement.
14. return $trans^*, Pid', P, PK_{cn_n}$.

When a drone d_j has passed the cross-domain authentication and entered the domain $Domain_i$, cn_i uses Algorithm 3 to publish a transaction for updating the task information of

d_j. Above all, cn_i obtains the task start time (TST_{dj}) of d_j in the domain. Then, cn_i calculates the expected time out of the domain by $Tout'_{dj} = TST_{dj} + TC^*$. After that, the *Mission* chain transaction including the task information of d_j is published, which can be expressed as $trans = \left(Mid,\ Did_{dj},\ CNid'_n,\ CNid_d,\ CNid'_c,\ Rout'_{dj},\ Tout'_{dj} \right)$. We consider that all the CNs in the CBC are trusted. Therefore, this paper uses the Raft consensus mechanism to package the transactions and generate new blocks. In addition, the Pid'_{dj}, P_{dj}, and PK_{cn_n} generated in Algorithm 2 are sent to the d_j during the cross-domain authentication process.

Algorithm 3. Task information update algorithm.

Input: Mission information
Output: True or False
1. Initialise variable isUploaded = FALSE // Initialization variable, upload success or not.
2.   Get task start time (TST_{d_j}) of d_j in the domain.
3.   Compute $Tout'_{d_j} = TST_{d_j} + TC^*$ // Calculate the expected time out of the domain.
4. $trans = \left(Mid,\ Did_{d_j},\ CNid'_n,\ CNid_d,\ CNid'_c,\ Rout'_{d_j},\ Tout'_{d_j} \right)$ // Generate transaction including task information of d_j.
5.   Publish the *trans* to the *Mission* chain.
6.   isUploaded = TRUE   // Upload successfully.
7. return isUploaded.

---

5. Performance Evaluation

5.1. Experimental Settings

We analyzed the performance of the proposed scheme by conducting simulation experiments. The performance of the method proposed in this paper was measured in terms of computational overhead, communication overhead, and cross-domain authentication time cost. The configuration of the PC for the experiments is: CPU: Intel Core i7-8550, RAM: 8 GB, OS: Ubuntu 18.04, 64-bit. Hyperledger Fabric is an open source project from the Linux Foundation. We used Hyperledger Fabric v1.4 to build the blockchain, and the consensus on the consortium blockchain was reached through the Raft algorithm. Additionally, we used the JPBC v2.0 bilinear pair cryptography library from Italy GAS Lab to generate the public and private keys, and to encrypt and decrypt messages and ciphertext, respectively. The applied elliptic curve is a Type A elliptic curve with an order length of 160 bits ($y^2 = x^3 + x$). Raspberry Pi, as an embedded single-board computer (SBC) from Uk Raspberry Pi Foundation that is easy to use for coding and other implementations, is widely used in the existing studies. To further evaluate the feasibility of the proposed scheme, we used Raspberry Pi 4B SBCs to simulate the drones. The configuration of the Raspberry Pi 4B is: CPU: Quad-core Cortex-A72, RAM: 8 GB, OS: Ubuntu 18.04, 64-bit. We also compared the proposed method with existing methods [25–27] that use a ground station as a trusted third party for identity authentication, as well as existing methods [31–33] for identity authentication through ground stations and blockchain architectures.

5.2. Computational Overhead

5.2.1. Materials and Methods

To evaluate the computational overhead of the proposed framework, we analyzed the computational operations required by each entity in different phases of tasks. Simple operations, such as integer addition and concatenation operation, were not taken into consideration because of their low computational expense. Specific notations are listed as follows. CN: A consortium node in charge of a domain; d_j: A drone in a drone group or a single drone; d_l: The group leader of a drone group or a single drone; RG: Registration mechanism; TR: Task-release mechanism; SC: Single-drone cross-domain authentication method; TA: Two-way authentication strategy; NG: Method of building a new group; JG: Method of joining a drone group; DGC: The drone group cross-domain authentication method; NMD: Notification mechanism between domains; DAT: Operation of determining the acceptable task type of a drone; BC: One reading or writing operation on blockchain; GKP: Operation of generating a public–private key pair for d_j; CAS: One asymmetric encrypt/decrypt operation; LD: One reading or writing operation on local database; PP: One path planning operation; HO: One hash operation; N: The number of drones to compose a group or in a group; and GL: Operation of generating, updating, or distributing a group member list.

Table 3 shows the computational overhead that each entity needs to undertake in different task model phases. For example, in the process of DGC, the total computational cost that a CN needs to undertake is 4CAS + BC + HO + (N − 1) × (3CAS + HO). Specifically, it denotes the total overhead of performing four asymmetric encrypt/decrypt operations, one blockchain operation, one hash operation, and N − 1 times (3CAS + HO). The function 3CAS + HO represents the computational overhead required by the CN for the cross-domain authentication of an ordinary drone in the drone group. Additionally, the computational overhead that the CN needs to undertake for the cross-domain authentication of the group leader (d_l) is 4CAS + BC + HO, which is the same as the cost of the CN in the process of SC. In addition, the computational overhead of d_j in the process of DGC is CAS + HO, which is reduced by two asymmetric encrypt/decrypt operations compared to that of d_j in the process of SC. In the process of NG, although the computational cost of d_j is (N − 1) × TA, the overall computational overhead of the drone group is $\frac{N(N-1)}{2}$TA because only one TA is required between two drones.

Table 3. The computational overhead in different phases of tasks.

	RG	TR	SC	TA	NG	JG	DGC	NMD
CN	DAT + GKP + BC + CAS + LD	PP + 3BC + CAS	4CAS + BC + HO	-	-	-	4CAS + BC + HO + (N−1) × (3CAS + BC + HO)	4BC + PP
d_j	-	-	3CAS + HO	4CAS + BC + HO	(N − 1) × TA	2TA + GL	CAS + HO	
d_l	-	-	-	4CAS + BC + HO	(N − 1) × TA + GL	TA + GL	3CAS + HO	

Note: "-" means no relevant operation.

To evaluate the efficiency of the proposed cross-domain authentication scheme, we firstly evaluated the computational time cost of the single-drone cross-domain (SC) authentication and the drone group cross-domain (DGC) authentication. Additionally, we evaluated the computational time consumption of the CN side and the UAV side in different situations. Secondly, we compared the computational time cost of our method with that of existing methods [25–27,31–33]. To further illustrate the advantages of our method, we also evaluated the variation in the computational time overhead with the number of drones, and compared it with that of existing methods [25,31].

5.2.2. Results and Discussion

In the cases of SC and DGC, the computational time of the main operations is shown in Table 4. The time cost in the table is the average time cost of executing the corresponding operation 100 times on the corresponding platform.

Table 4. The computational time of the main operations.

Notation	Description	CN	UAV
T_{EE}	ECC encrypt operation	2.43 ms	7.24 ms
T_{ED}	ECC decrypt operation	3.61 ms	9.31 ms
T_{HO}	Hash function	0.03 ms	0.32 ms
T_{BC}	Blockchain query	0.17 ms	0.63 ms

The computational time consumption of the CN side and the UAV side in different situations is shown in Figure 9. In the SC case, the time consumption of the CN side is $\{2 \times 3.61 + 2 \times 2.43 + 0.17 + 0.03\} = 12.28$ ms, and the time consumption of the UAV side is $\{2 \times 9.31 + 7.24 + 0.32\} = 26.18$ ms. In the case of DGC, the computational time cost needs to consider the scale of the drones. The time consumption of the UAV group leader is $\{2 \times 9.31 + 7.24 + 0.32\} = 26.18$ ms, and the time cost of an ordinary UAV in the group is $\{9.31 + 0.32\} = 9.63$ ms. The minimum average time consumption of the CN side is the time consumption when dealing with ordinary UAVs, i.e., $\{2 \times 3.61 + 2.43 + 0.17 + 0.03\} = 9.85$ ms. The maximum average time consumption on the CN side is when there are only two drones, that is, $\{(12.28 + 9.85)/2\} = 11.07$ ms. Therefore, the average computational time consumption interval of the CN side is (9.85 ms, 11.07 ms).

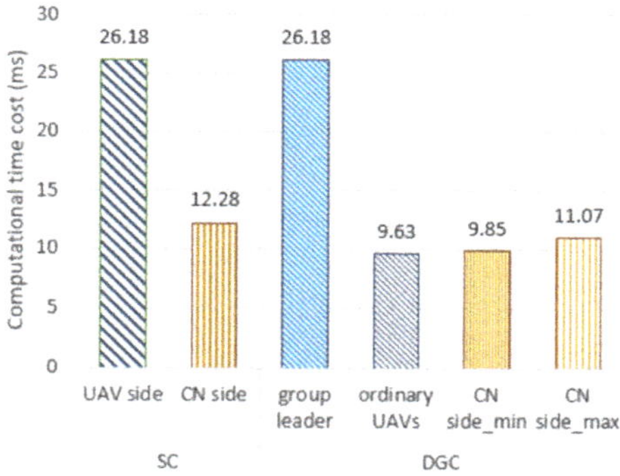

Figure 9. The computational time cost of SC and DGC.

The computational time cost of the proposed method and existing methods are shown in Figure 10. In the SC case, the computational time cost of our method is $\{12.28 + 26.18\} = 38.46$ ms. The figure also shows the maximum average computational time cost in the case of DGC, which is the average computational time cost required for each drone to cross domains when two drones perform DGC. The time cost is calculated by $\{(26.18 + 9.63 + 9.85 + 12.28)/2\} = 28.97$ ms. The existing methods for authenticating identity through a ground station as a trusted third party, as reported by Wazid et al. [25], Srinivas et al. [26], and Tanveer et al. [27], require 42.36 ms, 39.32 ms, and 38.12 ms, respectively. The existing methods for identity authentication through ground stations and blockchain architectures, as reported by Feng et al. [31], Shen et al. [32], and Gauhar et al. [33], require 32.93 ms, 36.87 ms, and 34.52 ms, respectively. Although the computational time cost of SC is not significantly different from that of existing methods [25–27,31–33], that of DGC is lower than that of other methods. Therefore, it can be considered that the DGC method can reduce the computational time cost of UAV cross-domain authentication. Figure 11 shows the computational time cost of cross-domain authentication when the number of drones increases. The computational time cost of DGC can be expressed as $\{26.18 + 12.28 + (N - 1)(9.63 + 9.85)\} = 19.48N + 18.98$ ms. As shown in

Figure 11, the time cost of each method increases linearly as the number of drones increases. Compared with the existing methods [25,31], the DGC method proposed in this paper has significant advantages when the number of drones is large.

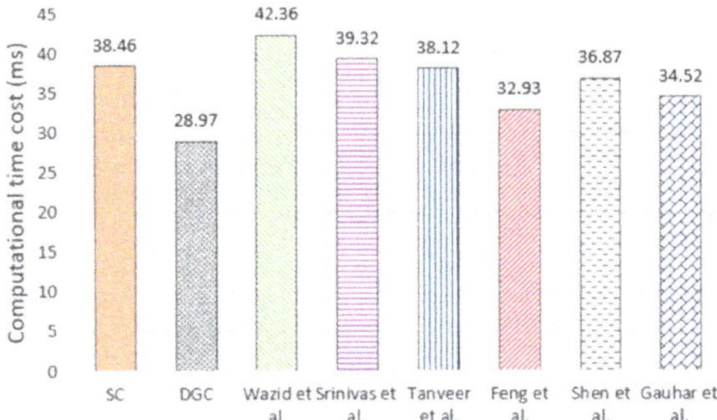

Figure 10. The comparison of computational time cost of different methods [25–27,31–33].

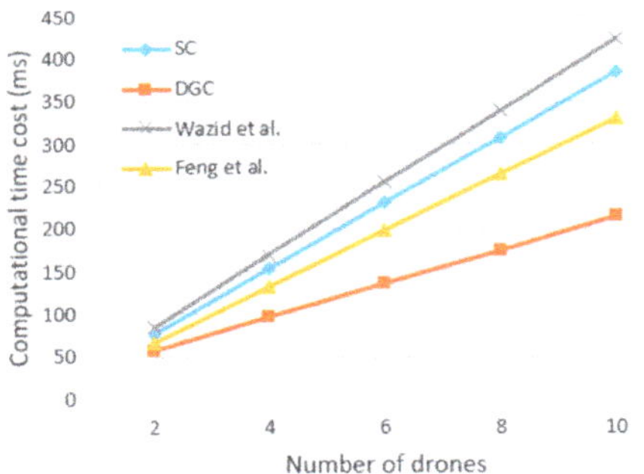

Figure 11. The computational time cost with increasing number of drones [25,31].

5.3. Communication Overhead

5.3.1. Materials and Methods

To evaluate the communication overhead of the proposed framework, we analyzed the number of communicated messages (bits) transmitted in different task model phases and compared it with existing advanced authentication schemes. In the SC case, the communicated messages are: $CRM : \left\{ Crequest, Enc_{SK_{cn_i}}\left(Did_{d_j}\right) \right\}$, $ack : \left\{ Enc_{PK_{d_j}}(x) \right\}$, $rsp_j : \left\{ Enc_{PK_{cn_n}}(x+1|| y) \right\}$, and $rsp_{cn} : \left\{ Enc_{PK_{d_j}}(y+1) \right\}$. The length of the CRM, ack, rsp_j, and rsp_{cn} is $\{64 + 4026\} = 4090$ bits, 1094 bits, 1094 bits, and 1094 bits, respectively. Thus, the total communication cost of the SC is 7372 bits. In the two-way authentication (TA) case, the communicated messages are: $M_{d_j} : \left\{ Taut, Enc_{SK_{cn_i}}\left(Pid_{d_j}\right), PBH_{d_j} \right\}$, $M_{d_n} : \left\{ Taut, Enc_{SK_{cn_i}}(Pid_{d_n}), PBH_{d_n} \right\}$, $ack : \left\{ Enc_{PK_{d_j}}(x) \right\}$, $rsp_j : \left\{ Enc_{PK_{d_n}}(x+1||y) \right\}$, and $rsp_n : \left\{ Enc_{PK_{d_j}}(y+1) \right\}$. The length of the M_{d_j}, M_{d_n}, ack, rsp_j, and rsp_n is $\{32 + 2350 + 128\} = 2510$ bits, 2510 bits, 1094 bits, 1094 bits, and 1094 bits, respectively. Thus, the total communication cost of the SC is 8302 bits. The communicated messages in the DGC case can be divided into two parts: (a) the communicated messages in the group leader authentication process, and (b) the communicated messages in an ordinary group member authentication process. The communicated messages in the group leader authentication process are: $GCM_{d_l} : \left\{ GCrequest, Enc_{SK_{cn_i}}\left(Did_{d_l}\right), GList_{d_l} \right\}$, $ack : \left\{ Enc_{PK_{d_l}}(x) \right\}$, $rsp_j :$

$\{Enc_{PK_{cn_n}}(x+1\|\,y)\}$, and $rsp_{cn} : \{Enc_{PK_{d_l}}(y+1)\}$. The length of the GCM_{d_l}, *ack*, rsp_j, and rsp_{cn} is $\{64 + 4026 + N \times 64\} = 4090 + 64N$ bits, 1094 bits, 1094 bits, and 1094 bits, respectively. Thus, the total communication cost of (a) is $7372 + 64N$ bits, where N is the number of members in the drone group. The communicated messages in an ordinary group member authentication process are: $GCM_{d_j} : \{GCrequest, Enc_{PK_{cn_n}}(Pid_{d_j}, x_j), Enc_{SK_{cn_i}}(Did_{d_j})\}$, and $rsp_{cn}^j = Enc_{PK_{d_j}}(y_j)$. The length of the GCM_{d_j} and rsp_{cn}^j is $\{32 + 2570 + 4026\} = 6628$ bits and 1094 bits, respectively. Thus, the total communication cost of (b) is 7722 bits. For a drone group with N members, the total communication cost is $\{a) + (N-1)b)\} = \{7372 + 64N + 7722N - 7722\} = 7786N\text{-}350$ bits. The average cost per drone is $7786 - 350/N$ bits.

5.3.2. Results and Discussion

Figure 12 shows the communication overhead required at different stages of tasks. The average minimum overhead for the DGC case is in the situation when two drones compose a group (7611 bits). To evaluate the cross-domain communication overhead, we compared our method with the existing novel methods, as shown in Figure 13. Among them, the methods proposed by Wazid et al. [25], Srinivas et al. [26], and Tanveer et al. [27], authenticating identity through a ground station as a trusted third party, require 6642 bits, 5938 bits, and 5522 bits, respectively. The above methods [25–27] mainly guarantee the credibility of the identity by performing multiple process parameter calculations on the device, registration center, and control center. Therefore, the communication overhead of the above methods is relatively lower than that of our method, but it increases the computational burden. The methods for identity authentication through ground stations and blockchain architectures proposed by Feng et al. [31], Shen et al. [32], and Gauhar et al. [33] require 7168 bits, 9280 bits, and 7680 bits, respectively. We noticed that the communication cost of DGC is higher than that of SC, and the maximum communication overhead difference between DGC and SC is $7786 - 7372 = 414$ bits. Additionally, Figure 10 shows that the computational time cost of SC is 38.46 ms, and the maximum average computational time cost of DGC is 28.97 ms. SC is 9.49 ms slower than DGC, and the difference increases as the number of drones increases. Therefore, it can be concluded that when the communication rate is higher than $414/9.49 = 43.6 \text{ b/ms} = 43.6$ kbps, DGC outperforms SC. As far as we know, a communication rate of 43.6 kbps is easily achievable. Therefore, while ensuring the overall efficiency of cross-domain authentication, it is feasible to consider increasing a portion of communication overhead to ensure security.

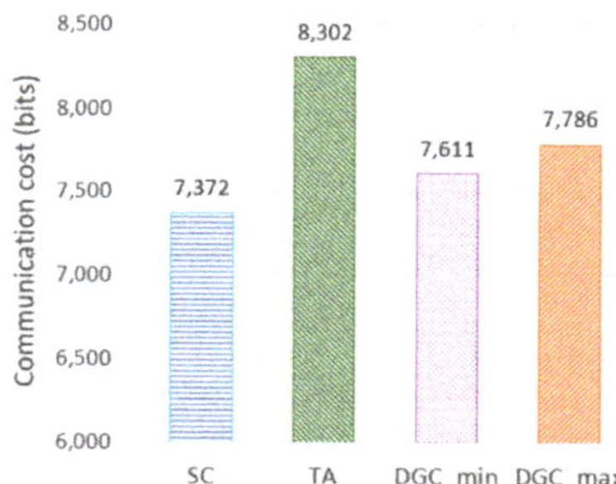

Figure 12. The communication cost in different task model phases.

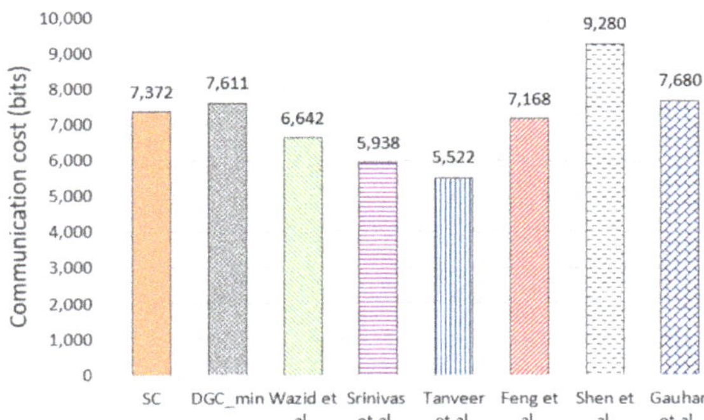

Figure 13. The comparison of communication cost of different methods [25–27,31–33].

The total communication time cost mainly includes transmission delay and propagation delay. The transmission delay can be expressed as $Sizedata/Tr$, where $Sizedata$ represents the size of the transmission data, and Tr represents the transmission rate of the channel. According to different transmission frequencies and communication bandwidths, Tr varies from tens of Kb to tens of Mb per second. Figure 14 shows how the transmission delay in the communicated messages changes with the transmission rate from 200 Kbps to 10 Mbps. We selected the minimum communication cost (Tanveer et al. [27]) and the maximum communication cost (Shen et al. [32]) among the comparison methods to compare them with our method. When the transmission rate is 5 Mps, the transmission time taken by SC, DGC_max, Tanveer et al. [27], and Shen et al. [32] is 1.43 ms, 1.51 ms, 1.07 ms, and 1.8 ms, respectively. The propagation delay can be expressed as $Dis/Velwave$, where Dis is the distance between the two sides of the communication and $Velwave$ is the propagation speed of the wave in the vacuum (about 3×10^5 km/s). Generally, the range of the domain is around 1 km. Therefore, propagation delay can be ignored.

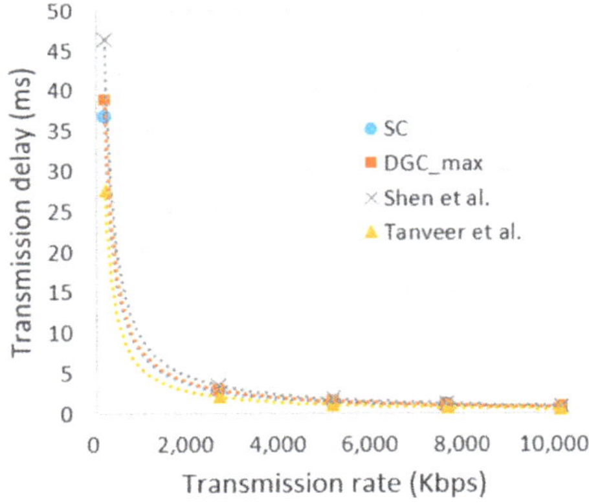

Figure 14. The transmission delay with increasing transmission rate [27,32].

5.4. Cross-Domain Authentication Time Cost

The total cross-domain time includes the computational time and communication time. Based on the experimental results in Section 5.2 and 5.3, we further evaluated the total cross-domain time taken by the method proposed in this paper by comparing it with the existing novel methods [25–27,31–33].

The experimental results are shown in Figure 15. The communication time cost of each scheme is that recorded when the transmission rate is 5 Mps. The total cross-domain time taken by the DGC proposed in this paper can be expressed as {19.48N + 18.98 + 1.55N − 0.07} = 21.03N + 18.91 ms, where N is the number of members in the drone group. In the case of cross-domain authentication for one drone, the SC method, Wazid et al. [25], Srinivas et al. [26], Tanveer et al. [27], Feng et al. [31], Shen et al. [32], and Gauhar et al. [33] require 39.89 ms, 43.69 ms, 40.51 ms, 39.19 ms, 34.37 ms, 38.67 ms, and 36.06 ms, respectively. The communication time cost of DGC in the figure (30.49 ms) is the average value for two drones. It can be seen that DGC has a better cross-domain authentication performance compared with the other methods [25–27,31–33]. Figure 16 shows the cross-domain authentication time cost when the number of drones increases. It can be seen that the DGC method proposed in this article has significant advantages when the number of drones is large.

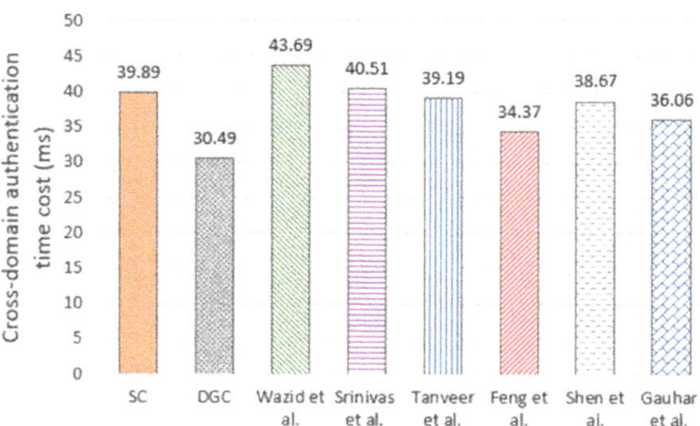

Figure 15. The comparison of cross-domain authentication time cost of different methods [25–27,31–33].

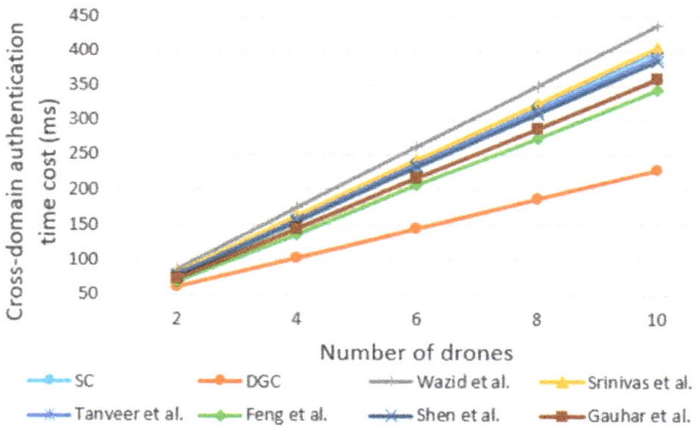

Figure 16. The cross-domain authentication time cost with increasing number of drones [25–27,31–33].

6. Security Analysis

We used the widely used Dolev and Yao (DY) threat model [35] to evaluate the security of the proposed method. In the DY threat model, a malicious attacker (MA) can inject, delete, eavesdrop, forge, or modify the exchanged messages over a public channel [36]. In this way, an MA can perform various security attacks on drones or CNs. The possible attacks and descriptions are as follows:

(1) Replay attack: An MA replays authentication messages to deceive the CN.
(2) Forgery attack: An MA generates an illegal or false ID to deceive the CN.
(3) Impersonation attack: An MA obtains authentication messages by impersonating terminals or eavesdropping on a channel, and impersonates a legitimate device to deceive the CN.

(4) Man-in-the-middle attack: An MA captures authentication messages and spoofs both parties of the communication.

(5) Database tampering: An MA attempts to tamper with the identity information in the database to pass the authentication.

Additionally, we made two assumptions: (a) The private keys of the CNs and drones are not revealed; and (b) An MA cannot deduce the private key from the public key, or it takes a lot of time. Considering the potential threats, we analyzed and compared the proposed scheme with the existing cross-domain authentication methods in terms of mutual authentication, cross-domain authentication, decentralization, anonymity, task path untraceability, path planning in advance, resilience to replay attacks, resilience to forgery attacks, resilience to impersonation attacks, resilience to man-in-the-middle attacks, and resilience to database tampering. The security analysis and comparison results are shown in Table 5.

Table 5. Security analysis and comparison results.

Features	[25]	[26]	[27]	[31]	[32]	[33]	Ours
Mutual authentication	Yes	Yes	Yes	Yes	Yes	Yes	Yes
Cross-domain authentication	Yes	Yes	Yes	Yes	Yes	Yes	Yes
Decentralization	No	No	No	Yes	Yes	Yes	Yes
Anonymity	Yes	Yes	Yes	Yes	No	No	Yes
Task path untraceability	No	No	No	No	No	No	Yes
Path planning in advance	No	No	No	No	No	No	Yes
Resilient to replay attack	Yes	Yes	Yes	Yes	Yes	Yes	Yes
Resilient to forgery attack	Yes	Yes	Yes	Yes	Yes	Yes	Yes
Resilient to impersonation attack	Yes	Yes	Yes	Yes	No	Yes	Yes
Resilient to man-in-the-middle attack	Yes	Yes	Yes	Yes	Yes	No	Yes
Resilient to database tampering	No	No	No	Yes	Yes	Yes	Yes

All the methods can well support mutual authentication and cross-domain authentication functions. At the same time, due to the use of blockchain technology and temporary intradomain ID methods, our method also has good decentralization and anonymity. Drones have different temporary IDs in different domains, and their device IDs on the CBC and all mission information can only be queried by the consortium nodes. Therefore, an MA cannot obtain the complete flight path of the drone, namely, task path untraceability. The notification mechanism between domains designed in this paper allows CNs to plan their paths in advance, which can improve their perception of the overall network situation. For possible attacks, we make the following analysis:

(1) Resilience to replay attacks: During the process of cross-domain authentication, the CNs and drones use PKI and a challenge–response mechanism to perform identity authentication and generate a session key. An MA cannot obtain useful information through this attack.

(2) Resilience to forgery attacks: The CNs need to query the *Devices* chain and the *Mission* chain transaction to confirm identity, and an MA cannot forge identity on the consortium chain.

(3) Resilience to impersonation attacks: Unregistered drones cannot obtain a legal *Did*, public key, and private key. In the process of the challenge–response game, an MA cannot decrypt the ciphertext to complete the verification. Therefore, it is difficult to implement an impersonation attack.

(4) Resilience to man-in-the-middle attacks: The communication data are encrypted by a public key or session key, which solves the problem of private data leakage. Even if the data are captured, the MA cannot decrypt the ciphertext to obtain the message.

(4) Resilience to database tampering: The important data are stored on the consortium blockchain. Only when the MA holds more than 51% of the nodes can it change the data in the blockchain, which is impracticable.

7. Conclusions

During long-distance flights for cargo transportation, drones need to apply cross-domain authentication mechanisms to enter the next domain. However, due to public wireless communication channels, drones are vulnerable to various security attacks in the process of cross-domain authentication. When facing a large number of cross-domain requests from drones, a CN requires significant computational and time overhead, which may lead to long waiting times for the cross-domain authentication of drones. To address this problem, we proposed BCDAIoD, an efficient blockchain-based cross-domain authentication scheme for the Internet of Drones. The BCDAIoD method includes a single-drone cross-domain authentication method, an establishment mechanism of drone groups, a drone group cross-domain authentication method, and a notification mechanism between domains. By taking advantage of blockchain, PKI, and the challenge–response game, BCDAIoD can ensure the authenticity and integrity of data, and can effectively prevent various attacks on drones and CNs. Furthermore, BCDAIoD uses the CBC and notification mechanism between domains to enable CNs to plan paths for drones in advance, which can further improve the efficiency of drone cross-domain authentication and task execution. The main contribution of this article is that BCDAIoD can improve the efficiency and security of the cross-domain authentication of drones. Experiment results show that the cross-domain authentication time cost and computational overhead of BCDAIoD are significantly lower than those of the existing state-of-the-art methods when facing a large number of drones.

Nevertheless, there are still limitations when applying BCDAIoD. First, blockchain brings additional communication and storage costs to the drone network. For example, drones in the IoD communicate with each other and update their local blockchains. Second, a small number of drones flying to the same destination or drones being far apart from each other may lead to drone group establishment failure. Hence, to address the above limitations, we seek to further simplify storage data in the block and design block pruning algorithms for the PBC to reduce communication and storage costs in future extensions of this work. At the same time, we will also attempt to design an optimization algorithm that dynamically adjusts between single-drone and drone group cross-domain methods based on the current state of IoD.

Author Contributions: Conceptualization, G.Q. and Y.Z.; methodology, G.Q. and T.Y.; software, T.Y. and G.Q.; investigation, G.Q. and Y.Q.; validation, G.Q., Y.Z. and T.Y.; result analysis, T.Y., Y.Q.; writing—original draft preparation, G.Q.; writing—review and editing, Y.Z. and G.Q.; supervision, Y.Z.; funding acquisition, Y.Z. All authors have read and agreed to the published version of the manuscript.

Funding: This research was funded by the National Natural Science Foundation of China (General Program) under Grant No. 61572253.

Data Availability Statement: The data presented in this study are available on request from the corresponding author.

Conflicts of Interest: The authors declare no conflict of interest.

References

1. Hassan, M.A.; Javed, A.R.; Hassan, T.; Band, S.S.; Sitharthan, R.; Rizwan, M. Reinforcing communication on the internet of aerial vehicles. *IEEE Trans. Green Commun. Netw.* **2022**, *6*, 1288–1297. [CrossRef]
2. Salah, K.; Rehman, M.H.U.; Nizamuddin, N.; Al-Fuqaha, A. Blockchain for AI: Review and open research challenges. *IEEE Access* **2019**, *7*, 10127–10149. [CrossRef]
3. Farah, M.F.; Mrad, M.; Ramadan, Z.; Hamdane, H. Handle with Care: Adoption of Drone Delivery Services. In Proceedings of the Advances in National Brand and Private Label Marketing: Seventh International Conference, Barcelona, Spain, 17–20 June 2020; pp. 22–29.
4. Makhdoom, I.; Zhou, I.; Abolhasan, M.; Lipman, J.; Ni, W. PrivySharing: A blockchain-based framework for privacy-preserving and secure data sharing in smart cities. *Comput. Secur.* **2020**, *88*, 101653. [CrossRef]
5. Li, X.; Wang, Y.; Vijayakumar, P.; He, D.; Kumar, N.; Ma, J. Blockchain-based mutual-healing group key distribution scheme in unmanned aerial vehicles ad-hoc network. *IEEE Trans. Veh. Technol.* **2019**, *68*, 11309–11322. [CrossRef]
6. Qiu, J.; Grace, D.; Ding, G.; Yao, J.; Wu, Q. Blockchain-Based Secure Spectrum Trading for Unmanned-Aerial-Vehicle-Assisted Cellular Networks: An Operator's Perspective. *IEEE Internet Things J.* **2020**, *7*, 451–466. [CrossRef]
7. Bera, B.; Chattaraj, D.; Das, A.K. Designing secure blockchain-based access control scheme in IoT-enabled Internet of Drones deployment. *Comput. Commun.* **2020**, *153*, 229–249. [CrossRef]
8. Yapıcı, Y.; Rupasinghe, N.; Güvenç, I.; Dai, H.; Bhuyan, A. Physical layer security for NOMA transmission in mmWave drone networks. *IEEE Trans. Veh. Technol.* **2021**, *70*, 3568–3582. [CrossRef]
9. Asheralieva, A.; Niyato, D. Distributed dynamic resource management and pricing in the IoT systems with blockchain-as-a-service and UAV-enabled mobile edge computing. *IEEE Internet Things J.* **2020**, *7*, 1974–1993. [CrossRef]
10. Li, T.; Liu, W.; Wang, T.; Ming, Z.; Li, X.; Ma, M. Trust data collections via vehicles joint with unmanned aerial vehicles in the smart Internet of Things. *Trans. Emerg. Telecommun. Technol.* **2022**, *33*, e3956. [CrossRef]
11. Nakamura, S.; Enokido, T.; Takizawa, M. Information flow control based on the CapBAC (capability-based access control) model in the IoT. *Int. J. Mob. Comput. Multimed. Commun.* **2019**, *10*, 13–25. [CrossRef]
12. Ali, G.; Ahmad, N.; Cao, Y.; Ali, Q.E.; Azim, F.; Cruickshank, H. BCON: Blockchain based access CONtrol across multiple conflict of interest domains. *J. Netw. Comput. Appl.* **2019**, *147*, 102440. [CrossRef]
13. Wang, Y.; Wang, H.; Wei, X.; Zhao, K.; Fan, J.; Chen, J.; Jia, R. Service Function Chain Scheduling in Heterogeneous Multi-UAV Edge Computing. *Drones* **2023**, *7*, 132. [CrossRef]
14. Jha, S.; Sural, S.; Atluri, V.; Vaidya, J. Specification and verification of separation of duty constraints in attribute-based access control. *IEEE Trans. Inf. Forensics Secur.* **2017**, *13*, 897–911. [CrossRef]
15. Sandhu, R.S.; Coyne, E.J.; Feinstein, H.L.; Youman, C.E. Role-based access control models. *Computer* **1996**, *29*, 38–47. [CrossRef]
16. Xu, S.; Ning, J.; Li, Y.; Zhang, Y.; Xu, G.; Huang, X.; Deng, R.H. Match in my way: Fine-grained bilateral access control for secure cloud-fog computing. *IEEE Trans. Dependable Secur. Comput.* **2022**, *19*, 1064–1077. [CrossRef]
17. Wang, K.; Zhang, X.; Qiao, X.; Li, X.; Cheng, W.; Cong, Y.; Liu, K. Adjustable Fully Adaptive Cross-Entropy Algorithms for Task Assignment of Multi-UAVs. *Drones* **2023**, *7*, 204. [CrossRef]
18. Abdel-Malek, M.A.; Akkaya, K.; Bhuyan, A.; Ibrahim, A.S. A proxy Signature-Based swarm drone authentication with leader selection in 5G networks. *IEEE Access* **2022**, *10*, 57485–57498. [CrossRef]
19. Fysarakis, K.; Soultatos, O.; Manifavas, C.; Papaefstathiou, I.; Askoxylakis, I. XSACd-Cross-domain resource sharing & access control for smart environment. *Future Gener. Comput. Syst.* **2018**, *80*, 572–582.
20. Nakamoto, S. Bitcoin: A peer-to-peer electronic cash system. In *Decentralized Business Review*; Scholastica: Seoul, Korea, 2008; p. 21260.
21. Mehta, P.; Gupta, R.; Tanwar, S. Blockchain envisioned drone networks: Challenges, solutions, and comparisons. *Comput. Commun.* **2020**, *151*, 518–538. [CrossRef]
22. Al-Hilo, A.; Samir, M.; Assi, C.; Sharafeddine, S.; Ebrahimi, D. Cooperative content delivery in UAV-RSU assisted vehicular networks. In Proceedings of the 2nd ACM MobiCom Workshop on Drone Assisted Wireless Communications for 5G and Beyond, London, UK, 21–25 September 2020; pp. 73–78.
23. Arafeh, M.; El Barachi, M.; Mourad, A.; Belqasmi, F. A blockchain based architecture for the detection of fake sensing in mobile crowdsensing. In Proceedings of the 2019 4th International Conference on Smart and Sustainable Technologies (SpliTech), Split, Croatia, 18–21 June 2019; pp. 1–6.
24. García-Magariño, I.; Lacuesta, R.; Rajarajan, M.; Lloret, J. Security in networks of unmanned aerial vehicles for surveillance with an agent-based approach inspired by the principles of blockchain. *Ad Hoc Netw.* **2019**, *86*, 72–82. [CrossRef]
25. Wazid, M.; Das, A.K.; Kumar, N.; Alazab, M. Designing authenticated key management scheme in 6G-enabled network in a box deployed for industrial applications. *IEEE Trans. Ind. Inform.* **2020**, *17*, 7174–7184. [CrossRef]
26. Srinivas, J.; Das, A.K.; Wazid, M.; Vasilakos, A.V. Designing secure user authentication protocol for big data collection in IoT-based intelligent transportation system. *IEEE Internet Things J.* **2020**, *8*, 7727–7744. [CrossRef]
27. Tanveer, M.; Alkhayyat, A.; Naushad, A.; Kumar, N.; Alharbi, A.G. RUAM-IoD: A robust user authentication mechanism for the Internet of Drones. *IEEE Access* **2022**, *10*, 19836–19851. [CrossRef]
28. Jan, S.U.; Abbasi, I.A.; Algarni, F.; Khan, A.S. A verifiably secure ECC based authentication scheme for securing IoD using FANET. *IEEE Access* **2022**, *10*, 95321–95343. [CrossRef]

29. Rajamanickam, S.; Vollala, S.; Ramasubramanian, N. EAPIOD: ECC based authentication protocol for insider attack protection in IoD scenario. *Secur. Priv.* **2022**, *5*, e248. [CrossRef]
30. Ever, Y.K. A secure authentication scheme framework for mobile-sinks used in the internet of drones applications. *Comput. Commun.* **2020**, *155*, 143–149. [CrossRef]
31. Feng, C.; Liu, B.; Guo, Z.; Yu, K.; Qin, Z.; Choo, K.K.R. Blockchain-based cross-domain authentication for intelligent 5G-enabled internet of drones. *IEEE Internet Things J.* **2021**, *9*, 6224–6238. [CrossRef]
32. Shen, M.; Liu, H.; Zhu, L.; Xu, K.; Yu, H.; Du, X.; Guizani, M. Blockchain-assisted secure device authentication for cross-domain industrial IoT. *IEEE J. Sel. Areas Commun.* **2020**, *38*, 942–954. [CrossRef]
33. Ali, G.; Ahmad, N.; Cao, Y.; Khan, S.; Cruickshank, H.; Qazi, E.A.; Ali, A. xDBAuth: Blockchain based cross domain authentication and authorization framework for Internet of Things. *IEEE Access* **2020**, *8*, 58800–58816. [CrossRef]
34. Zhang, H.; Chen, X.; Lan, X.; Jin, H.; Cao, Q. BTCAS: A blockchain-based thoroughly cross-domain authentication scheme. *J. Inf. Secur. Appl.* **2020**, *55*, 102538. [CrossRef]
35. Dolev, D.; Yao, A. On the security of public key protocols. *IEEE Trans. Inf. Theory* **1983**, *29*, 198–208. [CrossRef]
36. Yu, S.; Das, A.K.; Park, Y.; Lorenz, P. SLAP-IoD: Secure and lightweight authentication protocol using physical unclonable functions for internet of drones in smart city environments. *IEEE Trans. Veh. Technol.* **2022**, *71*, 10374–10388. [CrossRef]

Article

Machine Learning Methods for Inferring the Number of UAV Emitters via Massive MIMO Receive Array

Yifan Li [1], Feng Shu [1,2,*], Jinsong Hu [3], Shihao Yan [4], Haiwei Song [5], Weiqiang Zhu [5], Da Tian [5], Yaoliang Song [1] and Jiangzhou Wang [6]

[1] School of Electronic and Optical Engineering, Nanjing University of Science and Technology, Nanjing 210094, China
[2] School of Information and Communication Engineering, Hainan University, Haikou 570228, China
[3] College of Physics and Information Engineering, Fuzhou University, Fuzhou 350116, China
[4] School of Science and Security Research Institute, Edith Cowan University, Perth, WA 6027, Australia
[5] 8511 Research Institute, China Aerospace Science and Industry Corporation, Nanjing 210007, China
[6] School of Engineering, University of Kent, Canterbury CT2 7NT, UK
* Correspondence: shufeng0101@163.com

Abstract: To provide important prior knowledge for the direction of arrival (DOA) estimation of UAV emitters in future wireless networks, we present a complete DOA preprocessing system for inferring the number of emitters via a massive multiple-input multiple-output (MIMO) receive array. Firstly, in order to eliminate the noise signals, two high-precision signal detectors, the square root of the maximum eigenvalue times the minimum eigenvalue (SR-MME) and the geometric mean (GM), are proposed. Compared to other detectors, SR-MME and GM can achieve a high detection probability while maintaining extremely low false alarm probability. Secondly, if the existence of emitters is determined by detectors, we need to further confirm their number. Therefore, we perform feature extraction on the the eigenvalue sequence of a sample covariance matrix to construct a feature vector and innovatively propose a multi-layer neural network (ML-NN). Additionally, the support vector machine (SVM) and naive Bayesian classifier (NBC) are also designed. The simulation results show that the machine learning-based methods can achieve good results in signal classification, especially neural networks, which can always maintain the classification accuracy above 70% with the massive MIMO receive array. Finally, we analyze the classical signal classification methods, Akaike (AIC) and minimum description length (MDL). It is concluded that the two methods are not suitable for scenarios with massive MIMO arrays, and they also have much worse performance than machine learning-based classifiers.

Keywords: unmanned aerial vehicle (UAV); massive MIMO; threshold detection; emitter number detection; machine learning; information criterion

Citation: Li, Y.; Shu, F.; Hu, J.; Yan, S.; Song, H.; Zhu, W.; Tian, D.; Song, Y.; Wang, J. Machine Learning Methods for Inferring the Number of UAV Emitters via Massive MIMO Receive Array. *Drones* **2023**, *7*, 256. https://doi.org/10.3390/drones7040256

Academic Editor: Emmanouel T. Michailidis

Received: 14 March 2023
Revised: 6 April 2023
Accepted: 8 April 2023
Published: 10 April 2023

1. Introduction

With the advantages of high mobility and low cost, unmanned aerial vehicles (UAVs) play important roles in wireless networks for implementing tasks like weather monitoring, traffic control, emergency search, communication relaying, etc. [1]. However, unlike traditional ground-to-ground (G2G) communications, UAV communications have some special characteristics and challenges, e.g., the high mobility leads to the UAV communication channels changing much faster, the high flight altitude requires the ground base stations to provide larger 3D signal coverage for UAVs, and the line of sight (LoS) paths between UAVs and base stations are vulnerable to interference from ground users over the same frequency [2]. As is known to us, massive multiple-input multiple-output (MIMO) is a key technology in 5G or future 6G systems [3,4]; it can make significant improvements in system capacity, reliability, and spectral efficiency by using techniques such as spatial multiplexing, diversity, and beamforming [5]. Compared to small arrays, the higher array gain of massive

MIMO arrays can make a great extension of signal coverage [6], and experimental results in [7] showed massive MIMO works well with LoS mobile channels. So in view of the problems that UAV communications face, it is natural to consider the combination of UAVs and massive MIMO technology [8]. In [9], a nonstationary 3D geometry-based model was proposed for UAV-to-ground massive MIMO channels; this model considered the realistic scenarios and discussed the impact of some important UAV parameters such as altitude and flight velocity, so it can give some inspiration for future research on 6G standard UAV channel models. As UAVs often appear as clusters, the potential of massive MIMO ground station communication with UAV swarms was explored in [10], and a realistic geometric model was also developed.

Because of the high mobility of UAVs, it is necessary for ground base stations to obtain direction-of-arrival (DOA) information of UAVs in a timely manner for channel estimation and communication security. For most DOA estimation algorithms, such as MUSIC and ESPRIT, the number of emitters is required prior knowledge, but the number is usually unknown [11]. So inferring the number of emitters has been an active topic in array processing for a few decades [12]. In recent years, the potential of massive MIMO technology in array processing has also been gradually discovered [13], as the larger number of antennas can decrease the beamwidth and then increase the angular resolution of the arrays [14]. Therefore, considering the realistic needs of UAV communications and the advantages of massive MIMO technology in array processing, we will study the methods for inferring the number of UAV emitters via a massive MIMO receive array in this work.

In general, the solutions for inferring the number of emitters can be divided into two main categories. The first is based on the information-theoretic criteria and another is based on the analysis of the covariance matrices. Since detecting the number of signal sources can be viewed as a typical model order selection problem, Akaike firstly proposed a method focusing on finding the minimum Kullback–Leibler (KL) discrepancy between the probability density function (PDF) of obtained data and that of models for selection [15], and this method is now called AIC. Schwarz introduced Bayesian information criterion (BIC) based on Akaike's work [16], and Rissanen also derived a similar criterion called MDL [17]. Ref. [18] provided a good summary of these classical information criteria. In the last decade, Lu and Zoubir proposed the generalized Bayesian information criterion (GBIC) [19] and flexible detection criterion (FDC) [20], which effectively improved the performance on source enumeration. The other basic method for enumerating the number of sources is performing analysis on the covariance matrices of signals received by arrays. Williams and Johnson proposed a sphericity test for source enumeration in [21], which was based on a hypothesis test for the covariance matrix. Ref. [22] gave a bootstrap-based method to estimate the null distributions of the test statistics. Wax and Adler solved this problem by performing signal subspace matching [23].

Signal detection is another technique adopted in this work. In order to reduce the interference of the noise to the detection of signal number, some good methods were proposed, such as classic signal detection algorithms containing energy detection [24], matched-filter detection [25], cyclostationarity-based detection [26], etc. On the basis of these methods, Zeng and Liang proposed two eigenvalue-based algorithms in [27], Zhang et al. used the generalized likelihood ratio test (GLRT) approach to improve detection performance [28], and an eigenvalue-based LRT algorithm was also given in [29].

In recent years, machine learning (ML) has played an important role in the fields of array signal processing [30] and UAV communications [31], and now the ML-based methods used in 5G mainly include supervised learning, unsupervised learning, and reinforcement learning [32]. Thilina et al. compared the performance of unsupervised learning approaches and supervised learning approaches for cooperative spectrum sensing [33]. A machine learning-based DOA measurement method was also proposed in [34], and ref. [35] used a neural network for power allocation in a wireless communication network.

In this paper, we will combine the techniques mentioned above for inferring the number of UAV emitters via massive MIMO receive array. First, the pure noise signals are separated by threshold detectors, and then the feature vectors are extracted from the sample covariance matrices of the remaining signals. Finally, the ML-NN and other machine learning methods are used to classify the signals for determining the number of emitters. Therefore, our main contributions are summarized as follows:

1. A DOA preprocessing system is proposed for obtaining the number of UAV emitters via a massive MIMO array. The main steps of this system include signal detection and inferring the number of emitters. The received signals are first inputted into signal detectors. If the detection result shows the presence of emitters, this signal is further transmitted to signal classifiers to determine the number of emitters.

2. Two high-precision signal detectors, the square root of the maximum eigenvalue times the minimum eigenvalue (SR-MME) and the geometric mean (GM), are proposed in Section 3. Their thresholds and probability of detection are also derived with the aid of random matrix theories. The simulation results show that SR-MME and GM have significant improvement in detection performance compared with the MME detector proposed in [27] and the M-MME detector proposed in [36], even though the SNR is very low and the number of samples is small. The simulation results also show that SR-MME and GM can maintain a low false alarm probability while achieving a high detection probability.

3. Since the existence of emitters is known, we innovatively introduce machine learning-based classifiers to infer their number, including multi-layer neural networks (ML-NNs), support vector machine (SVM), and naive Bayesian classifier (NBC). Important features which make up feature vectors are also extracted from eigenvalue sequences of signals' sample covariance matrices. The results show that machine learning methods are very suitable for performing signal classification, especially neural networks, because they can achieve a classification accuracy of 70%, even under extreme conditions. Finally, we validate the classification performance of AIC and MDL under different SNR and number of receive antennas. We show that they are unapplicable to scenarios with low SNR and massive MIMO receive arrays compared to machine learning-based methods.

The rest of the paper is organized as follows. In Section 2, we present a specific system model and assumptions. Two high-precision signal detectors are given in Section 3. Section 4 shows how to perform feature extraction on received signals and classify them by machine learning methods. Then, the advantages of the proposed detectors and classifiers are presented through simulation results in Section 5. Finally, Section 6 draws conclusions.

Notation: Matrices, vectors, and scalars are denoted by letters of bold upper case, bold lower case, and lower case, respectively. Signs $(\cdot)^T$, $(\cdot)^*$, and $(\cdot)^H$ represent transpose, conjugate, and conjugate transpose. $\mathbf{I}_M$ denotes the $M \times M$ identity matrix. diag$\{\cdot\}$ stands for diagonal matrix.

2. System Model

As the system shown in Figure 1, we consider a scenario with K far-field UAV emitters and one massive MIMO receiver equipped with an M-element uniform linear array (ULA). The signals transmitted by the kth UAV are denoted by $s_k(t)e^{j2\pi f_c t}$, where $s_k(t)$ is the baseband signal and f_c is the carrier frequency. Referring to [37], the received signals at the mth antenna are given by

$$y_m(t) = \sum_{k=1}^{K} s_k(t)e^{j2\pi f_c t}e^{-j2\pi f_c \tau_{k,m}} + v_m(t),\tag{1}$$

where $v_m(t) \sim \mathcal{CN}(0, \sigma_v^2)$ represents the additive white Gaussian noise (AWGN) term, and $\tau_{k,m}$ denotes the propagation delay from the kth UAV to the mth antenna, expressed by

$$\tau_{k,m} = \tau_0 - \frac{(m-1)d\sin\theta_k}{c},\tag{2}$$

where τ_0 is the propagation delay from the UAV to the reference point on the receive array, θ_k is the angle of signal incidence from the kth UAV, $d = \lambda/2$ represents the space between array elements, and c denotes the speed of light. Then received signals go through ADC and down converter, and we obtain

$$y_m(n) = \sum_{k=1}^{K} e^{-j2\pi(m-1)d\sin\theta_k/\lambda} s_k(n) + v_m(n), \tag{3}$$

and by combining all the M antennas, we obtain

$$\mathbf{y}(n) = \sum_{k=1}^{K} \mathbf{a}(\theta_k) s_k(n) + \mathbf{v}(n), \tag{4}$$

where $\mathbf{v}(n) = [v_1(n),\ldots,v_M(n)]^T$ denotes the noise vector and

$$\mathbf{a}(\theta_k) = [1, e^{-j2\pi d\sin\theta_k/\lambda}, \ldots, e^{-j2\pi(M-1)d\sin\theta_k/\lambda}]^T, \tag{5}$$

is the array manifold.

Initially, it is not clear whether the UAVs exist, so we should consider two situations, including the signals' presence and only noise [38]. By turning (4) to matrix form, we obtain

$$H_0 : \mathbf{y}(n) = \mathbf{v}(n) \quad H_1 : \mathbf{y}(n) = \mathbf{As}(n) + \mathbf{v}(n), \tag{6}$$

where $\mathbf{s}(n) = [s_1(n),\ldots,s_K(n)]^T$, $\mathbf{A} = [\mathbf{a}(\theta_1),\ldots,\mathbf{a}(\theta_K)]$. Then the covariance matrix of the received signal can be expressed by

$$\mathbf{Q_y} = \mathbf{AQ_sA}^H + \sigma_\mathbf{v}^2\mathbf{I}_M = \sum_{k=1}^{K} \sigma_{\mathbf{s},k}^2 \mathbf{a}(\theta_k)\mathbf{a}^H(\theta_k) + \sigma_\mathbf{v}^2\mathbf{I}_M. \tag{7}$$

where $\mathbf{Q_s} = \mathrm{E}[\mathbf{S}(n)\mathbf{S}^H(n)] = \mathrm{diag}\{\sigma_{\mathbf{s},1}^2,\ldots,\sigma_{\mathbf{s},K}^2\}$.

Since the base station is equipped with a massive array, $M \gg K$ and $\mathrm{rank}(\mathbf{A}) = K$. Then the eigenvalues of $\mathbf{Q_y}$ satisfy the following properties

$$\underbrace{\lambda_1 \geq \lambda_2 \geq \ldots \geq \lambda_K}_{\text{signal subspace}} > \underbrace{\lambda_{K+1} = \ldots = \lambda_M = \sigma_\mathbf{v}^2}_{\text{noise subspace}}, \tag{8}$$

and

$$\lambda_m = \rho_m + \sigma_\mathbf{v}^2, \tag{9}$$

where $\rho_1 \geq \ldots \geq \rho_K > \rho_{K+1} = \ldots = \rho_M = 0$ are the eigenvalues of $\mathbf{AQ_sA}^H$.

In practice, the covariance matrix of received signal $\mathbf{y}$ cannot be obtained accurately, so the sample covariance matrix of the received signal is usually used to approximate it:

$$\hat{\mathbf{Q}}_\mathbf{y} = \frac{1}{N}\sum_{n=1}^{N} \mathbf{y}(n)\mathbf{y}^H(n) = \frac{1}{N}\mathbf{YY}^H, \tag{10}$$

where

$$H_0 : \mathbf{Y} = \mathbf{V} \quad H_1 : \mathbf{Y} = \mathbf{AS} + \mathbf{V}, \tag{11}$$

and $\mathbf{S} = [\mathbf{s}(1),\mathbf{s}(2),\ldots,\mathbf{s}(N)]$, $\mathbf{V} = [\mathbf{v}(1),\mathbf{v}(2),\ldots,\mathbf{v}(N)]$.

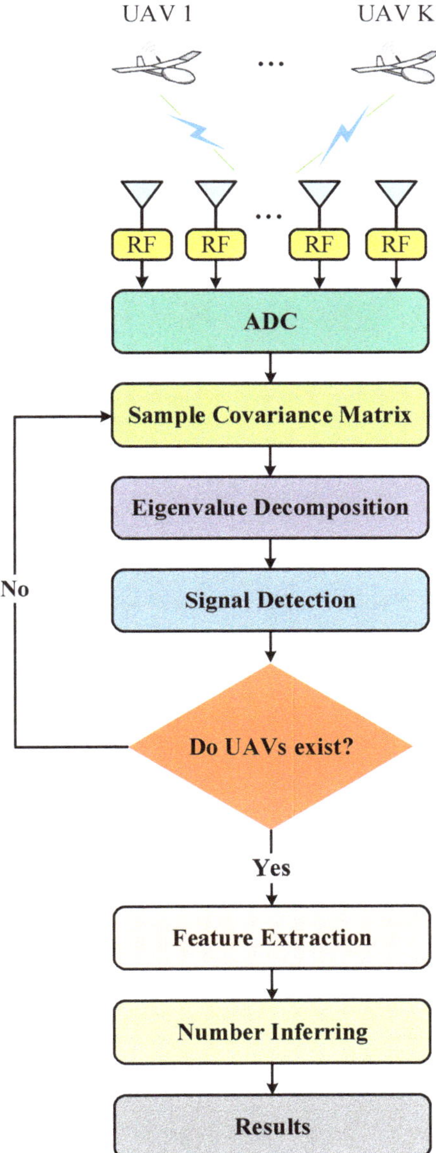

Figure 1. The procedure of proposed system for inferring the number of UAV emitters by massive MIMO receive array.

3. Signal Detectors

As shown in Figure 1, after the sample covariance matrix of the received signal is obtained, we take eigenvalue decomposition (EVD) on it. For the two situations in (11), eigenvalues are represented by $\lambda_1(\hat{\mathbf{Q}}_{\mathbf{y},H_0}) \geq \cdots \geq \lambda_M(\hat{\mathbf{Q}}_{\mathbf{y},H_0})$ and $\lambda_1(\hat{\mathbf{Q}}_{\mathbf{y},H_1}) \geq \cdots \geq$

$\lambda_M(\hat{\mathbf{Q}}_{\mathbf{y},H_1})$, respectively. For convenience, we consider moving the constant $1/N$ to the left-hand side of (10). Assuming $\sigma_{\mathbf{v}}^2 = 1$, we obtain

$$\mathbf{R}_{H_0} = \mathbf{V}\mathbf{V}^H, \tag{12a}$$

$$\mathbf{R}_{H_1} = N\mathbf{A}\hat{\mathbf{Q}}_{\mathbf{S}}\mathbf{A}^H + \mathbf{R}_{H_0}, \tag{12b}$$

where $\mathbf{R}_{H_0}$ is a Wishart matrix and $\hat{\mathbf{Q}}_{\mathbf{S}}$ is the sample covariance matrix of $\mathbf{S}$. The eigenvalues of $\mathbf{R}_{H_0}$ and $\mathbf{R}_{H_1}$ can also be expressed as $\lambda_1(\mathbf{R}_{H_0}) \geq \ldots \geq \lambda_M(\mathbf{R}_{H_0})$ and $\lambda_1(\mathbf{R}_{H_1}) \geq \ldots \geq \lambda_M(\mathbf{R}_{H_1})$, where $\lambda_m(\mathbf{R}_{H_0}) = N\lambda_m(\hat{\mathbf{Q}}_{\mathbf{y},H_0})$ and $\lambda_m(\mathbf{R}_{H_1}) = N\lambda_m(\hat{\mathbf{Q}}_{\mathbf{y},H_1})$. Since $\mathbf{R}_{H_0}$ is a complex Gaussian Wishart matrix, its largest eigenvalue should follow Tracy–Widom distribution of order 2 [39]:

$$\frac{\lambda_{\max}(\mathbf{R}_{H_0}) - \mu}{\nu} \xrightarrow{d} \mathcal{TW}_2, \tag{13}$$

where

$$\mu = (\sqrt{M} + \sqrt{N})^2, \tag{14a}$$

$$\nu = \sqrt{\mu}\left(\frac{1}{\sqrt{M}} + \frac{1}{\sqrt{N}}\right)^{1/3}, \tag{14b}$$

are center and scaling parameters. Then the cumulative distribution function (CDF) of the largest eigenvalue, i.e., $F(x)$, can be approximated as

$$F(x) \approx F_2\left(\frac{x - \mu}{\nu}\right), \tag{15}$$

where $F_2(x)$ denotes the distribution function of $\mathcal{TW}_2$. Referring to [40,41], it is defined as

$$F_2(x) = \exp\left\{-\int_x^\infty (a - x)q^2(a)da\right\}, \tag{16}$$

where $q(a)$ is the solution of Painlevé II differential equation

$$q''(a) = aq(a) + 2q^3(a). \tag{17}$$

with boundary condition $q(a) \sim Ai(a)$ as $a \to \infty$, where $Ai(a)$ represents the Airy function [42]. The value of $F_2(x)$ can be computed by using software packages such as [43].

In addition, for the Wishart matrix $\mathbf{R}_{H_0}$, if $\lim_{N \to +\infty} \frac{M}{N} = z$ ($z \in [0,1]$), its maximum and minimum eigenvalues can be approximated as $(\sqrt{N} + \sqrt{M})^2$ and $(\sqrt{N} - \sqrt{M})^2$, respectively. Next we will present several high-performance signal detectors based on the knowledge given earlier.

3.1. Proposed SR-MME Detector

The SR-MME detector is defined as the square root of the maximum eigenvalue times the minimum eigenvalue, and is given by

$$\sqrt{\lambda_{\max}(\hat{\mathbf{Q}}_{\mathbf{y}})\lambda_{\min}(\hat{\mathbf{Q}}_{\mathbf{y}})} \underset{H_0}{\overset{H_1}{\gtrless}} \gamma_1, \tag{18}$$

where $\lambda_{\max}(\hat{\mathbf{Q}}_{\mathbf{y}})$, $\lambda_{\min}(\hat{\mathbf{Q}}_{\mathbf{y}})$ are maximum and minimum eigenvalues, respectively, of sample covariance matrix $\hat{\mathbf{Q}}_{\mathbf{y}}$, and γ_1 denotes the judgment threshold.

At the end of judgment, there will be four possible results: true positive (TP), false positive (FP), true negative (TN), false negative (FN). From a probabilistic perspective, we know $P_{TP} + P_{TN} = 1$ and $P_{FP} + P_{FN} = 1$, where the probability of FP is also called false alarm (FA) probability, so only TP and FP situations need to be addressed. Therefore, P_{FA} of the SR-MME detector is defined as

$$
\begin{aligned}
P_{FA} &= P\left(\sqrt{\lambda_{\max}(\hat{\mathbf{Q}}_{\mathbf{y},H_0})\lambda_{\min}(\hat{\mathbf{Q}}_{\mathbf{y},H_0})} > \gamma_1 \right) \\
&= P\left(\lambda_{\max}(\mathbf{R}_{H_0}) > \frac{(N\gamma_1)^2}{\lambda_{\min}(\mathbf{R}_{H_0})} \right) \\
&= P\left(\frac{\lambda_{\max}(\mathbf{R}_{H_0}) - \mu}{\nu} > \frac{\left(\frac{N\gamma_1}{\sqrt{N}-\sqrt{M}}\right)^2 - \mu}{\nu} \right) \\
&= 1 - F_2\left(\frac{\left(\frac{N\gamma_1}{\sqrt{N}-\sqrt{M}}\right)^2 - \mu}{\nu} \right).
\end{aligned}
\tag{19}
$$

Then the threshold can be derived as

$$
\gamma_1 = \frac{\sqrt{N} - \sqrt{M}}{N} \sqrt{\nu F_2^{-1}(1 - P_{FA}) + \mu}.
\tag{20}
$$

When the signal exists, the sample covariance matrix (12b) is no longer a Wishart matrix. As shown in [27], its maximum and minimum eigenvalues can be approximated as

$$
\lambda_{\max}(\mathbf{R}_{H_1}) \approx N\rho_1 + \lambda_{\max}(\mathbf{R}_{H_0}),
\tag{21a}
$$
$$
\lambda_{\min}(\mathbf{R}_{H_1}) \approx N\rho_M + \sqrt{N}(\sqrt{N} - \sqrt{M}),
\tag{21b}
$$

The detection probability P_D, i.e., P_{TP}, is given by

$$
\begin{aligned}
P_D &= P\left(\sqrt{\lambda_{\max}(\hat{\mathbf{Q}}_{\mathbf{y},H_1})\lambda_{\min}(\hat{\mathbf{Q}}_{\mathbf{y},H_1})} > \gamma_1 \right) \\
&= P\left(\lambda_{\max}(\mathbf{R}_{H_1}) > \frac{(N\gamma_1)^2}{\lambda_{\min}(\mathbf{R}_{H_1})} \right) \\
&= P\left(\frac{\lambda_{\max}(\mathbf{R}_{H_0}) - \mu}{\nu} > \frac{\frac{(N\gamma_1)^2}{N\rho_M + N - \sqrt{MN}} - N\rho_1 - \mu}{\nu} \right) \\
&= 1 - F_2\left(\frac{\frac{(N\gamma_1)^2}{N\rho_M + N - \sqrt{MN}} - \rho_1 - \mu}{\nu} \right).
\end{aligned}
\tag{22}
$$

3.2. Proposed GM Detector

The geometric mean (GM) detector is defined as

$$
\sqrt[M]{\prod_{m=1}^{M} \lambda_m(\hat{\mathbf{Q}}_{\mathbf{y}})} \underset{H_0}{\overset{H_1}{\gtrless}} \gamma_2,
\tag{23}
$$

where $\lambda_m(\hat{\mathbf{Q}}_{\mathbf{y}})$ is the eigenvalue of the sample covariance matrix and γ_2 represents the judgment threshold of this detector. Similar to SR-MME detector, the false alarm probability of the GM detector is given by

$$
\begin{aligned}
P_{FA} &= P\left(\sqrt[M]{\prod_{m=1}^{M} \lambda_m(\hat{\mathbf{Q}}_{\mathbf{y},H_0})} > \gamma_2 \right) \\
&= P\left(\lambda_{\max}(\mathbf{R}_{H_0}) > \gamma_2^M \frac{\lambda_{\max}(\mathbf{R}_{H_0})}{\det(\hat{\mathbf{Q}}_{\mathbf{y},H_0})} \right) \\
&= P\left(\frac{\lambda_{\max}(\mathbf{R}_{H_0}) - \mu}{\nu} > \frac{\gamma_2^M \frac{(\sqrt{N}+\sqrt{M})^2}{\det(\hat{\mathbf{Q}}_{\mathbf{y},H_0})} - \mu}{\nu} \right) \\
&= 1 - F_2\left(\frac{\gamma_2^M \frac{(\sqrt{N}+\sqrt{M})^2}{\det(\hat{\mathbf{Q}}_{\mathbf{y},H_0})} - \mu}{\nu} \right),
\end{aligned}
\tag{24}
$$

and the threshold is

$$
\gamma_2 = \sqrt[M]{\frac{\left(\nu F_2^{-1}(1 - P_{FA}) + \mu \right)\det(\hat{\mathbf{Q}}_{\mathbf{y},H_0})}{(\sqrt{N} + \sqrt{M})^2}}.
\tag{25}
$$

Finally, the detection probability of the GM detector can be expressed by

$$
\begin{aligned}
P_D &= P\left(\sqrt[M]{\prod_{m=1}^{M} \lambda_m(\hat{\mathbf{Q}}_{\mathbf{y},H_1})} > \gamma_2 \right) \\
&= P\left(\lambda_{\max}(\mathbf{R}_{H_0}) > \gamma_2^M \frac{\lambda_{\max}(\mathbf{R}_{H_0})}{\det(\hat{\mathbf{Q}}_{\mathbf{y},H_1})} \right) \\
&= 1 - F_2\left(\frac{\gamma_2^M \frac{(\sqrt{N}+\sqrt{M})^2}{\det(\hat{\mathbf{Q}}_{\mathbf{y},H_1})} - \mu}{\nu} \right).
\end{aligned}
\tag{26}
$$

4. Proposed Classifiers for Inferring the Number of UAV Emitters

The detectors proposed in Section 3 are designed for detecting whether the signals received by the base station are from UAV emitters or noise only. If the UAVs are present, we need to further determine their number. Therefore, a multi-layer neural network (ML-NN) classifier is given in the following. Support vector machine (SVM) classifier and naive Bayesian classifier (NBC) are also discussed as benchmarks.

4.1. Feature Selection and Extraction

As can be seen in Figure 1, after the sampling of the received signal, taking eigenvalue decomposition on the sample covariance matrix $\hat{\mathbf{Q}}_{\mathbf{y}}$, we can obtain eigenvalues $\hat{\lambda}_1 \geq \hat{\lambda}_2 \geq \ldots \geq \hat{\lambda}_M$. Although the sample covariance matrix is only an approximation of the actual received signal covariance matrix, its eigenvalues also approximately satisfy (8) if the sample number N is large enough, i.e., the maximum K eigenvalues belong to signal subspace. Therefore, this character can be used to determine the number of signal emitters.

Firstly, the following features of $\{\hat{\lambda}_m\}_{m=1}^{M}$ are selected to construct the feature space of received signal $\mathbf{Y}$, where

$$
\begin{cases}
\hat{\lambda}_{\max}, \ \hat{\lambda}_{\min} \\
\bar{\lambda} = \dfrac{1}{M} \sum_{m=1}^{M} \hat{\lambda}_m, \ \tilde{\lambda} = \left(\prod_{m=1}^{M} \hat{\lambda}_m \right)^{1/M} \\
\sigma_{\hat{\lambda}} = \sqrt{\dfrac{\sum_{m=1}^{M} (\hat{\lambda}_m - \bar{\lambda})^2}{M}}.
\end{cases}
\tag{27}
$$

As the number of emitters grows, the features also increase. In order to enlarge the discrimination between the different signals, we perform log operations on them. Then, the feature vector of any received signal is given by

$$
\mathbf{x} = \left(\log(\hat{\lambda}_{\max}), \log(\hat{\lambda}_{\min}), \log(\bar{\lambda}), \log(\tilde{\lambda}), \log(\sigma_{\hat{\lambda}}) \right).
\tag{28}
$$

Since the signal received by the base station is derived from different emitters, and it is a typical multiclass problem, machine learning-based methods are very suitable. Assuming there are most K emitters in the coverage area of the base station, we can obtain a K-elements classifier based on the existing training data and then substitute the signal to be detected into this classifier for classification. Then we will introduce several high-performance classification algorithms.

4.2. Proposed Multi-Layer Neural Network Classifier

We first take a set of received signals for training, such as $\mathbf{X} = \{(\mathbf{x}_i, \mathbf{g}_i)\}_{i=1,2,\ldots}$, where $\mathbf{g}_i = [g_{i,1}, \ldots, g_{i,k}, \ldots, g_{i,K}]$ is the corresponding output vector. It is a unit vector if signal i belongs to class k, $g_{i,k} = 1$. As is shown in Figure 2, the input of this neural network is a feature vector defined in (28), and the input layer is constructed of five neurons. Since most K emitters are in the coverage area of the base station, the number of neurons in the output layer is also K and the outputs of these neurons are denoted by $\{\hat{g}_1, \hat{g}_2, \ldots, \hat{g}_K\}$. Assuming there are a total s hidden layers in this network, these hidden layers contain $q_1, q_2, \ldots, q_s$ neurons, respectively. Therefore, referring to [44], the input received by the j_1th neuron of hidden layer 1 can be represented as

$$
\alpha_{1,j_1} = \sum_{h=1}^{5} v_{h,j_1} \mathbf{x}(h),
\tag{29}
$$

where v_{h,j_1} is the connection coefficient between the hth neuron of the input layer and the j_1th neuron of hidden layer 1. Then, the output of this neuron is given by

$$
z_{j_1}^1 = f(\alpha_{1,j_1} - \delta_{1,j_1}),
\tag{30}
$$

where δ_{1,j_1} denotes the threshold of the jth neuron of hidden layer 1. $f(\cdot)$ is the activation function, and usually a sigmoid function is adopted, which can be defined as

$$
\mathrm{sigmoid}(x) = \frac{1}{1 + e^{-x}}.
\tag{31}
$$

We can deduce the input and output of the rest of the hidden layers from hidden layer 1, and the output from the j_sth neuron of hidden layer s is given as

$$
\begin{aligned}
z_{j_s}^s &= f(\alpha_{s,j_s} - \delta_{s,j_s}) \\
&= f\left(\sum_{j_{s-1}=1}^{q_{s-1}} u_{j_{s-1},j_s} z_{j_{s-1}}^{s-1} - \delta_{s,j_s} \right),
\end{aligned}
\tag{32}
$$

where u_{j_{s-1},j_s} represents the connection coefficient between the j_{s-1}th neuron of hidden layer $s-1$ and the j_sth neuron of hidden layer s. Since the output of the last hidden layer is transmitted to the output layer, the final output of this network is

$$\hat{g}_k = f(\beta_k - \varepsilon_k) = f\left(\sum_{j_s=1}^{q_s} w_{j_s,k} z_{j_s}^s - \varepsilon_k\right), \tag{33}$$

where $w_{j_s,k}$ is the connection coefficient between hidden layer s and the output layer, and ε_k is the threshold of the kth neuron of the output layer.

When the input signal is $\mathbf{x}_1$, the ideal output is $\mathbf{g}_i$. However, the actual output of this neural network is $\hat{\mathbf{g}}_i = [\hat{g}_{i,1}, \ldots, \hat{g}_{i,k}, \ldots, \hat{g}_{i,K}]$, then the mean squared error (MSE) between ideal output and actual output is derived as

$$E_i = \frac{1}{K} \sum_{k=1}^{K} (\hat{g}_{i,k} - g_{i,k})^2, \tag{34}$$

Based on the classification error, we can update all the $(5q_1 + \sum_{t=1}^{s-1} q_t q_{t+1} + q_s K)$ connection coefficients and $(\sum_{t=1}^{s} q_t + K)$ thresholds of this neural network. Taking the j_sth neuron of hidden layer s as an example, we obtain

$$w_{j_s,k}^{l+1} = w_{j_s,k}^{l} + \Delta w_{j_s,k}^{l}, \tag{35a}$$

$$\delta_{s,j_s}^{l+1} = \delta_{s,j_s}^{l} + \Delta \delta_{s,j_s}^{l}, \tag{35b}$$

where l represents the number of iterations. According to the gradient descent method, the update terms are defined as

$$\begin{aligned}
\Delta w_{j_s,k}^{l} &= -\eta \frac{\partial E_i}{\partial w_{j_s,k}^{l}} \\
&= -\eta \frac{\partial E_i}{\partial \hat{g}_{i,k}} \cdot \frac{\partial \hat{g}_{i,k}}{\partial \beta_k} \cdot \frac{\partial \beta_k}{\partial w_{j_s,k}^{l}} \\
&= -\frac{2\eta}{K} z_{j_s}^s \cdot G_{i,k},
\end{aligned} \tag{36}$$

and

$$\begin{aligned}
\Delta \delta_{s,j_s}^{l} &= -\eta \frac{\partial E_i}{\partial \delta_{s,j_s}^{l}} \\
&= -\eta \sum_{k=1}^{K} \frac{\partial E_i}{\partial \hat{g}_{i,k}} \cdot \frac{\partial \hat{g}_{i,k}}{\partial \beta_k} \cdot \frac{\partial \beta_k}{\partial z_{j_s}^s} \cdot \frac{\partial z_{j_s}^s}{\partial \delta_{s,j_s}^{l}} \\
&= -\frac{2\eta}{K} z_{j_s}^s (1 - z_{j_s}^s) \cdot \sum_{k=1}^{K} w_{j_s,k}^{l} G_{i,k},
\end{aligned} \tag{37}$$

where η is the learning rate and

$$G_{i,k} = \hat{g}_{i,k}(1 - \hat{g}_{i,k})(\hat{g}_{i,k} - g_{i,k}). \tag{38}$$

All the parameters in the neural network are updated in each iteration until the parameters change less than a certain threshold or a certain number of iterations is reached. Therefore, the final classification result for signal i is given by

$$C_i = \arg\max_{k} \hat{g}_{i,k}^{L}, \tag{39}$$

where $C_i \in \{1, 2, \ldots, K\}$.

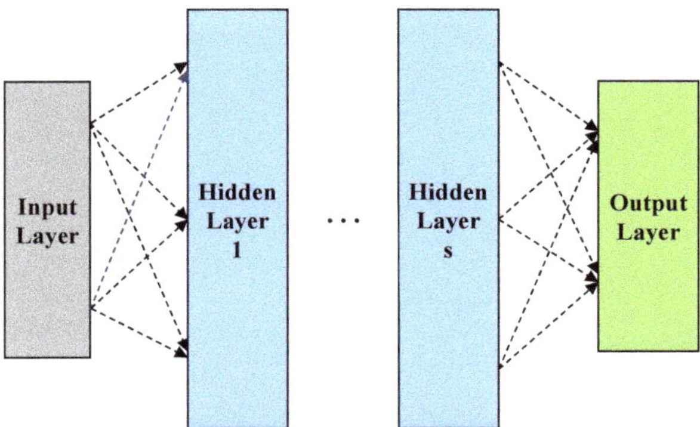

Figure 2. Multi-layer neural network.

4.3. Support Vector Machine Classifier

Since determining the number of signal sources is a K-elements classification problem, it can be decomposed into $K(K-1)/2$ binary classification problems and each of these binary classification problems can be solved by the support vector machine (SVM) method. Given a training sample set $D = \{(\mathbf{x}_1, g_1), (\mathbf{x}_2, g_2), \ldots, (\mathbf{x}_s, g_s)\}$, where $g_i = \{-1, +1\}$, $g_i = -1$ denotes that signal i belongs to class 1 and $g_i = +1$ denotes that this signal belongs to class 2. The separable hyperplane for the sample space can be expressed by

$$\mathbf{w}^T \mathbf{x} + b = 0, \tag{40}$$

where $\mathbf{w}$ is the normal vector which determines the direction of this hyperplane, and b denotes the bias which is defined as the distance from the hyperplane to the original point. Therefore, the separable hyperplane can be denoted as $(\mathbf{w}, b)$.

Assuming the samples can be classified by hyperplane $(\mathbf{w}, b)$ accurately, if $g_i = -1$, we can obtain $\mathbf{w}^T \mathbf{x}_i + b < 0$, and if $g_i = +1$, we obtain $\mathbf{w}^T \mathbf{x}_i + b > 0$. Then the following conditions should be satisfied:

$$\begin{cases} \mathbf{w}^T \mathbf{x}_i + b \geq +1, \ g_i = +1 \\ \mathbf{w}^T \mathbf{x}_i + b \leq -1, \ g_i = -1. \end{cases} \tag{41}$$

The samples closest to the separable hyperplane make the equalities in (41) hold, and they are support vectors. The sum of the distance from the two heterologous support vectors to the hyperplane is called the margin, and it is defined as $\delta = \frac{2}{\|\mathbf{w}\|}$. For maximizing the margin of the separable hyperplane, the optimization problem can be designed as

$$\min_{\mathbf{w}, b} \ \frac{1}{2} \|\mathbf{w}\|^2 \tag{42a}$$

$$\text{s.t.} \ g_i(\mathbf{w}^T \mathbf{x}_i + b) \geq 1. \tag{42b}$$

Actually, the training samples can hardly be linearly separated in the current sample space. Firstly, we map the samples to a higher-dimensional feature space. Then the model of the separable hyperplane is modified as

$$f(\mathbf{x}) = \mathbf{w}^T \phi(\mathbf{x}) + b. \tag{43}$$

Secondly, to avoid overfitting, we introduce the concept of soft margin. This concept allows SVM to make errors in the classification of some samples, i.e., these samples can not satisfy constraint $g_i(\mathbf{w}^T\phi(\mathbf{x}_i) + b) \geq 1$. Consequently, the optimization problem (42) is transformed to maximize the margin while minimizing the classification error:

$$\min_{\mathbf{w},b,\xi_i} \quad \frac{1}{2}\|\mathbf{w}\|^2 + C\sum_{i=1}^{s}\xi_i, \tag{44a}$$

$$\text{s.t.} \quad g_i(\mathbf{w}^T\phi(\mathbf{x}_i) + b) \geq 1 - \xi_i, \tag{44b}$$

$$\xi_i \geq 0. \tag{44c}$$

where $C > 0$ is the regularization constant, $\xi_i \geq 0$ is a slack variable, and $\xi_i \geq 1$ means sample $\mathbf{x}_i$ is misclassified.

Obviously, (44) is a quadratic programming (QP) problem, and it can be solved by the Lagrangian multiplier method. Therefore, the Lagrangian of (44) is given by

$$\begin{aligned}
L(\mathbf{w}, b, \xi, \alpha, \beta) = {} & \frac{1}{2}\|\mathbf{w}\|^2 + C\sum_{i=1}^{s}\xi_i - \sum_{i=1}^{s}\beta_i\xi_i \\
& + \sum_{i=1}^{s}\alpha_i\Big[1 - \xi_i - g_i(\mathbf{w}^T\phi(\mathbf{x}_i) + b)\Big],
\end{aligned} \tag{45}$$

where $\alpha_i \geq 0$ and $\beta_i \geq 0$ are Lagrangian multipliers. Computing the partial derivatives of $\mathbf{w}, b, \xi_i$, we obtain

$$\mathbf{w} = \sum_{i=1}^{s}\alpha_i g_i\phi(\mathbf{x}_i), \tag{46a}$$

$$\sum_{i=1}^{s}\alpha_i g_i = 0, \tag{46b}$$

$$C = \alpha_i + \beta_i. \tag{46c}$$

Taking them into Equation (45), the dual problem of (44) is derived as

$$\max_{\alpha_i} \quad \sum_{i=1}^{s}\alpha_i - \frac{1}{2}\sum_{i=1}^{s}\sum_{j=1}^{s}\alpha_i\alpha_j g_i g_j \kappa(\mathbf{x}_i, \mathbf{x}_j), \tag{47a}$$

$$\text{s.t.} \quad (46b), \tag{47b}$$

$$0 \leq \alpha_i \leq C, \tag{47c}$$

where $\kappa(\mathbf{x}_i, \mathbf{x}_j) = \phi(\mathbf{x}_i)^T\phi(\mathbf{x}_j)$ is the kernel function.

Since (44) contains the inequality constraint, the above optimization procedure must satisfy the KKT conditions

$$\begin{cases}
\alpha_i \geq 0, \ \beta_i \geq 0 \\
g_i f(\mathbf{x}_i) - 1 + \xi_i \geq 0 \\
\alpha_i(g_i f(\mathbf{x}_i) - 1 + \xi_i) = 0 \\
\xi_i \geq 0, \ \beta_i\xi_i = 0.
\end{cases} \tag{48}$$

4.4. Naive Bayesian Classifier

As given in (28), three features of the ith signal are considered in our problem. We assume that the five features are independent of each other, then according to Bayes' theorem, the probability that the ith signal belongs to a certain class is

$$P(c_k|\mathbf{x}_i) = \frac{P(c_k)P(\mathbf{x}_i|c_k)}{P(\mathbf{x}_i)} = \frac{P(c_k)P(\mathbf{x}_i|c_k)}{\sum_{k=1}^{K}P(\mathbf{x}_i|c_k)P(c_k)}, \tag{49}$$

where c_k, $k \in D = \{1, 2, \ldots, K\}$ is the label for classification. Therefore, the NBC for our problem can be verified as

$$h(\mathbf{x}_i) = \arg\max_{k \in D} P(c_k)P(\mathbf{x}_i|c_k). \tag{50}$$

The training process is based on the training set to estimate the class prior probability $P(c_k)$ and conditional probability $P(\mathbf{x}_i|c_k)$. Since the features in (28) are continuous, we can suppose $P(\mathbf{x}_i|c_k) \sim \mathcal{N}(\mu_k, \Sigma_k)$, where μ_k and Σ_k are the mean and covariance matrix of feature vectors for all training samples that belong to class k. Therefore, the conditional probability can be represented by its PDF as

$$P(\mathbf{x}_i|c_k) = \frac{1}{(\sqrt{2\pi})^5|\Sigma|^{1/2}}e^{-\frac{1}{2}(\mathbf{x}_i-\mu_k)^T\Sigma_k^{-1}(\mathbf{x}_i-\mu_k)}. \tag{51}$$

Then, we can compute the logarithm of (50). Finally, the NBC can be transformed as

$$\begin{aligned}
h(\mathbf{x}_i) &= \arg\max_{k \in D} \ln(P(c_k)P(\mathbf{x}_i|c_k)) \\
&= \arg\max_{k \in D} \left(\ln P(c_k) - \frac{5}{2}\ln 2\pi - \frac{1}{2}\ln|\Sigma_k| \right. \\
&\quad \left. - \frac{1}{2}(\mathbf{x}_i - \mu_k)^T\Sigma_k^{-1}(\mathbf{x}_i - \mu_k) \right).
\end{aligned} \tag{52}$$

5. Simulation Results

In this section, representative simulation results are given to show the high performance of signal detectors and classifiers proposed in this paper. Next, we will compare the two proposed signal detectors with existing detectors.

5.1. Signal Detectors

Firstly, it is assumed that there are three UAV emitters in the coverage area of the base station, i.e., $K = 3$ and the signals used in this simulation are randomly generated signals. After sampling the received signal, we can obtain the sample covariance matrix. The largest eigenvalue of the noise-only sample covariance matrix ($\mathbf{R}_{H_0}$) follows Tracy–Widom distribution of order 2, so we want to use its statistical properties to derive P_{FA}, P_D, and γ of the signal detectors. However, (17) is difficult to evaluate, since we cannot obtain the CDF of $\mathcal{TW}_2$. Fortunately, M. Prähofer and H. Spohn fitted this function and gave tables for the CDF of the Tracy–Widom distribution in [45]. We may select a part of the values and put them in Table 1. To highlight the advantages of our proposed signal detectors, we also introduce two existing detectors for comparison. The two detectors, M-MME and MME [27], are defined as

$$\textbf{M-MME}: \quad \frac{\lambda_{\max}(\hat{\mathbf{Q}}_\mathbf{y}) + \lambda_{\min}(\hat{\mathbf{Q}}_\mathbf{y})}{2} \underset{H_0}{\overset{H_1}{\gtrless}} \gamma_4, \tag{53a}$$

$$\textbf{MME}: \quad \frac{\lambda_{\max}(\hat{\mathbf{Q}}_\mathbf{y})}{\lambda_{\min}(\hat{\mathbf{Q}}_\mathbf{y})} \underset{H_0}{\overset{H_1}{\gtrless}} \gamma_3. \tag{53b}$$

As can be seen in Figure 3, the relationship between SNR and probability of detection is plotted, where the probability of false alarm $P_{FA} = 10^{-4}$, the number of receive antennas $M = 64$, the number of snapshots $N = 100$, and the final results are obtained from 5000 Monte-Carlo simulations. Among these four detectors, SR-MME has the best performance across all SNR values, especially in the low-SNR region. In extremely poor communication conditions, i.e., SNR in the range from -30 dB to -20 dB, M-MME and MME can hardly detect the presence of the signal sources, while SR-MME can keep the detection probability above 85%, so we can say that SR-MME is the best signal detector for the low-SNR situations. For the GM detector, its detection probability is slightly less than

SR-MME in the low-SNR situation, but it still has a great improvement compared to the other two detectors.

Table 1. Numerical table for the Tracy–Widom distribution of order 2.

t	-3.70	-2.90	-1.80	-0.60	-0.23	0.49	1.32	2.06	2.68
$F_2(t)$	0.01	0.1	0.5	0.9	0.95	0.99	0.999	0.9999	0.99999

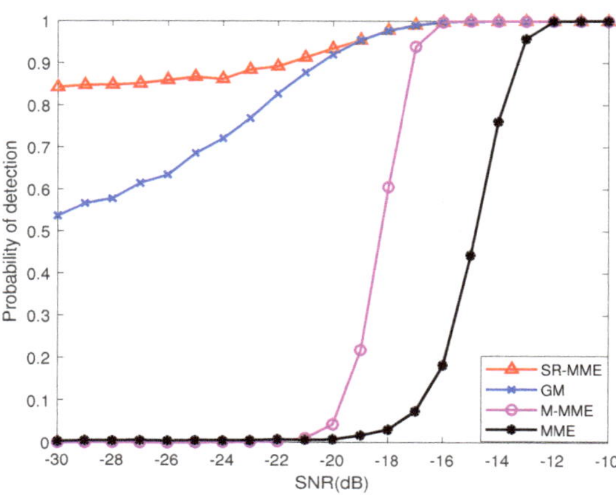

Figure 3. Probability of detection versus SNR, $P_{FA} = 10^{-4}$, $N = 100$.

Figure 4 presents the detection probability of these four signal detectors with the number of samples, where $M = 64$, $P_{FA} = 10^{-4}$ and SNR $= -20$ dB. The overall trend of the curves in this figure is similar to Figure 3, with SR-MME still the best performing of these four signal detectors and achieving a detection probability of at least 93%. The detection performance of the GM detector also improves as the number of samples increases, especially when N ranges between 100 and 200. GM has a significant improvement compared with M-MME and MME. Therefore, the robust performance of SR-MME and GM at a lower number of samples can help us save lots of time and spatial resources, and not at the cost of a loss of detection performance.

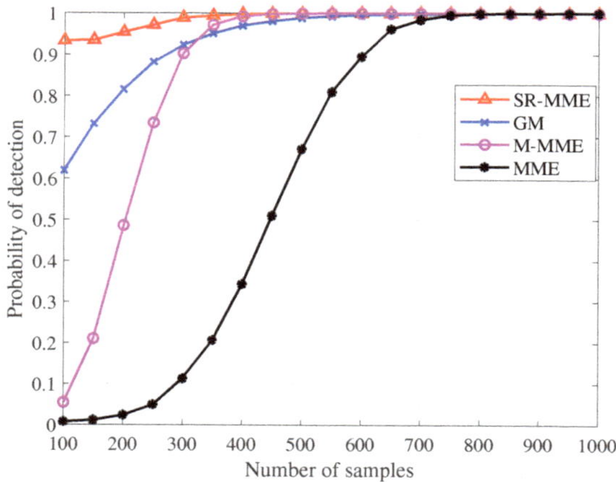

Figure 4. Probability of detection versus number of samples, SNR $= -20$ dB, $P_{FA} = 10^{-4}$.

Figure 5 shows the most commonly used indicator in the field of threshold detection, the Receiver Operating Characteristic (ROC) curve. It evaluates a detector comprehensively in terms of both detection probability and false alarm probability. The parameters involved in this simulation are $M = 64$, $N = 200$, and SNR = -20 dB. The ROC curve of SR-MME is above the other three curves, so it is the best detector for the overall performance. Correspondingly, the MME has the worst performance. For GM and M-MME, due to a cross-over of their ROC curves, the area under ROC curve (AUC) is introduced for comparing their performance. Since the axes in this figure employ scientific counting, after converting it to ordinary coordinates, the AUC value of M-MME is larger than GM. From this perspective, M-MME performs better than GM. However, in practice, we would prefer a relatively low false alarm probability, so GM will be more useful, since it can guarantee a low false alarm probability while maintaining a high detection probability.

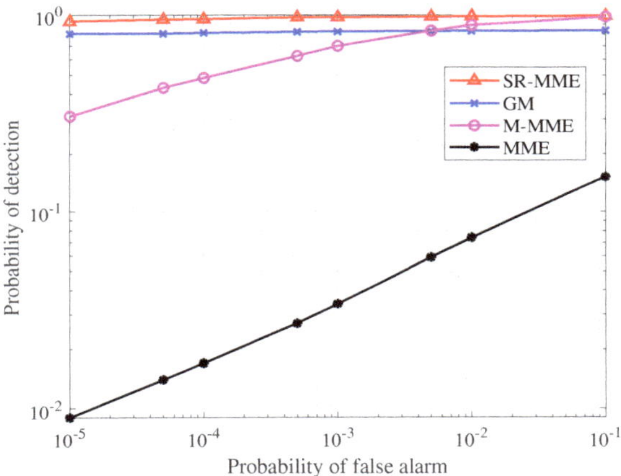

Figure 5. ROC curve, SNR = -20 dB, $N = 200$.

5.2. Signal Classifiers

After the presence of the emitters is determined by the signal detectors, we need to further determine the number of emitters. According to the three machine learning-based signal classifiers, the first step is to design an appropriate training set. As mentioned in Section 4, the feature vector of received signals is given by (28), so the training set is defined as

$$\{\mathbf{X}_1, \ldots, \mathbf{X}_k, \ldots, \mathbf{X}_K\}, \tag{54}$$

where

$$\mathbf{X}_k = \left\{ (\mathbf{x}_{k,1}, k), (\mathbf{x}_{k,2}, k), \ldots, (\mathbf{x}_{k,i}, k), \ldots \right\}, \tag{55}$$

and $K \in \{1, 2, 3\}$. For the training of ML-NN, the epoch size is 400, and the learning rate is set as 0.01. The input layer and output layer have five neurons and three neurons, respectively, and the hidden layer size of the three-layer NN is 10; the four-layer NN has two hidden layers, and their sizes are 7 and 5.

In order to compare the complexity of the ML-based methods mentioned in our work, Table 2 gives the training duration of each classifier at different amounts of training data. The neural network takes more training time as the number of training samples is small. When the amount of training data reaches 50, the average training duration of SVM exceeds the three-layer neural network. Unlike other classifiers, the change in the number of training samples has less impact on NBC.

Table 2. Average training duration of different classifiers.

Classifiers	Number of Training Samples					
	10	**20**	**30**	**40**	**50**	**100**
4-layer Neural Network	0.734149	0.809213	0.936686	1.038361	1.133686	1.660306
3-layer Neural Network	0.629034	0.705787	0.799842	0.875255	0.949917	1.356083
SVM	0.221015	0.333413	0.520857	0.753500	1.007692	3.077889
NBC	0.090488	0.092070	0.093222	0.094849	0.095326	0.113129

Figure 6 plots the relationship between the classification accuracy of the four classifiers and SNR, where $M = 64$, $N = 200$, and K = 3 in the test. This figure shows that ML-NNs have much stronger performance than NBC in all the SNR regions, and the accuracy of SVM is obviously lower than ML-NNs when SNR ≥ -18 dB. Since neural networks have strong learning ability, the deeper networks will cause overfitting and result in the decrease in classification accuracy; we only consider 3L-NN and 4L-NN in this work.

By observing the curves of the signal detectors and the signal classifiers about SNR in Figures 3 and 6, we can find when SNR = -20 dB and $P_{FA} = 10^{-4}$; the P_D of SR-MME can achieve 95%. Since $P_{FA} + P_{AN} = 1$, SR-MME almost separates all the noise while ensuring a high signal detection rate. However, for the optimal neural network-based signal classifier, its classification accuracy at SNR = -20 dB is also only about 70%, that is, if the noise is directly added to the classification process, nearly 30% of the noise will be misclassified as signals. Therefore, we believe that adding the step of signal detection is necessary. Moreover, the time required to perform one signal detection was approximately 0.04 s, and the training duration required for the four-layer neural network after adding noise is also increased to about 1.02 s when the number of training sample is 10. Therefore, using the signal detectors can also save time.

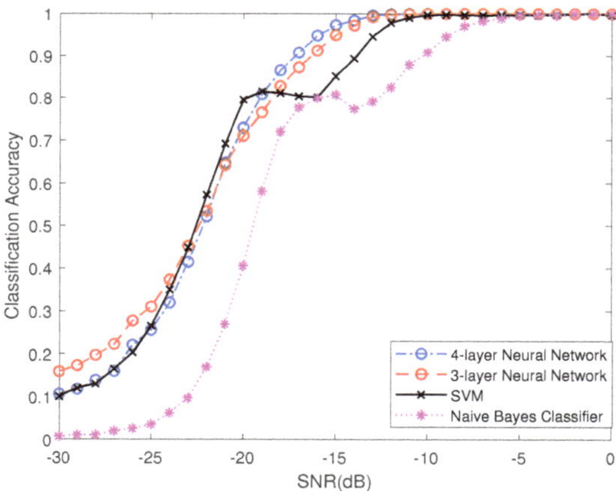

Figure 6. Classification accuracy versus SNR, $M = 64$.

In Figure 7, we show classification accuracy varying with the number of received antennas when SNR = -15 dB, and other conditions are the same as Figure 6. In general, the array containing 64 antennas or more can be called a massive array. Therefore, as can be seen in this figure, the classification accuracy of neural networks can approach nearly 100% when a massive receive array is adopted. The performance of SVM and NBC is worse than the neural network with a massive receive array.

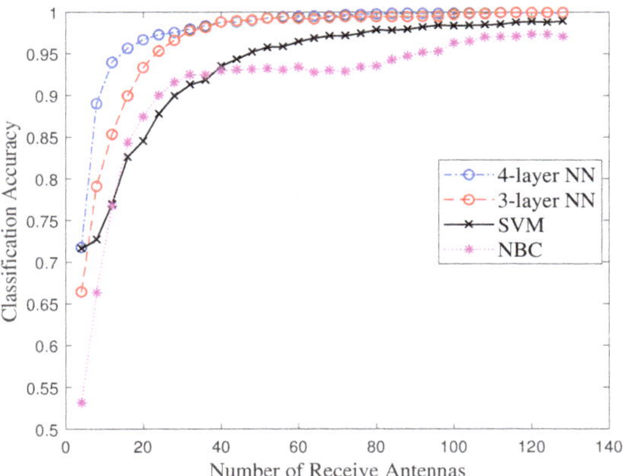

Figure 7. Classification accuracy versus number of receive antennas, SNR $= -15$ dB.

5.3. Analysis of Classic Classifiers

AIC and MDL are two classic information-theoretic criteria for model selection, which were proposed by Akaike [15,46], Schwartz [16], and Rissanen [17]. In Akaike's works, the AIC criterion is defined as

$$\text{AIC}(m) = -2\log L_m^{(M-m)N} + 2m(2M - m), \tag{56}$$

where $m \in \{0, 1, \dots, M - 1\}$ and

$$L_m = \frac{\prod_{i=m+1}^{M} \hat{\lambda}_i^{1/(M-m)}}{\frac{1}{M-m} \sum_{i=m+1}^{M} \hat{\lambda}_i}. \tag{57}$$

The classification results of received signals are determined by AIC criterion as

$$\text{AIC}(C) = \min(\text{AIC}(0), \text{AIC}(1), \dots, \text{AIC}(M - 1)), \tag{58}$$

where C is the number of emitters.

Similarly, the definition of the MDL criterion is given as

$$\text{MDL}(m) = -2\log L_m^{(M-m)N} + \frac{1}{2}m(2M - m)\log N. \tag{59}$$

MDL modified the bias term based on AIC, leading to improved classification performance. The classification result of MDL is

$$\text{MDL}(C) = \min(\text{MDL}(0), \text{MDL}(1), \dots, \text{MDL}(M - 1)). \tag{60}$$

The former papers only verified the work performance of AIC and MDL with a small-sized receiving array, such as arrays with around eight antennas. To find out whether these two methods can maintain good performance with a massive receive array, we present a curve between their classification accuracy and the number of receive antennas. Unfortunately, as shown in Figure 8, AIC and MDL can only achieve good performance when the number of receive antennas is between 8 and 36. Once the number of receive antennas exceeds 36, their classification accuracy drops sharply until the number of emitters is completely inaccessible at 44 antennas. By analyzing the definitions of AIC and MDL, since the number of receive antennas is equal to the number of possible classifications, the corresponding model complexity increases when the number of antennas increases. If the model is too complex, the values of AIC and MDL will increase, and this will result in overfitting. Thus, we can conclude that AIC and MDL are not applicable for scenarios using massive receive arrays.

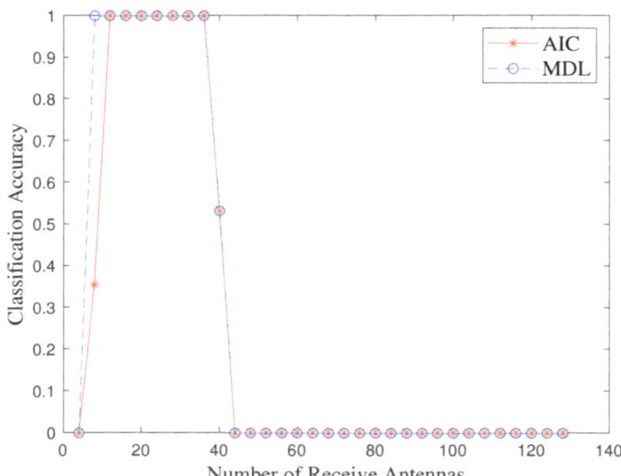

Figure 8. Classification accuracy versus number of receive antennas for AIC and MDL, SNR = 0 dB.

To compare the performance differences between traditional and machine learning-based methods, we plot the classification accuracy of these methods with SNR in Figure 9, where $M = 32$. Although this is not a massive array scenario, the machine learning-based method still has higher classification accuracy than the AIC and MDL. Therefore, machine learning-based signal classifiers are robust and are applicable to a broader SNR range and array size.

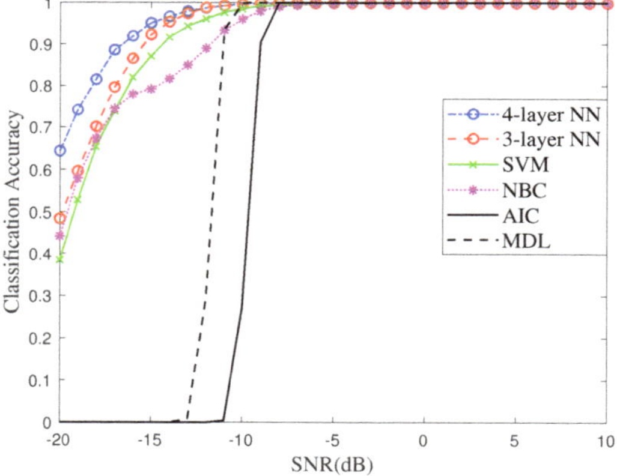

Figure 9. Classification accuracy versus SNR, $M = 32$.

6. Conclusions

In order to provide the vital prior knowledge for DOA estimation, a DOA preprocessing system containing signal detectors and ML-based signal classifiers has been proposed for inferring the number of UAV emitters in a massive MIMO system. Two high-precision signal detectors, i.e., SR-MME and GM, can quickly and accurately judge the presence of the signal emitters based on the statistical characteristics of the received signals and the threshold detection theory. Simulation results showed that the proposed SR-MME and GM have much better detection performance than existing detectors like MME and M-MME, especially in the low-SNR region and situations with a small number of samples.

After determining the presence of signals, the specific number of emitters can be further determined by ML-based classifiers including ML-NN, SVM, and NBC. Compared to traditional methods, like AIC and MDL, the proposed methods can work well with a massive MIMO array and have higher accuracy when SNR is low. In conclusion, we believe that the proposed system and method will be helpful for the future implementation of UAV massive MIMO communications.

Author Contributions: Conceptualization, Y.L.; Methodology, Y.L.; Software, Y.L.; Validation, Y.L.; Investigation, J.H., S.Y., W.Z., D.T. and Y.S.; Resources, H.S.; Writing—review & editing, J.W.; Project administration, F.S. All authors have read and agreed to the published version of the manuscript.

Funding: This work was supported in part by the National Natural Science Foundation of China (Nos. U22A2002 and 62071234), the Major Science and Technology plan of Hainan Province under Grant ZDKJ2021022, and the Scientific Research Fund Project of Hainan University under Grant KYQD(ZR)-21008. This work was also supported in part by the National Natural Science Foundation of China under Grant 62001116, the Natural Science Foundation of Fujian Province under Grant 2020J05106.

Conflicts of Interest: The authors declare no conflict of interest.

References

1. Zeng, Y.; Zhang, R.; Lim, T.J. Wireless communications with unmanned aerial vehicles: Opportunities and challenges. *IEEE Commun. Mag.* **2016**, *54*, 36–42. [CrossRef]
2. Huang, Y.; Wu, Q.; Lu, R.; Peng, X.; Zhang, R. Massive MIMO for cellular-connected UAV: Challenges and promising solutions. *IEEE Commun. Mag.* **2021**, *59*, 84–90. [CrossRef]
3. Wang, C.X.; Haider, F.; Gao, X.; You, X.H.; Yang, Y.; Yuan, D.; Aggoune, H.M.; Haas, H.; Fletcher, S.; Hepsaydir, E. Cellular architecture and key technologies for 5G wireless communication networks. *IEEE Commun. Mag.* **2014**, *52*, 122–130. [CrossRef]
4. Saad, W.; Bennis, M.; Chen, M. A vision of 6G wireless systems: Applications, trends, technologies, and open research problems. *IEEE Netw.* **2019**, *34*, 134–142. [CrossRef]
5. Zhang, Z.; Xiao, Y.; Ma, Z.; Xiao, M.; Ding, Z.; Lei, X.; Karagiannidis, G.K.; Fan, P. 6G wireless networks: Vision, requirements, architecture, and key technologies. *IEEE Veh. Technol. Mag.* **2019**, *14*, 28–41. [CrossRef]
6. Chandhar, P.; Larsson, E.G. Massive MIMO for connectivity with drones: Case studies and future directions. *IEEE Access* **2019**, *7*, 94676–94691. [CrossRef]
7. Harris, P.; Malkowsky, S.; Vieira, J.; Bengtsson, E.; Tufvesson, F.; Hasan, W.B.; Liu, L.; Beach, M.; Armour, S.; Edfors, O. Performance characterization of a real-time massive MIMO system with LOS mobile channels. *IEEE J. Sel. Areas Commun.* **2017**, *35*, 1244–1253. [CrossRef]
8. Geraci, G.; Garcia-Rodriguez, A.; Azari, M.M.; Lozano, A.; Mezzavilla, M.; Chatzinotas, S.; Chen, Y.; Rangan, S.; Di Renzo, M. What will the future of UAV cellular communications be? A flight from 5G to 6G. *IEEE Commun. Surv. Tuts.* **2022**, *24*, 1304–1335. [CrossRef]
9. Bai, L.; Huang, Z.; Cheng, X. A Non-Stationary Model with Time-Space Consistency for 6G Massive MIMO mmWave UAV Channels. *IEEE Trans. Wireless Commun.* **2022**, *22*, 2048–2064. [CrossRef]
10. Chandhar, P.; Danev, D.; Larsson, E.G. Massive MIMO for communications with drone swarms. *IEEE Trans. Wireless Commun.* **2017**, *17*, 1604–1629. [CrossRef]
11. Huang, L.; Qian, C.; So, H.C.; Fang, J. Source enumeration for large array using shrinkage-based detectors with small samples. *IEEE Trans. Aerosp. Electron. Syst.* **2015**, *51*, 344–357. [CrossRef]
12. Krim, H.; Viberg, M. Two decades of array signal processing research: the parametric approach. *IEEE Signal Process. Mag.* **1996**, *13*, 67–94. [CrossRef]
13. Aquino, S.; Vairavel, G. A Review of Direction of Arrival Estimation Techniques in Massive MIMO 5G Wireless Communication Systems. In *Proceedings of the Fourth International Conference on Communication, Computing and Electronics Systems: ICCCES 2022*; Springer: Berlin/Heidelberg, Germany, 2023; pp. 15–34.
14. Björnson, E.; Sanguinetti, L.; Wymeersch, H.; Hoydis, J.; Marzetta, T.L. Massive MIMO is a reality—What is next? Five promising research directions for antenna arrays. *Digit. Signal Process.* **2019**, *94*, 3–20. [CrossRef]
15. Akaike, H. A new look at the statistical model identification. *IEEE Trans. Autom. Control* **1974**, *19*, 716–723. [CrossRef]
16. Schwarz, G. Estimating the dimension of a model. *Ann. Stat.* **1978**, *6*, 461–464. [CrossRef]
17. Rissanen, J. Modeling by shortest data description. *Automatica* **1978**, *14*, 465–471. [CrossRef]
18. Stoica, P.; Selen, Y. Model-order selection: a review of information criterion rules. *IEEE Signal Process. Mag.* **2004**, *21*, 36–47. [CrossRef]
19. Lu, Z.; Zoubir, A.M. Generalized Bayesian information criterion for source enumeration in array processing. *IEEE Trans. Signal Process.* **2012**, *61*, 1470–1480. [CrossRef]
20. Lu, Z.; Zoubir, A.M. Flexible detection criterion for source enumeration in array processing. *IEEE Trans. Signal Process.* **2012**, *61*, 1303–1314. [CrossRef]

21. Williams, D.B.; Johnson, D.H. Using the sphericity test for source detection with narrow-band passive arrays. *IEEE Trans. Acoust. Speech Signal Process.* **1990**, *38*, 2008–2014. [CrossRef]
22. Brcich, R.F.; Zoubir, A.M.; Pelin, P. Detection of sources using bootstrap techniques. *IEEE Trans. Signal Process.* **2002**, *50*, 206–215. [CrossRef]
23. Wax, M.; Adler, A. Detection of the Number of Signals by Signal Subspace Matching. *IEEE Trans. Signal Process.* **2021**, *69*, 973–985. [CrossRef]
24. Cabric, D.; Mishra, S.M.; Brodersen, R.W. Implementation issues in spectrum sensing for cognitive radios. In Proceedings of the Conference Record of the Thirty-Eighth Asilomar Conference on Signals, Systems and Computers, Pacific Grove, CA, USA, 7–10 November 2004; Volume 1, pp. 772–776.
25. Cabric, D.; Tkachenko, A.; Brodersen, R.W. Spectrum sensing measurements of pilot, energy, and collaborative detection. In Proceedings of the Milcom 2006-2006 IEEE Military Communications Conference, Washington, DC, USA, 23–25 October 2006; pp. 1–7.
26. Gardner, W.A. Exploitation of spectral redundancy in cyclostationary signals. *IEEE Signal Process. Mag.* **1991**, *8*, 14–36. [CrossRef]
27. Zeng, Y.; Liang, Y.C. Eigenvalue-based spectrum sensing algorithms for cognitive radio. *IEEE Trans. Commun.* **2009**, *57*, 1784–1793. [CrossRef]
28. Zhang, R.; Lim, T.J.; Liang, Y.C.; Zeng, Y. Multi-antenna based spectrum sensing for cognitive radios: A GLRT approach. *IEEE Trans. Commun.* **2010**, *58*, 84–88. [CrossRef]
29. Liu, C.; Li, H.; Wang, J.; Jin, M. Optimal eigenvalue weighting detection for multi-antenna cognitive radio networks. *IEEE Trans. Wirel. Commun.* **2016**, *16*, 2083–2096. [CrossRef]
30. Yang, M.; Ai, B.; He, R.; Huang, C.; Ma, Z.; Zhong, Z.; Wang, J.; Pei, L.; Li, Y.; Li, J. Machine-learning-based fast angle-of-arrival recognition for vehicular communications. *IEEE Trans. Veh. Technol.* **2021**, *70*, 1592–1605. [CrossRef]
31. Bithas, P.S.; Michailidis, E.T.; Nomikos, N.; Vouyioukas, D.; Kanatas, A.G. A survey on machine-learning techniques for UAV-based communications. *Sensors* **2019**, *19*, 5170. [CrossRef]
32. Jiang, C.; Zhang, H.; Ren, Y.; Han, Z.; Chen, K.C.; Hanzo, L. Machine learning paradigms for next-generation wireless networks. *IEEE Wirel. Commun.* **2016**, *24*, 98–105. [CrossRef]
33. Thilina, K.M.; Choi, K.W.; Saquib, N.; Hossain, E. Machine learning techniques for cooperative spectrum sensing in cognitive radio networks. *IEEE J. Sel. Areas Commun.* **2013**, *31*, 2209–2221. [CrossRef]
34. Zhuang, Z.; Xu, L.; Li, J.; Hu, J.; Sun, L.; Shu, F.; Wang, J. Machine-learning-based high-resolution DOA measurement and robust directional modulation for hybrid analog-digital massive MIMO transceiver. *Sci. China Inf. Sci.* **2020**, *63*, 1–18. [CrossRef]
35. Shu, F.; Liu, L.; Yang, L.; Jiang, X.; Xia, G.; Wu, Y.; Wang, X.; Jin, S.; Wang, J.; You, X. Spatial Modulation: an Attractive Secure Solution to Future Wireless Network. *arXiv* **2021**, arXiv:2103.04051.
36. Jie, Q.; Zhan, X.; Shu, F.; Ding, Y.; Shi, B.; Li, Y.; Wang, J. High-performance Passive Eigen-model-based Detectors of Single Emitter Using Massive MIMO Receivers. *arXiv* **2021**, arXiv:2108.02011.
37. Zhang, R.; Shim, B.; Wu, W. Direction-of-Arrival Estimation for Large Antenna Arrays With Hybrid Analog and Digital Architectures. *IEEE Trans. Signal Process.* **2021**, *70*, 72–88. [CrossRef]
38. Chen, C.E.; Lorenzelli, F.; Hudson, R.E.; Yao, K. Stochastic maximum-likelihood DOA estimation in the presence of unknown nonuniform noise. *IEEE Trans. Signal Process.* **2008**, *56*, 3038–3044. [CrossRef]
39. Chiani, M. Distribution of the largest eigenvalue for real Wishart and Gaussian random matrices and a simple approximation for the Tracy–Widom distribution. *J. Multivar. Anal.* **2014**, *129*, 69–81. [CrossRef]
40. Wei, L.; Tirkkonen, O. Analysis of scaled largest eigenvalue based detection for spectrum sensing. In Proceedings of the 2011 IEEE International Conference on Communications (ICC), Kyoto, Japan, 5–9 June 2011; pp. 1–5.
41. Tracy, C.A.; Widom, H. The distributions of random matrix theory and their applications. In *Proceedings of the New Trends in Mathematical Physics: Selected Contributions of the XVth International Congress on Mathematical Physics*; Springer: Berlin/Heidelberg, Germany, 2009; pp. 753–765.
42. Fokas, A.S.; Its, A.R.; Novokshenov, V.Y.; Kapaev, A.A.; Kapaev, A.I.; Novokshenov, V.I. *Painlevé Transcendents: The Riemann-Hilbert Approach*; Number 128; American Mathematical Society: Providence, RI, USA, 2006.
43. Perry, P.; Johnstone, I.; Ma, Z.; Shahram, M. Rmtstat: Distributions and Statistics from Random Matrix Theory. R Software Package Version. 2009. Available online: https://cran.rstudio.com/web/packages/RMTstat/index.html (accessed on 6 April 2023).
44. Hagan, M.T.; Demuth, H.B.; Beale, M. *Neural Network Design*; PWS Publishing Co.: Boston, MA, USA, 1997.
45. Prähofer, M.; Spohn, H. Exact scaling functions for one-dimensional stationary KPZ growth. *J. Stat. Phys.* **2004**, *115*, 255–279. [CrossRef]
46. Akaike, H. Information theory and an extension of the maximum likelihood principle. In *Selected Papers of Hirotugu Akaike*; Springer: Berlin/Heidelberg, Germany, 1998; pp. 199–213.

 **_drones_**

MDPI

Article

Joint Communication and Action Learning in Multi-Target Tracking of UAV Swarms with Deep Reinforcement Learning

Wenhong Zhou [1], Jie Li [2] and Qingjie Zhang [1,*]

[1] Aviation University of Air Force, Changchun 130022, China
[2] College of Intelligence Science and Technology, National University of Defense Technology, Changsha 410073, China
* Correspondence: nudtzhang@hotmail.com

Abstract: Communication is the cornerstone of UAV swarms to transmit information and achieve cooperation. However, artificially designed communication protocols usually rely on prior expert knowledge and lack flexibility and adaptability, which may limit the communication ability between UAVs and is not conducive to swarm cooperation. This paper adopts a new data-driven approach to study how reinforcement learning can be utilized to jointly learn the cooperative communication and action policies for UAV swarms. Firstly, the communication policy of a UAV is defined, so that the UAV can autonomously decide the content of the message sent out according to its real-time status. Secondly, neural networks are designed to approximate the communication and action policies of the UAV, and their policy gradient optimization procedures are deduced, respectively. Then, a reinforcement learning algorithm is proposed to jointly learn the communication and action policies of UAV swarms. Numerical simulation results verify that the policies learned by the proposed algorithm are superior to the existing benchmark algorithms in terms of multi-target tracking performance, scalability in different scenarios, and robustness under communication failures.

Keywords: UAV swarms; reinforcement learning; cooperation; communication; policy gradient

Citation: Zhou, W.; Li, J.; Zhang, Q. Joint Communication and Action Learning in Multi-Target Tracking of UAV Swarms with Deep Reinforcement Learning. *Drones* **2022**, *6*, 339. https://doi.org/10.3390/drones6110339

Academic Editor: Oleg Yakimenko

Received: 23 September 2022
Accepted: 28 October 2022
Published: 2 November 2022

Publisher's Note: MDPI stays neutral with regard to jurisdictional claims in published maps and institutional affiliations.

1. Introduction

Multi-target tracking (MTT) is an important application of unmanned aerial vehicle (UAV) swarms, which is widely applied to environmental monitoring, border patrol, antiterrorism, emergency response, etc. [1–3]. However, due to constraints, such as flight distance, endurance, sensor coverage, etc., the individual abilities are usually insufficient to meet the task requirements, so UAVs need to communicate to achieve information sharing and better cooperation [4,5], then improve the MTT capability.

Currently, the communication between UAVs mainly follows the manually designed communication protocol, and UAVs transmit specific messages in accordance with specific formats and prescriptions [6–8]. However, the design of the communication protocol requires prior knowledge and is highly task-relevant [9], and manually customized protocols may bring side effects, such as insufficient flexibility and versatility, which may affect the communication capabilities of UAVs and are not conducive to their efficient cooperation in highly dynamic environments.

With the development of multi-agent deep reinforcement learning (MADRL), many works using MADRL to learn the complex cooperative action policies of UAVs have appeared. This also provides a new idea for learning cooperative communication, that is, applying this advanced artificial intelligence technique to learn the effective communication between UAVs to achieve efficient cooperation. Different from those methods using manually customized communication protocols, such as value function decomposition [10,11] and reward shaping [8,12,13], communication learning is a more general and exploratory cooperation enhancement method. It empowers UAVs to learn how to actively share knowledge to achieve cooperation without requiring expert domain knowledge and experience [14,15]. In addition, the learned communication policy enables the UAV to independently decide the content according to its real-time status, so as to improve the autonomy and adaptability of the UAVs. Therefore, this method can easily be extended to different multi-agent systems, such as unmanned transportation networks, logistics robots, etc., but is not limited to the MTT scenarios in this paper.

This paper no longer follows the traditional idea of manually designing the communication protocol, but adopts a new data-driven idea to model the communication protocol as the communication policy, then uses a deep neural network (DNN) to approximate and fit the policy. On that basis, this paper proposes an MADRL algorithm to learn the communication and action policies of a UAV simultaneously; thus, the UAV learns how to communicate with others for better cooperation, thereby improving the overall MTT capability of UAV swarms. Then, the effectiveness of the proposed algorithm is verified through numerical simulation experiments, and the performance of the learned policies is further tested.

The major contributions and innovations of this paper include:

(1) Different from the manually designed communication protocol, the communication learning in this paper enables the UAV to independently decide the message content to be published according to its current state and endows the UAV with the ability of active communication and autonomous cooperation.
(2) The communication policy of a UAV is parameterized as a function from its input variable to the published message. Then, two neural networks based on an attention mechanism are designed to approximate the communication and action policies, respectively, which can not only automatically distinguish the important messages received but also scale to the dynamic changes of the local communication topology.
(3) To maximize the rewards of neighboring UAVs, a gradient optimization procedure for deterministic communication policy over continuous space is derived. Then, a MADRL algorithm for UAV swarms is proposed to jointly learn the continuous communication and discrete action policies of the UAVs.

The paper is organized as follows. Section 2 summarizes the related works. In Section 3, the background and some definitions about reinforcement learning are introduced. Section 4 analyzes the MTT problem and establishes the mathematician models. Then, the communication settings of UAV swarms are configured. Next, the specific methods are proposed in Section 5, including the models of communication and action policies, the derivations of policy gradient , and the corresponding algorithm. Then, numerical simulation experiments are implemented in Section 6 to verify the effectiveness of the proposed algorithm. A discussion of the proposed algorithm and numerical simulation experiments is presented in Section 7. Finally, Section 8 gives the summary and outlook of the paper.

2. Related Works

As an emerging research hotspot, the MADRL-based communication learning research in recent years can be classified into several categories, including communication protocol, communication structure, communication object, and communication timing, etc.

2.1. Communication Protocol

The communication protocol specifies the textual content that agents communicate with each other. Foerster et al. [16] firstly proposed two communication learning methods: reinforced inter-agent learning (RIAL) and differentiable ning (DIAL) to learn the communication protocol between two agents. Although they can only learn the simple low-dimensional communication protocols between two agents, their findings inspired a lot of follow-up works. Similarly, grounded semantic network (GSN) [15] was proposed to encode high-dimensional observation information and transmit it to other agents to realize information sharing. Experiments verified that GSN can reduce the limitations caused by the individual partial observability and improve the cooperation between agents. Pesce and Emanuele [17] proposed a memory-shared communication mechanism in which each agent can generate a belief state about its local observation and store it in a shared memory, and all agents can access and update the memory to achieve message passing between agents. However, in complex and drastically dynamic scenarios, the belief states generated by different agents may be all kinds of strange, which is not conducive to establishing a stable cooperative relationship between agents.

2.2. Communication Structure

Communication structure focuses on how the communication messages flow between agents. Peng et al. [18] modeled the communication link between agents as the bidirectionally-coordinated nets (BiCNet), which can not only transfer information between agents but also store local memory. However, the chain relationship in BiCNet is not necessarily suitable and accurate to capture the interactions between agents. In addition, BiCNet can be extremely complex and fragile when the scale of the agents is large. Therefore, BiCNet cannot be scaled well to the large-scale and highly dynamic UAV swarms. CommNet [14] assumed that each agent can globally receive and average the messages from the hidden layers of all other agents' neural networks. It can scale well to the population changes of agents but cannot distinguish the importance of the messages from different agents, which may overwhelm some important ones. Moreover, global communication is usually impractical for swarms. With the introduction of graph neural networks (GNNs), communication learning methods based on graph attention network (GAT) have been proposed, such as ATOC [19], GA-Comm and GA-AC [20]. The graph attention network can adaptively assign the weight of neighbor nodes, which improves the flexibility and adaptability of the communication of agents.

2.3. Communication Object

In the study of communicating object, an agent learns to choose which adjacent agent(s) to communicate with peer-to-peer rather than broadcast. Ding et al. [21] proposed the individually inferred communication (I2C) algorithm to train a neural network that maps an agent's local observation to others' index codes to determine who to communicate with. Similarly, targeted multi-agent communication (TARMAC) [22] was proposed to learn the communication objects of each agent and the message to be sent. The simulation verified that TARMAC can learn effective communication in a simple discrete environment, enabling effective cooperation among agents.

2.4. Communication Timing

In some competition and confrontation scenarios, an agent may only need to communicate with neighbors at certain important moments, thereby reducing the communication frequency and bandwidth requirements. To learn when to communicate, the individualized controlled continuous communication model (IC3Net) [23] assumed that each agent's action variable set includes a physical movement and a discrete communication switch signal. The later one is modeled as a gating unit that controls whether the agent publishes its communication message to the outside.

Although there are many related studies on communication learning, there are few works applicable to UAV swarms. Aiming at the MTT problem of UAV swarms, how to learn the efficient, scalable and robust communication between UAVs to achieve active cooperation and improve the MTT capability of UAVs is the focus of this paper.

3. Preliminary

3.1. Decentralized Partially Observable Markov Decision Process (Dec-POMDP)

Dec-POMDP [24] is a model of a Markov decision process (MDP) for multi-agents in which each one can only partially observe the environment and make its action decision accordingly. For n agents, each one is indexed by $i \in [1, n]$; the Dec-POMDP at every step (the subscript t is omitted for convenience) can be described as:

$$(N, S, \mathcal{A}, O, Z, T, R, \gamma), \tag{1}$$

where N is the collective set of all agents, S is the global state space denoting all agents' and the environment's configurations, and $s \in S$ denotes the current and specific state. The joint action space of all agents is denoted as $\mathcal{A} : \mathcal{A}^1 \times \cdots \times \mathcal{A}^n$ in which $a^i \in \mathcal{A}^i$ is agent i's specific action; $O : (O^1, \cdots, O^n)$ denotes all agents' joint observation space; $Z : o^i = Z(s, i)$ denotes the individual observation model of agent i given the global state s, and $o^i \in O^i$ is agent i's local observation. $T : P(s' \mid s, a) \to [0, 1]$ denotes the probability of s transiting to new state s' executing joint action $a : (a^1, \cdots, a^n)$; R is the reward function; $\gamma \in [0, 1]$ is the constant discount factor.

In Dec-POMDP, each agent makes its action decision following individual policy $\pi^i : O^i \mapsto A^i$, and the joint policy is denoted as $\pi : (\pi^{(1)}, \cdots, \pi^{(n)})$. Then, all agents execute the joint action to refresh the environment. Given a specific joint observation o and all agents' joint policy π, if each agent can access its private reward r_t^i at every time step t, $V_\pi(o) = E_\pi[\sum_{t=0}^{\infty} \sum_{i=1}^{N} \gamma^t r_t^i \mid o_{t=0} = o]$ denotes the state-value function of all agents. Furthermoremore, executing the joint action a, their action-value function is denoted as $Q_\pi(o, a) = E_\pi[\sum_{t=0}^{\infty} \sum_{i=1}^{N} \gamma^t r_t^i \mid (o, a)_{t=0} = (o, a)]$.

3.2. Actor–Critic (AC)

AC combines the policy gradient and value function approximation methods in which each actor is a policy function to predict the agent's action, and each critic is a value function to evaluate the performance of the policy function [25]. Thus, the policy function π_θ, which is parameterized with θ, can be optimized via maximizing the value function, and the policy gradient with respect to θ is:

$$\nabla_\theta J(\theta) = \mathbb{E}_{s,a \sim \pi_\theta(s)}[\nabla_\theta \log \pi_\theta(a \mid s) Q_\pi(s, a)], \tag{2}$$

where the value function can be optimized via minimizing the square of the temporal-difference (TD) error [25].

3.3. Deep Deterministic Policy Gradient (DDPG)

DDPG is an extended version of AC in which the policy function directly outputs a deterministic action value ($a = \pi_\theta(s)$) instead of a probability distribution over the action space ($a \sim \pi_\theta(a \mid s)$). Then the gradient of the policy function is:

$$\nabla_\theta J(\theta) = \mathbb{E}_s \left[\nabla_\theta \pi_\theta(s) \nabla_a Q_\pi(s, a) \mid_{a = \pi_\theta(s)} \right]. \tag{3}$$

The value function in DDPG is updated with the frozen network trick, and in addition to the two networks appearing in AC, the target-policy function and the target-value function are used to improve training stability [26].

4. Problem Formulation

4.1. Problem Description

The research focus of this paper is to explore a communication and action policies joint learning method to achieve swarm cooperation. To reduce the learning difficulty, we make reasonable assumptions and simplifications of the models of both the UAV and the target. As shown in Figure 1, a large number of homogeneous small fixed-wing UAVs track an unknown number of moving targets on the ground. Each UAV can only perceive the targets below it but cannot distinguish the specific identities or indices of the tracked targets. It is assumed that the UAVs move at a uniform constant speed in a two-dimensional plane and rotate their headings according to the local communication messages and observation information. However, since the targets are non-cooperative and there is no explicit target assignment, a single UAV may track multiple aggregated targets, or multiple UAVs may cooperatively track one or multiple targets simultaneously. Therefore, the UAVs should cooperate in a decentralized manner to keep targets within their field of view and track as many targets as possible. In addition, the UAVs should also satisfy the safety constraints, such as avoiding collisions, crossing boundaries, etc.

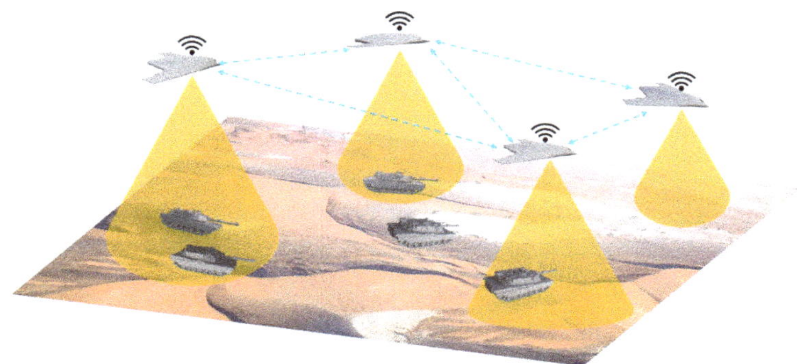

Figure 1. MTT scenario for UAV swarms.

4.1.1. Kinematic Model

There are n UAVs and m targets in the two-dimensional mission area. The motions of these UAVs and targets can be modeled with two-dimensional plane motion models. For any UAV i, $i \in [1, n]$, its speed is denoted as v_U, the heading angular is denoted as θ_U, and the control variable is its heading angular rate $\dot{\theta}_U$. Then, the kinematic model is described by its position and heading, that is:

$$\begin{cases} x^i_{U,t+1} = x^i_{U,t} + v^i_U \cos \theta^i_{U,t} \Delta t, & 0 \le x^i_{U,t} \le x_{max} \\ y^i_{U,t+1} = y^i_{U,t} + v^i_U \sin \theta^i_{U,t} \Delta t, & 0 \le y^i_{U,t} \le y_{max} \\ \theta^i_{U,t+1} = \theta^i_{U,t} + \dot{\theta}^i_{U,t} \Delta t, & -\dot{\theta}_{max} \le \dot{\theta}^i_{U,t} \le \dot{\theta}_{max} \end{cases} \tag{4}$$

where the subscription t is denoted as the current time, Δt is the discrete time step, $\dot{\theta}_{max}$ is the UAV's maximum heading angular rate, and x_{max} and y_{max} are the maximum boundaries.

Similarly, for any target k, $k \in [1, m]$, its kinematic model can also be described with the position $[x^k_T, y^k_T]$ and heading angular θ^k_T, and the difference is that the target's heading angular rate $\dot{\theta}^k_T$ is assumed to be a bounded random variable.

4.1.2. Target Observation Model

Shown in Figure 2, each UAV can only observe these targets in a circle with radius d_o below it and can resolve the position, speed and other information of the tracked targets from the raw observation but cannot identify their specific indexes. The ground projection distance between UAV i and target k is denoted as $d^{i,k}$, and target k is tracked by UAV i when $d^{i,k} \le d_o$. The observation is denoted as $o^k_T = [x^k_T, y^k_T, v^k_{x_T}, v^k_{y_T}]$. Furthermore, the observation information o^k_T should be transformed from a global coordinate to UAV i's local coordinate considering partial observability, denoted as $o^{i,k}_T$.

Suppose UAV i can obtain the relative location information o^i_B between itself and the boundaries of the task area through its GEO-fencing system and its partial observation of the targets. Then, the environment is denoted as $o^i = \{ o^i_B, \{o^{i,k}_T\} \mid \forall k \in [1, m], d^{i,k} \le d_o \}$.

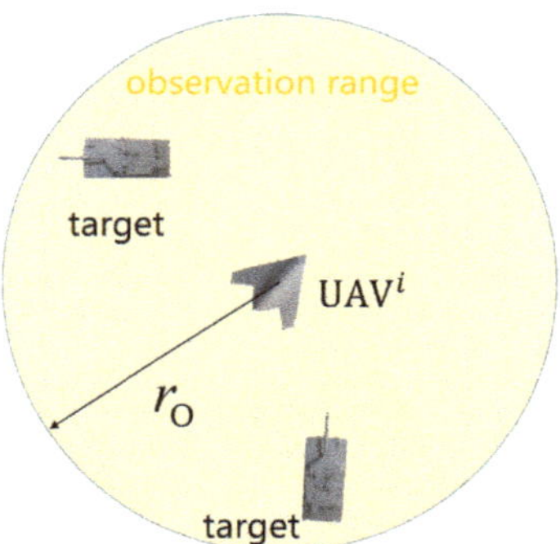

Figure 2. Target observation diagram.

4.1.3. Action Space

The purpose of this paper is to learn the cooperative policy of UAV swarms rather than the precise control of each individual. To facilitate the learning process, the action space of each UAV can be discretized into a limited number of action primitives as follows:

$$\dot{\theta}_{\mathrm{U},t} = \frac{2n_a - N_a - 1}{N_a - 1}\dot{\theta}_{\max}, n_a \in [1, N_a], \tag{5}$$

where N_a is the cardinality of the discrete action set.

4.1.4. Reward Shaping

In MTT, UAVs are expected to track as many targets as possible. Therefore, each UAV should keep the tracked targets within its field of view as much as possible, while maximizing observation benefits by avoiding observation outside the boundaries and repeated tracking. Thus, the reward of each UAV i is shaped as the sum of multiple items, including:

(1) **Target Tracking Reward**: Since the observation range of a UAV is limited, a naive idea is that the target should be as close as possible to the UAV's observation center. Accordingly, the target tracking reward of UAV i to target k is defined as:

$$r_{\mathrm{tar}}^{i,k} = \begin{cases} 1 + \left(r_{\mathrm{o}} - d^{i,k}\right)/r_{\mathrm{o}} & d^{i,k} \le r_{\mathrm{o}}, \\ 0 & \text{else} . \end{cases} \tag{6}$$

When UAV i tracks multiple targets, its target tracking reward is $r_{\mathrm{tar}}^i = \sum_{k=1}^{m} r_{\mathrm{tar}}^{i,k}$. Specifically, the constant bias 1 in Equation (6) can encourage the UAV to track more targets rather than just obsessing over a single target. For example, when tracking two targets, $r_{\mathrm{tar}}^i \ge 2$, but when tracking a single target, $r_{\mathrm{tar}}^i < 2$.

(2) **Repeated Observation Penalty**: Repeated observation of a target by multiple UAVs may not increase the number of tracked targets but may increase the risk of collision due to the proximity of the UAVs. Therefore, to improve the observation efficiency and track more targets, a penalty item is defined to guide the UAV i and $j, j \ne i$ to avoid repeated observations, that is:

$$r_{\mathrm{rt}}^{i,j} = \begin{cases} -0.5 \times \exp\left((2 \times r_{\mathrm{o}} - d^{i,j})/(2 \times r_{\mathrm{o}})\right) & d^{i,j} \le 2 \times r_{\mathrm{o}}, \\ 0 & \text{else} , \end{cases} \tag{7}$$

and $r_{\mathrm{rt}}^{i,j} = r_{\mathrm{rt}}^{j,i}$. In Equation (7), if $d^{i,j} > 2 \times r_{\mathrm{o}}$, there is no observational overlap between UAV i and j, and UAV i's repeated observation penalty is $r_{\mathrm{rt}}^i = \sum_{j=1,j \ne i}^{n} r_{\mathrm{rt}}^{i,j}$.

(3) **Boundary Penalty**: To effectively capture and track targets, UAV i's observation area should always be within the boundaries. When the observation range is outside

the boundaries, the outside part is invalid. To this end, the minimum distance from UAV i to all boundaries is d_{bound}^i, and the boundary penalty item is defined as:

$$
r_{\text{bound}}^i =
\begin{cases}
-0.5 \times \left(r_{\text{o}} - d_{\text{bound}}^i \right) / r_{\text{o}} & d_{\text{bound}}^i < d_{\text{o}} \\
0 & \text{else} .
\end{cases}
\tag{8}
$$

To sum up, the individual reward of UAV i is shaped as:

$$
r^i = r_{\text{tar}}^i + r_{\text{rt}}^i + r_{\text{bound}}^i .
\tag{9}
$$

4.2. Communication Settings

To cooperate among UAVs, they need to follow certain communication protocols to exchange information, and communication within a UAV swarm should meet the following requirements:

(1) **Local communication:** In a large-scale UAV swarm, each one is both the communication receiving and output nodes, and all the nodes constitute a complex network. Considering the limitation of communication power, each one only communicates with the neighbors within its maximum communication range, which can effectively reduce the complexity of the communication network;

(2) **Direct communication:** The MTT problem requires high timeliness of communication between UAVs. Therefore, to reduce the communication delay, it is assumed that each UAV only communicates with adjacent ones in a single-hop, and multi-hop bridge communication with ones outside the communication range is not considered;

(3) **Broadcast communication:** To reduce bandwidth requirements and avoid communication congestion, each UAV broadcasts the same message to its neighbors once, instead of sending one-to-one multiple times;

(4) **Dynamic communication:** The rapid movement of UAVs leads to dramatic changes in communication network and asymmetry between uplink and downlink. To this end, it is assumed that all neighbors within the communication range can receive the messages sent by a UAV to improve the dynamics and reliability of the communication network;

(5) **Autonomous communication:** In complex scenarios, UAVs should be able to autonomously decide the content of messages to be sent based on their local observations, so as to promote efficient cooperation between them;

(6) **Safe communication:** To improve the survivability of UAVs in the confrontation scenarios, the anti-jamming and anti-interception capabilities of communication should be improved to protect communication messages from being deciphered by non-receivers and improve communication security, etc.

5. Methods

5.1. Communication and Action Policies Modeling

Based on the above settings, the set of UAV i's neighbors that can communicate locally with it at time t is denoted as $\mathcal{N}_t^i$. Its communication and action decision-making processes is shown in Figure 3. Specifically, $j \in \mathcal{N}_t^i$, a_t^i is its heading angular rate $\dot{\theta}_{\text{U},t}^i$; m_t^i indicates the continuous and deterministic message that is about to be published to the neighbors. Here, UAV i can receive the messages from itself and all neighbors in the last moment; c_t^i is denoted as $c_t^i = \left\{ m_{t-1}^i, m_{t-1}^j \mid \forall j \in \mathcal{N}_t^i \right\}$. UAV i makes its action and communication decisions based on its local observation and the messages received. Then, the action policy is defined as:

$$
a_t^i \sim \pi_{\text{a}}\left(a \mid o_t^i, c_t^i \right),
\tag{10}
$$

and the communication policy is defined as:

$$
m_t^i = \pi_{\text{c}}\left(o_t^i, c_t^i \right).
\tag{11}
$$

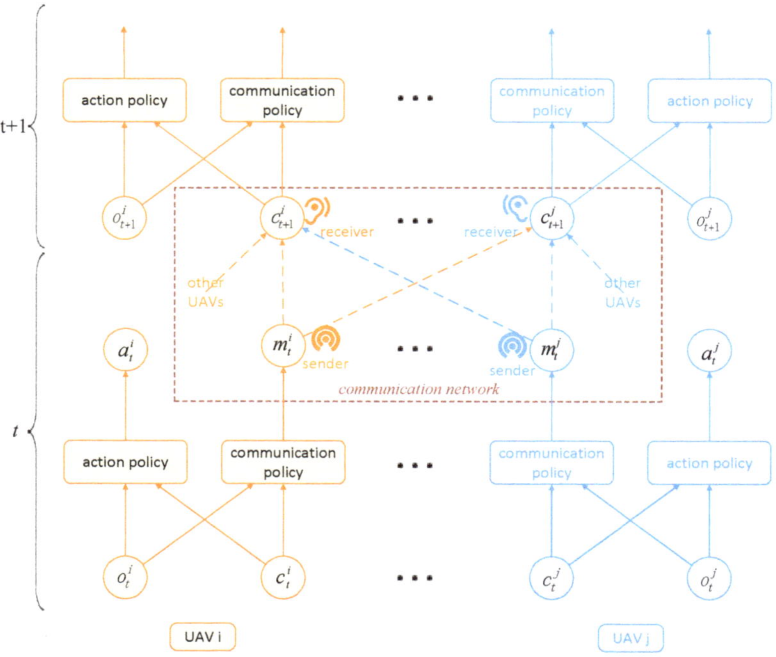

Figure 3. Communication and action decision-making processes.

The communication and action decision process of UAV i is as follows:

(1) At each time t, UAV i accesses its local observation o_t^i and receives message set c_t^i;
(2) Input o_t^i and c_t^i into both Equations (10) and (11) to output its action a_t^i and message m_t^i;
(3) Execute joint action $(a_t^1, \cdots, a_t^i, \cdots, a_t^n)$ to refresh the environment and publish message m_t^i to the neighboring UAVs, then receive the reward r_t^i from the environment;
(4) $t = t + 1$, and continue to step (1).

The input variables of both UAV i's action and communication policies are the local observation and the received messages. As the UAVs and targets move continually, both the number of objects observed and the number of messages received by UAV i are dynamically changing accordingly. However, the input dimension of a neural network is usually fixed at initialization, and input variables with uncertain cardinality cannot be directly input into the neural network.

In MTT, each UAV can interpret the precise physical features of the tracked targets, such as their speeds, positions, etc. These explicit feature sets can be encoded as a dimension-determined input variable using feature embedding methods in [27]. Unfortunately, the message received from a neighbor is usually high-dimensional and often cryptic, i.e., its content composition may be time-varying, depending on the context of the sender, and has no definite physical properties. Therefore, the received messages cannot be easily encoded as a fixed-dimensional feature embedding. To this end, we adopt the graph attention mechanism [28] (GAT) to aggregate the received messages for each UAV, and its ability to extract and aggregate variable-length messages has been verified by [29,30].

Thus, the communication and action policies of UAV i can be approximated with neural networks. Take communication policy as an example, the overview of its neural network is shown in Figure 4, and the aggregation process of its communication messages in the dashed box on the right is as follows:

(1)   At time t, transform the communication messages with function F whose parameters can be learned to obtain the high-level feature [28], and denote $F(m^i_{t-1})$ as *query*, which represents the prior knowledge of UAV i, while $\{F(m^j_{t-1}) \mid \forall j \in \mathcal{N}^i_t\}$ are the set of *sources*, and each one indicates the received message to be aggregated;

(2)   The correlation coefficient from any adjacent UAV $j, j \in \mathcal{N}^i_t$ to the central UAV i is defined as:

$$e_{ij} = \langle F(m^i_{t-1}), F(m^j_{t-1}) \rangle, j \in \mathcal{N}^i_t \tag{12}$$

the inner product represents a parameter-free calculation, which outputs a scalar that measures the correlation;

(3)   Use the softmax function to normalize the similarity set $\{e_{ij} \mid \forall j \in \mathcal{N}^i_t\}$ to obtain the weight set $\{w_{ij} \mid \forall j \in \mathcal{N}^i_t\}$ in which

$$w_{ij} = \frac{\exp(e_{ij})}{\sum_{j \in \mathcal{N}^i_t} \exp(e_{ij})} \tag{13}$$

(4)   Weighted summation over the *source* set yields the aggregated message $\hat{c}^i_t$:

$$\hat{c}^i_t = \sum_{j \in \mathcal{N}^i_t} w^{ij} F\left(m^j_{t-1}\right) \tag{14}$$

Then, o^i_t and $\hat{c}^i_t$ are concatenated and input into the following hidden layers to calculate the output message m^i_t, and Equation (11) is redefined as:

$$m^i_t = \pi_c\left(o^i_t, \mathrm{GAT}\left(\left\{m^i_{t-1}, m^j_{t-1} \mid \forall j \in \mathcal{N}^i_t\right\}\right); \theta_c\right) \tag{15}$$

where θ_c is the parameter of the communication neural network, and the GAT component is a part of the network.

Similarly, the action policy could also be approximated by a neural network, only the output layer should be modified accordingly. Then, the discrete actions of each UAV i obey the distribution:

$$a^i_t \sim \pi_a\left(a \mid o^i_t, \mathrm{GAT}\left(\left\{m^i_{t-1}, m^j_{t-1} \mid \forall j \in \mathcal{N}^i_t\right\}\right); \theta_a\right) \tag{16}$$

where θ_a is the parameter of the action neural network.

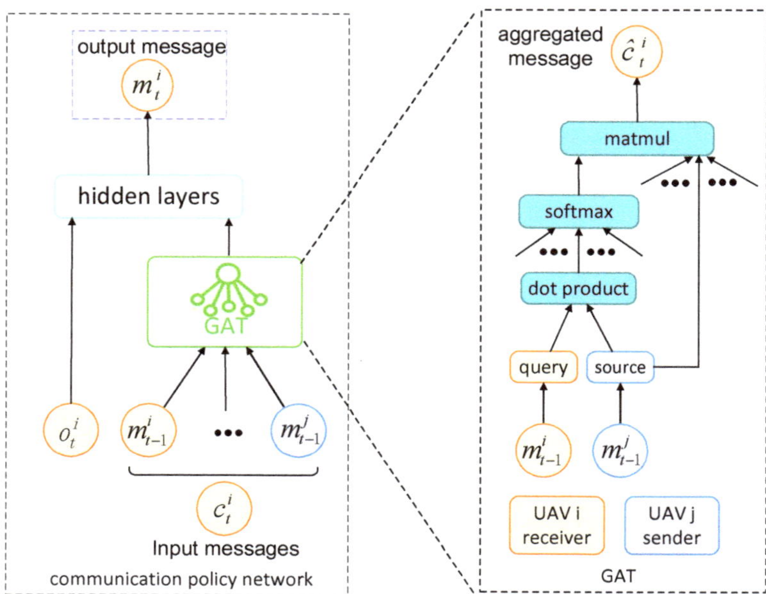

Figure 4. The overview of communication policy neural network.

5.2. Policy Gradient Optimization

Assuming that the action and communication policies of a UAV are independent of each other, the latter is frozen when training the action neural network and vice versa.

5.2.1. Action Policy Gradient

To learn the action policy π_a, the action-value function is defined as $Q^i(o^i, c^i, a^i; \phi_Q)$, and ϕ_Q is its parameter, which is updated by minimizing the following loss function:

$$\mathcal{L}_Q^i(\phi_Q) = \mathbb{E}_{\pi_a}[\frac{1}{2}(y^i - Q(o^i, c^i, a^i; \phi_Q))^2] \tag{17}$$

where $y^i = r^i + \gamma Q(o^{i'}, c^{i'}, a^{i'}; \phi_Q^-) |_{a^{i'} \sim \pi_a(a|o^{i'}, c^{i'}; \theta_a)}$, ϕ_Q^- is the parameter of the corresponding target network, $a^{i'}$ is the next action, $o^{i'}$ and $c^{i'}$ are the local observations and the set of received messages at the next moment, respectively. The time-difference(TD) error is denoted as $\delta^i = r^i + \gamma Q(o^{i'}, c^{i'}, a^{i'}; \phi_Q^-) - Q(o^i, c^i, a^i; \phi_Q)$, and the gradient of this loss function with respect to ϕ_Q performing gradient descent is:

$$\nabla_{\phi_Q} \mathcal{L}_Q^i(\phi_Q) = -\delta \nabla_{\phi_Q} Q(o^i, c^i, a^i; \phi_Q) \tag{18}$$

Then, the action policy is updated via maximizing the action-value function:

$$J_a^i(\theta_a) = \mathbb{E}_{\pi_a}[Q(o^i, c^i, a^i; \phi_Q)|_{a^i \sim \pi_a(a|o^i, c^i; \theta_a)}] \tag{19}$$

and the policy gradient is:

$$\nabla_{\theta_a} J_a^i(\theta_a) = \nabla_{\theta_a} \log \pi(a^i \mid o^i, c^i; \theta_a) \delta^i \tag{20}$$

5.2.2. Communication Policy Gradient

In local communication topography, all the adjacent UAVs receive the message m_t^i that is an input variable of their next action and communication decisions. Given the action policy π_a, the parameter of the action-value function ϕ_Q, and UAV i's current input variables (o_t^i, c_t^i), the communication objective is denoted as:

$$J_c^i(\theta_c) = \frac{1}{|\mathcal{N}^i|} \sum_{j \in \mathcal{N}^i} \mathbb{E}_{\pi_c}[Q(o_{t+1}^j, (c_{t+1}^{j/i}, m_t^i), a_{t+1}^j; \phi_Q) \tag{21}$$

$$\Big|_{m_t^i = \pi_c(o_t^i, c_t^i; \theta_c), a_{t+1}^j \sim \pi(a | o_{t+1}^j, (c_{t+1}^{j/i}, m_t^i); \theta_a)}\Big],$$

where $c_{t+1}^{j/i}$ is the set of UAV j's received messages except m_t^i.

Then, the communication policy gradient is derived according to the policy gradient theorem and the chain derivation rule as:

$$\nabla_{\theta_c} J_c^i(\theta_c) = \frac{1}{|\mathcal{N}^i|} \sum_{j \in \mathcal{N}^i} \mathbb{E}_{\pi_c}[\nabla_{\theta_c} \pi_c(o_t^i, c_t^i; \theta_c)$$

$$\cdot \nabla_{m_t^i} \log \pi_a(a | o_{t+1}^j, (c_{t+1}^{j/i}, m_t^i); \theta_a) Q(o_{t+1}^j, (c_{t+1}^{j/i}, m_t^i), a_{t+1}^j; \phi_Q) \tag{22}$$

$$+ \nabla_{\theta_c} \pi_c(o_t^i, c_t^i; \theta_c) \nabla_{m_t^i} Q(o_{t+1}^j, (c_{t+1}^{j/i}, m_t^i), a_{t+1}^j; \phi_Q)].$$

For simplicity, the conditional term in Equation (21) is omitted. Thus, given the input variables of UAV i at the current moment t and that of all adjacent UAVs $\mathcal{N}^i$ at the next moment $t + 1$, the communication policy gradient can be calculated via Equation (22). Then, the policy can be updated accordingly.

Note that the objective functions, Equations (17), (19) and (21), are non-convex when using neural networks to approximate them, respectively. The common optimizers, such as MBSGD (mini-batch stochastic gradient descent) or Adam (adaptive moment estimation) in PyTorch, are usually adopted to solve these optimization problems.

5.2.3. Joint Communication and Action Policies Learning

As mentioned above, when calculating UAV i's action policy gradient, its observation and messages received are required. However, for the communication policy gradient, in addition to these variables, it is necessary to further obtain the relevant variables of each adjacent UAV at the next moment. Employing the experience-sharing training mechanism [27] to train the communication and action policy neural networks, the action experience is denoted as $e_a^i = (o^i, c^i, a^i, r^i, o^{i'}, c^{i'})$, and the communication experience is denoted as $e_c^i = (o^i, c^i, \{o^{j'}, c^{j'}, a^{j'} \mid \forall j \in \mathcal{N}^i\})$. Utilizing the centralized-training decentralized-execution (CTDE) framework, an algorithm for jointly learning the communication and action policies for UAV swarms is proposed, and its pseudo-code is as follows.

In Algorithm 1, the two policy networks are not coupled with each other. During centralized training, they both have private experience buffers, and when one network is updated, the other one is frozen. However, communication policy gradient can backpropagate across UAVs, which enables closed-loop feedback updates of the communication policy. In decentralized execution, each UAV can decide its action and what to publish to its adjacent UAVs based on its own observation and received messages.

Algorithm 1 Joint communication-action multi-agent deep reinforcement learning

---

Initialize: Neural network parameter: action policy, θ_a; communication policy, θ_c; action-value function and its target function, ϕ_Q and ϕ_Q^-. Action experience buffer: D_1. Communication experience buffer: D_2

//Centralized-Training:

1: **for** epi = 1: episodes **do**
2:     Environment Reset
3:     **for** $t = 1 : T$ **do**
4:         **for** i = 1: n **do**
5:             Access observation o^i and message set c^i, and execute policy π_a and policy π_c to output action a^i and message m^i, respectively
6:         **end for**
7:         Execute joint action $\{a^1, a^2, \cdots, a^n\}$ to update immediate rewards $\{r^1, r^2, \cdots r^n\}$ and joint observation at next moment $\{o^{1'}, o^{2'}, \cdots, o^{n'}\}$
8:         **for** i = 1: n **do**
9:             Publish message m^i, receive the neighbors' messages to form $c^{i'}$
10:            Push action experience $(o^i, c^i, a^i, r^i, o^{i'}, c^{i'})$ into buffer D_1
11:            Push communication experience $(o^i, c^i, \{o^{j'}, c^{j'}, a^{j'} \mid \forall j \in \mathcal{N}^i\})$ into buffer D_2
12:         **end for**
13:         Randomly sample B_1 experiences from D_1
14:         Minimize loss function $\mathcal{L}_Q$ to update the action-value function:

$$\mathcal{L}_Q(\phi_Q) = \frac{1}{2B_1} \sum_{k=1}^{B_1} [(y^k - Q(o^k, c^k, a^k; \phi_Q))^2];$$

15:         **if** Update target network **then**

$$\phi_Q^- \leftarrow lr^- \phi_Q + (1 - lr^-)\phi_Q^-;$$

16:         **end if**
17:         Update parameter θ_a with gradient:

$$\nabla_{\theta_a} J_a(\theta_a) \approx \frac{1}{B_1} \sum_{k=1}^{B_1} \nabla_{\theta_a} \log \pi(a^k \mid o^k, c^k; \theta_a) \delta^k$$

18:         Randomly sample B_2 experiences from D_2
19:         Perform Equation (22), and update communication policy network parameter with gradient:

$$\nabla_{\theta_c} J_c(\theta_c) \approx \frac{1}{B_2} \sum_{k=1}^{B_2} \nabla_{\theta_c} J_c^k(o^k, c^k; \theta_c)$$

20:     **end for**
21: **end for**

//Decentralized-Execution:

22: Environment Reset
23: Load shared action policy π_a and communication policy π_c for each UAV
24: **for** $t = 1 : T$ **do**
25:     For each UAV i, access o^i and c^i, and execute π_a and π_c to output its action a^i and message m^i, respectively
26:     Execute joint action $\{a^1, a^2, \cdots, a^n\}$ to update environment, and each UAV i publishes message m^i
27: **end for**

---

6. Experiments

6.1. Benchmark Algorithms

In this paper, the proposed Algorithm 1 is named Att-Message for simplification, and we hardly see from the existing literature that techniques other than MADRL can achieve the similar goal of solving communication and action policies for large-scale UAV swarms to cooperate. Thus, we select and adopt several benchmark algorithms that are commonly used by researchers from [14,19], and the non-communication one , including:

(1) **No-Comm.** Literally, in No-Comm, each UAV can only receive local observation and selfishly maximize its individual rewards. There is no clear communication channel between UAVs and naturally no explicit cooperation or competition. Thus, the communication policy is $\pi_c = Null$, and the action policy is:

$$a^i \sim \pi_a(a \mid o^i; \theta_a). \tag{23}$$

(2) **Local-CommNet.** In CommNet [14], it is assumed that each agent can receive all agents' communication messages. It should be adapted to the local communication configuration of UAV swarms in this paper, named Local-CommNet. Specifically, each UAV publishes the hidden layer information h of its action policy network to its adjacent UAVs, i.e., $m^j_{t-1} = h^j_{t-1}$, Then, the messages received by UAV i are denoted as:

$$c^i_t \doteq \{h^j_{t-1} \mid \forall j \in \mathcal{N}^i_t\}. \tag{24}$$

Next, c^i_t is aggregated using the average pooling method to obtain:

$$\hat{c}^i_t = \frac{1}{\mid \mathcal{N}^i_t \mid} \sum_{j \in \mathcal{N}^i_t} h^j_{t-1}. \tag{25}$$

(3) **Att-Hidden.** In addition to the average pooling method, the GAT can also be used to aggregate c^i_t [12,19]. Then:

$$\hat{c}^i_t = \mathrm{GAT}(\{h^i_{t-1}, h^j_{t-1} \mid \forall j \in \mathcal{N}^i_t\}). \tag{26}$$

The message of each UAV is its hidden layer information of the action policy network, and there is no separate communication policy network. So GAT, as an encoder, could be a component of the action policy network. The network can be updated according to the input variable $(o^i_t, \{h^i_{t-1}, h^j_{t-1} \mid \forall j \in \mathcal{N}^i_t\})$ following Equation (20).

6.2. Settings

In this section, the effectiveness of the proposed algorithm is verified by numerical simulation experiments. According to the problem description (Section 4.1), the training environmental parameters are set in Table 1. These parameters have been used in our previous work [8,27], and the rationality has been verified. During testing, the environment size and the numbers of UAVs and targets may change. The hyper-parameters of those algorithms are configured in Table 2.

Table 1. Environmental parameters.

Object	Parameter	Value
Environment	Shape	Square
	Size	2 km × 2 km
UAV	Quantity (n)	10
	Communication Range (d_c)	500 m
	Observation Range (d_o)	200 m
	Flight Speed (v_U)	20 m/s
	Max Heading Angular Rate ($\dot{\theta}_{max}$)	30°/s
	Cardinality of Action Set (N_a)	7
Target	Quantity (m)	10
	Moving Speed (v_T)	5 m/s
	Max Steering Angular Rate	30°/s

Table 2. Hyper-parameters configuration.

Hyper-Parameter	Value
Iteration Episode	2×10^3
Replay Buffer	5×10^5
Max Step	200
Batch Size	64
Target Network Update Interval	100
Action Policy Learning Rate	1×10^{-4}
Communication Policy Learning Rate	5×10^{-5}
Communication Policy Output Dimension	100
Discount Factor	0.95

To evaluate the tracking performance of UAV swarms, the following metrics are defined:

(1) **Average Reward:**

$$\frac{1}{Tn} \sum_{t=1}^{T} \sum_{i=1}^{n} r_t^i, \tag{27}$$

where r_t^i has been defined in Equation (9), which comprehensively evaluates the performance of UAV swarms from the aspects of target tracking, repeated observation, safe flight, etc.

(2) **Average Target:**

$$\frac{1}{Tn} \sum_{t=1}^{T} \sum_{i=1}^{n} \sum_{k=1}^{m} \mathbb{1}(i,k), \mathbb{1}(i,k) = \begin{cases} 1, & d^{(i,k)} \leqslant d_o; \\ 0, & else. \end{cases} \tag{28}$$

which evaluates the number of targets tracked from the perspective of each UAV.

(3) **Collective Target:**

$$\frac{1}{T} \sum_{t=1}^{T} \sum_{k=1}^{m} \mathbb{1}(k), \mathbb{1}(k) = \begin{cases} 1, & \exists i \in [1,n], s.t.\ d^{(i,k)} \leqslant d_o; \\ 0, & else. \end{cases} \tag{29}$$

which evaluates the number of targets tracked by all the UAVs.

(4) **Coverage:**

$$\frac{1}{Tm} \sum_{t=1}^{T} \sum_{k=1}^{m} \mathbb{1}(k), \mathbb{1}(k) = \begin{cases} 1, & \exists i \in [1,n], s.t.\ d^{(i,k)} \leqslant d_o; \\ 0, & else. \end{cases} \tag{30}$$

which is denoted as the proportion of the tracked targets to all targets. Furthermore, the coverage rate, as a normalized indicator, can evaluate the tracking capability of UAV swarms in different scenarios from the perspective of targets.

6.3. Validity Verification

The neural networks in the four algorithms are randomly initialized and trained, and the curves of the defined metrics are plotted in Figure 5. Overall, the MTT performance of the Att-Message is the best, followed by Att-Hidden and Local-CommNet; the last one is No-Comm. Again, there is no explicit communication and cooperation between the UAVs in No-Comm, and each UAV greedily maximizes its private interest.

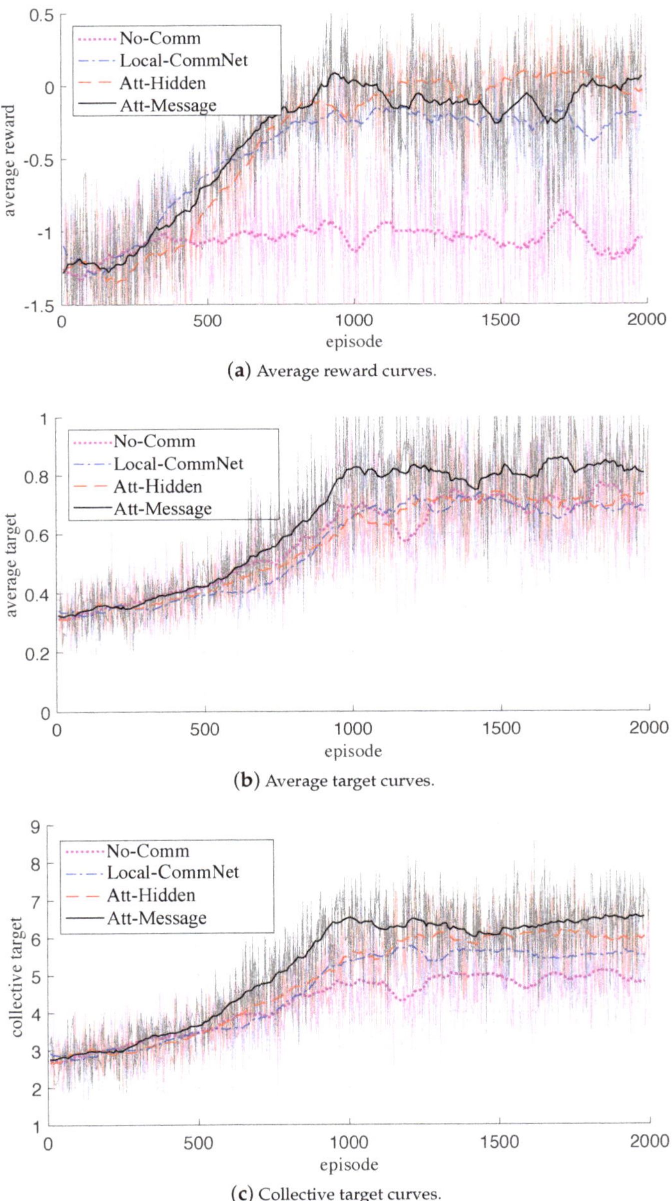

(**a**) Average reward curves.

(**b**) Average target curves.

(**c**) Collective target curves.

Figure 5. The metric curves during the training process of the four algorithms.

Looking at specifics, (1) the three algorithms using explicit communication outperform No-Comm without communication, which indicates that communication can effectively promote the cooperation between UAVs, thereby improving the tracking performance of UAV swarms; (2) the comprehensive performance of Att-Hidden using GAT is better than that of Local-CommNet, but the UAV in both algorithms transmits the hidden layer of the action policy network. The reason may be that GAT can better aggregate the received messages, then effectively improve the action policy of UAVs and the cooperation between them; (3) furthermore, the comprehensive performance of the Att-Message is superior to that of Att-Hidden, indicating that compared with the hidden layer of the action policy neural network, the communication message can better capture the information that is helpful for cooperation. It is also proved that the communication policy in Att-Message can be optimized based on feedback from other UAVs to facilitate cooperation between UAVs.

Furthermore, Figure 6 intuitively visualizes the tracking process of the UAVs using the four algorithms, respectively, and the snapshots verify the previous conclusions again. In addition, it can be seen that executing the policies learned by Att-Message, the UAVs emerge with obvious cooperative behaviors. For example, when a target escapes the observation range of a UAV, the adjacent UAVs can quickly track and recapture the target again. Alternatively, there is a tendency to avoid getting too close between the UAVs to avoid repeated tracking as much as possible and to improve the observation coverage to capture more targets.

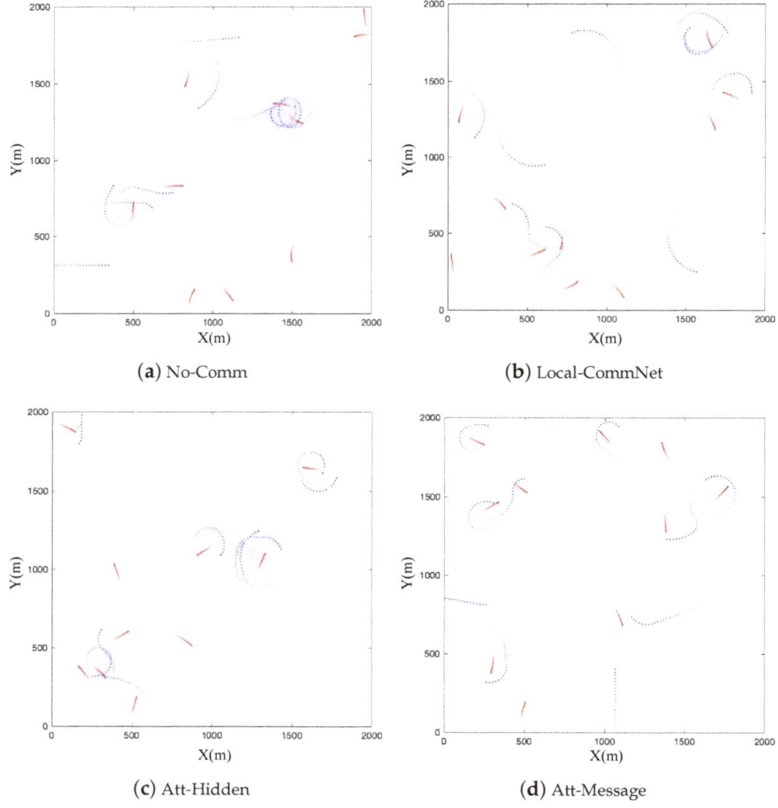

(**a**) No-Comm

(**b**) Local-CommNet

(**c**) Att-Hidden

(**d**) Att-Message

Figure 6. Visualization of MTT executing different algorithms.

6.4. Scalability Test

To test whether the policies learned by the above four algorithms can be scaled to other scenarios beyond the training one, these policies were executed for 100 rounds in different scenarios, and the metrics of single-step are counted in Table 3.

Table 3. Result statistics of scalability test.

Map Size	$\frac{n}{m}$	Metrics	No-Comm	Local-CommNet	Att-Hidden	Att-Message
1000	$\frac{5}{5}$	Average Reward	−1.3108	−0.8653	**−0.2912**	−0.3554
		Average Target	1.0626	1.1393	1.0513	**1.2109**
		Coverage	0.6555	0.7370	0.7626	**0.7915**
1000	$\frac{10}{10}$	Average Reward	−4.1760	−2.5640	−1.6756	**−1.4816**
		Average Target	**1.9919**	1.9792	1.4915	1.6382
		Coverage	0.7692	0.8278	0.8508	**0.8722**
2000	$\frac{10}{10}$	Average Reward	−0.8157	−0.0425	0.0645	**0.0776**
		Average Target	0.7190	0.8166	0.7589	**0.8777**
		Coverage	0.5357	0.6207	0.6275	**0.6714**
2000	$\frac{20}{20}$	Average Reward	−2.5680	−1.1612	−0.5765	**−0.5166**
		Average Target	1.2339	**1.3396**	1.0992	1.2432
		Coverage	0.6581	0.7523	0.7586	**0.8026**
2000	$\frac{50}{50}$	Average Reward	−6.9769	−6.1045	−3.3002	**−3.0900**
		Average Target	**2.5686**	2.5803	1.9871	2.1014
		Coverage	0.8562	0.8475	0.9100	**0.9183**
5000	$\frac{100}{100}$	Average Reward	−2.2617	−1.1111	−0.9921	**−0.5119**
		Average Target	1.1118	1.1958	1.1712	**1.0925**
		Coverage	0.6170	0.7174	0.7297	**0.7542**
5000	$\frac{200}{200}$	Average Reward	−4.5219	−3.7763	−2.2895	**−2.0386**
		Average Target	1.8413	**1.9586**	1.4805	1.6496
		Coverage	0.7743	0.8177	0.8392	**0.8705**
10,000	$\frac{1000}{1000}$	Average Reward	−6.0707	−5.3054	−3.4510	**−3.1223**
		Average Target	2.2993	**2.3369**	1.7754	1.9648
		Coverage	0.8242	0.8406	0.8814	**0.9043**

The statistical results generally indicate that the average reward and coverage of the three algorithms that introduce explicit communication in different scenarios are significantly better than No-Comm without communication, which once again verify the effectiveness of the communication. Specifically, Att-Message performs better than other algorithms in terms of average reward and coverage, which directly reflects that the UAVs adopting the action and communication policies learned with Att-Message can better cooperate to track more targets in different scenarios. However, in the scenario with dense UAVs and targets, the average targets of No-Comm and Local-CommNet are higher, indicating that the individual performance of a single UAV is excellent, while the cooperation between UAVs is much lower. This also reveals the importance of cooperation for the emergence of swarm intelligence.

Combined with the visualization in Figure 6, the numerical results verify that UAV swarms can learn more efficient communication and action policies by using Att-Message and can scale these policies to different scenarios and achieve better performance.

6.5. Communication Failure Assessment

The above experiments assume perfect communication between UAVs; that is, the messages can always be received correctly by the adjacent UAVs. However, how does the performance of the UAVs change if there is a communication failure, while the UAVs still execute the policies learned while communicating perfectly? In this paper, two communication failures are assumed here: message loss and message error. The former refers to the message not being received due to communication disconnection or delay; and the later refers to the received message being inconsistent with the sent one due to electromagnetic interference or other reasons. Here, the error message is assumed to be a random noise signal. Under different failure probabilities, such as $\{0, 0.1, \cdots, 1.0\}$, the UAVs execute the policies learned by the four algorithms, respectively.

The variation trends of the metrics with the failure probability under the two failures are plotted in Figures 7 and 8, respectively. At first glance, the corresponding statistical metric curves in both the two failure cases have similar trends; that is, when the probability gradually increases, the average reward and collective target curves of the three explicit communication-based algorithms gradually decrease, while the average target curve gradually increases, and the metric curves of No-Comm (without communication) is approximately flat. For the same failure probability, the descending order of comprehensive performance of the four algorithms is: Att-Message > Att-Hidden > Local-CommNet > No-Comm, which is consistent with the training results.

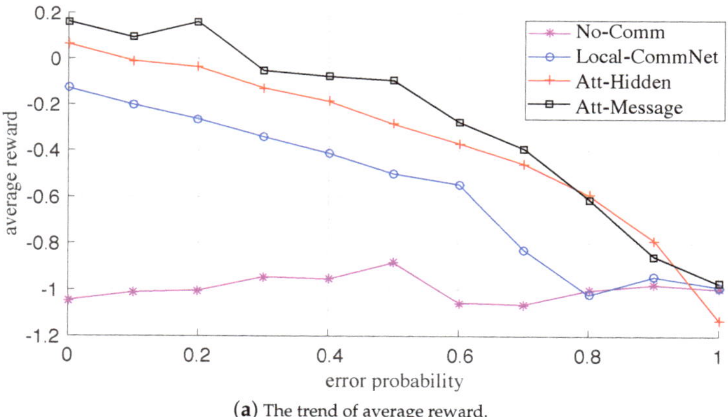

(**a**) The trend of average reward.

Figure 7. *Cont.*

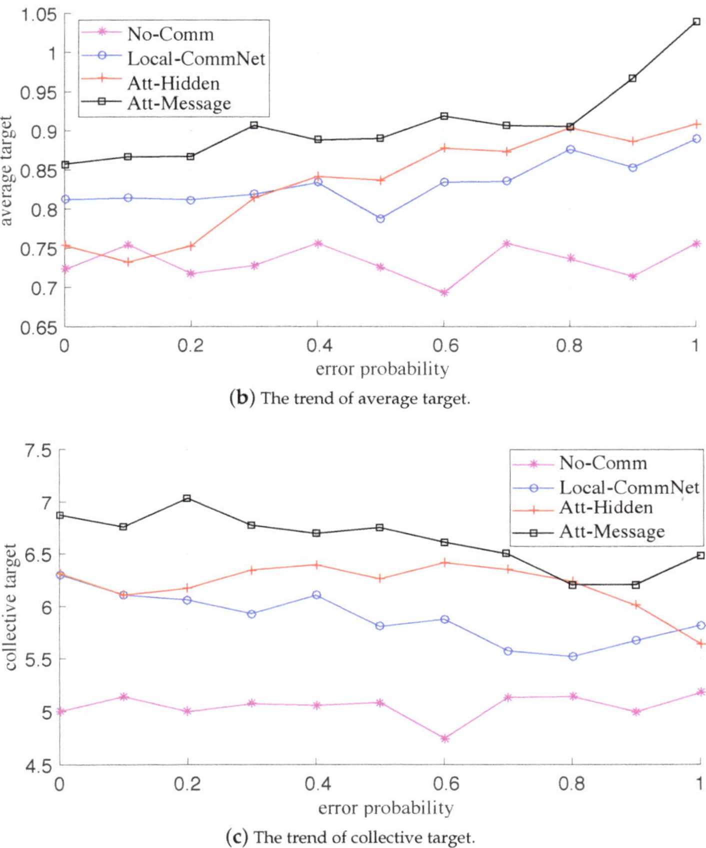

(**b**) The trend of average target.

(**c**) The trend of collective target.

Figure 7. The variation trends of the metrics with communication error probability.

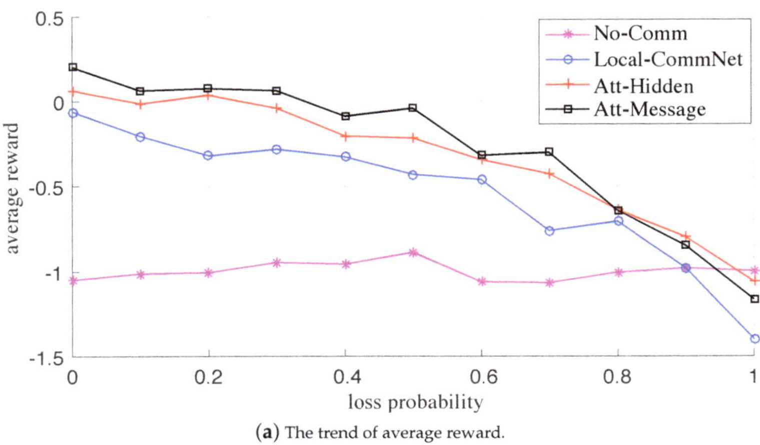

(**a**) The trend of average reward.

Figure 8. *Cont.*

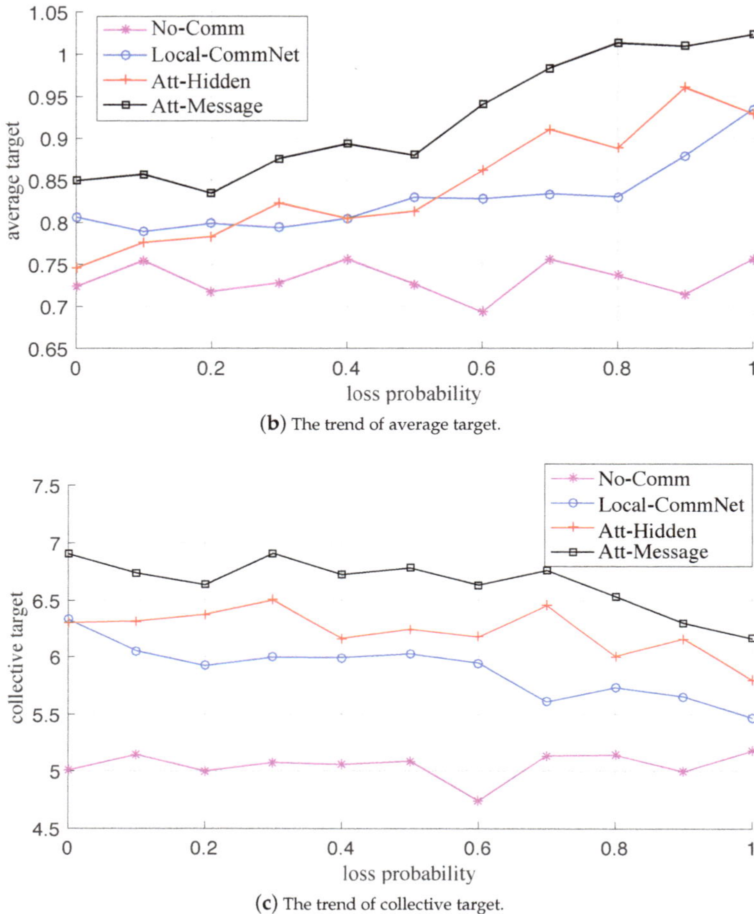

(**b**) The trend of average target.

(**c**) The trend of collective target.

Figure 8. The variation trends of the metrics with communication loss probability.

It is conceivable that when the probability increases, the available messages gradually dwindle, and the comprehensive performance of the former three algorithms with communication gradually deteriorates. The reason is that the reduction of useful messages leads to increased conflicts between UAVs and a decrease in cooperation. Moreover, as the probability increases, the average targets of the former three algorithms gradually increase, indicating that the UAVs shortsightedly maximize the number of targets tracked by individuals. In addition, when the communication is paralyzed, each UAV makes a completely independent decision. It can be seen that the comprehensive performance of Att-Message is the best, which reveals that while learning the communication policy, the UAVs can also learn a better action policy for tracking targets. Therefore, even the communication fails, and the improvement of the individual MTT capability can also feed back the overall capability of the swarm to a certain extent.

In summary, when there is a communication failure, such as message loss or error, the comprehensive performance of the communication and action policies learned by the proposed algorithm would be affected to a certain extent, but it is also better than the other three benchmark algorithms. The numerical results also demonstrate the robustness of the learned policies.

7. Discussion

As mentioned earlier, the research object of this paper is large-scale UAV swarms in which each UAV can only communicate and interact locally with the adjacent ones when making decisions. In local topology and ignoring other factors, the computational complexity of the action and communication policies for processing (aggregating) a message is assumed to be a unit, denoted as $O(1)$, and the average cardinal number of the message set is denoted as $|c|$.

Then, in the decentralized execution, the computational complexity of the action and communication policies of each UAV is $O(|c|)$ according to Equations (16) and (15), respectively. In centralized training, the computational complexity of updating action policy is also $O(|c|)$ according to Equation (20), and that of the communication policy is $O(2|c|^2)$ since a message can influence the communication and action decisions at the next step of all adjacent UAVs according to Equation (22).

Therefore, similar to most MADRL algorithms adopting the CTDE framework, the training of the proposed algorithm requires more computational resources than execution, which is suitable for offline implementation. The offline training in this paper was deployed on the computer equipped with Intel (R) Xeon E5 CPU (Manufacturer: Intel Corporation, Santa Clara, CA, USA) and GTX Titan X GPU (Manufacturer: ASUS,Taiwan) , the operating system was Ubuntu 16.04 LTS, and the algorithm was implemented by Pytorch. Then the learned policies can be performed online without retraining. The specific requirement of computational resources should comprehensively take the constraints, such as computing platform, neural network design and optimization, decision frequency and so on, into consideration.

Moreover, in the observation of a target, we only consider the simple numerical information, such as its location and speed, but not the real-time image, and the communication policy can also compress and encode the high-dimensional information to realize lightweight embedding interaction. These can further improve the feasibility of the algorithm in real-world scenarios.

8. Conclusions and Future Works

Communication is an important medium for transferring information and realizing cooperation between UAVs. This paper adopts a data-driven approach to learn the cooperative communication and action policies of UAV swarms and improve their communication and MTT capabilities. Specifically: (1) The communication policy of a UAV is mapped from the input variables to the message sent out, so that the UAV can autonomously decide the content of the message according to its real-time status. (2) The neural networks based on the attention mechanism are designed to approximate the communication and action policies, where the attention mechanism can distinguish the importance of different messages and aggregate the variable number of messages into a fixed-length code to adapt to the dynamic changes of the local communication topology. (3) To maximize the cumulative reward of the adjacent UAVs, the gradient optimization process of the continuous communication policy is derived. (4) Based on the CTDE framework, a reinforcement learning algorithm is proposed to jointly learn the communication and action policies of UAV swarms. The numerical simulation verifies that the proposed algorithm can learn effective cooperative communication and action policies to conduct the cooperation of UAV swarms, thereby improving their MTT ability, and the learned policies are robust to communication failures.

Although the communication policy in this paper can extract the message that is beneficial to cooperation, the physical meaning of the message cannot be explicitly parsed. Therefore, the interpretability of the message remains to be further explored. How to reasonably set the output dimension of the communication policy neural network, that is, the length of the message, also needs to be further solved. If the output dimension is too small, it may limit the communication capability of the UAV; otherwise, it may increase the difficulty of learning, which is not conducive to learning an effective communication policy. In addition, it is necessary to take more complex scenarios into consideration and establish more accurate models to investigate how the physical aspects of both the UAVs and targets would affect the MTT performance.

Author Contributions: Conceptualization, W.Z.; data curation, J.L.; formal analysis, W.Z.; funding acquisition, J.L.; investigation, W.Z.; methodology, W.Z.; project administration, Q.Z.; software, W.Z.; supervision, J.L.; validation, W.Z.; visualization, W.Z.; writing—original draft, W.Z.; writing—review and editing, W.Z., J.L. and Q.Z. All authors have read and agreed to the published version of the manuscript.

Funding: This research was funded by the Science and Technology Innovation 2030-Key Project of "New Generation Artificial Intelligence" under Grant 2020AAA0108200.

Data Availability Statement: Not applicable.

Conflicts of Interest: The authors declare no conflict of interest.

References

1. Goldhoorn, A.; Garrell, A.; Alquezar, R.; Sanfeliu, A. Searching and tracking people in urban environments with static and dynamic obstacles. *Robot. Auton. Syst.* **2017**, *98*, 147–157. [CrossRef]
2. Senanayake, M.; Senthooran, I.; Barca, J.C.; Chung, H.; Kamruzzaman, J.; Murshed, M. Search and tracking algorithms for swarms of robots: A survey. *Robot. Auton. Syst.* **2016**, *75*, 422–434. [CrossRef]
3. Abdelkader, M.; Güler, S.; Jaleel, H.; Shamma, J.S. Aerial Swarms: Recent Applications and Challenges. *Curr. Robot. Rep.* **2021**, *2*, 309–320. [CrossRef] [PubMed]
4. Emami, Y.; Wei, B.; Li, K.; Ni, W.; Tovard, E. Joint Communication Scheduling and Velocity Control in Multi-UAV-Assisted Sensor Networks: A Deep Reinforcement Learning Approach. *IEEE Trans. Veh. Technol.* **2021**, *9545*, 1–13. [CrossRef]
5. Maravall, D.; de Lope, J.; Domínguez, R. Coordination of Communication in Robot Teams by Reinforcement Learning. *Robot. Auton. Syst.* **2013**, *61*, 661–666. [CrossRef]
6. Kriz, V.; Gabrlik, P. UranusLink—Communication Protocol for UAV with Small Overhead and Encryption Ability. *IFAC-PapersOnLine* **2015**, *48*, 474–479. [CrossRef]
7. Khuwaja, A.A.; Chen, Y.; Zhao, N.; Alouini, M.S.; Dobbins, P. A Survey of Channel Modeling for UAV Communications. *IEEE Commun. Surv. Tutor.* **2018**, *20*, 2804–2821. [CrossRef]
8. ZHOU, W.; LI, J.; LIU, Z.; SHEN, L. Improving multi-target cooperative tracking guidance for UAV swarms using multi-agent reinforcement learning. *Chin. J. Aeronaut.* **2022**, *35*, 100–112. [CrossRef]
9. Bochmann, G.; Sunshine, C. Formal Methods in Communication Protocol Design. *IEEE Trans. Commun.* **1980**, *28*, 624–631. [CrossRef]
10. Rashid, T.; Samvelyan, M.; Schroeder, C.; Farquhar, G.; Foerster, J.; Whiteson, S. QMIX: Monotonic value function factorisation for deep multi-agent reinforcement learning. In Proceedings of the 35th International Conference on Machine Learning, Stockholm, Sweden, 10–15 July 2018; Volume 80, pp. 4295–4304.
11. Son, K.; Kim, D.; Kang, W.J.; Hostallero, D.; Yi, Y. QTRAN: Learning to factorize with transformation for cooperative multi-agent reinforcement learning. In Proceedings of the 36th International Conference on Machine Learning, Long Beach, CA, USA, 9–15 June 2019; pp. 10329–10346. Available online: https://arxiv.org/abs/1905.05408 (accessed on 20 September 2022).
12. Wu, S.; Pu, Z.; Qiu, T.; Yi, J.; Zhang, T. Deep Reinforcement Learning based Multi-target Coverage with Connectivity Guaranteed. *IEEE Trans. Ind. Inf.* **2022**, *3203*, 1–12. [CrossRef]
13. Xia, Z.; Du, J.; Wang, J.; Jiang, C.; Ren, Y.; Li, G.; Han, Z. Multi-Agent Reinforcement Learning Aided Intelligent UAV Swarm for Target Tracking. *IEEE Trans. Veh. Technol.* **2022**, *71*, 931–945. [CrossRef]
14. Sukhbaatar, S.; Szlam, A.; Fergus, R. Learning multiagent communication with backpropagation. In Proceedings of the 30th Conference on Neural Information Processing Systems (NIPS 2016), Barcelona, Spain, 5–10 December 2016. [CrossRef]
15. Hausknecht, M.; Stone, P. Grounded semantic networks for learning shared communication protocols. In Proceedings of the International Conference on Machine Learning (Workshop), New York City, NY, USA, 19–24 June 2016.
16. Foerster, J.; Assael, Y.M.; de Freitas, N.; Whiteson, S. Learning to Communicate with Deep Multi-Agent Reinforcement Learning. *Adv. Neural Inf. Process. Syst.* **2016**, *29*, 2137–2145.
17. Pesce, E.; Montana, G. Improving coordination in small-scale multi-agent deep reinforcement learning through memory-driven communication. *Mach. Learn.* **2020**, *109*, 1–21. [CrossRef]
18. Peng, P.; Wen, Y.; Yang, Y.; Yuan, Q.; Tang, Z.; Long, H.; Wang, J. Multiagent Bidirectionally-Coordinated Nets: Emergence of Human-Level Coordination in Learning to Play StarCraft Combat Games. 2017. Available online: https://arxiv.org/abs/1703.10069 (accessed on 20 September 2022).
19. Jiang, J.; Lu, Z. Learning attentional communication for multi-agent cooperation. In Proceedings of the 32nd International Conference on Neural Information Processing Systems, Montréal, QC, Canada, 3–8 December 2018; Volume 18, pp. 7265–7275.

20. Liu, Y.; Wang, W.; Hu, Y.; Hao, J.; Chen, X.; Gao, Y. Multi-agent game abstraction via graph attention neural network. In Proceedings of the AAAI 2020—34th AAAI Conference on Artificial Intelligence, New York, NY, USA, 7–12 December 2020; pp. 7211–7218. [CrossRef]
21. Ding, G.; Huang, T.; Lu, Z. Learning Individually Inferred Communication for Multi-Agent Cooperation. In Proceedings of the 34th Conference on Neural Information Processing Systems (NeurIPS2020), Vancouver, BC, Canada, 6–12 December 2020; Volume 33, pp. 22069–22079.
22. Das, A.; Gervet, T.; Romoff, J.; Batra, D.; Parikh, D.; Rabbat, M.; Pineau, J. TarMAC: Targeted multi-agent communication. In Proceedings of the 36th International Conference on Machine Learning, Long Beach, CA, USA, 9–15 June 2019; Volume 97, pp. 1538–1546.
23. Singh, A.; Jain, T.; Sukhbaatar, S. Learning when to Communicate at Scale in Multiagent Cooperative and Competitive Tasks. *arXiv* **2018**, arXiv:1812.09755.
24. Dibangoye, J.S.; Amato, C.; Buffet, O.; Charpillet, F. Optimally Solving Dec-POMDPs as Continuous-State MDPs. *J. Artif. Intell. Res.* **2016**, *55*, pp.443–497. [CrossRef]
25. Sutton, R.S.; Barto, A.G. Temporal-difference learning. In *Reinforcement Learning: An Introduction*; MIT Press: Cambridge, MA, USA, 1998; pp. 133–160.
26. Lillicrap, T.P.; Hunt, J.J.; Pritzel, A.; Heess, N.; Erez, T.; Tassa, Y.; Silver, D.; Wierstra, D. Continuous control with deep reinforcement learning. *arXiv* **2015**, arXiv:1509.02971.
27. Zhou, W.; Liu, Z.; Li, J.; Xu, X.; Shen, L. Multi-target tracking for unmanned aerial vehicle swarms using deep reinforcement learning. *Neurocomputing* **2021**, *466*, 285–297. [CrossRef]
28. Veličković, P.; Cucurull, G.; Casanova, A.; Romero, A.; Lio, P.; Bengio, Y. Graph attention networks. *arXiv* **2017**, arXiv:1710.10903.
29. Lee, J.B.; Rossi, R.A.; Kim, S.; Ahmed, N.K.; Koh, E. Attention Models in Graphs: A Survey. *ACM Trans. Knowl. Discov. Data* **2019**, *13*, 1–25. [CrossRef]
30. Zhou, J.; Cui, G.; Hu, S.; Zhang, Z.; Yang, C.; Liu, Z.; Wang, L.; Li, C.; Sun, M. Graph neural networks: A review of methods and applications. *AI Open* **2020**, *1*, 57–81. [CrossRef]

Article

A Lightweight Authentication Protocol for UAVs Based on ECC Scheme

Shuo Zhang, Yaping Liu *, Zhiyu Han and Zhikai Yang

CIAT, Guangzhou University, Guangzhou 510006, China; szhang18@gzhu.edu.cn (S.Z.);
2111906006@e.gzhu.edu.cn (Z.H.); 2112006047@e.gzhu.edu.cn (Z.Y.)
* Correspondence: ypliu@gzhu.edu.cn

Abstract: With the rapid development of unmanned aerial vehicles (UAVs), often referred to as drones, their security issues are attracting more and more attention. Due to open-access communication environments, UAVs may raise security concerns, including authentication threats as well as the leakage of location and other sensitive data to unauthorized entities. Elliptic curve cryptography (ECC) is widely favored in authentication protocol design due to its security and performance. However, we found it still has the following two problems: inflexibility and a lack of backward security. This paper proposes an ECC-based identity authentication protocol LAPEC for UAVs. LAPEC can guarantee the backward secrecy of session keys and is more flexible to use. The time cost of LAPEC was analyzed, and its overhead did not increase too much when compared with other authentication methods.

Keywords: UAV; internet of drones; authentication protocol; key agreement

Citation: Zhang, S.; Liu, Y.; Han, Z.; Yang, Z. A Lightweight Authentication Protocol for UAVs Based on ECC Scheme. *Drones* **2023**, *7*, 315. https://doi.org/10.3390/drones7050315

Academic Editor: Emmanouel T. Michailidis

Received: 7 March 2023
Revised: 29 April 2023
Accepted: 30 April 2023
Published: 9 May 2023

1. Introduction

Unmanned aerial vehicles (UAVs) have experienced rapid developments in recent years and have attracted the interest of researchers [1]. They have been deployed for many applications and missions such as data transmission, surveillance, cellular service provisioning, package delivery, firefighting, traffic monitoring, military operations, agriculture, etc. [2,3]. Here, a common UAV scenario (target surveillance as an example) is shown in Figure 1.

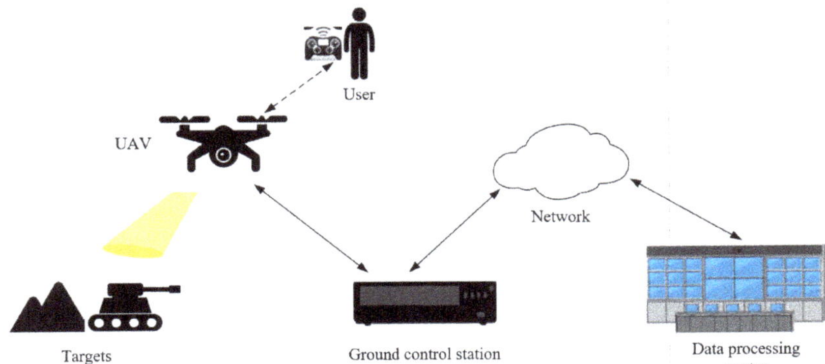

Figure 1. A common UAV scenario (target surveillance as an example).

In the above scenario, a drone is controlled by the user. After it has received the control signal from the user, it collects the data of the targets (e.g., video, photo) and sends the data to the closest ground control station (GCS). Since the GCS connects to the data processing center (DPC) through the network, it can send the data of the targets to the DPC. Finally, the DPC utilizes the data from the GCS to analyze the behavior of the targets.

UAV communication relies on wireless channels, which makes UAVs vulnerable to many attacks such as replay attacks, man-in-the-middle attacks, and masquerading attacks. These attacks can have serious consequences, which can lead to commercial and non-commercial losses. Attackers may also aim to exploit these UAVs to eavesdrop on sensitive information, tamper with data, or cause malicious interference [4,5].

With the rapid development of the internet of drones (IoD), the security of the IoD is becoming more and more important. Among many, authentication is one of the research hotspots in the field of IoD security. Because most drones have shortcomings (such as low computing power, small storage space, etc.), it is difficult to directly apply traditional identity authentication and key agreement protocols within the IoD [2]. Thus, it is necessary to design identity authentication and key agreement protocols suitable for the IoD [3]. The traditional security provisioning applicable to distributed networks fails to give similar results for UAVs [6]. The large-scale deployment of UAVs is hindered due to these many security challenges [7,8].

Aiming at the lack of a pre-registration process and the backward security of session keys in the EDHOC (ephemeral Diffie–Hellman over COSE) protocol, an ECC (elliptic curve cryptography)-based authentication protocol for the IoD (called LAPEC) is proposed in this paper, which can achieve high security and an acceptable time overhead. A formal security proof is given for the proposed LAPEC protocol to demonstrate its security properties. At the end of the paper, a time cost analysis and a comparison of LAPEC with other protocols are carried out.

2. Background and Related Works

In order to deal with the authentication issues of the IoD, some researchers have proposed various solutions in recent years. Due to the versatility of RFID technology [9], which is ideal for identifying and tracking objects, some researchers use it in the identification and authentication of UAVs [10] in both commercial and/or military scenarios. In this case, these drones can be equipped with RFID tags and be required to pass through a reader checkpoint whereby the tag is scanned and its credentials sent to a secure server unit for verification. Either the drone is authenticated, or if not, it can be intercepted [11]. Authentication methods based on RFID are simple and easy to use, but securing an RFID-based system is a challenging task due to the computational capability of RFID tags being very limited [12,13].

To address this issue of RFID-based authentication methods, the concept of physically unclonable function (PUF) technology [14–16] has been introduced. A PUF is a function derived from a physical characteristic and is used to produce a device-specific output for any input such as with a fingerprint. With the inclusion of PUF, RFID can ensure hardware security. However, these methods can deal with problems related to one-to-one authentication, but they fail to provide solutions for dynamic and large-scale networks [6].

2.1. Elliptical Curve Cryptography Scheme

Some researchers [4,17] have presented authentication protocols based on elliptic curve cryptography (ECC). Although they increase the level of security, their techniques are far from being scalable.

ECC is an asymmetric public key cryptography [18], whose theoretical basis comes from elliptic curves. Compared with traditional public key cryptography (such as RSA, etc.), ECC requires less computation and uses a shorter key length to achieve the same key strength as RSA.

The design of using ECC for a public key cryptosystem is based on the following two mathematical problems about chaotic maps: (1) the discrete logarithm problem based on ECC (ECDLP); (2) the computational Diffie–Hellman problem based on ECC (ECDHP).

Ever [19] proposed an authentication framework for the IoD using elliptic curve cryptosystems, but it still has some of the inherent issues of ECC. Tao [20] has proposed a two-way identity authentication scheme based on the SM2 algorithm and adopted the pre-shared secret information to improve the efficiency of authentication, but how to securely pre-share the secret is also an issue. Lin [21] proposed a certificate signing based on elliptic curve multiple authentication schemes, but it still has inherent issues with certificate mechanisms.

Some related work can be found in [22–32], and we will compare ours with them when analyzing the time cost in Section 5.

2.2. EDHOC Scheme

The EDHOC protocol is one of the most practical used for the IoT, so we will discuss it and compare it with our proposed protocol. The EDHOC protocol is based on the SIGMA (SIGn and Mac) protocol structure, which is a series of theoretical protocols with a large number of variants. The EDHOC protocol uses digital signatures for authentication. Similar protocols include the IKEv2 (RFC7296) [33] and TLS 1.3 (RFC8446) [34] protocols. EDHOC implements the SIGMA-I variant as Mac-then-Sign. EDHOC consists of three messages (message_1, message_2, message_3) that map directly to the three messages in SIGMA-I. The scheme of EDHOC is shown in Figure 2, showing it needs three messages to finish the authentication. Message_1 is composed of method (authentication method), SUITES_I (array of cipher suites which the Initiator supports), G_X (the ephemeral public key of the Initiator), C_I (variable length connection identifier), and EAD_1 (external authorization data). Message_2 is composed of G_Y_CIPHERTEXT_2 (the concatenation of G_Y and CIPHERTEXT_2) and C_R (variable length connection identifier). Message_3 is composed of CIPHERTEXT_3.

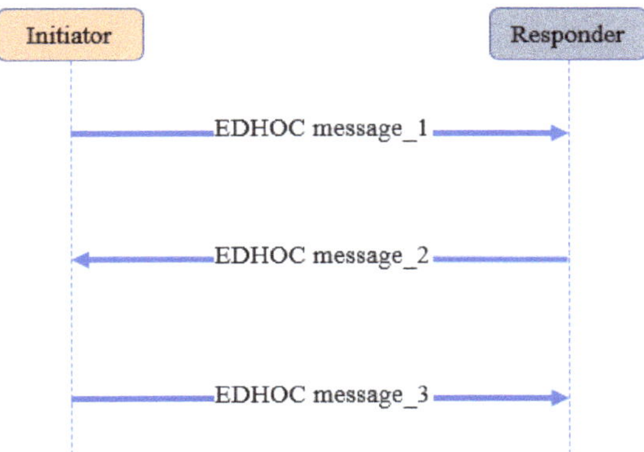

Figure 2. Interaction flow of the EDHOC Scheme.

2.3. Problem Analysis

The EDHOC protocol and some other protocol solutions for limited UAV devices have taken into account the characteristics of insufficient computing and storage space of UAV devices. They have carried out lightweight optimizations in the protocol process, encryption, and decryption. However, the EDHOC protocol has the following shortcomings in terms of deployment flexibility and security:

- Public key preset problem.

In the process of the EDHOC protocol, since the signature and verification operations of the certificate are not required, the burden of the device is greatly reduced. However, in actual use, the two parties who authenticate by default have the public key of the other party. Therefore, this requires EDHOC to preset the other party's public key in the implementation, which will cause the UAV devices to face extreme inflexibility during large-scale deployment and use. It may bring a lot of inconvenience.

- Session key backward security.

Backward security, also known as future security or post-compromise security (PCS), was formally defined by Katriel et al. [35]. Backward security means that after the long-term key or session key is leaked or compromised, the security of messages after the session can still be guaranteed.

The scheme of EDHOC relies on the automatic update of the symmetric session key after completing the authentication and key negotiation process. Therefore, EDHOC needs to use the symmetric session key to secure the subsequent messages. Once the session key is leaked or compromised, the subsequent messages will face significant security risks, that is to say, backward secrecy cannot be guaranteed.

3. Proposed Scheme

3.1. Design Principles

Aiming at the problems of inflexibility and security of EDHOC, this section proposes an enhanced elliptic-curve-based lightweight authentication protocol for IoD, which is named LAPEC (lightweight authentication protocol over elliptic curve). The main design ideas are as follows:

(1) In view of the inflexibility of EDHOC use, the corresponding pre-registration steps are designed to reduce the use of public key certificates of both parties, and users do not need to configure the public key in advance, which is flexible in large-scale deployment and use.

(2) For backward security, based on the non-interactive zero-knowledge proof protocol, a corresponding session key update mechanism is designed to ensure the security of message communication. Even if the session key is leaked, the attacker cannot complete the zero-knowledge proof, so the key cannot be modified and session-backward security is guaranteed. In the session key update phase, the Schnorr zero-knowledge proof is introduced to design the session key update process.

In the LAPEC protocol, (1) a pre-registration process is added, which is before the authentication process, and (2) a new session key update process is designed using the zero-knowledge proof to increase the backward security of the session key.

Therefore, the LAPEC protocol consists of three phases: the pre-registration phase, the authentication phase, and the session key update phase. Figure 3 shows the general process of interaction flow of a LAPEC message.

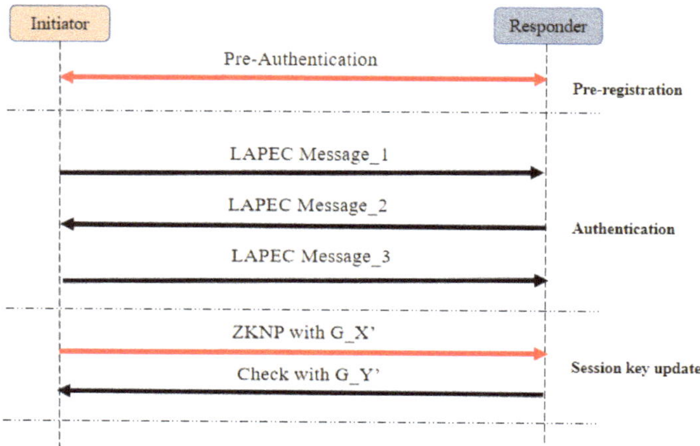

Figure 3. The general process of interaction flow of a LAPEC message.

3.2. Symbols and Meanings

This section describes the overall design of the LAPEC protocol message structure. LAPEC mainly includes three processes: a pre-registration process, an authentication and key negotiation process, and a session key update process. The parameters and meanings used in the LAPEC protocol are shown in Table 1:

Table 1. Symbols in LAPEC.

Symbols	Meaning
DEV, *GWN*	the UAV *DEV*, its ground control station (gateway) *GWN*
P_A, P_B	Ephemeral public key for device A and B
P_D, P_G	Authentication public key of the device and the gateway
ID_CRED_D, ID_CRED_G	The public key identifier of the device and the gateway
AEAD(*K*;(Plaintext))	Additional data are encrypted with authentication using a key *K* derived from the shared secret
Extract	Pseudorandom key generation function
Expand	Symmetric key generation function
MAC	Message authentication code
t_D	The current timestamp of the device
t_G	The current timestamp of the gateway
Δt	Maximum time interval allowed
H_m	Hash of message data
H(*)	Collision resistant hash function
||	Connect operation
$\oplus$	XOR operation

The proposed LAPEC scheme mainly includes the pre-registration phase, the authentication and key negotiation phase, and the key update phase. Figure 4 shows the interaction messages during the scheme process.

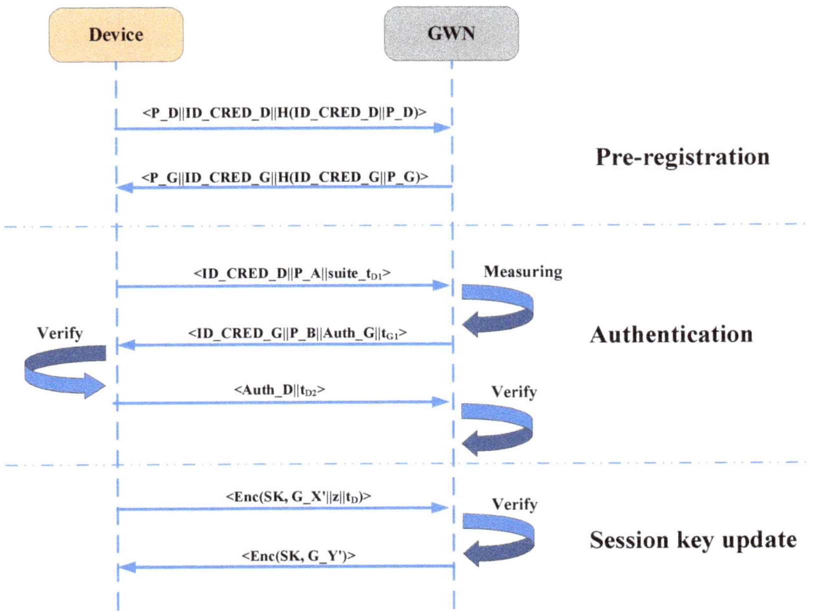

Figure 4. The interaction messages of LAPEC process.

3.3. Pre-Registration Phase

UAV devices and a ground station (serviced as a gateway) respectively hold their own authentication public and private key pairs: <D, P_D> and <G, P_G>. Among them, D and G represent the private key for both parties to authenticate. Key pairs generate as follows: P_D = DP, P_G = GP. Among them, P is the base point of the elliptic curve recognized by both parties.

At the same time, both parties also need to use ID_CRED_D and ID_CRED_G as the identifiers of the above authentication keys. Both parties calculate the following results: C_D = H(ID_CRED_D||P_D), C_G = H(ID_CRED_G||P_G).

In the pre-registration phase shown in Figure 4, the device sends the first message to GWN, which is formatted as <P_D||ID_CRED_D||H(ID_CRED_D||P_D)>. After receiving the first message from the device, GWN will respond with a reply message formatted as <P_G||ID_CRED_G||H(ID_CRED_G||P_G)>.

3.4. Authentication Phase

The interaction process during authentication is shown in Figure 5.

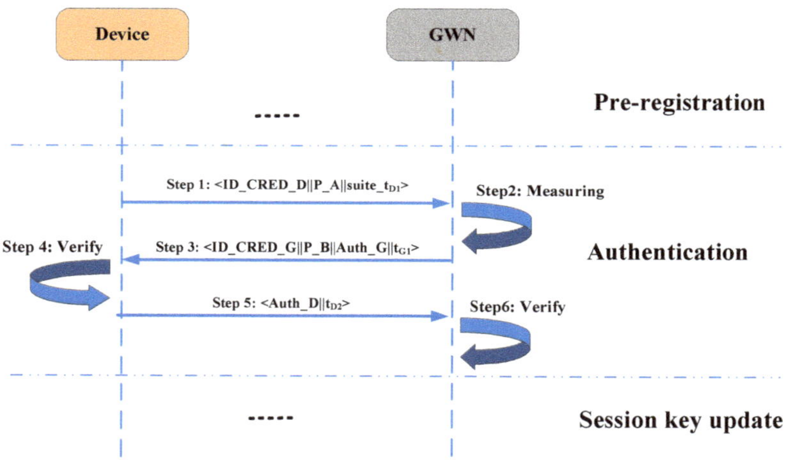

Figure 5. Authentication Phase of interaction process showing the authentication process.

(1) Step 1

Firstly, the UAV device generates the current timestamp t_{D1} to determine the freshness of the message, selects a random number A, and calculates the ephemeral public key: P_A = A × P.

Secondly, the device needs to determine the cipher suite suite_D. The function of the suite parameter is to ensure that both parties use the same cipher algorithm in the next protocol process, especially to determine the AEAD algorithm that both parties need to use and the parameters required by the Extract and Expand functions to generate a key.

Finally, the device connects the above parameters and sends Message_1 (Step 1 shown in Figure 5) to the ground station *GWN* via the open channel:

$$Message\_1 = ID\_CRED\_D||P\_A||suite\_D||t_{D1}$$

(2) Step 2

After the ground station, *GWN* receives the first message, it first needs to extract and verify the parameters (Step 2: Measuring shown in Figure 5). It mainly checks whether the time when *GWN* receives the message meets the timeliness and whether it supports the cipher suite suite_D contained in Message_1. For timeliness, it records the current timestamp t_{G1}, and judges: $|t_{G1} - t_{D1}| < \Delta t$?

If the above decoding or verification process fails, *GWN* must send back an authentication error message and abort the process. If *GWN* does not support the selected cipher suite, it will return the parameter suite_G containing its own supported cipher suites.

(3) Step 3

After successfully decoding Message_1, the ground station *GWN* selects a random number B, calculates the ephemeral public key, and saves it as its own temporary public–private key pair: P_B = B × P.

In the process of identity authentication and key generation, corresponding cryptographic algorithms are required to encrypt plaintext or decrypt ciphertext. The Extract and Expand functions are used with a hash algorithm in the selected cipher suite to derive the key. Extract is used to derive a uniform pseudorandom key (PRK) of fixed length from the shared secret. Expand is used to derive other key material from PRK. The process of generating the intermediate key *PRK* is as follows: *PRK* = Extract(*salt*, *IKM*), where *salt* is the added salt value, and *IKM* represents the input key material. The Extract function is specifically determined by the suite parameter in Step 1.

The keys used in LAPEC are derived from PRK using Expand function, and the process of generating the symmetric key *K* is as follows: *K* = Expand(*PRK*, *H*). Among them, *PRK* is the pseudo-random intermediate key generated by the above Extract function, and *H* represents the text hash value of a certain message.

The ground station *GWN* first needs to calculate the shared secret P_AB according to P_A and B: P_AB = B × P_A. *GWN* uses P_AB to calculate the first and the second *PRK*: *PRK_1* = Extract(t_{D1}, P_AB), *PRK_2* = Extract(*PRK_1*, P_GA). Among them, P_GA is the shared secret calculated from P_A and G: P_GA = G × P_A.

After the generation of the *PRK* is completed at the ground station *GWN*, the generation of the symmetric key *K* used for authentication needs to be performed. *GWN* first needs to generate *K_1* using the Expand function described in Step 2, the generated *PRK_1*, and text hash *H_1*. The calculations of *H_1* and *K_1* are as follows:

$$H\_1 = H(\text{Message\_1} \,||\, t_{G1} \,||\, \text{P\_B}).$$

$$K\_1 = \text{Expand}(PRK\_1, H\_1).$$

Similarly, *GWN* also generates the symmetric key *K_2*:

$$K\_2 = \text{Expand}(PRK\_2, H\_1).$$

Next, *GWN* constructs a message authentication code (MAC), which is calculated using the AEAD algorithm in the selected cipher suite. AEAD constructs an additional piece of auxiliary authentication data during encryption to ensure that after decryption using the symmetric key, it can be judged whether the symmetric key used is correct. The AEAD algorithm is used to encrypt the auxiliary authentication data *external_aad_G* with the key *K_2* generated above:

$$external\_aad\_G = AEAD(H\_1 \,||\, \text{P\_G} \,||\, t_{G1})$$

$$MAC\_2 = AEAD(K\_2, external\_aad\_G).$$

Finally, *GWN* uses another key *K_1* generated above to perform XOR encryption with *MAC_2* to obtain the authentication data segment of Message_2: Auth_G = K_1⊕MAC_2.

Then, *GWN* connects the authentication data segment with other parameters to get Message_2 (Step 3 shown in Figure 5), and sends it to the device for authentication via the open channel:

$$\text{Message\_2} = \text{ID\_CRED\_G} \,||\, \text{P\_B} \,||\, \text{Auth\_G} \,||\, t_{G1}$$

(4)   Step 4

After receiving message_2, the UVA device should handle message_2 (Step 4: Verify shown in Figure 5) as follows

1.   Decode message_2 and record the timestamp to determine the freshness of the message.
2.   XOR the Auth_G with the key K_1 to decrypt the Auth_G field.
3.   Verify MAC_2 using the algorithm in the selected cipher suite.

If the timestamp or AEAD algorithm fails to verify the authentication packet of MAC_2, an error message is returned and the protocol process is aborted.

The UAV device also needs to calculate the shared secret P_AB according to P_B and A. The calculation process is P_AB = A $\times$ P_B. Next, similar to Step 3, the UAV device also uses P_AB to calculate PRK: PRK_1 = Extract(t_{D1}, P_AB), PRK_2 = Extract(PRK_1, P_GA), PRK_3 = Extract(PRK_2, P_DB), where P_DB = D $\times$ P_B.

The UAV device also needs to generate K_1':

$$H\_1 = H(\text{Message\_1} \mid\mid t_{G1} \mid\mid \text{P\_B})$$

$$K\_1' = \text{Expand}(PRK\_1, H\_1).$$

Similarly, the device side generates the symmetric key K_2':

$$K\_2' = \text{Expand}(PRK\_2, H\_1).$$

After the key is generated, the verification process can be performed. The UAV device first performs the following XOR decryption for the Auth_G part:

$$MAC\_2' = K\_1' \oplus \text{Auth\_G}.$$

Then, it uses the generated K_2' as the key to decrypt the MAC_2':

$$external\_aad\_G' = AEAD\_dec(K\_2', MAC\_2')$$

where $AEAD\_dec(K, M)$ is a decryption function that uses the key K to decrypt and verify the encrypted message M. AEAD determines whether the key K_2 is correct or not by comparing the decrypted auxiliary authentication data:

$$external\_aad\_G' = external\_aad\_G?$$

(5)   Step 5

After the UAV device completes the processing of the authentication data packet to the ground station, if the authentication is passed, it constructs Message_3. During the verification process in Step 3, the UAV device has completed the calculation of the pseudo-random keys PRK_1, PRK_2, and PRK_3, as well as the keys K_1 and K_2 used for verification. In order to construct the authentication data packet MAC_3, the UAV device first calculates the text hash value H_2 as follows: $H\_2 = H(H\_1 \mid\mid Auth\_G \mid\mid \text{P\_B} \mid\mid t_{G1})$. K_3 is constructed using H_2 and pseudo-random key PRK_3 as follows: $K\_3 = \text{Expand}(PRK\_3, H\_2)$.

Similar to Step 3, the additional authentication data of MAC_3 are constructed as follows:

$$external\_aad\_D = H\_1 \mid\mid \text{P\_D} \mid\mid t_{G1}$$

At this point, the UAV device can construct MAC_3 as:

$$MAC\_3 = AEAD(K\_3, external\_aad\_D)$$

Finally, the UAV device calculates the encryption key K_4 of Auth_D:

$$K\_4 = \text{Expand}(PRK\_2, H\_2)$$

$$\text{Auth\_D} = AEAD(K\_4, MAC\_3 \mid\mid t_{G1} \mid\mid H\_2).$$

The UAV device connects the generated Auth_D with the timestamp to get the final Message_3 (Step 5 shown in Figure 5) and sends it to *GWN*.

$$\text{Message\_3} = \text{Auth\_D} || t_{D2}$$

(6)   Step 6

After the ground station, *GWN* receives the corresponding Message_3. It first needs to authenticate the device (Step 6: Verify shown in Figure 5). The intermediate pseudo-random key has been calculated in Step 3. At this time, the gateway needs to calculate H_2, K_3', and K_4':

$$H\_2 = H(H\_1 || \text{Auth\_G} || \text{P\_B} || t_{G1})$$

$$K\_3' = Expand(PRK\_3, H\_2)$$

$$K\_4' = Expand(PRK\_2, H\_2).$$

Similarly, *GWN* uses AEAD to decrypt the authentication packet contained in ciphertext_3:

$$MAC\_3' || t_{G1} || H\_2 = AEAD\_dec(K\_4', \text{Auth\_D})$$

$$external\_aad\_D' = AEAD\_dec(K\_3', MAC\_3' || t_{G1} || H\_2).$$

AEAD determines whether the key K_3 is correct or not by comparing the decrypted auxiliary authentication data, that is, verifying $external\_aad\_D' = external\_aad\_D$?

If the verification is successful, *GWN* also considers whether the UAV device's identity is legal and can construct the session key. If unsuccessful, the UAV device identity authentication fails, and *GWN* immediately terminates the authentication process and returns an authentication failure message.

If the UAV device is authenticated, both parties can calculate the session key separately by first calculating the text hash value H_3:

$$H\_3 = H(H\_2, \text{Auth\_D})$$

$$SK = Expand(PRK\_3, H\_3)$$

Both parties encrypt subsequent messages and communicate via *SK*.

3.5. Session Key Update Phase

Figure 6 shows the message interaction process in the session key update phase, which is part of Figure 4.

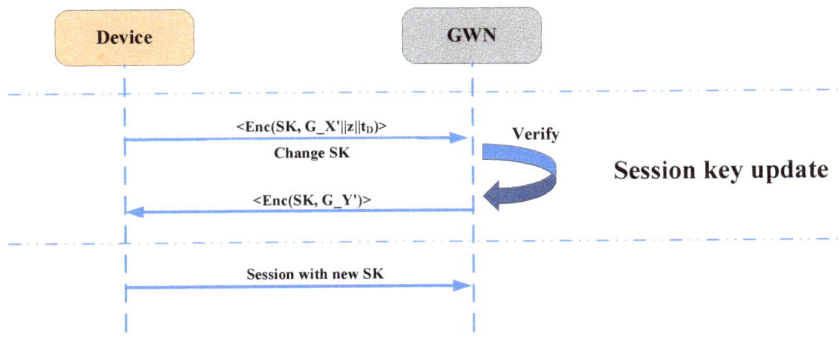

Figure 6. Session Key Update Phase.

After successfully completing the mutual authentication and key negotiation process, both parties should communicate by sharing the secret session key SK. If the session key needs to be updated (i.e., the session key has a valid time), either party will initiate a key update request.

The entity that needs to initiate the update of the session key (presumably the D party) first selects a random number X' and calculates the temporary public–private key pair $G\_X' = X' \times P$. Party D calculates the following results and sends them to GWN: $c = H(P\_D \mid\mid G\_X')$.

The UAV device constructs the following response based on challenge c: $z = X' + c \times D$, where D is the authentication public key of the UAV device, and c is the challenge result calculated by the above formula. The device constructs and sends a session key change request message (step *Change SK* in Figure 6): $Message\_ChangeSK = Enc(SK, G\_X' \mid\mid z \mid\mid t_D)$.

After GWN receives the session key change request, it checks the following steps (step *Verify* in Figure 6):

1. Decode the message and obtain and check the freshness of the message.
2. Calculate random challenges.
3. Calculate and check:

$$z \times P = G\_X' + c \times P\_D?$$

If not, the receiver aborts the session key update procedure and returns an update failure error message. If so, GWN considers that the identity of the requester for updating the session key is legitimate, and the receiver generates the updated session key according to the following steps:

$$P\_GX' = X' \times P\_G$$

Next, both parties calculate:

$$PRK\_x = Extract(PRK\_3, P\_GX')$$

$$H\_4 = H(Message\_ChangeSK)$$

$$SK' = Expand(PRK\_x, H\_4)$$

Both parties can then communicate via updated encrypted session key SK' in follow-up messages.

4. Security Analysis

4.1. Security Properties Analysis

In this section, the security properties of LAPEC are discussed. The LAPEC protocol has five security attributes: backward security, anti-replay attack, forward security, anti-masquerade attack, and session key confidentiality. However, the EDHOC protocol has four security attributes, which are shown in Table 2:

Table 2. Security properties of protocols.

Security Properties	LAPEC	EDHOC
Backward Security	✓	✗
Anti-replay Attack	✓	✓
Forward Security	✓	✓
Session Key Confidentiality	✓	✓
Anti-Camouflage Attack	✓	✓

4.2. Security Properties Proof

This section will formally prove the backward security, anti-replay attack, forward security, anti-masquerading attack, and session key secrecy of LAPEC.

Theorem 1. *The LAPEC protocol can inherit the anti-replay attack, forward security, anti-masquerading attack, and session key confidentiality of EDHOC in the authentication negotiation process.*

Proof. Since the pre-registration process added to the LAPEC protocol does not change the key calculation in the authentication protocol phase, the LAPEC protocol can inherit the security properties of EDHOC during the authentication phase. According to the formal analysis of EDHOC using Tamarin tools, LAPEC can at least inherit the following security properties: forward security, session key independence, anti-replay attack, and anti-masquerading attack. □

Lemma 1. *The LAPEC protocol has the security properties of anti-replay attack, anti-masquerading attack, and session key confidentiality.*

Proof. According to Theorem 1, the LAPEC protocol can inherit the relevant security properties of EDHOC in the authentication negotiation process, so the LAPEC protocol has the security properties of anti-replay attack, anti-masquerading attack, and session key confidentiality. □

Suppose that attacker A can launch different attacks by interrogating the oracles as shown in Table 3.

Table 3. Oracles and description.

Oracle	Description
Creat (D, r, G)	Create a new session oracle with peer G as D's identity r
Send (D, i, M)	Execute and return the result at the *ith* session oracle of D
Corrupt (C)	Leak C's long-term key
Test-session (s)	If b = 1, C outputs the current session key *SK*. If b = 0, C returns a random number. If no session key is generated, returns null.
Randomness (C, i)	Leak the random number in the *ith* session of C
Session-key (s)	Leaked session key *SK*
Hsm (C)	Hardware security module for C
Guess (b)	End game

Definition 1. *After receiving the last expected message M3, C will generate a session key and enter the accept state. All communication messages M1, M2, and M3 are concatenated in sequence to form a session identifier.*

Definition 2. *If D and G meet the following conditions, they are defined as a **partnership**: (1) D and G are both in the accepted state; (2) D and G authenticate each other and share the same session ID.*

Definition 3 (Semantic Security). *The correct probability of an adversary A guessing coin b is an advantage of its authentication scheme Semantic Security (AKE):*

$$Adv_C^{AKE} = |2Pr\,[Succ(A)] - 1| = |2Pr\,[b = b'] - 1|.$$

Definition 4. *Attacker A has the following equation for the ECDLP problem within time t_A:*

$$Adv_A^{ECDLP}(t_A) \le \varepsilon,\ \varepsilon > 0$$

ε is the advantage of A for the semantic safety of the ECDLP problem within time t_A.

Theorem 2. *The LAPEC protocol has session key backward security.*

Let A be a polynomial-time adversary whose running time upper limit is t_A. In order to destroy the backward security of the protocol, A can perform at most Hash Oracle queries,

Send queries and Execute queries q_H, q_S, q_E times, and session-key queries, respectively. Then, for A, we have:

$$Adv_C^{PCS} \le 2q_H/2^{lH} + 10q_S/2^{lr} + 4q_S Adv_A^{ECDLP}(t_A)$$

Proof. Game 0, Game 1, Game 2, Game 3, Game 4, Game 5 are a defined set of games, and $Succ_i$ is the probability of correctly guessing coin b in Game i. $\square$

Game 0: Assume that Game 0 is the same as the actual scheme in the random oracle, with:

$$Adv_C^{PCS} = |2Pr[Succ_0] - 1|$$

Game 1: Query the oracle in Game 1. Since Game 0 and Game 1 are indistinguishable, there are:

$$Pr[Succ_0] = Pr[Succ_1]$$

Game 2: Game 2 considers that the Hash function collides with the key update message. According to the birthday paradox, the probability of Hash query collision is at most $q_H/2^{lH}$, so there are:

$$|Pr[Succ2] - Pr[Succ_1]| \le q_H/2^{lH}$$

Game 3: The adversary tries to query the oracle machine to guess the random number directly from the message. The probability of guessing the random number will not exceed $2q_S/2^{lr}$. Therefore, there are:

$$|Pr[Succ3] - Pr[Succ_2]| \le 2q_S/2^{lr}$$

Game 4: Adversary A will guess attack by asking about corruption.

C1: Adversary A attempts to use the advantages of ECDLP to crack the session key after updating without the participation of the oracle H. Since two random numbers are required for ECDH exchange during the process of updating the session key, the probability will not exceed $2q_S Adv_A^{ECDLP}(t_A)$.

C2: Since random number participation and a zero-knowledge proof are required in the process of updating the session key, and the parameter guessing of zero-knowledge proof is similar to random number guessing, the probability will not exceed $3q_S/2^{lr}$.

In summary, we can get:

$$|Pr[Succ_4] - Pr[Succ_3]| \le 3q_S/2^{lr} + 2q_S Adv_A^{ECDLP}(t_A)$$

After completing the game, adversary A has no more advantage in guessing b, so there is:

$$Pr[Succ4] = 1/2$$

From the triangle inequality, we can get:

$$|Pr[Succ_0] - 1/2| = |Pr[Succ_4]\text{-}Pr[Succ_1]| \le qH/2lH + 5qS/2lr + 2qSAdvAECDLP(tA)$$

Thus:

$$Adv_C^{PCS} \le 2q_H/2^{lH} + 10q_S/2^{lr} + 4q_S Adv_A^{ECDLP}(t_A)$$

The theorem is proved.

5. Time Cost Analysis

5.1. Computation Cost Analysis

In the process of identity authentication and key negotiation, the main overhead is concentrated on the encryption and decryption calculation, key storage, and message interaction of the cryptosystem. In terms of time overhead, related primitive operations and communication overhead are mainly considered [36–39]. The primitive operation and time overhead of the authentication protocol based on ECC are shown in Table 4:

Table 4. Computation cost of ECC-based schemes.

Scheme	Time Cost			Total
	User	GWN	UAV	
Xu [22]	-	-	-	$9T_H + 4T_{SM}$
Wu F [23]	$2T_{SM} + 13T_H$	$14T_H$	$2T_{SM} + 4T_H$	$22T_H$
Jiang [24]	$8T_H + 2T_{SM}$	$9T_H + T_{SM}$	$6T_H$	$23T_H + 3T_{SM}$
Li X [25]	$8T_H + 3T_{SM}$	$7T_H + T_{SM}$	$4T_H + 2T_{SM}$	$19T_H + 6T_{SM}$
Li X [26]	$2T_{SM} + 8T_H$	$T_{SM} + 9T_H$	$4T_H$	$3T_{SM} + 21T_H$
Chang [27]	$11T_H + 5T_{SM}$	$10T_H + 4T_{SM}$	$4T_H + T_{SM}$	$25T_H + 10T_{SM}$
Lu [28]	-	-	-	$6T_{SM} + 13T_H + 4T_S$
Saeed [29]	$T_{SM} + 3T_H + 2T_S$	$2T_{SM} + 3T_H$	$2T_{SM} + 3T_H + 2T_S$	$5T_{SM} + 9T_H + 4T_S$
Bander [30]	$3T_{SM} + 6T_H + 3T_S$	$T_{SM} + 9T_H + 7T_S$	$2T_{SM} + 5T_H + 3T_S$	$6T_{SM} + 21T_H + 13T_S$
Deebak [31]	-	-	-	$19T_H + 12T_{EX}$
LAPEC	-	$3T_{SM} + 6T_H + 2T_S$	$3T_{SM} + 6T_H + 2T_S$	$6T_{SM} + 12T_H + 4T_S$

In the table, User represents the user of the UAV, while GWN and UAV represent the ground control station (gateway) and the UAV, respectively. T_{SM} represents the overhead of the ECC scalar multiply operation, T_A represents the overhead of the point-add operation, T_H represents the overhead of the hash operation, T_S represents the overhead of symmetric encryption/decryption, and T_{EX} represents the exponential function to execute the computational complexity.

In terms of communication overhead, the LAPEC protocol only needs to perform the interaction in the pre-registration phase when the LAPEC protocol is connected for the first time and it is quite small. The pre-registration phase only performs $2T_H$ which costs almost 10% of the authentication phase computation cost ($6T_{SM} + 12T_H + 4T_S$). After the second connection, only the overhead of the authentication phase and the session key update phase is considered.

- For the authentication phase:

In order to facilitate the time cost comparison without the hardware platform, refer to the experimental results of Roy et al. [32]. The overhead of hash operations and symmetric encryption and decryption operations is about 8% and 14% of elliptic curve scalar multiplication operations. As it is shown in Table 4, LAPEC has a computational overhead similar to most schemes in the authentication phase (for example, schemes such as Lu [28], Bander [30], Deebak [31], etc.). However, the computation cost of LAPEC is a little higher than the scheme of Saeed [29]. What is more, LAPEC is better than some ECC-based schemes.

- For the session key update phase:

Since some schemes do not design corresponding key update steps, this paper uses the default key Diffie–Hellman exchange for comparison.

As we can see, LAPEC needs to complete the zero-knowledge proof in the key update phase, so one more scalar multiplication operation T_M is required. We perform a zero-knowledge proof session key update phase after five traditional update processes. In this update method, the phase only increases the computational overhead by about 8% but still maintains backward security.

5.2. Communication Cost Analysis

The EDHOC protocol has great advantages in the number of message exchanges (3 messages) and the computational overhead in the authentication negotiation stage. Therefore, we mainly compare the LAPEC protocol with the EDHOC protocol to analyze the performance overhead. In terms of communication overhead, the protocol is divided into the following three stages for analysis:

- Message Interaction Cost

Messaging cost refers to the number of message interactions and the latency of the communication channel. In fact, the channel delay occupies a large overhead in the authentication protocol.

Since both EDHOC and LAPEC conduct authentication negotiation through three messages, it can be considered that the message channel delay and the number of interactions are the same. Similarly, the session key update phase does not add new message interactions, so LAPEC's update phase is also the same as EDHOC.

For the pre-registration phase, LAPEC adds two messages. However, as mentioned in the previous section, the pre-registration phase is only performed on the first connection, and the subsequent authentication and update phases have significantly more messages than the pre-registration phase.

- Message Size Cost

In the pre-registration stage, the LAPEC protocol needs to send two pre-registration messages; the message sizes are 32 bytes, respectively.

In the authentication negotiation stage, the LAPEC protocol needs to send three authentication negotiation messages; the message sizes are 36, 65, and 128.

In the session key update phase, the LAPEC protocol needs to send two session key update messages; the message sizes are 64 and 32, respectively. Meanwhile, the EDHOC protocol needs to send two session key update messages. The message sizes are 32, respectively.

Assuming that the network bandwidth is the same as M, the analysis results are shown in Table 5.

Table 5. Message size cost of EDHOC AND LAPEC.

Phase	LAPEC (Bytes)	EDHOC (Bytes)
Pre-registration	32 + 32	0
Authentication	36 + 65 + 128	38 + 66 + 129
Key Update	64 + 32	32 + 32

From Table 5, LAPEC adds message overhead in the pre-authentication phase, but it only needs to be considered when connecting for the first time, and it is only a small part of the overall connection process (in the experiment, less than 10%).

For the key update phase, it increases the message size by about 50%. However, it is only about 14% compared to the authentication phase messages. Considering that most of the actual overhead is the channel delay of message exchange, these increases are acceptable as long as the number of message exchanges during the update phase is guaranteed to be equal.

At the same time, it can be seen that in the process of protocol implementation, the number of public key operations such as elliptic curve scalar multiplication between the two parties should be minimized, and the number of message exchanges should be controlled.

6. Conclusions

This paper proposed an ECC-based identity authentication protocol LAPEC for UAVs. We introduced the interaction process of the LAPEC protocol in detail, and we proved that it has session key backward security. In the end, we compared the LAPEC protocol with other authentication protocols and found that the time overhead of the LAPEC protocol is small. However, due to the need to increase the backward security in the key update phase, the time overhead in the session key update phase only increased by about 8%. Since the pre-authentication phase is only required when connecting for the first time, the extra overhead added to the pre-authentication phase was only about 10% of the entire authentication process.

In the future, we will continue to optimize the LAPEC protocol and apply it in multiple scenarios such as the authentication between UAV–UAV communications.

Author Contributions: Conceptualization, S.Z. and Z.H.; methodology, S.Z.; validation, S.Z., Z.H., and Z.Y.; formal analysis, S.Z.; writing—original draft preparation, S.Z. and Z.H.; writing—review and editing, S.Z.; supervision, Y.L. All authors have read and agreed to the published version of the manuscript.

Funding: This research was supported by the Major Key Project of PCL (Grant No. PCL2022A03, PCL2021A02, PCL2021A09) and the Key-Area Research and Development Program of Guangdong Province (Grant No. 2019B010137005).

Institutional Review Board Statement: Not applicable.

Informed Consent Statement: Not applicable.

Data Availability Statement: The data presented in this paper will be made available on request via the author's email with appropriate justification.

Conflicts of Interest: The authors declare no conflict of interest.

References

1. Mozaffari, M.; Saad, W.; Bennis, M.; Nam, Y.-H.; Debbah, M. A tutorial on UAVs for wireless networks: Applications, challenges, and open problems. *IEEE Commun. Surv. Tutor.* **2018**, *21*, 2334–2360. [CrossRef]
2. Hayat, S.; Yanmaz, E.; Muzaffar, R. Survey on Unmanned Aerial Vehicle Networks for Civil Applications: A Communications Viewpoint. *IEEE Commun. Surv. Tutor.* **2016**, *18*, 2624–2661. [CrossRef]
3. Motlagh, N.H.; Taleb, T.; Arouk, O. Low-Altitude Unmanned Aerial Vehicles-Based Internet of Things Services: Comprehensive Survey and Future Perspectives. *IEEE Internet Things J.* **2016**, *3*, 899–922. [CrossRef]
4. Jangirala, S.; Das, A.K.; Kumar, N.; Rodrigues, J. Tcalas: Temporal credential-based anonymous lightweight authentication scheme for internet of drones environment. *IEEE Trans. Veh. Technol.* **2019**, *68*, 6903–6916.
5. Li, B.; Fei, Z.; Zhang, Y.; Guizani, M. Secure UAV Communication Networks over 5G. *IEEE Wirel Commun.* **2019**, *26*, 114–120. [CrossRef]
6. Gaurang, B.; Naren, N.; Vinay, C.; Biplab, S. SHOTS: Scalable Secure Authentication-Attestation Protocol Using Optimal Trajectory in UAV Swarms. *IEEE Trans. Veh. Technol.* **2022**, *71*, 5827–5836.
7. Kaufman, C.; Hoffman, P.; Nir, Y.; Eronen, P.; Kivinen, T. *RFC 7296: Internet Key Exchange Protocol Version 2 (IKEv2)*; RFC Editor; IETF: Fremont, CA, USA, 2014.
8. Rescorla, E. *RFC 8446: The Transport Layer Security (TLS) Protocol Version 1.3*; RFC Editor; IETF: Fremont, CA, USA, 2018.
9. Zhong, C.; Yao, J.; Xu, J. Secure uav communication with cooperative jamming and trajectory control. *IEEE Commun. Lett.* **2018**, *23*, 286–289. [CrossRef]
10. Zeng, Y.; Zhang, R. Energy-efficient uav communication with trajectory optimization. *IEEE Trans. Wirel. Commun.* **2017**, *16*, 3747–3760. [CrossRef]
11. Grover, A.; Berghel, H. A survey of RFID deployment and security issues. *Inf. Process. Syst.* **2011**, *7*, 561–580. [CrossRef]
12. Gope, P.; Sikdar, B. An efficient privacy-preserving authenticated key agreement scheme for edge-assisted internet of drones. *IEEE Trans. Veh. Technol.* **2020**, *69*, 13621–13630. [CrossRef]
13. Gope, P.; Millwood, O.; Saxena, N. A provably secure authentication scheme for RFID-enabled UAV applications. *Comput. Commun.* **2021**, *166*, 19–25. [CrossRef]
14. Khattab, A.; Jeddi, Z.; Amini, E.; Bayoumi, M. *RFID Security Threats and Basic Solutions*; Springer International Publishing: Cham, Switzerland, 2017; pp. 27–41.
15. Lopez, P.P.; Hernandez-Castro, J.C.; Estevez-Tapiador, J.M.; Ribagorda, A. *RFID Systems: A Survey on Security Threats and Proposed Solutions*; Springer: Berlin/Heidelberg, Germany, 2006; pp. 159–170.
16. Suh, G.; Devadas, S. Physical unclonable functions for device authentication and secret key generation. In Proceedings of the Design Automation Conference (DAC '07), San Diego, CA, USA, 4–6 June 2007.
17. Sung, J.Y.; Ashok, K.D.; Youngho, P.; Pascal, L. SLAP-IoD: Secure and lightweight authentication protocol using physical unclonable functions for internet of drones in smart city environments. *IEEE Trans. Veh. Technol.* **2022**, *71*, 10374–10388.
18. Bansal, G.; Sikdar, B. S-MAPS: Scalable Mutual Authentication Protocol for Dynamic UAV Swarms. *IEEE Trans. Veh. Technol.* **2021**, *70*, 12088–12100. [CrossRef]
19. Wazid, M.; Das, A.K.; Kumar, N.; Vasilakos, A.V.; Rodrigues, J.J. Design and analysis of secure lightweight remote user authentication and key agreement scheme in internet of drones deployment. *IEEE Internet Things J.* **2018**, *6*, 3572–3584. [CrossRef]
20. Ever, Y.K. A secure authentication scheme framework for mobile-sinks used in the Internet of Drones applications. *Comput. Commun.* **2020**, *155*, 143–149. [CrossRef]

21. Tao, X.; Jun, H. An Identity Authentication Scheme Based on SM2 Algorithm in UAV Communication Network. *Wirel. Commun. Mob. Comput.* **2022**, *4*, 1–10.
22. Lin, L.; Xiao, F.L.; Yu, L.W.; Tan, L. CSECMAS: An Efficient and Secure Certificate Signing Based Elliptic Curve Multiple Authentication Scheme for Drone Communication Networks. *Appl. Sci.* **2022**, *12*, 9203. [CrossRef]
23. Hankerson, D.; Vanstone, S.; Menezes, A.J. *Guide to Elliptic Curve Cryptography*; Springer Science & Business Media: Berlin/Heidelberg, Germany, 2006.
24. Cohn-Gordon, K.; Cremers, C.; Garratt, L. On post-compromise security. In Proceedings of the 2016 IEEE 29th Computer Security Foundations Symposium (CSF), Lisboa, Portugal, 27 June–1 July 2016.
25. He, Y.X.; Sun, F.J.; Li, Q.A.; He, J.; Wang, L.M. A survey on public key mechanism in wireless sensor networks. *Jisuanji Xuebao/Chin. J. Comput.* **2020**, *43*, 381–408.
26. Huang, Z.; Wang, Q. A PUF-based unified identity verification framework for secure IoT hardware via device authentication. *World Wide Web* **2020**, *23*, 1057–1088. [CrossRef]
27. Li, X.; Liu, J.; Ding, B.; Li, Z.; Wu, H.; Wang, T. A SDR-based verification platform for 802.11 PHY layer security authentication. *World Wide Web* **2020**, *23*, 1011–1034. [CrossRef]
28. Shao, S.; Chen, F.; Xiao, X.; Gu, W.; Lu, Y.; Wang, S.; Tang, W.; Liu, S.; Wu, F.; He, J.; et al. IBE-BCIOT: An IBE based cross-chain communication mechanism of blockchain in IoT. *World Wide Web* **2021**, *24*, 1665–1690. [CrossRef]
29. Xu, X.; Zhu, P.; Wen, Q.; Jin, Z.; Zhang, H.; He, L. A secure and efficient authentication and key agreement scheme based on ECC for telecare medicine information systems. *J. Med. Syst.* **2014**, *38*, 1–7. [CrossRef]
30. Wu, F.; Li, X.; Sangaiah, A.K.; Xu, L.; Kumari, S.; Wu, L.; Shen, J. A lightweight and robust two-factor authentication scheme for personalized healthcare systems using wireless medical sensor networks. *Future Gener. Comput. Syst.* **2018**, *82*, 727–737. [CrossRef]
31. Jiang, Q.; Ma, J.; Wei, F.; Tian, Y.; Shen, J.; Yang, Y. An untraceable temporal-credential-based two-factor authentication scheme using ECC for wireless sensor networks. *J. Netw. Comput. Appl.* **2016**, *76*, 37–48. [CrossRef]
32. Li, X.; Niu, J.; Bhuiyan, M.Z.A.; Wu, F.; Karuppiah, M.; Kumari, S. A robust ECC-based provable secure authentication protocol with privacy preserving for industrial Internet of Things. *IEEE Trans. Industr. Inform.* **2018**, *14*, 3599–3609. [CrossRef]
33. Li, X.; Niu, J.; Kumari, S.; Wu, F.; Sangaiah, A.K.; Choo, K.-K.R. A three-factor anonymous authentication scheme for wireless sensor networks in IoT environments. *J. Netw. Comput. Appl.* **2018**, *103*, 194–204. [CrossRef]
34. Chang, I.P.; Lee, T.F.; Lin, T.H.; Liu, C.M. Enhanced two-factor authentication and key agreement using dynamic identities in wireless sensor networks. *Sensors* **2015**, *15*, 29841–29854. [CrossRef] [PubMed]
35. Lu, Y.R.; Xu, G.Q.; Li, L.X.; Yang, Y. Anonymous three-factor authenticated key agreement for wireless sensor networks. *Wirel. Netw.* **2019**, *25*, 1461–1475. [CrossRef]
36. Chatterjee, S.; Roy, S.; Das, A.K.; Chattopadhyay, S.; Kumar, N.; Vasilakos, A.V. Secure biometric-based authentication scheme using chebyshev chaotic map for multi-server environment. *IEEE Trans. Dependable Secur. Comput.* **2018**, *15*, 824–839. [CrossRef]
37. Saeed, U.J.; Irshad, A.A.; Fahad, A.; Adnan, S.K. A Verifiably Secure ECC Based Authentication Scheme for Securing IoD Using FANET. *IEEE Access* **2022**, *10*, 95321–95343.
38. Bander, A.A.; Ahmed, B.; Shehzad, A.C. A Resource-Friendly Authentication Protocol for UAV-Based Massive Crowd Management Systems. *Secur. Commun. Netw.* **2021**, *2021*, 3437373. [CrossRef]
39. Deebak, B.D.; Al-Turjman, F. A smart lightweight privacy preservation scheme for IoT-based UAV communication systems. *Comput. Commun.* **2020**, *162*, 102–117. [CrossRef]

Article

Connectivity-Maintenance UAV Formation Control in Complex Environment

Liangbin Zhu [1], Cheng Ma [2,*], Jinglei Li [2], Yue Lu [2] and Qinghai Yang [2]

[1] School of Information and Electronics, Beijing Institute of Technology, Beijing 100081, China; 3420205018@bit.edu.cn
[2] School of Telecommunications Engineering, Xidian University, Xi'an 710071, China
* Correspondence: chengma@stu.xidian.edu.cn

Abstract: Cooperative formation control is the research basis for various tasks in the multi-UAV network. However, in a complex environment with different interference sources and obstacles, it is difficult for multiple UAVs to maintain their connectivity while avoiding obstacles. In this paper, a Connectivity-Maintenance UAV Formation Control (CMUFC) algorithm is proposed to help multi-UAV networks maintain their communication connectivity by changing the formation topology adaptively under interference and reconstructing the broken communication topology of a multi-UAV network. Furthermore, through the speed-based artificial potential field (SAPF), this algorithm helps the multi-UAV formation to avoid various obstacles. Simulation results verify that the CMUFC algorithm is capable of forming, maintaining, and reconstructing multi-UAV formation in complex environments.

Keywords: multi-UAV network; connectivity maintenance; formation control; complex environment

1. Introduction

With the rapid development of UAV technology, the multi-UAV network is widely used in civil and military fields such as disaster rescue, air reconnaissance, etc. [1–5]. The communication topology of the multi-UAV network affects its work efficiency, and its connectivity can be maintained by controlling the topology of the multi-UAV formation. Therefore, it is fundamental to carry out research on the formation topology control of multi-UAVs. During the actual flight, the limited communication range of UAVs and different environmental factors, such as obstacles and interference sources, will affect the connectivity of the multi-UAV network. Thus, it is necessary to maintain connectivity of the entire multi-UAV network by controlling multi-UAV formations in complex environments [6].

Scholars at home and abroad have conducted many studies on connectivity maintenance between UAVs. An UAV formation control law was proposed to generate a leader–follower structure based on consistency under the balance of control constraints and communication constraints, so as to avoid collisions and maintain connectivity between UAVs [7]. The authors in [8] proposed using the graph coalition formation game to model the cooperation between UAVs, which can quickly restore the required connectivity between UAV networks. In [9], the connectivity methods were compared in four application scenarios, mainly by increasing or decreasing the communication links between UAVs to increase or decrease the connectivity of UAV clusters. A connectivity tracking algorithm was proposed to track the connectivity distribution over time, and the results are analyzed. The authors in [10] used the second-order integral characteristic to solve the time-varying formation tracking control problem of multiple UAVs. We consider the correspondence between multi-UAV connectivity and formation control and maintain the connectivity of multi-UAV networks through formation control in complex environments. These papers also consider the problem of UAV formation flight

Citation: Zhu, L.; Ma, C.; Li, J.; Lu, Y.; Yang, Q. Connectivity-Maintenance UAV Formation Control in Complex Environment. *Drones* **2023**, *7*, 229. https://doi.org/10.3390/drones7040229

Academic Editors: Zhihong Liu, Shihao Yan, Kehao Wang and Yirui Cong

Received: 28 February 2023
Revised: 20 March 2023
Accepted: 23 March 2023
Published: 26 March 2023

in the case of limited communication. The authors in [11] studied the formation control problem of multiple agents in the noise environment and transformed the formation control problem into the convergence problem of the infinite product of general random matrix sequences. A flight strategy was proposed to improve the multi-UAV cooperative search ability under the condition of limited resources. A multi-UAV cooperative search model was established. The optimization function of the model considers communication cost and formation benefit to ensure multi-UAV Effectiveness of Human–Machine Search [12]. A new adaptive formation control method was proposed for UAVs with limited leader information and communication. The method was extended to replace the leader with adjacent UAVs, where the leader can convey location and direction information [13].

In addition, the formation obstacle avoidance problem of UAVs needs to be considered in the process of formation flight. The aim is the formation and maintenance of a specific configuration to adapt to mission requirements and friendly aircrafts. Currently widely used strategies include the leader–follower method [14], virtual structure method [15], behavior-based control method [16], and the consensus algorithm [17]. Among them, the algorithm based on consensus theory emphasizes the synchronization, cooperation, and substitutability among individuals. This algorithm meets the characteristics of decentralization, autonomy, and autonomy of UAVs; it thus gradually became the main method and research direction to solve in the formation control of UAVs. In addition, the obstacle avoidance problem of UAVs needs to be considered in the process of formation flight. The artificial potential field (APF) algorithm proposed by Khatib [18] in 1986 stands out among many obstacle avoidance algorithms because of its simple structure, easy real-time control, and rapid response to environmental changes. An observer-based memory consensus protocol was proposed in [19] for achieving the consensus of nonlinear multi-agent systems with Markov switching topologies. This approach was applicable for an observer-based nonlinear multi-agent system which was described by switched undirected topologies. In [20], the authors solved the consensus problem in multi-agent systems with Markov jump, time-varying delay, and uncertainties. In [21], the authors developed a consistent algorithm to decompose the motion of UAV into three directions, but the constraint processing of instructions in the algorithm convergence process is too cumbersome, which is not conducive to engineering implementation. The authors in [22] introduced a particle swarm optimization algorithm to deal with static and dynamic obstacles. They added UAV formation configuration requirements to the consensus algorithm. An adaptive distributed control algorithm was proposed to realize the problem of cooperative formation of heterogeneous vertical take-off and landing UAVs under the condition of parameter uncertainty in [23]. In [24], the authors developed a novel decentralized adaptive consensus formation control method. Each UAV sets a coordinate and controls its relative position with adjacent UAVs to obtain the desired formation. A multi-UAV formation system based on the leader–follower model was proposed in [25]. The follower predicts the state of the leader, maintains a relative position in the formation, and finally reaches a consensus with the leader. A topology control algorithm was proposed in [26] to complete the distributed communication maintenance and formation configuration of four quadrotor UAVs. However, the security requirements for the long-running machine in the cluster are very high.

In this paper, aiming at connectivity maintenance of a multi-UAV network and obstacle avoidance of multi-UAV formation, we design a formation control algorithm to overcome the connectivity maintenance and obstacle avoidance problem. The main challenge is to design an excellent formation control algorithm to ensure the connectivity and security of the multi-UAV network during the actual flight due to the limited communication range of UAVs and the existence of different environmental factors, such as obstacles and interference sources. Specifically, the formation switching of the multi-UAV network or the failure of some communication networks will cause the system connectivity to be destroyed. Therefore, the designed formation control algorithm requires the ability to maintain the system connectivity. At the same time, considering the flying speed of a UAV, the designed algorithm requires timely and safe obstacle avoidance. Thus, it is quite necessary to maintain the connectivity of an entire multi-UAV network by controlling multi-UAV formations in complex environments.

This paper proposes a Connectivity-Maintenance UAV Formation Control algorithm, called CMUFC, for the multi-UAV network to complete tasks in complex environments with various obstacles and interference sources. This algorithm considers the communication range and kinematics constraints of UAVs and overcomes the problem of maintaining connectivity when a multi-UAV network is disturbed and avoids obstacles. This paper has the following contributions:

- The CMUFC algorithm maintains the connectivity of the multi-UAV network through adaptive scaling formation, and the UAV changes its relative position with other UAVs to maintain the stability of the entire system in the case of interference.
- A speed-based artificial potential field (SAPF) algorithm, which helps UAVs avoid obstacles safely in the process of rapid flight, is proposed. Combined with the SAPF and the consensus formation control algorithm, it overcomes the problem of local minimum and solves the problem that APF cannot make UAVs tend to the specified formation.
- Aiming at the situation that the formation of a multi-UAV network is forced to change in order to avoid obstacles, a recursive self-repairing formation algorithm based on layering is used to enable the multi-UAV to complete the formation reconstruction and maintain the connectivity of the multi-UAV network.

The rest of this paper is organized as follows. Section 2 describes the system model. Section 3 introduces the Connectivity-Maintenance UAV Formation Control algorithm. Section 4 verifies and analyzes our algorithm. Finally, concluding remarks are provided in Section 5.

2. System Model

As shown in Figure 1, this paper considers the formation control problem of multi-UAV connectivity maintenance in complex environments where there are K interference sources of different interference powers and O obstacles of different sizes. In the considered multi-UAV scenario, we construct the model in a 3D Cartesian coordinate system. Among them, M UAVs are modeled as discs with a radius l_{min}. Let $p_i^{uav}(t) = [p_{ix}^{uav}(t), p_{iy}^{uav}(t), H]$, $i \in [1, 2, \ldots, M]$, $t \in [1, 2, \ldots, T_i]$ denote the 3D position of the UAV, where H is the altitude of the UAV,which is assumed to be fixed; T_i denotes the time for UAV i to complete its mission. The o-th obstacle is modeled as a disk with radius r_o, $o \in [1, 2, \ldots, O]$, and its position is $p_o^{obs}(t) = [p_{ox}^{obs}(t), p_{oy}^{obs}(t)]$. The position of the interference source k is $p_k^{int}(t) = [p_{kx}^{int}(t), p_{ky}^{int}(t)]$, and its transmission power is P_k^{int}, $k \in [1, 2, \ldots, K]$. The target location of the multi-UAV network is $p^{tar} = [p_x^{tar}, p_y^{tar}]$.

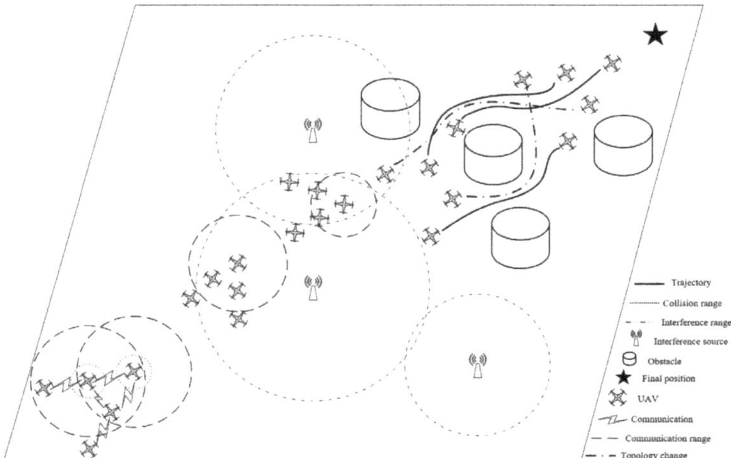

Figure 1. System model.

2.1. Formation Control Model

When multiple UAVs form a formation and fly jointly, they must maintain a fixed geometric shape with each other. At the same time, they must meet mission requirements and adapt to surrounding environmental constraints, such as obstacle avoidance. This paper adopts the virtual structure method. That is, a virtual structure is established in the formation, and each UAV only needs to follow a certain point or a certain edge in the virtual structure to realize formation control. The task of the multi-UAV network we consider is to maintain communication connectivity and generate the formation of UAVs while avoiding obstacles. The dynamic system of UAV can be abstracted as a double integral dynamic system

$$\dot{p}_i = v_i, \ \dot{v}_i = u_i, i = 1, \ldots, M \tag{1}$$

where $\dot{p}_i$ and $\dot{v}_i$ represent the derivatives of p_i^{uav} and v_i, respectively. $u_i \in R^m$ is the acceleration and control inputs for UAV i. The control input u_i helps UAVs form a designated formation. According to the double-integral dynamic system in Equation (1), the consensus method for a multi-UAV network consisting of M UAVs is expressed as

$$u_i = -\sum_{j=1}^{M} a_{ij}(t)\left[(p_i^{uav} - p_j^{uav}) + \gamma(t)(v_i - v_j)\right], \ i = 1, 2, \ldots, M \tag{2}$$

where $\gamma(t)$ is a positive number and a_{ij} is the (i, j)-th term in the Laplacian matrix of an undirected graph G_M. The consensus formation control algorithm of a double integral dynamic system makes the relative position between UAVs tend to the set value by controlling the input u_i, so as to form the formation of multiple UAVs. In addition, the speed and acceleration of the UAV must be less than its maximum limit

$$v_i \leq v_{\max}, u_i \leq a_{\max} \tag{3}$$

where $v_i, a_{\max}$ are the maximum speed and maximum acceleration of the UAV, respectively.

2.2. Communication Model

2.2.1. Topology Model

In terms of multi-UAV formation control, directed graphs and undirected graphs have different effects on the stability, convergence speed, and robustness of formation control. In general, directed graphs require more complex control algorithms and coordination strategies. In order to reduce the impact of communication delay and other factors on the multi-UAV network topology control, this paper considers two-way communication between UAVs to transmit information such as position and speed. Thus, the topology of the multi-UAV network is represented by an undirected graph $G_M \equiv (Q_M, E_M, W_M)$, where $Q_M = \{1, 2, \ldots, M\}$ denotes a non-empty finite set of UAVs. $E_M \subseteq Q_M \times Q_M$ is the edge set of the communication links connecting two UAVs. If there is a reachable communication link between UAV i and UAV j, it means that there is an edge E_{ij} in the undirected graph G_M, and Q_i can obtain the information consisting of position and speed from Q_j. $W_M \subseteq Q_M \times Q_M$ represents the weight matrix of communication links between UAVs in the network topology, and we think that the communication between UAVs is symmetric, i.e., $W_{ij} = W_{ji}, \forall E_{ij}$. Specifically, W_M is described as the communication quality matrix, where W_{ij} represents the communication weight between UAV i and UAV j, which is related to the communication distance between two UAVs. An undirected graph is connected if there is an undirected path between any two different UAVs in the undirected graph G_M.

Figure 2 shows the correspondence between the formation structure and communication topology of the multi-UAV network. By controlling the relative position between two UAVs, the distance between them satisfies the communication requirements, $p_1 - p_2 \leq R_c$, $E_{12} = 1$. That is, the multi-unmanned systems maintain connectivity.

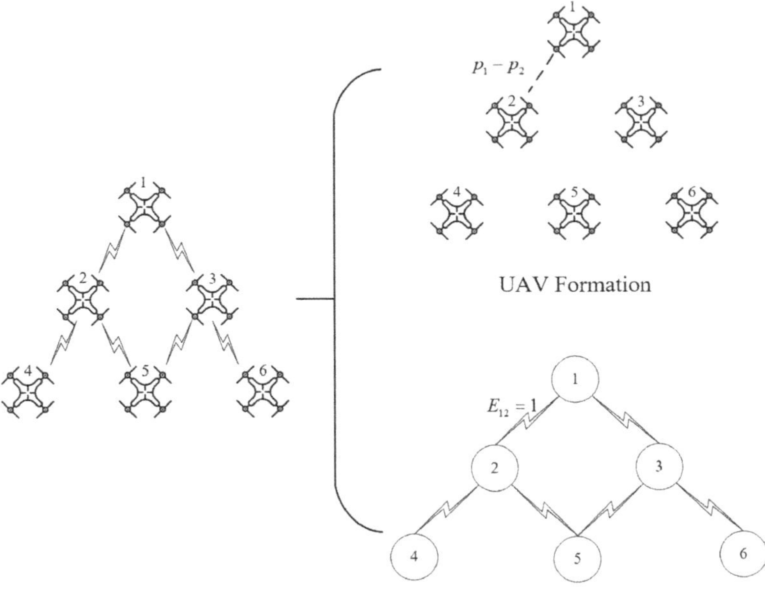

Figure 2. Multi-UAV network.

As shown in Figure 3, the network topologies considered in this paper include string type, ring type, tree type, and star type. There is at least one undirected path between every two UAVs in the multi-UAV network to ensure the connectivity of the system.

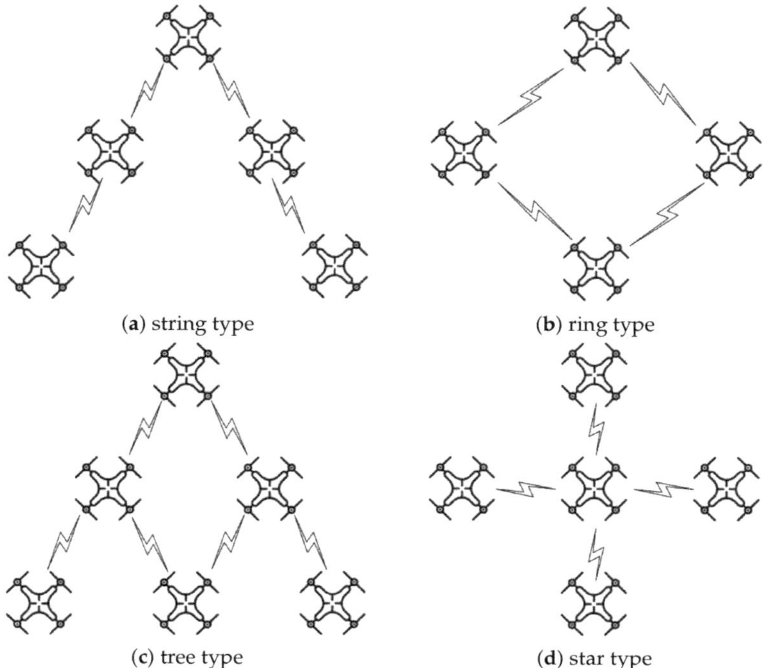

(**a**) string type   (**b**) ring type

(**c**) tree type   (**d**) star type

Figure 3. Communication topology.

2.2.2. Channel Model

1. UAV–UAV Channel Model

   In this paper, the communication between UAVs in the time-varying channel is considered, and only the average path loss is considered. The power of the signal transmitted by UAV i to UAV j is expressed as

$$P_j = \left(\frac{\lambda}{4\pi d_{ij}}\right)^{\alpha} \frac{G_i G_j}{L_m} P_i \tag{4}$$

   where G_i is the transmitting antenna gain of UAV i, G_j is the receiving antenna gain of UAV j, λ is the wavelength, d_{ij} indicates the distance between UAV i and UAV j, α denotes the average path loss constant, L_m is the loss factor, and P_i is the signal transmission power of UAV i.

2. UAV–Interference source Channel Model

   The transmission scenario in an urban area is considered, where the elevation angle-dependent probability LoS channel model is considered between the UAV and the ground interference source [27]. The instantaneous interference from ground interference source k to UAV i is as follows

$$P_{i,k} = P_k(P_{LoS}(\theta)\beta_0 d_{i,k}^{-\alpha} + (1 - P_{LoS}(\theta))\kappa\beta_0 d_{i,k}^{-\alpha}) \tag{5}$$

   where $d_{i,k}$ is the distance between UAV i and interference source k, $\beta_0 = (\lambda/4\pi)^2$ is the path loss at a reference distance of $1m$ under LoS conditions, λ is the carrier, $\kappa < 1$ is the additional attenuation factor due to NLoS propagation, and α is the path loss exponent, which is modeled as a monotonically decreasing function of the height H of the UAV. The probability $P_{LoS}(\theta)$ of having a LoS environment is modeled as

$$P_{LoS}(\theta) = \frac{1}{1 + a \cdot \exp(-b(\theta - a))} \tag{6}$$

Among them, a and b are modeling parameters, and θ is the elevation angle from interference source k to UAV i, namely

$$\theta = \arcsin(H/d_{i,k}(t)) \tag{7}$$

where H is the height of the UAV. The probability of an NLoS environment is given by

$$P_{NLoS}(\theta) = 1 - P_{LoS}(\theta) \tag{8}$$

The instantaneous interference received by UAV i from all ground interference sources is

$$P_{i,K} = \sum_{k} P_{i,k} \tag{9}$$

Therefore, the maximum transmission distance R_c between UAVs is expressed as

$$R_c = \frac{\lambda}{4\pi} \left(\frac{P_i}{\sigma^2 + P_{i,K} \cdot 10^{\frac{SINR_{th}}{10}}} \right)^{\frac{1}{\alpha}} \tag{10}$$

where σ^2 is the average power of the noise in the wireless channel and $SINR_{th}$ is the signal to interference plus noise ratio (SINR) threshold. In order to ensure the connectivity of the multi-UAV network, there is at least one undirected link between every two UAVs; the communication between adjacent UAVs in the undirected path needs to meet its maximum transmission distance.

3. Connectivity-Maintenance UAV Formation Control Algorithm

The multi-UAV network maintains connectivity. That is, there is at least one undirected path between every two UAVs, and the communication between adjacent UAVs in this undirected path needs to meet its maximum transmission distance. In areas where there are interference sources, the CMUFC algorithm helps UAVs adaptively change the formation structure to maintain the connectivity of multi-UAV communication topology. In addition, this algorithm combines the SAPF and the consensus formation control algorithm to help the multi-UAV formation to fly to the target position and avoid obstacles, while making the flight distance between the UAVs meet the connectivity requirements.

3.1. Connectivity Maintenance of Multi-UAV Network under Interference

Figure 4 shows the collision zone and communication interaction zone around the UAV, where R_c is the maximum transmission distance of signals between UAVs, l_o is the maximum range of influence of obstacles on the UAV, and $l_{\min}$ is the radius of the UAV. In order to maintain system connectivity, the distance between two adjacent UAVs d_{ij} in the undirected path cannot be greater than the maximum transmission distance R_c. In this paper, the effects of interference sources and obstacles on the connectivity of multi-UAV networks are considered.

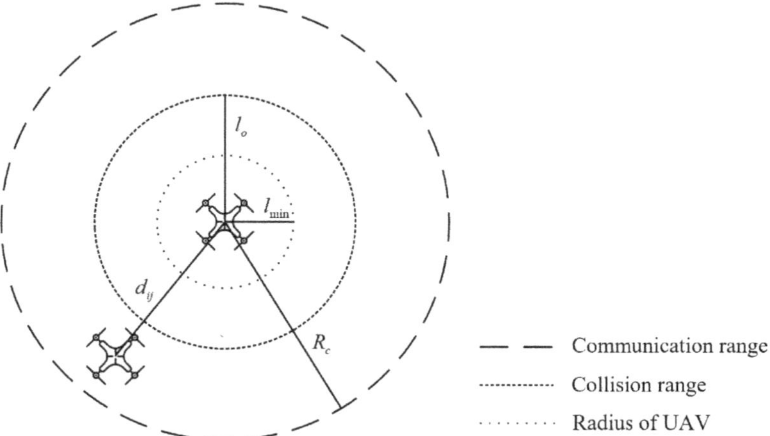

Figure 4. Area classification around UAV.

The situation of the interference of a UAV is shown in Figure 5. When the multi-UAV network is interfered by an interference source, the maximum transmission distance of the UAV signal is reduced. The closer the UAV is to the interference source, the smaller the communication range. This situation reflects the actual UAV formation. That is, the distance between UAVs is scaled adaptively to maintain the connectivity of the system.

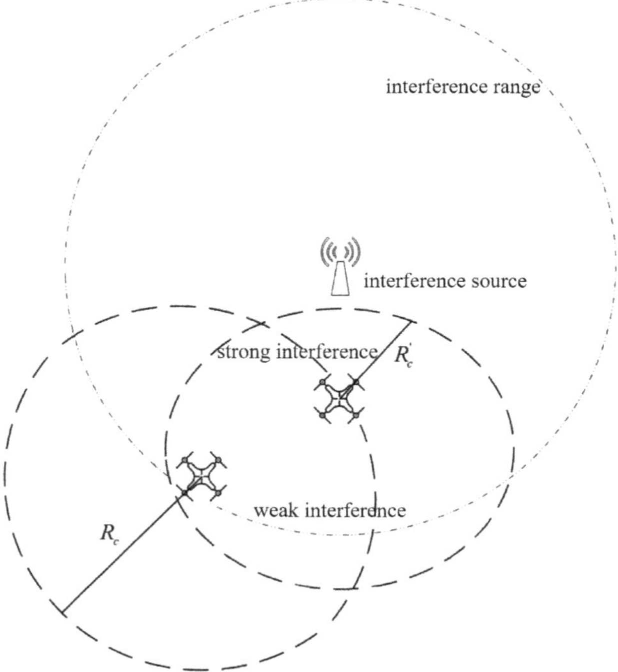

Figure 5. Influence of interference source for UAV communication range.

3.2. Multi-UAV Formation Control under Obstacle Avoidance

In this paper, the SAPF algorithm and consensus control algorithm are combined to help UAV formation avoid obstacles and keep formation. SAPF is used to help UAVs fly to the target and avoid obstacles, and the consensus control algorithm is used to help multiple UAVs form a specified communication topology. The two algorithms are combined to simultaneously ensure that no collision occurs between UAVs and communication interaction can be maintained. That is,

$$l_{\min} \leq d_{ij} \leq R_c \tag{11}$$

The SAPF algorithm establishes an attractive potential field for the target and a repulsive potential field for the obstacle. The two potential fields are combined to avoid the collision between the UAV and the obstacle in the process of flying to the target position. The attractive and repulsive potential fields are expressed as

$$U_{att}(p) = \frac{1}{2}k_{att} \cdot l^2(p^{uav}, p^{tar}) \tag{12}$$

$$U_{rep}(p) = \begin{cases} \frac{1}{2}k_{rep}\left(\dfrac{1}{l(p^{uav}, p^{obs})} - \dfrac{1}{l_o}\right)^2, l(p^{uav}, p^{obs}) \leq l_o \\ 0, l(p^{uav}, p^{obs}) > l_o \end{cases} \tag{13}$$

where k_{att} is the attraction gain factor, k_{rep} is the repulsive force gain coefficient, $l(p^{uav}, p^{tar})$ denotes the vector distance between UAV and target position, $l(p^{uav}, p^{obs})$ is the vector distance between the UAV and the obstacle, i.e., the Euclidean distance between two points. l_o is a constant that represents the maximum range over which the obstacle can affect the UAV. The attractive and repulsive forces are the negative gradients of the attractive and repulsive potential fields, respectively, and the attractive and repulsive force functions are expressed as

$$F_{att}(p) = -\nabla(U_{att}(p)) = -k_{att} \cdot l(p^{uav}, p^{tar}) \tag{14}$$

$$F_{rep}(p) = \begin{cases} k_{rep}\left(\dfrac{1}{l(p^{uav}, p^{obs})} - \dfrac{1}{l_o}\right) \cdot \dfrac{1}{l^2(p^{uav}, p^{obs})} \cdot \dfrac{\partial(l(p^{uav}, p^{obs}))}{\partial(p)}, l(p^{uav}, p^{obs}) \leq l_o \\ 0, l(p^{uav}, p^{obs}) > l_o \end{cases} \tag{15}$$

Then, adding the speed steering force to solve the local minimum problem, the speed steering force is expressed as

$$F^v_{rep} = \begin{cases} k^v_{rep}\left(\dfrac{1}{l(p^{uav}, p^{obs})} - \dfrac{1}{l_o}\right) \cdot v, l(p^{uav}, p^{obs}) \leq l_o \\ 0, l(p^{uav}, p^{obs}) > l_o \end{cases} \tag{16}$$

where k^v_{rep} is the speed repulsion force gain coefficient, v is the speed of the UAV, and the direction of F^v_{rep} is perpendicular to v. Therefore, the resultant repulsive force is expressed as

$$F^{sum}_{rep} = F_{rep}(p) + F^v_{rep} \tag{17}$$

In addition, this paper adopts the formation control mode of the virtual pilot. Then, the consensus algorithm, according to the double integral dynamic system shown in Equation (2), is further expressed as

$$u_i = -\sum_{j=1}^{n} a_{ij}(t)(c_1(p_i^{uav} - p_j^{uav} - \Delta h_{ij}) + c_2(v_i - v_j)) - f_r, \; i = 1, 2, \ldots, n \tag{18}$$

$$f_r = c_3(p_i^{uav} - p_\nabla) + c_4(v_i - v_\nabla) \tag{19}$$

where c_1 and c_3 are stiffness gains, c_2 and c_4 are damping gains, $a_{ij}(t)$ represents the adjacency matrix of each UAV communication topology in a multi-UAV network, and Δh_{ij} is the relative position of UAV i and UAV j. p_∇ and v_∇ are the speed and position of the virtual leader. That is, f_r represents the tracking item of the virtual leader by the UAV.

Therefore, based on the SAPF generated forces derived from each UAV's current position and speed and environmental conditions, combined with the control inputs generated by the consensus control algorithm, the control inputs for UAV i in the multi-UAV network are as follows

$$F_{sum}^i = F_{att}^i + F_{rep}^{sum,i} + u_i \cdot m_i \tag{20}$$

where m_i is the mass of UAV i. In summary, the formation control algorithm of multi-UAV network controls the flight direction and speed of UAV i by controlling the input F_{sum}^i to solve the obstacle avoidance problem of multi-UAV network.

3.3. Formation and Connectivity Restoration of Multi-UAV Network

As shown in Figure 6, UAVs move away from the formation in order to avoid obstacles during flight. UAV 3 loses connection with the formation to avoid obstacles, and UAV 5 restores connectivity to the multi-UAV network as a repair UAV. In this paper, a layer-based recursive self-healing formation algorithm is used for the situation that a multi-UAV network cannot maintain connectivity when UAVs have to stay away from the system in order to avoid obstacles during flight. When the topology of multi-UAV network formation is forced to change, the algorithm can maintain the connectivity of the system network and complete formation reconstruction without changing the network topology relationship of UAVs. The proposed algorithm block diagram is shown in Figure 7.

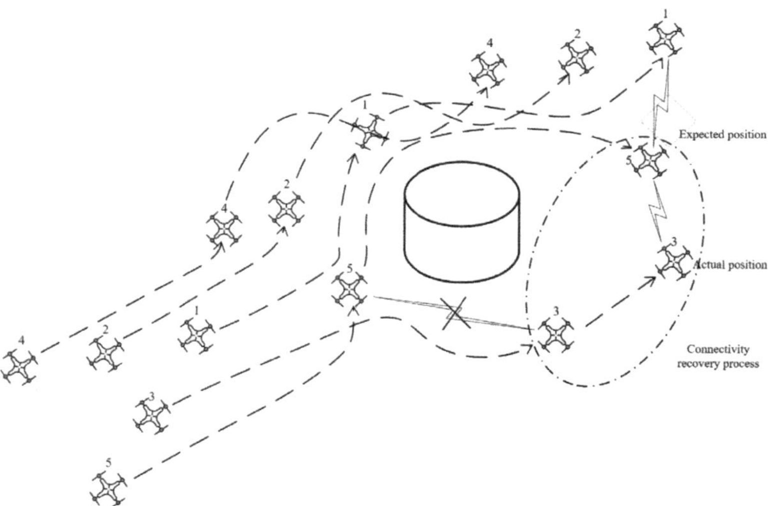

Figure 6. Influence of obstacle for UAV topology.

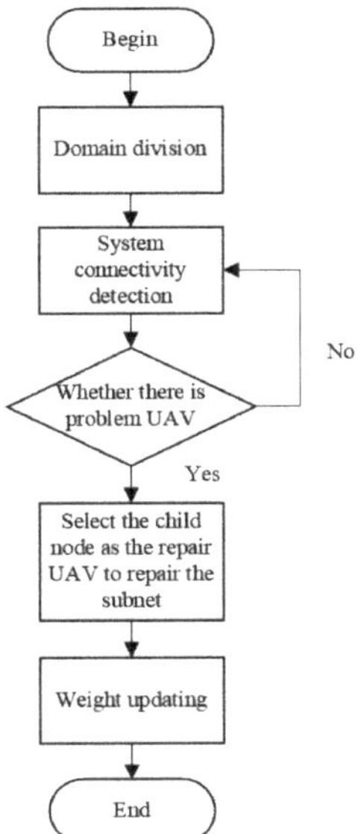

Figure 7. Algorithm block diagram of topology reconfiguration.

Before the departure of the multi-UAV network, UAVs are divided into layers according to the number of communication links of them. If there is a communication link between UAV i and UAV j, then $E_{ij} = 1$. Select UAV i, satisfying $\arg\max \sum_j E_{ij}, \forall i$, as the first layer of the multi-UAV network. If there are UAVs with the same number of links, select the UAV close to the target position. Then, the UAVs that have a communication link with the UAVs on the first layer are used as the second layer, and the division method of the third layer and subsequent layers is the same as above. Then, number each UAV in order from top to bottom and from left to right and assign weights to UAVs according to the position difference of each UAV in the expected formation. Generally, the multi-UAV expected formation is divided into three layers from top to bottom according to the principle of hierarchical division, and the basic formation configuration is obtained. The position of the first UAV in the initial formation is generally at the center UAV of the first layer. The numbering method of the second layer specifies the relative position of each UAV in order from left to right. The naming method of the third layer and subsequent layers is the same as that of the second layer. After layering, two control mechanisms, hierarchical weight β_q and intra-layer position weight β_p, were established by setting the corresponding weight coefficients to ensure the stability of the UAV reconstruction formation. The UAVs in the first layer have the largest β_q, which decrease according to the increase of the number of layers; the position weights β_q within the layer decrease in order from left to right. For V-shaped formations, each UAV β_q within the same layer is equal, β_p is not equal, and $\beta_q >> \beta_p$.

When a UAV is damaged or forced to leave the system, the child UAV of the problem UAV is used as the repair UAV. The multi-UAV formation is traversed down along the communication link until the entire UAV formation is traversed. Then, the repair subnet is established. If there are multiple child UAVs, the child UAV that can reach the expected position of the problem UAV the fastest is judged as the repair UAV according to the position, speed, and acceleration of each child UAV at the current moment. If there are multiple problematic UAVs, select the child UAV of the problematic UAV with a larger weight to repair the missing position. The repair UAV first flies to the desired position of the problem UAV, so as to establish connectivity with other child UAVs of the original problem

UAV. The repair UAV is within the maximum communication link range with the root UAV of the problem UAV and approaches the movement direction of the problem UAV when it leaves the team. It then restores the connection with the problem UAV as much as possible. If the connection with the problem UAV cannot be restored, the sub-UAV of the repair also approaches the problem UAV to form a serial link to expand the communication range.

After the repair subnetwork is established, the weights of the sub-UAVs of the problem UAVs are updated. First, each UAV recalculates the current weights according to the formation in the repair subnetwork. It then sends the new weights to the UAVs through the link. Human–machine and the repair UAV sums the new weight and its own weight to realize the weight update.

3.4. Connectivity-Maintenance Formation Control Algorithm

Algorithm 1 shows the pseudocode of the CMUFC algorithm. First, initialize the position of M UAVs in the multi-UAV network, as well as the radius, maximum speed, acceleration and other parameters of the UAVs. Randomly initialize the positions of O obstacles and K interference sources. Set the transmit power of interference sources, the influence range of obstacles, etc. (lines 1–2). Let T_i denote the mission completion time of UAV i, where $i \in \{1, 2, \ldots, M\}$. Divide each task duration T_i in the discrete time domain into multiple time steps t according to a fixed time interval Δt. That is, t is represented as the t-th time period Δt. Second, at each time slot t, for each UAV i, first calculate the communication distance according to the interference power and path loss of the interference source. Then, calculate the distance between it and other UAVs with communication links in the system. Calculate the distance between it and the target position and obstacle position. After that, calculate the resultant force generated by the multi-UAV formation obstacle avoidance control algorithm according to the distance. Finally, calculate the position of the UAV under the constraints of speed and acceleration at time $t + 1$ (lines 5–8). Third, judge whether there is an undirected path in the multi-UAV network at time $t + 1$ to satisfy the system connectivity. If it exists, continue looping. If it does not exist according to the CMUFC algorithm, the problem UAV is set as the root UAV, and its child UAVs are used as the repair UAV. Then, let it fly to the expected position to restore system connectivity (lines 9–13).

Algorithm 1 CMUFC

  1: Initialize the physical parameters of M UAVs
  2: Initialize the physical parameters of O obstacles and K interference sources
  3: **for** $t = 1, \ldots, T$ **do**
  4:     **for** $i = 1, \ldots, M$ **do**
  5:         Calculate the communication distance of UAV i in Equation (10)
  6:         Calculate the distance between UAV i and neighboring UAVs in the undirected path
  7:         Calculate the resultant force of UAV i in Equation (20)
  8:         Calculate the position of UAV i under the constraints at time $t + 1$
  9:         **if** there is an undirected path in the multi-UAV network **then**
10:            Continue the cycle
11:         **else**
12:            Repair system connectivity
13:         **end if**
14:     **end for**
15: **end for**

---

4. Simulation Results

In this section, we simulate a V-formation multi-UAV network and analyze the simulation results. The relevant parameters of the simulation are shown in Table 1.

Table 1. Simulation Parameters.

Parameter	Value
Number of UAVs	$M = 5$
Transmitting power of UAVs	$p^{uav} = 36$ dBm
Maximum speed of UAV	$v_{max} = 30$ m/s
Maximum acceleration of UAV	$a_{max} = 30$ m/s^2
Number of interference sources	$K = 3.1$
Power of interference sources	$p^{int} = 10\text{–}36$ dBm
Number of obstacles	$O = 10$
Obstacle size	$r_o = 30\text{–}50$ m
Position attractive force coefficient	$k_{att} = 0.1$
Position repulsive force coefficient	$k_{rep} = 1500$
Speed repulsive force coefficient	$k_{rep}^{v} = 100$
Radius of influence of obstacles	$l_o = 100$ m
Safe radius of the UAV	$l_{min} = 10$ m

We simulate the performance of the CMUFC algorithm in different scenarios and compare it with the traditional formation control algorithm by APF. The scenario where there are interference sources is shown in Figure 8, and the powers of these three interference sources are 23 dBm, 26 dBm, and 10 dBm, respectively. The CMUFC algorithm helps UAVs to fly to areas with less interference, thereby maintaining the connectivity of the multi-UAV network. Figure 9 shows the communication range of UAVs. When the multi-UAV network is closer to the interference source, the communication range is smaller. Among them, UAV1 and UAV4 are the UAV communication ranges calculated by our proposed algorithm, and UAV1′ and UAV4′ are the communication ranges calculated by the traditional formation control algorithm. Figure 10 shows the distance between two UAVs. The proposed algorithm can help the multi-UAV network adaptively reduce the formation distance to maintain the connectivity of the entire network when the communication range decreases. In addition, as shown in Figure 10, the farthest distance between two UAVs in the multi-UAV network, UAV1 and UAV4, satisfies the communication requirements of UAVs shown in Figure 9. However, the traditional formation control algorithm makes the distance between UAVs far greater than its communication distance, resulting in the inability of the multi-UAV network to maintain connectivity.

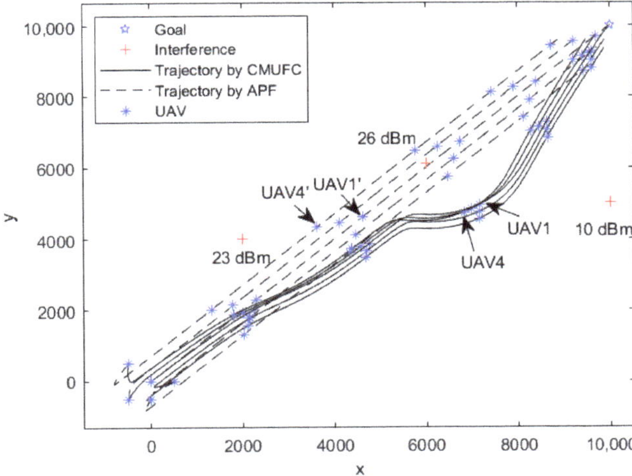

Figure 8. Multi-UAV flight under interference.

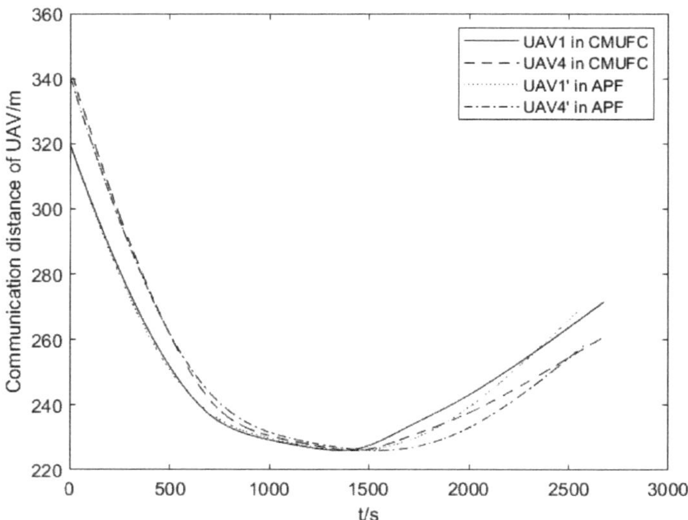

Figure 9. Communication distance of UAVs under interference.

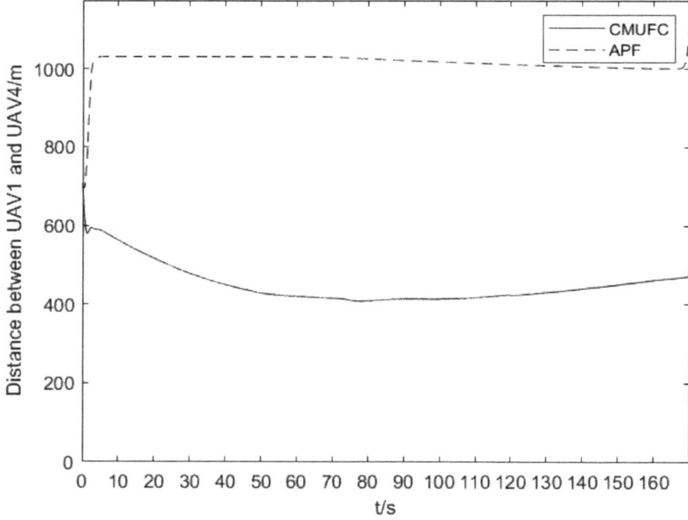

Figure 10. Distance between UAVs under interference.

Furthermore, we simulate a complex environment with obstacles and interference sources as shown in Figure 11, where the power of the interference source is 36 dBm, and the radius of the obstacle is 30–50 m. The CMUFC algorithm helps the multi-UAV network fly away from the interference source to maintain connectivity. Due to the existence of interference, the communication distance between UAVs is reduced, making it easier for UAVs to collide with obstacles. Compared with the traditional formation control algorithm, our algorithm keeps the multi-UAV network away from obstacles and improves safety. The distance between UAV1 and UAV4 when flying in a complex environment is shown in Figure 12. The CMUFC algorithm helps UAVs shorten the distance between them without colliding with each other to maintain the connectivity of the entire system when UAVs are interfered. Figure 13 shows the communication distance of UAVs in complex environments. Compared with the traditional formation control algorithm, the CMUCF algorithm controls the flight distance between the UAVs to be less than its communication distance in their entire flight.

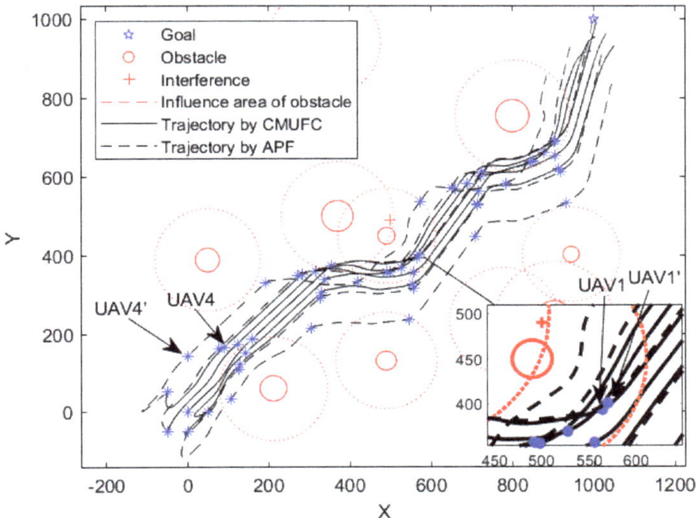

Figure 11. Multi-UAV flight in complex environment.

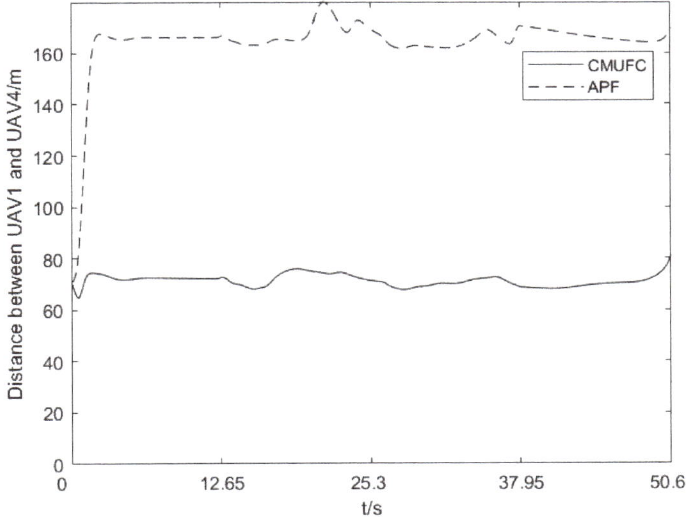

Figure 12. Distance between UAVs in complex environment.

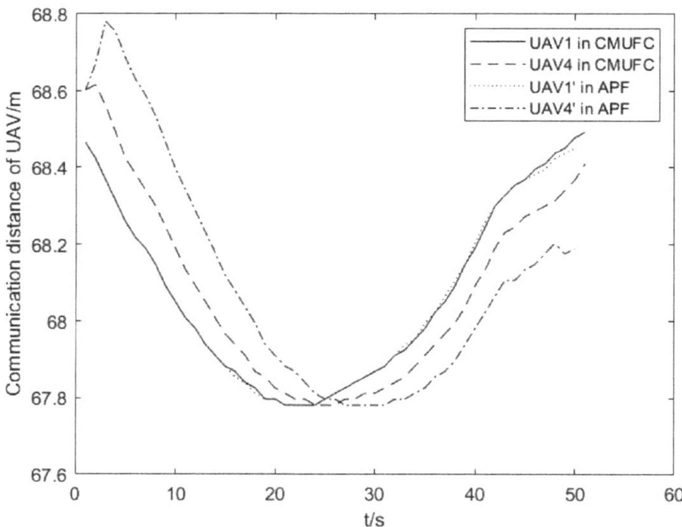

Figure 13. Communication distance of UAVs in complex environment.

5. Conclusions

We investigated the problem of maintaining the connectivity of multi-UAV networks in complex environments. For complex environments with obstacles and interference sources, the CMUFC algorithm helps multi-UAV networks safely avoid obstacles and maintain connectivity during flight. In order to solve the problem that UAVs may collide with obstacles during fast flight, the traditional APF is improved, and SAPF is proposed to help UAVs avoid obstacles more safely. In addition, in order to solve the situation that UAVs leave the team and are forced to change the communication topology during the obstacle avoidance process, the proposed method helps the multi-UAV network to perform formation reconstruction. The simulation results show that the CMUFC algorithm is helpful for multiple UAVs to form, maintain, and reconstruct the formation during their flight.

Author Contributions: Conceptualization, L.Z.; Methodology, C.M.; Software, J.L.; Validation, Y.L.; Writing–review, Q.Y. All authors have read and agreed to the published version of the manuscript.

Funding: This research was supported by the Natural Science Basis Research Plan in Shaanxi Province of China (2023JCYB555).

Institutional Review Board Statement: Not applicable.

Informed Consent Statement: Not applicable.

Data Availability Statement: Not applicable.

Conflicts of Interest: The authors declare no conflict of interest.

References

1. Wu, E.; Sun, Y.; Huang, J.; Zhang, C.; Li, Z. Multi UAV Cluster Control Method Based on Virtual Core in Improved Artificial Potential Field. *IEEE Access* **2020**, *8*, 131647–131661. [CrossRef]
2. Yue, X.; Zhang, W. UAV Path Planning Based on K-Means Algorithm and Simulated Annealing Algorithm. In Proceedings of the 37th Chinese Control Conference (CCC), Wuhan, China, 25–27 July 2018; pp. 2290–2295.
3. Li, Z.; Han, R. Unmanned Aerial Vehicle Three-dimensional Trajectory Planning Based on Ant Colony Algorithm. In Proceedings of the 37th Chinese Control Conference (CCC), Wuhan, China, 25–27 July 2018; pp. 9992–9995.
4. Lin, Y.; Saripalli, S. Sampling-Based Path Planning for UAV Collision Avoidance. *IEEE Trans. Intell. Transp. Syst.* **2017**, *18*, 3179–3192. [CrossRef]

5.   Wei, Z.; Meng, Z.; Lai, M.; Wu, H.; Han, J.A.; Feng, Z. Anti-collision Technologies for Unmanned Aerial Vehicles: Recent Advances and Future Trends. *IEEE Internet Things J.* **2022**, *9*, 7619–7638. [CrossRef]
6.   Tegicho, B.E.; Bogale, T.E.; Eroglu, A.; Edmonson, W. Connectivity and Safety Analysis of Large Scale UAV Swarms: Based on Flight Scheduling. In Proceedings of the 2021 IEEE 26th International Workshop on Computer Aided Modeling and Design of Communication Links and Networks (CAMAD), Porto, Portugal, 25–2 October 2021; pp. 1–6.
7.   Mukherjee, S.; Namuduri, K. Formation Control of UAVs for Connectivity Maintenance and Collision Avoidance. In Proceedings of the IEEE National Aerospace and Electronics Conference (NAECON), Dayton, OH, USA, 15–19 July 2019; pp. 126–130.
8.   Huang, Y.T.; Qi, N.; Huang, Z.Q.; Jia, L.L.; Wu, Q.H.; Yao, R.G.; Wang, W.J. Connectivity Guarantee Within UAV Cluster: A Graph Coalition Formation Game Approach. *IEEE Open J. Commun. Soc.* **2023**, *4*, 79–90. [CrossRef]
9.   Trimble, J.; Pack, D.; Ruble, Z. Connectivity Tracking Methods for a Network of Unmanned Aerial Vehicles. In Proceedings of the IEEE 9th Annual Computing and Communication Workshop and Conference (CCWC), Las Vegas, NV, USA, 7–9 January 2019; pp. 0440–0447.
10.   Ma, Z.; Qi, J.; Wang, M.; Wu, C.; Guo, J.; Yuan, S. Time-Varying Formation Tracking Control for Multi-UAV Systems with Directed Graph and Communication Delays. In Proceedings of the 40th Chinese Control Conference (CCC), Shanghai, China, 26–28 July 2021; pp. 5436–5441.
11.   Li, Z.; Huang, T.; Tang, Y.; Zhang, W. Formation Control of Multiagent Systems With Communication Noise: A Convex Analysis Approach. *IEEE Trans. Cybern.* **2021**, *51*, 2253–2264. [CrossRef] [PubMed]
12.   Fei, B.; Bao, W.; Zhu, X.; Liu, D.; Men, T.; Xiao, Z. Autonomous Cooperative Search Model for Multi-UAV With Limited Communication Network. *IEEE Internet Things J.* **2022**, *9*, 19346–19361. [CrossRef]
13.   Baldi, S.; Sun, D.; Zhou, G.; Liu, D. Adaptation to Unknown Leader Velocity in Vector-Field UAV Formation. *IEEE Trans. Aerosp. Electron. Syst.* **2022**, *58*, 473–484. [CrossRef]
14.   Roldao, V.; Cunha, R.; Cabecinhas, D.; Silvestre, C.; Oliveira, P. A leader-following trajectory generator with application to quadrotor formation flight. *Robot. Auton. Syst.* **2014**, *62*, 1597–1609. [CrossRef]
15.   Davidi, A.; Berman, N.; Arogeti, S. Formation flight using multiple Integral Backstepping controllers. In Proceedings of the IEEE 5th International Conference on Cybernetics and Intelligent Systems (CIS), Qingdao, China, 17–19 September 2011; pp. 317–322.
16.   Song, M.; Wei, R.X.; Hu, M.L. Unmanned aerial vehicle formation contorl for reconnaissance task based on virtual leader. *Syst. Eng. Electron.* **2010**, *32*, 2412–2415.
17.   Li, C.X.; Liu, Z.; Yin, H. Cooperative motions control method guided by virtual formations for multi-UAVs. *Syst. Eng. Electron.* **2012**, *34*, 1220–1224.
18.   Khatib, O. Real-time obstacle avoidance for manipulators and mobile robots. In Proceedings of the 1985 IEEE International Conference on Robotics and Automation, St. Louis, MO, USA, 25–28 March 1985.
19.   Parivallal, A.; Sakthivel, R.; Amsaveni, R.; Alzahrani, F.; Saleh Alshomrani, A. Observer-based memory consensus for nonlinear multi-agent systems with output quantization and Markov switching topologies. *Phys. Stat. Mech. Appl.* **2020**, *551*, 123949. [CrossRef]
20.   Parivallal, A.; Sakthivel, R.; Wang, C. Guaranteed cost leaderless consensus for uncertain Markov jumping multi-agent systems. *J. Exp. Theor. Artif. Intell.* **2023**, *35*, 257–273. [CrossRef]
21.   Wu, Y.; Liang, T.J. Improved consensus-based algorithm for unmanned aerial vehicle formation control. *Acta Aeronaut. Astronaut. Sin.* **2020**, *41*, 172–190.
22.   Wu, Y.; Guo, J.Z.; Hu, X.T.; Huang, Y.T. A new consensus theory-based method for formation control and obstacle avoidance of UAVs. *Aerosp. Sci. Technol.* **2020**, *107*, 106332. [CrossRef]
23.   Zou, Y.; Zhang, H.; He, W. Adaptive Coordinated Formation Control of Heterogeneous Vertical Takeoff and Landing UAVs Subject to Parametric Uncertainties. *IEEE Trans. Cybern.* **2022**, *52*, 3184–3195. [CrossRef] [PubMed]
24.   Tran, V.P.; Santoso, F.; Garratt, M.A.; Petersen, I.R. Distributed Formation Control Using Fuzzy Self-Tuning of Strictly Negative Imaginary Consensus Controllers in Aerial Robotics. *IEEE/ASME Trans. Mechatron.* **2021**, *26*, 2306–2315. [CrossRef]
25.   Wang, Y.; Cheng, Z.; Xiao, M. UAVs' Formation Keeping Control Based on Multi-Agent System Consensus. *IEEE Access* **2020**, *8*, 49000–49012. [CrossRef]
26.   Qin, T.; Yu, H.; Lv, Y.; Guo, Y. Artificial Potential Field Based Distributed Cooperative Collision Avoidance for UAV Formation. In Proceedings of the 3rd International Conference on Unmanned Systems (ICUS), Harbin, China, 27–28 November 2020; pp. 897–902.
27.   Zeng, Y.; Wu, Q.; Zhang, R. Accessing from the sky: A tutorial on UAV communications for 5G and beyond. *Proc. IEEE* **2019**, *107*, 2327–2375. [CrossRef]

Article

Consensus Control of Large-Scale UAV Swarm Based on Multi-Layer Graph

Taiqi Wang [1], Shuaihe Zhao [1,2], Yuanqing Xia [1,*], Zhenhua Pan [1] and Hanwen Tian [1]

[1]   School of Automation, Beijing Institute of Technology, Beijing 100081, China
[2]   Aerospace Shenzhou Aerial Vehicle Ltd., Tianjin 300300, China
*   Correspondence: xia_yuanqing@bit.edu.cn

Abstract: An efficient control of large-scale unmanned aerial vehicle (UAV) swarm to establish a complex formation is one of the most challenging tasks. This paper investigates a novel multi-layer topology network and consensus control approach for a large-scale UAV swarm moving under a stable configuration. The proposed topology can make the swarm remain robust in spite of the large number of UAVs. Then a potential function-based controller is developed to control the UAVs in realizing autonomous configuration swarming under the consideration of mutual collision, and the stability of the controller from the individual UAV to the entire swarm system is analyzed by a Lyapunov approach. Afterwards, a yaw angle adjustment approach for the UAVs to reach consensus is developed for the multi-layer swarm, then the direction state of each UAV converges with a fast rate. Finally, simulations are performed on the large-scale UAV swarm system to demonstrate the effectiveness of the proposed scheme.

Keywords: multi-layer graph; potential function; consensus control; UAV swarm

Citation: Wang, T.; Zhao, S.; Xia, Y.; Pan, Z.; Tian, H. Consensus Control of Large-Scale UAV Swarm Based on Multi-Layer Graph. *Drones* **2022**, 6, 402. https://doi.org/10.3390/drones6120402

Academic Editors: Zhihong Liu, Shihao Yan, Yirui Cong and Kehao Wang

Received: 25 October 2022
Accepted: 2 December 2022
Published: 7 December 2022

Publisher's Note: MDPI stays neutral with regard to jurisdictional claims in published maps and institutional affiliations.

1. Introduction

Over the past few decades, the investigation of large-scale swarm has received extensive attention in different fields, such as biology, physics, medicine, sociology, engineering, et al. [1–3]. Swarm refers to a super large-scale isomorphic individual, based on group dynamics and information perception, supported by efficient and safe collaborative interaction between individuals, with the emergence of swarm intelligence as the core, and based on a comprehensive integration of open architecture. It is a complex system with the advantages of invulnerability, adaptive dynamic configuration, functional distribution, and intelligent features. The Boid model is the first model established by Reynold to simulate group behavior [4], and three heuristic rules are introduced in this model, namely separation, cohesion and alignment. On the basis of these three rules, many scholars have conducted in-depth investigation on the swarming movement [5,6]. For example, Olfati-Saber R [7] proposes a theoretical framework of distributed swarm algorithms, and swarm in free space and multiple obstacles avoidance are also considered. Inspired by the above work, Su H et al. [8] revisits the problem of multiagent system in the absence of the above two assumptions.

Combining robot technology with swarm algorithms is one of the hotspots [9,10]. In particular, Enrica Soria et al. [11] published in the journal Nature combines the local principles of potential field methods into an objective function and optimizes those principles with knowledge of the agent's dynamics and environment, resulting in improving drone swarm speed, order and safety. In [12], a multi-layer grouping coordination methodology is proposed to achieve different shape configuration for a large-scale agents. In [13], a new topology approach based on multilevel construction is adopted to present swarm robots of different shapes in the desired region. A novel multi-layer graph is presented by [14] for multi-agent systems to enable scalability of the interaction networks, and the model predictive control method is applied for tracking trajectories; In [15], a multi-layer formation control scheme and a layered distributed finite-time estimator is designed for agents, which impels them to reach the desired positions and velocities according to the the information of agents in their prior layers. In practice, many issues need to be considered in order to implement formation control approaches successfully, such as the avoidance of the obstacles and collisions.

The artificial potential field (APF) models provides an effective solution for practical applications, which attracts the agent to the target and repulses it for avoidance, and can be executed quickly and provides a viable solution [16]. In [17], a rotating potential field is introduced, which makes the UAVs can escape from the oscillations and ensures that the follower-leader maintains the desired angles and distances. Based on the APF approach in [18], a novel automatic vehicles motion planning and tracking framework is presented, and the effectiveness is validated in real experiment. In [19], an adaptive synchronized tracking control based on the neural network is applied to boat by combining with APF and robust $H\infty$ methods, and the artificial potential method is used to guarantee the boat maintaining desired distance with obstacles. In [20], different forms of potential field functions are used for repulsion, velocity alignment and interaction with walls and obstacles, and the proposed model is validated on a self-organized swarm of 30 drones.

Consensus control of multiagent systems is also a hotspot now, which means all the agents in the system converge to the same state by the specific control law. In [21], a distributed active anti-disturbance cooperative control method with a finite-time disturbance observer is proposed to achieve the consensus in finite time for the agents. In [22], the consensus control problem is investigated under an event-triggered mean-square consensus control law for a class of discrete time-varying stochastic multi-agent system. There are three approaches proposed by [23] for consensus control of the multi-agent systems on directed graphs, and some correlative examples are presented to validate the effectiveness. In [24], the synergistic trajectory tracking problem of UAVs formation is investigated, both the position tracking to the desired position and the attitude tracking to the command attitude signal are achieved with the stability analysis and simulations validation.

The main challenges that impede the solving of the configuration and consensus problem for the swarm are the large-sclae of the community and the chronological order of configuration and consensus. Therefore, we have carried out the following research to solve these problems. In order to improve the scalability of the network topology under the large size of the swarm situation, based on the concept of [14], a multi-layer network graph model is proposed for the large-scale UAV swarm, which allows the configuration to be more adjustable and robust. After the configuration of the swarm is completed, to make each UAV in the swarm reach an agreement, a multi-layer recursive consensus control concept is designed for the UAV swarm, so that the yaw angles of UAVs in each layer tend to be consistent.

The remainder of the paper is organized as follows. Section 2 describes some preliminaries and formulates the problem to be investigated in this paper. In Section 3, the multi-layer UAVs swarm configuration control strategy and the consensus concept are proposed. The effectiveness of the proposed methodologies is illustrated by numerical analysis in Section 4. Finally, the results of our work are briefly summarized in Section 5.

2. Preliminaries and Problem Statements

2.1. Graph Theory

In this subsection, some introductions of the graph theory are listed. Firstly, we define undirected graph $\mathcal{G} = (v, \varepsilon)$ as the interaction topology which consists of $v = (1, 2, \ldots, n)$ a list of vertices , whose elements represent individual UAV in the swarm, and $\varepsilon \subseteq ((i, j) : i, j \in v, i \neq j;)$ a list of edges, containing unordered pairs of vertices. An edge $(i, j) \in \varepsilon$ of the undirected graph $\mathcal{G}$ means that the UAV i and UAV j can exchange information. For the undirected graph $\mathcal{G}$, the adjacency matrix is given by $A = \left[a_{ij}\right] \in R^{N \times N}$ with $a_{ij} = 0 \Leftrightarrow (i,j) \in \varepsilon$, $a_{ij} = a_{ji}$. The neighboring set of agent is denoted in [25]:

$$N_i = \{j \in v : a_{ij} \neq 0\} = \{j \in v : (i,j) \in \varepsilon\} \tag{1}$$

2.2. Problem Descriptions

The consistency problem aims at designing appropriate protocols such that the group of UAVs can reach consensus, exploiting only local information exchange among neighbors and unreliable information exchange and dynamically changing interaction topologies. In this paper, our target is to regulate the entire swarm (from each agent to multilayer) move at a same velocity (the same magnitude and direction)and maintain constant distances between the same agent layers. Based on the vicsek model, our hypothesis implies that each agent in the same layer adjusts its velocity by adding to it a weighted average of the differences of its velocity with the others. Then the potential function is necessary to proposed for maintaining a constant distance of each UAV and each layer such that their potentials become minima. In the next section, we will describe the control strategy for our multi-layer grouping swarm specifically.

For brevity, two assumptions is given as follows

Assumption 1. *We assume the Large-scale UAVs swarm consisting n UAVs with the same dynamic characteristics flying in a same altitude space. Therefore, the working environment of each UAV can be consider a two-dimensional space.*

Assumption 2. *In the case of controlling large-scale swarm, we assume each UAV as a point mass, which means the influence of the size and shape of each UAV can be ignored.*

3. Multi-Layer Consensus Control Architecture

3.1. Multi-Layer Graph Model

The proposed multi-layer UAVs swarm model is a multipartite network, which is composed of a series of similar layer structures. In each layer, a certain number of subgroups form a higher layer network by the corresponding control law. Note that each layer is strictly follows the same network characteristics, such as position distribution, potential function, velocity consistency, and so on. Based on the above rules, a hierarchical network structure is constructed.

When the multi-layer structure is considered, the first layer characterized by the position-based interaction forms a primary formation configuration. We assume that the whole swarm includes n UAVs, l layers, and there can only be N_o neighbors from the independent UAVs to each subgroups. We assume that n is divisible by $N_o + 1$. Therefore, the undirect graph of the first layer can be defined as $\mathcal{G}_1 = (v_1, \varepsilon_1)$, where $v_1 = (1, 2, \ldots, n/(N_o + 1))$ and $\varepsilon_1 \subseteq ((i,j) : i, j \in v_1, i \neq j;)$; The second layer undirected graph consists of the first layer, which is defined as $\mathcal{G}_2 = (v_2, \varepsilon_2)$, where $v_2 = \left(1, 2, \ldots, n/(N_o + 1)^2\right)$ and second list of edges $\varepsilon_2 \subseteq ((i,j) : i, j \in v_2, i \neq j;)$; Based on the above rules, we denote $\mathcal{G}_m = (v_m, \varepsilon_m)$, $m = 1, 2, \ldots l$ as the interaction network topology to characterize the underlying information flow among the UAVs in the mth layer, where $v_m = (1, 2, \ldots, n/(N_o + 1)^m)$ and $\varepsilon_m \subseteq ((i,j) : i, j \in v_m, i \neq j;)$. Then the neighboring set from the first layer to the mth layer can be denoted as $N_i^1, N_i^2, \ldots, N_i^m$.

3.2. Swarm Configuration

Based on the above conclusions, in order to realize the multi-layer grouping configuration of the whole swarm, firstly we define the dynamic of each UAV as follows:

$$\begin{cases} \dot{x}_i = v_i \\ \dot{v}_i = u_i \end{cases} \quad i = 1, 2, \ldots, n \tag{2}$$

From single agent to multi-agent system, the dynamic protocol of the UAV swarm in each layer is described as follows:

$$
\begin{cases}
first\ layer \begin{cases} \dot{x}_i^1 = v_i^1 \\ \dot{v}_i^1 = u_i^1 = f_{sum}^1 - k_1 \dot{x}_i^1 \end{cases} & i \in v_1 \\[2em]
second\ layer \begin{cases} \dot{x}_i^2 = v_i^2 \\ \dot{v}_i^2 = u_i^2 = f_{sum}^2 - k_1 \dot{x}_i^2 \end{cases} & i \in v_2 \\[2em]
\quad\quad\vdots \\[1em]
mth\ layer \begin{cases} \dot{x}_i^m = v_i^m \\ \dot{v}_i^m = u_i^m = f_{sum}^m - k_1 \dot{x}_i^m \end{cases} & i \in v_m
\end{cases}
\tag{3}
$$

where $x_i^1, v_i^1 \in R^n$ are respectively the position and velocity of each UAV in the subgroup of first layer, $u_i^1 \in R^n$ is the control input acting on it, f_{sum}^1 is the resultant force contains obstacle avoidance force and collision avoidance force between UAVs. For the second layer, $x_i^2, v_i^2 \in R^{k_2}$ and $u_i^2 \in R^{k_2}$ are respectively the position, velocity and the control input of the UAV in the subgroup of second layer, where $k_2 = n/(N_0 + 1)$ is the element number of the v_2; and f_{sum}^2 is the resultant force contains not only mutual forces from each UAV but also has potential field force from other subgroups in the second layer. Silimarly, $x_i^m, v_i^m \in R^{k_m}$ and $u_i^m \in R^{k_m}$ are respectively the position, velocity and the control input of the UAV in the subgroup of mth layer, where $k_m = n/(N_0 + 1)^{(m-1)}$ is the element number of the v_m; and the resultant force f_{sum}^m contains the mutual forces from each UAV in the whole global and the potential field forces from the second layer to the mth layer. Furthermore, k_1 is a positive constant for damping action.

By analyzing the dynamic model of the UAV (2), we design the corresponding control law to make UAVs reach their desired configuration. Two forces will be engendered based on the designed potential functions to drive all the UAVs move into the desired position and avoid mutual collisions.

The mathematical expression of potential function is as follows

$$
V_{ij}(d_{ij}) = \begin{cases} -\xi \frac{d_{ij}}{r_0} \ln(\frac{d_{ij}}{r_0}) + \frac{d_{ij}}{r_0} & x_i \in N_i^1 \\ 0 & otherwise \end{cases}
\tag{4}
$$

where ξ is the positive control coefficient, $d_{ij} = \|x_i - x_j\|$ is the distance between agent i and agent j, r_0 is the desired radius between each UAV.

Differentiating (3) with respect to d_{ij} yields a potential force as

$$
f_{ij} = -\nabla V_{ij}(d_{ij}) = \begin{cases} \xi \ln(\frac{d_{ij}}{r_0}) & x_i \in N_i^1 \\ 0 & otherwise \end{cases}
\tag{5}
$$

In another case, when UAV i and UAV j are not well-defined neighbors, both can be regarded as obstacles to each other. Therefore, another potential function to avoid obstacles is necessary to proposed as follows

$$
V_o(d_{io}) = \begin{cases} \eta(r_0 - d_{io})\frac{x_i - x_j}{d_{io}} & d_{io} < r_0 \\ 0 & d_{io} \geq r_0 \end{cases}
\tag{6}
$$

where η is the positive control gain, x_o is the position of the obstacle o. $d_{io} = \|x_i - x_o\|$ is the distance between UAV i and the obstacles.

Then we define the set of the obstacles as

$$
O_i = \{j \notin N_i | d_{io} < r_0\}
\tag{7}
$$

The corresponding force obtained from V_o is

$$f_{io} = -\nabla V_o(d_{io}) = \begin{cases} \frac{\eta}{d_{io}^2} r_0 (x_i - x_o) & d_{io} < r_0 \\ 0 & d_{io} \geq r_0 \end{cases} \tag{8}$$

Based on the above two forces, the resultant force f_{sum}^1 for the first layer is expressed as follows

$$f_{sum}^1 = \sum_{j \in N_i^1} f_{ij} + \sum_{o \in O_i} f_{io} \tag{9}$$

the control input can be described as follows

$$\begin{aligned} u_i^1 &= \sum_{j \in N_i^1} f_{ij} + \sum_{o \in O_i} f_{io} - k_1 \dot{x}_i^1 \\ &= -\sum_{j \in N_i^1} \nabla V_{ij}(d_{ij}) - \sum_{o \in O_i} \nabla V_o(d_{io}) - k_1 \dot{x}_i^1 \end{aligned} \tag{10}$$

For the second layer, in addition to the mutual force between the individual UAV, the swarm are also affected by the potential field force between the subgroups. We define the potential function of the second layer as follows

$$V_{ij}^2(d_{ij}^2) = \begin{cases} -\zeta \frac{d_{ij}^2}{r_0^2} \ln(\frac{d_{ij}^2}{r_0^2}) + \frac{d_{ij}^2}{r_0^2} & x_i^2 \in N_i^2 \\ 0 & otherwise \end{cases} \tag{11}$$

where $d_{ij}^2 = \left\| x_i^2 - x_j^2 \right\|$, r_0^2 is the desired distance of the second layer. Then the corresponding potentional force is expressed as follows

$$f_{ij}^2 = -\nabla V_{ij}^2(d_{ij}^2) \tag{12}$$

At the same time, each UAV in the swarm has gathered within a fixed area, then the force to avoid obstacles disappears. Therefore, resultant force f_{sum}^2 are combined as follows

$$f_{sum}^2 = \sum_{j \in N_i^1} f_{ij} + \sum_{j \in N_i^2} f_{ij}^2 \tag{13}$$

The control law u_i^2 of the second layer can be describe as follows

$$\begin{aligned} u_i^2 &= \sum_{j \in N_i^1} f_{ij} + \sum_{j \in N_i^2} f_{ij}^2 - k_1 \dot{x}_i^2 \\ &= -\sum_{j \in N_i^1} \nabla V_{ij}(d_{ij}) - \sum_{j \in N_i^1} \nabla V_{ij}^2(d_{ij}^2) - k_1 \dot{x}_i^2 \end{aligned} \tag{14}$$

For the mth layer, we assume it as the last layer of the whole swarm, then each UAV in mth layer is subject to global forces. The potential function is described as follows

$$V_{ij}^m(d_{ij}^m) = \begin{cases} -\zeta \frac{d_{ij}^m}{r_0^m} \ln(\frac{d_{ij}^m}{r_0^m}) + \frac{d_{ij}^m}{r_0^m} & x_i^m \in N_i^m \\ 0 & otherwise \end{cases} \tag{15}$$

where $d_{ij}^m = \left\| x_i^m - x_j^m \right\|$, r_0^m is the desired distance of the second layer. Then the corresponding potentional force is expressed as follows

$$f_{ij}^m = -\nabla V_{ij}^m(d_{ij}^m) \tag{16}$$

Different from (13), the resultant force f_{sum}^m combines all the forces from the first layer to the mth layer in the global, and the form is as follows

$$f_{sum}^m = \sum_{j \in N_i^1} f_{ij} + \sum_{j \in N_i^2} f_{ij}^2 + \ldots + \sum_{j \in N_i^{m-1}} f_{ij}^{m-1} + \sum_{j \in N_i^m} f_{ij}^m \tag{17}$$

The global control law u_i^m is list as follows

$$
\begin{aligned}
u_i^m &= \sum_{j \in N_i^1} f_{ij} + \sum_{j \in N_i^2} f_{ij}^2 + \ldots + \sum_{j \in N_i^{m-1}} f_{ij}^{m-1} + \sum_{j \in N_i^m} f_{ij}^m - k_1 \dot{x}_i^m \\
&= -\sum_{j \in N_i^1} \nabla V_{ij}(d_{ij}) - \sum_{j \in N_i^2} \nabla V_{ij}^2(d_{ij}^2) - \ldots - \sum_{j \in N_i^{m-1}} \nabla V_{ij}^{m-1}(d_{ij}^{m-1}) \\
&\quad - \sum_{j \in N_i^m} \nabla V_{ij}^m(d_{ij}^m) - k_1 \dot{x}_i^m
\end{aligned} \tag{18}
$$

The control law of the entire UAV swarm are completed. Furthermore, the stability of the configuration needs to be analyzed.

Theorem 1. *Consider a subgroup of n UAVs with dynamics (2), under the control law (10), each UAV can stay at a desired position and the forces and velocity converge to zero finally.*

Proof of Theorem 1. Define a Lyapunov function candidate as

$$V_1 = \sum_{j \in N_i^1} V_{ij}(d_{ij}) + \sum_{o \in O_i} V_o(d_{io}) + \frac{1}{2} \dot{x}_i^{1^T} \dot{x}_i^1 \tag{19}$$

From the above conclusion we can get V_1 is non-negative. Differentiating (19) with respect to time and combining with (2), (3) and (10), we have

$$
\begin{aligned}
\dot{V}_1 &= \dot{x}_i^{1^T} \left(\sum_{j \in N_i^1} \nabla V_{ij}(d_{ij}) + \sum_{o \in O_i} \nabla V_o(d_{io}) + \ddot{x}_i^1 \right) \\
&= \dot{x}_i^{1^T} (-f_{sum}^1 + u_i^1) \\
&= -k_1 \dot{x}_i^{1^T} \dot{x}_i^1 \\
&\leq 0
\end{aligned} \tag{20}
$$

Thus the energy of each UAV i $(i = 1, 2, \ldots, n)$ monotonically decreasing. From the analysis we can conclude that the velocity of UAVs eventually converge as the same. $\square$

Theorem 2. *For the entire swarm with n agents, under the global control law (18), all the UAVs can arrive at the desired positions, the potential forces from the first layer to the mth layer and velocity converge to zero finally.*

Proof of Theorem 2. Define the global Lyapunov function candidate as

$$
\begin{aligned}
V_m = \sum_{i=1}^{n} \Big(&\sum_{j \in N_i^1} V_{ij}(d_{ij}) + \sum_{j \in N_i^2} V_{ij}^2(d_{ij}^2) + \ldots + \sum_{j \in N_i^{m-1}} V_{ij}^{m-1}(d_{ij}^{m-1}) \\
&+ \sum_{j \in N_i^m} V_{ij}^m(d_{ij}^m) + \tfrac{1}{2} \dot{x}_i^{m^T} \dot{x}_i^m \Big)
\end{aligned} \tag{21}
$$

From the (5), (11) and (15) we can get V_m is non-negative.

Differentiating (21) with respect to time and combining with (3) and (18), we have

$$\begin{aligned}
\dot{V}_m &= \sum_{i=1}^{n} \dot{x}_i^{mT} \Big(\sum_{j\in N_i^1} \nabla V_{ij}(d_{ij}) + \sum_{j\in N_i^2} \nabla V_{ij}^2(d_{ij}^2) \\
&\quad + \ldots + \sum_{j\in N_i^{m-1}} \nabla V_{ij}^{m-1}(d_{ij}^{m-1}) + \sum_{j\in N_i^m} \nabla V_{ij}^m(d_{ij}^m) + \ddot{x}_i^m \Big) \\
&= \sum_{i=1}^{n} \dot{x}_i^{mT} (-f_{sum}^m + u_i^m) \\
&= -k_1 \sum_{i=1}^{n} \dot{x}_i^{mT} \dot{x}_i^m \\
&\leq 0
\end{aligned} \tag{22}$$

Therefore, the total potential energy can approach the minimumwe and $\dot{x}_i^m \to 0$ as $t \to \infty$ for all the UAVs in the swarm, and so is $\ddot{x}_i^m$. As a result, the multi-layer configuration of the swarm is constructed. $\square$

3.3. Consensus Strategy

In this subsection, all the UAVs in the swarm have formed a fixed formation configuration based on the potential function control law. However, the yaw angle ψ of each UAV is still arbitrarily uncertain. In order to keep the flight states of all the UAVs in consensus, a yaw angle adjustment method based on the concept of the Vicsek model [26] is proposed for the multi-layer UAVs swarm.

For simplicity, we relabel each UAV in different layers. The edge of the first layer is $\varepsilon_1 \subseteq ((i_1, j_1) : i_1, j_1 \in v_1, i_1 \neq j_1;)$. For the second layer, $\varepsilon_2 \subseteq ((i_2, j_2) : i_2, j_2 \in v_2, i_2 \neq j_2;)$. For the mth layer, the edge is labeled as $\varepsilon_m \subseteq ((i_m, j_m) : i_m, j_m \in v_m, i_m \neq j_m;)$. Then the UAVs in each subgroup make corresponding updates according to the states of the previous subgroups, and finally achieve consensus.

For the first layer, the UAVs yaw angle adjustment strategy is as follows

$$\psi_{i_1}(t+1) = \arctan \frac{\sum_{j_1=1}^{N_o+1} \sin \psi_{j_1}(t)}{\sum_{j_1=1}^{N_o+1} \cos \psi_{j_1}(t)} \tag{23}$$

The attitude of the UAV i_1 can be updated according to the attitude of all the UAVs in the same subgroup. Therefore, the consensus of the first layer is achieved.

For the second layer, the UAVs yaw angle are adjusted by the following approach

$$\begin{aligned}
\psi_{i_2}(t+1) &= \arctan \frac{\sum_{j_2=1}^{N_o+1} \sin \psi_{j_2}(t)}{\sum_{j_2=1}^{N_o+1} \cos \psi_{j_2}(t)} \\
&= \arctan \frac{\sum_{j_2=1}^{N_o+1} \sin(\frac{1}{N_o+1} \sum_{j_1=1}^{N_o+1} \psi_{j_1}(t))}{\sum_{j_2=1}^{N_o+1} \cos(\frac{1}{N_o+1} \sum_{j_1=1}^{N_o+1} \psi_{j_1}(t))}
\end{aligned} \tag{24}$$

Therefore, ψ_{i_2} is obtained from the average of the yaw angles of the individual UAVs in all the subgroups for the first layer.

Based on the above strategy, the UAVs yaw angle adjustment strategy for the mth layer is as follows

$$\psi_{i_m}(t+1) = \arctan\frac{\sum\limits_{j_m=1}^{N_0+1}\sin\psi_{j_2}(t)}{\sum\limits_{j_m=1}^{N_0+1}\cos\psi_{j_2}(t)}$$

$$= \arctan\frac{\sum\limits_{j_m=1}^{N_0+1}\sin(\frac{1}{N_0+1}\sum\limits_{j_{m-1}=1}^{N_0+1}\psi_{j_{m-1}}(t))}{\sum\limits_{j_m=1}^{N_0+1}\cos(\frac{1}{N_0+1}\sum\limits_{j_{m-1}=1}^{N_0+1}\psi_{j_{m-1}}(t))} \tag{25}$$

In the above, we describe the consensus strategy between different layers, then all the UAVs in the swarm achieve consensus eventually. For the specific example of the UAVs swarm, as shown in Figure 1, assume $n = 9$, $N_0 = 2$, there are nine UAVs labeled as $A_1, \ldots, A_9$, then the whole swarm can be combined as three first layer subgroups named as $\mathcal{G}_1^1, \mathcal{G}_1^2, \mathcal{G}_1^3$, which constitute a second layer $\mathcal{G}_2$. Futhermore, $\mathcal{G}_1^1$ is composed of $A_1, A_2, A_3, \mathcal{G}_1^2$ and $\mathcal{G}_1^3$ are consisted of $A_4, A_5, A_6, A_7, A_8, A_9$, respectively. Then we set the communication between $\mathcal{G}_1^1$ and $\mathcal{G}_1^2$ are connected by A_2 and A_4, $\mathcal{G}_1^2$ and $\mathcal{G}_1^3$ are connected by A_6 and A_8, $\mathcal{G}_1^1$ and $\mathcal{G}_1^3$ are connected by A_3 and A_7. Firstly, the UAV swarm achieves the desired configuration through the forces between the UAV individuals and between the same layers. Taking A_2 as an example, A_2 is subjected to the forces of A_1 and A_3, namely $f_{A_2A_1}$ and $f_{A_2A_3}$, A_2 is also subject to $f^2_{\mathcal{G}_1^1\mathcal{G}_1^2}$ and $f^2_{\mathcal{G}_1^1\mathcal{G}_1^3}$, which are the components between $\mathcal{G}_1^1$ and $\mathcal{G}_1^2$ and between $\mathcal{G}_1^1$ and $\mathcal{G}_1^3$, respectively. After the whole swarm reaches the desired configuration, the resultant force of A_2 is zero. Furthermore, let $\mathcal{G}_1^1$, $\mathcal{G}_1^2$ and $\mathcal{G}_1^3$ achieve intra-group consensus through the yaw angle adjustment stragety (23), and reach the same yaw angle ψ_{A_1}, ψ_{A_4} and ψ_{A_7} respectively. For the second layer, $\mathcal{G}_2$ achieve the intra-group consistent yaw angle from the average of the ψ_{A_1}, ψ_{A_4} and ψ_{A_7}. Based on this rule, the entire swarm achieves consensus eventually.

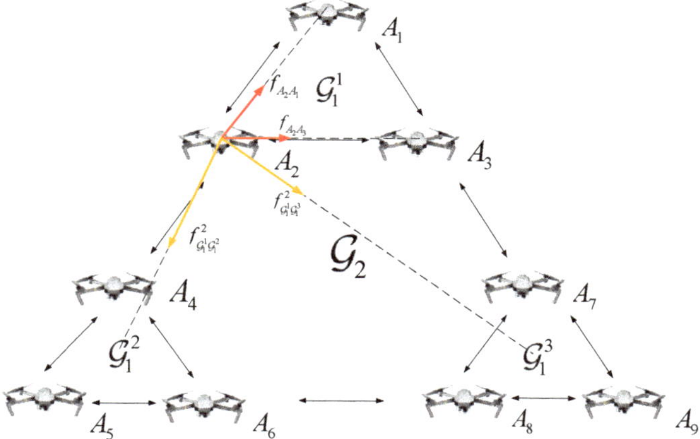

Figure 1. Communication topology with $n = 9$, $N_0 = 2$.

4. Simulation Study

To illustrate the effectiveness of the proposed multi-layer topology and the consensus algorithm, corresponding simulation results under different conditions are presented in this section. For the multi-layer UAVs swarm, we consider a group of networked UAVs with $n = 27$, $N_0 = 2$, which contains two layer subgroups. The control parameters are chosen as $r_0 = 2$ m, $r_0^2 = 4$ m, $\xi = 20$, $\eta = 5$.

4.1. Swarm Configuration

For the proposed multi-layer UAVs system, all the UAVs are randomly distributed in a fixed working area. Firstly, based on the adjacent principle, under the control law (10), all the neighbors in the UAVs swarm are assigned to establish the multi-layer network topology. All the UAVs move towards their desired location and keep the desired distances with their neighbours under the forces between the UAV individuals and the forces between layers at the same level, the repulsive forces makes UAVs avoid collisions and keep the desired distance between them, while the force between layers makes the UAV swarm achieve the desired configuration, then the resultant force from the artificial potential converges to zero. In different situations, the number of UAVs in the whole system and the number of their neighbors can be seted arbitrarily, so as to adjust the structure of the entire network topology. When all the UAVs complete the assignment of neighbors, the first layer topology within a set of subgroups is constructed, under the swarm configuration control law (14), all the subgroups can be treated as independent individuals, then the neighbors are assigned to these subgroups and the second layer network topology is established. Based on this rule, the subgroups will adjust their position and form a higher level group, until all the UAVs achieve the desired configuration. Here, we take 27 UAVs as an example to illustrate the effectiveness of control laws. As shown in Figure 2, in the initial state, the distance between UAV individuals is arbitrary, within the range of 25 m, then after the UAVs start to communicate with each other, in a short iteration step, the UAVs at any position converge quickly. When the step reaches around 150, the distance between UAVs is within 2 m, and the expected configuration is basically achieved. As a result, all UAVs in the swarm of each layer tend to be at the desired distance with the high formation configuration results after 800 steps.

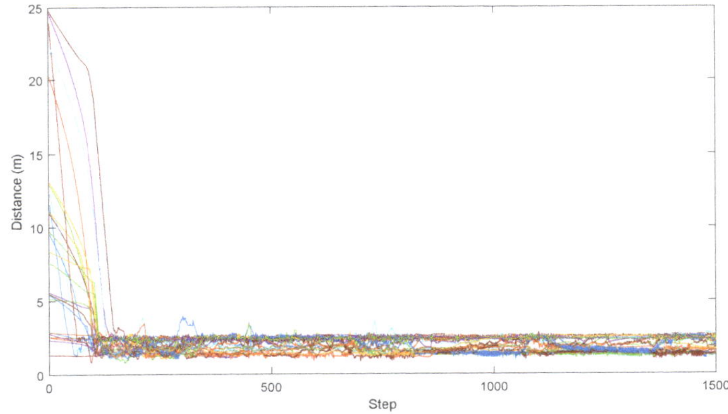

Figure 2. The distance between all neighbor UAVs in the configuration.

4.2. Consensus Control

When the swarm have achieved the desired configuration, the proposed consensus control approach will adjust the attitude of all the UAVs to achieve consensus. Figure 3 shows the process of achieving consensus from the initial yaw angle states. After the system completes the desired configuration, the UAVs have arbitrary yaw angles. Then, under the control law (23), the three UAVs in each group in the first layer adjust the yaw angles to reach a consistent state, and then control law (24) enables the unification of the yaw angle of UAVs between layers. It can be seen that when step = 150, the yaw angle of UAV basically reaches 3 degree. After step = 500, the swarm completes the unification of yaw angle. As a result, all the UAVs realize the motion consensus according to the proposed recursive consensus control concept, while maintaining the desired swarm configuration and moving in the same direction.

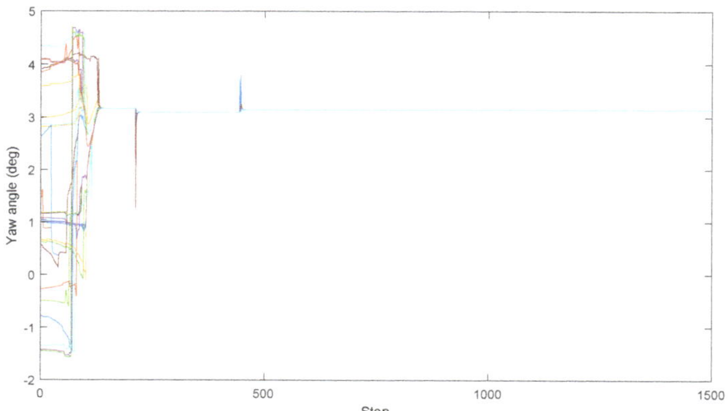

Figure 3. The yaw angle of each UAV in the configuration.

5. Conclusions

The current paper proposed a multilayer framework based on the multi-layer concept to deal with the multiagent problem with arbitrary number of UAVs. The primary contribution is that the designed multi-layer structure can be used to form the desired configuration and keep consensus under the context of large-scale UAVs swarm with Assumption 1 and Assumption 2, rather than moving into random positions. A potential function-based multi-layer controller is developed to drive all the UAVs to achieve the desired configuration precisely without collisions. Then all the UAVs reach an agreement through the consensus algorithm. The stability of the system is proved by the Lyapunov approach. The simulation studies demonstrated the effectiveness of the proposed methods for the UAVs swarm. In our future work, the trajectory tracking and the obstacle avoidance of the large-scale UAVs swarm will be investigated under the Active Disturbance Rejection Control approach.

Author Contributions: Conceptualization, formal analysis, investigation, writing, T.W.; Methodology, pre-processing, writing, S.Z.; Simulation validation, writing-review Z.P.; Supervision, Y.X.; Writing-review, H.T. All authors have read and agreed to the published version of the manuscript.

Funding: This paper is funded by National Natural Science Foundation of China (62203050) and National Natural Science Foundation of China (61720106010).

Institutional Review Board Statement: Not applicable.

Informed Consent Statement: Not applicable.

Data Availability Statement: Not applicable.

Acknowledgments: The authors would like to thank all coordinators and supervisors involved and the anonymous reviewers for their detailed comments that helped to improve the quality of this article.

Abbreviations

The following abbreviations are used in this manuscript:

UAV   Unmanned Aerial Vehicle
APF   Artificial Potential Field

References

1. Nasir, M.H.; Khan, S.A.; Khan, M.M.; Fatima, M. Swarm Intelligence inspired Intrusion Detection Systems—A systematic literature review. *Comput. Netw.* **2022**, *205*, 108708. [CrossRef]
2. Rosenberg, L.; Willcox, G.; Askay, D.; Metcalf, L.; Harris, E. Amplifying the social intelligence of teams through human swarming. In Proceedings of the 2018 First International Conference on Artificial Intelligence for Industries (AI4I), San Francisco, CA, USA, 26–28 September 2018; pp. 23–26.
3. Kiebert, L.; Joordens, M. Autonomous robotic fish for a swarm environment. In Proceedings of the 2016 11th System of Systems Engineering Conference (SoSE), Waurn Ponds, Australia, 12–16 June 2016; pp. 1–6.
4. Reynolds, C.W. Flocks, herds and schools: A distributed behavioral model. In Proceedings of the 14th Annual Conference on Computer Graphics and Interactive Techniques, Los Angeles, CA, USA, 27–31 July 1987; pp. 25–34.
5. Cucker, F.; Smale, S. Emergent behavior in flocks. *IEEE Trans. Autom. Control.* **2007**, *52*, 852–862. [CrossRef]
6. Saska, M.; Vakula, J.; Přeučil, L. Swarms of micro aerial vehicles stabilized under a visual relative localization. In Proceedings of the 2014 IEEE International Conference on Robotics and Automation (ICRA), Hong Kong Convention and Exhibition Center, Hong Kong, China, 31 May–7 June 2014; pp. 3570–3575.
7. Olfati-Saber, R. Flocking for multi-agent dynamic systems: Algorithms and theory. *IEEE Trans. Autom. Control.* **2006**, *51*, 401–420. [CrossRef]
8. Su, H.; Wang, X.; Lin, Z. Flocking of multi-agents with a virtual leader. *IEEE Trans. Autom. Control.* **2009**, *54*, 293–307. [CrossRef]
9. Fu, X.; Pan, J.; Wang, H.; Gao, X. A formation maintenance and reconstruction method of UAV swarm based on distributed control. *Aerosp. Sci. Technol.* **2020**, *104*, 1270–9638. [CrossRef]
10. Wu, Y.; Gou, J.; Hu, X.; Huang, Y. A new consensus theory-based method for formation control and obstacle avoidance of UAVs. *Aerosp. Sci. Technol.* **2020**, *107*, 1270–9638. [CrossRef]
11. Soria, E.; Schiano, F.; Floreano, D. Predictive control of aerial swarms in cluttered environments. *Nat. Mach. Intell.* **2021**, *3*, 545–554. [CrossRef]
12. Haghighi, R.; Cheah, C.C. Multi-group coordination control for robot swarms. *Automatica* **2012**, *48*, 2526–2534. [CrossRef]
13. Yan, X.; Chen, J.; Sun, D. Multilevel-based topology design and shape control of robot swarms. *Automatica* **2012**, *48*, 3122–3127. [CrossRef]
14. Pan, Z.; Sun, Z.; Deng, H.; Li, D. A Multilayer Graph for Multiagent Formation and Trajectory Tracking Control Based on MPC Algorithm. *IEEE Trans. Cybern.* **2021**, *52*, 13586–13597. [CrossRef]
15. Li, D.; Ge, S.S.; He, W.; Ma, G.; Xie, L. Multilayer formation control of multi-agent systems. *IEEE Trans. Cybern.* **2019**, *109*, 108558. [CrossRef]
16. Liu, X.; Ge, S.S.; Goh, C.H. Formation potential field for trajectory tracking control of multi-agents in constrained space. *Int. J. Control* **2017**, *90*, 2137–2151. [CrossRef]
17. Pan, Z.; Zhang, C.; Xia, Y.; Xiong, H.; Shao, X. An Improved Artificial Potential Field Method for Path Planning and Formation Control of the Multi-UAV Systems. *IEEE Trans. Circuits Syst. II Express Briefs* **2021**, *69*, 1129–1133. [CrossRef]
18. Huang, Y.; Ding, H.; Zhang, Y.; Wang, H.; Cao, D.; Xu, N.; Hu, C. A motion planning and tracking framework for autonomous vehicles based on artificial potential field elaborated resistance network approach. *IEEE Trans. Ind. Electron.* **2019**, *67*, 1376–1386. [CrossRef]
19. Wen, G.; Ge, S. S.; Tu, F.; Choo, Y.S. Artificial Potential-Based Adaptive H_∞ Synchronized Tracking Control for Accommodation Vessel. *IEEE Trans. Ind. Electron.* **2017**, *64*, 5640–5647. [CrossRef]
20. Vásárhelyi, G.; Virágh, C.; Somorjai, G.; Nepusz, T.; Eiben, A.E.; Vicsek, T. Optimized flocking of autonomous drones in confined environments. *Sci. Robot.* **2018**, *3*, eaat3536. [CrossRef]
21. Wang, X.; Li, S.; Yu, X.; Yang, J. Distributed active anti-disturbance consensus for leader-follower higher-order multi-agent systems with mismatched disturbances. *IEEE Trans. Autom. Control* **2016**, *62*, 5795–5801. [CrossRef]
22. Ma, L.; Wang, Z.; Lam, H.K. Event-triggered mean-square consensus control for time-varying stochastic multi-agent system with sensor saturations. *IEEE Trans. Autom. Control* **2016**, *62*, 3524–3531. [CrossRef]
23. Zhang, H.; Lewis, F.L.; Qu, Z. Lyapunov, adaptive, and optimal design techniques for cooperative systems on directed communication graphs. *IEEE Trans. Ind. Electron.* **2011**, *59*, 3026–3041. [CrossRef]
24. Zou, Y.; Meng, Z. Coordinated trajectory tracking of multiple vertical take-off and landing UAVs. *IEEE Trans. Ind. Electron.* **2019**, *99*, 33–40. [CrossRef]
25. Olfati-Saber, R.; Murray, R.M. Consensus problems in networks of agents with switching topology and time-delays. *IEEE Trans. Autom. Control.* **2004**, *49*, 1520–1533. [CrossRef]
26. Vicsek, T.; Czirók, A.; Ben-Jacob, E.; Cohen, I.; Shochet, O. Novel type of phase transition in a system of self-driven particles. *Phys. Rev. Lett.* **1995**, *75*, 1226. [CrossRef] [PubMed]

Article

Distributed Offloading for Multi-UAV Swarms in MEC-Assisted 5G Heterogeneous Networks

Mingfang Ma and Zhengming Wang *

College of Science, National University of Defense Technology, Changsha 410073, China
* Correspondence: wzm@nudt.edu.cn

Abstract: Mobile edge computing (MEC) is a novel paradigm that offers numerous possibilities for Internet of Things (IoT) applications. In typical use cases, unmanned aerial vehicles (UAVs) that can be applied to monitoring and logistics have received wide attention. However, subject to their own flexible maneuverability, limited computational capability, and battery energy, UAVs need to offload computation-intensive tasks to ensure the quality of service. In this paper, we solve this problem for UAV systems in a 5G heterogeneous network environment by proposing an innovative distributed framework that jointly considers transmission assessment and task offloading. Specifically, we devised a fuzzy logic-based offloading assessment mechanism at the UAV side, which can adaptively avoid risky wireless links based on the motion state of an UAV and performance transmission metrics. We introduce a multi-agent advantage actor–critic deep reinforcement learning (DRL) framework to enable the UAVs to optimize the system utility by learning the best policies from the environment. This requires decisions on computing modes as well as the choices of radio access technologies (RATs) and MEC servers in the case of offloading. The results validate the convergence and applicability of our scheme. Compared with the benchmarks, the proposed scheme is superior in many aspects, such as reducing task completion delay and energy consumption.

Keywords: unmanned aerial vehicle; heterogeneous networks; computation offloading; fuzzy logic; deep reinforcement learning

Citation: Ma, M.; Wang, Z. Distributed Offloading for Multi-UAV Swarms in MEC-Assisted 5G Heterogeneous Networks. *Drones* **2023**, *7*, 226. https://doi.org/10.3390/drones7040226

Academic Editor: Oleg Yakimenko

Received: 22 February 2023
Revised: 18 March 2023
Accepted: 20 March 2023
Published: 24 March 2023

1. Introduction

Unmanned aerial vehicles (UAVs) have become popular in recent years, thanks to their mobility, flexibility, and limited costs [1,2]. For instance, when UAV swarms are equipped with sensing devices, they can be candidates for rapid computing and communication in scenarios, such as reconnaissance, property surveillance, transportation, agriculture 4.0, etc. [3,4]. In the future, UAV swarms will play a more prominent role in enhancing existing services and enabling new ones [5]. However, novel services tend to be more computation-intensive, which is a significant challenge for UAVs with limited on-board computing power and battery capacity.

In order to alleviate the strain on device resources, multi-access edge computing (MEC) has received significant attention. By deploying powerful computing units at the edge of the network to serve devices, MEC can provide computation resources to UAV swarms in close proximity; as a result, the transmission delay between them as well as the energy consumed locally are greatly shortened [6]. In other words, by utilizing MEC, UAV swarms are able to offload task data to edge servers via wireless transmission to assist computing.

The transmission link and network node selections need to be seriously considered in the task-offloading process [7]. Previous studies [8,9] have mostly been offloaded via cellular networks, which can cause base station (BS) overload and network congestion in the face of large-scale UAV swarm connections. Fortunately, thanks to the recent popular 5G heterogeneous network architecture, which integrates different radio access technologies (RATs), UAVs can also offload tasks through access points (APs) of deployed Wi-Fi networks, thus reducing the pressure on a single cellular network and enhancing the exploits on available network resources [10]. As a result, different UAV tasks will face many choices when selecting the target network nodes to request services, and facilities in close proximity are not always the best choice. Notably, although the UAV link selection in a network-sparse environment is small and relatively fixed [11], a critical concern of this paper is how flying UAVs adaptively evaluate and select transmission links in a distributed manner to achieve flexible and stable offloading in hotspots with overlapped coverage of heterogeneous networks.

For task offloading, previous works have mainly focused on developing strategies under system certainty or used centralized approaches when faced with environmental dynamics. In most cases, they fall short of settling the multi-UAV offloading problem in unknown environments, particularly when multiple heterogeneous network nodes are deployed and UAVs fly arbitrarily. In addition, the heuristics or dynamic programming methods commonly used to achieve optimal task-offloading solutions may be time-consuming due to the large number of iterations required. As a result, these approaches may not be suitable for real-time offloading decision-making in dynamic environments. Accordingly, reinforcement learning (RL) has the potential to alleviate excessive computational demands, so as to enable learning for the agents. Previous online schemes based on RL have coped with system uncertainty to a certain extent, while offloading strategies are made centrally by the system or independently by each agent.

In this paper, we propose a distributed task-offloading scheme for multi-UAVs in MEC-assisted heterogeneous networks with the objective of maximizing the utilities of all UAVs for processing tasks through multi-UAV collaboration. To prevent UAVs from offloading via easily disconnected communication links and poorly performing service nodes, we propose an offloading assessment mechanism for UAV swarms based on fuzzy logic. In the framework, UAV velocity and transmission quality are jointly considered, and UAVs can make assessments locally and efficiently based on the perceived information. Subsequently, we designed an offloading algorithm by applying deep reinforcement learning (DRL), which adopts multi-agent advantage actor–critic (A2C) policy optimization to automatically and effectively work out the optimal solution, so as to reduce the task completion time and energy consumption of an UAV swarm in a MEC environment.

The contributions of this paper are summarized as follows:

- We introduce a multi-agent task-offloading model in a heterogeneous network environment (which is different from the existing works that consider single-network scenarios or independent devices). Moreover, the optimization problem is formulated as a Markov decision process (MDP), which is beneficial for solving the sequential offloading decision-making for UAV swarm in dynamic environments.
- To facilitate stable offloading of UAVs in any motion state, we devised a fuzzy logic-based offloading assessment mechanism. The mechanism is executed in a decentralized manner on the UAV with low complexity and can adaptively identify available offloading nodes that are prone to disconnection or have undesirable transmission quality.
- Based on the multi-agent DRL framework, we propose a distributed offloading scheme named DOMUS. DOMUS effectively enables each UAV to learn the joint optimal policy, such as determining the computing mode and selecting the RATs and MEC servers in the offloading case.
- We performed different numerical simulations to verify the rationality and efficiency of the DOMUS scheme. The evaluation results show that the DOMUS proposed is capable of rapidly converging to a stable reward, achieving the optimal offloading performance in energy consumption and delay by comparing with four other benchmarks.

The rest of the paper is structured as follows. Section 2 presents the related works on task offloading. Section 3 illustrates the system model, presents the mathematical presentation of the task computing model, formulates a utility model for the performance metrics of task computing, and defines the optimization problem. An offloading assessment mechanism based on fuzzy logic is devised in Section 4. In Section 5, we propose a distributed task-offloading algorithm by applying the multi-agent DRL framework. Finally, Section 6 demonstrates and compares the efficiency of the proposed scheme and Section 7 summarizes this paper. For ease of reference, the definitions of the key symbols are listed in Table 1.

Table 1. Key symbol definitions.

Symbols	Definition
$\mathcal{U} = \{1, ..., U\}$	Set of UAVs
$\mathcal{M} = \{1, ..., M\}$	Set of servers
κ_u	Task of UAV $u \in \mathcal{U}$
$d_{u,\kappa}$	Data size of κ_u
$c_{u,\kappa}$	Computation resources required by task κ_u
$\alpha^1_{u,\kappa},\ \alpha^2_{u,\kappa}$	Offloading decisions
λ_u	Computational capability of UAV $u \in \mathcal{U}$
λ_m	Computational capability of server $m \in \mathcal{M}$
$l^{loc}_{u,\kappa}$	Execution time in local computing
$e^{loc}_{u,\kappa}$	Energy consumption in local computing
ρ^e_u	Energy consumption coefficient per CPU cycle
$\zeta^c_{u,\kappa},\ \zeta^w_{u,\kappa}$	Transmission rate via cellular and Wi-Fi networks, respectively
$B^c_u,\ B^w_u$	Allocated bandwidth to the UAV u from cellular and Wi-Fi networks, respectively
$P^c_{u,\kappa},\ P^w_{u,\kappa}$	Transmission power of the UAV u via cellular and Wi-Fi connectivities, respectively
$G^c_{u,\kappa},\ G^w_{u,\kappa}$	Channel gain over cellular and Wi-Fi networks, respectively
$(\sigma^c_{u,\kappa})^2,\ (\sigma^w_{u,\kappa})^2$	Noise power of the channel over cellular and Wi-Fi networks, respectively
$d_{u,m}$	Distance between the UAV u and the server m
$l^{tr}_{u,\kappa,m}$	Task transmission time in the MEC offloading
$e^{mec}_{u,\kappa}$	Task execution time on the server
$e^{mec}_{u,\kappa}$	Transmission energy consumption in the MEC offloading
$l^{mec}_{u,\kappa}$	Total time in the MEC offloading
$\hat{e}_u$	Maximum energy constraint of the UAV u
$\hat{l}_{u,k},\ \hat{b}_{u,k},\ \hat{p}_{u,k}$	Tolerable upper bound values for delay, BER, and PLR, respectively
$w_{d,\kappa},\ w_{e,\kappa}$	Balance factors for delay and energy consumption, respectively
$\hat{C}_m$	Computation capacity of the server m
F_u	Utility of the UAV u
$fuzzy(\cdot)$	Fuzzy logic processor
$p^m_{u,\kappa}$	Packet loss rate generated in the data transmission
$b^m_{u,\kappa}$	Bit error rate generated in the data transmission
χ_m	Offloading probability

2. Related Work

Effective offloading of computer-intensive application tasks for smart devices, especially UAVs, is becoming more critical. Accordingly, many studies related to task offloading are being proposed. In this section, we briefly outline the related work.

Some studies consider centralized controllers to realize offloading decisions. Li et al. [12] considered that the tasks performed in Maritime environments have strict delay requirements; they designed a genetic-based offloading algorithm for energy-starved UAVs, which optimizes energy consumption under the task delay constraint. Guo et al. [13] studied task offloading in a MEC system and attempted to minimize the system overhead by expressing the offloading as a mixed-integer non-linear programming problem, proposing a heuristic algorithm based on a greedy policy. Zhang et al. [14] integrated latency and energy consumption to obtain the offloading utility, which was combined with simulated annealing to make offloading strategies in MEC, so as to enhance the utility. These efforts [12–14] have global coordination but require UAVs to upload private information related to the tasks executed and real-time status to enable centralized offloading decision-making. This will significantly increase the burden on the centralized controller when the scale of the UAV swarm increases.

As network environments become larger and more complex, distributed frameworks are becoming more popular in some computation-offloading efforts [15,16]. Dai et al. [1] developed a vehicle-assisted offloading architecture for UAVs in smart cities, where vehicles and UAVs are matched according to preferences and the offloading process of data are modeled as part of a bargaining game to enhance the offloading efficiency and optimize the system utility. Zhou et al. [17] modeled the interaction in offloading as part of a Stackelberg game and maximized the utility of the system. To select a suitable service provider for the offloaded task of UAV, Gu et al. [18] devised an evolutionary game-based offloading approach to make a trade-off between latency, energy, and cost. However, these methods require multiple interactions and iterations of all participants to reach a satisfactory optimal solution, and they are not always suitable for making real-time decisions due to the fact that UAVs have good maneuverability, which can lead to rapid changes in environmental states.

Swarm intelligence is a popular approach used in multi-UAV systems and can enable global behavior to emerge from UAV clusters through operations such as interactions. As a result, swarm intelligence algorithms have received more attention in the implementation of UAV offloading. You et al. [19] introduced a computation-offloading scheme based on particle swarm optimization, which can offload tasks to low-latency MEC servers and balance the load on the servers. Li et al. [20] constructed an offloading model, which aims to minimize the delay of whole UAVs under the constraint of consumed energy; they applied the bat algorithm to solve the model. In [21], Asaamoning et al. researched computing offloading in a networked control system consisting of UAVs and discussed the application of swarm intelligence approaches, such as ant colony optimization and bee colony optimization. In addition, these swarm intelligence approaches can help determine the optimal positions of drone base stations, which can provide support for drones to act as base stations in the next generation of the Internet of Things [22,23].

Some studies on computation offloading in MEC tend to leverage reinforcement learning because of its strength in adapting to dynamic environments. Chen et al. [24] constructed task-offloading architecture based on deep deterministic policy gradients to optimize the offloading performance. Different from these schemes, refs. [24–26], our work devises the distributed decision-making mechanism by leveraging the multi-agent DRL framework, which can collaboratively deal with the optimization of offloading policies for multi-UAVs with heterogeneous tasks. Although there are distributed approaches for computation offloading decision-making that applies reinforcement learning, such as Q-TOMEC [27], TORA [28], and a distributed offloading technique based on deep Q-learning [29], these approaches utilize parallel deep neural networks instead of considering collaboration among agents.

This paper considers the popular 5G heterogeneous network architecture rather than a single network, as considered in many papers [8,9,30,31]. Correspondingly, it is important to evaluate and choose the appropriate offloading link among many heterogeneous networks for UAVs with good maneuverability. This is because an improper offloading selection may lead to frequent service interruptions, network hand-offs, and transmission link failures. However, existing offloading schemes are only concerned with task deadlines, energy consumed, or a balance between the two. In order to make the offloading scheme effective, we propose an offloading assessment mechanism that jointly considers the effects of transmission quality and UAV mobility to ensure efficient data transmission. Furthermore, we designed the mechanism to be fully decentralized on the UAV side, so that the mechanism has great scalability. To our knowledge, this is the first attempt to research the link evaluation during the task offloading of an UAV swarm.

3. System Model and Problem Definition

In this section, we first depict the system model (Section 3.1), and present the mathematical formulation of task computing models (Section 3.2). Then we formulate a utility function to evaluate the critical attributes that can affect the decision-making of UAVs (Section 3.3). Finally, according to the system model and task computing models, we define the optimization problem to be solved in this paper (Section 3.4).

3.1. System Model

The system in Figure 1 shows a flying UAV swarm that is defined by a group of agents $\mathcal{U} = \{1, \ldots, U\}$. In addition, the wireless network is supported by heterogeneous RATs, including Wi-Fi APs and cellular BSs; each is equipped with a MEC server that can provide computational capability for the energy-constrained UAV swarms to process their tasks. We let $\mathcal{M} = \{1, \ldots, M\}$ indicate a set of geo-distributed MEC servers.

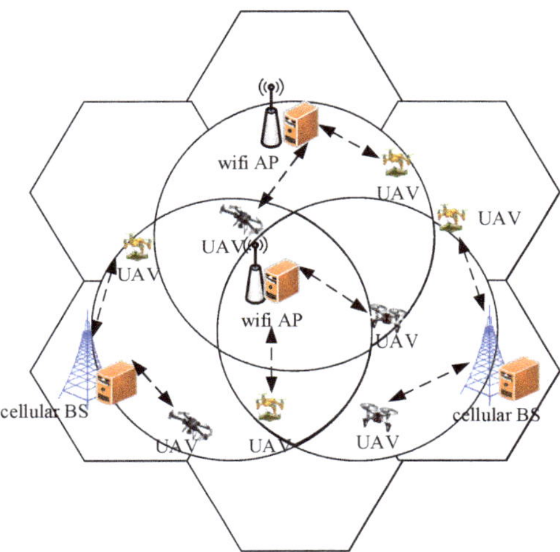

Figure 1. Task offloading for multi-UAV swarms in MEC-assisted 5G heterogeneous networks.

Furthermore, each UAV $u \in \mathcal{U}$ has a task to process at a certain time; we use a tuple $\kappa_u = \{d_{u,\kappa}, c_{u,\kappa}\}$ to express the UAV task u; the data size of the task κ_u is indicated by $d_{u,\kappa}$, and the total computation resources required to complete κ_u are denoted by $c_{u,\kappa}$. Moreover, the application tasks place tight requirements on the quality of service (QoS) attributes, such as delay, BER, and PLR when executing tasks.

In view of the above, each UAV in the system can perform its task κ_u by computing locally, offloading to a MEC server through cellular BS or a Wi-Fi AP. Correspondingly, when each UAV u performs its task κ_u, two binary variables ($\alpha_{u,\kappa}^1$ and $\alpha_{u,\kappa}^2$) are used to characterize the decisions made by the UAV u; we provide the following explanations for them.

$$\alpha_{u,\kappa}^1 = \begin{cases} 0 & \text{local computing} \\ 1 & \kappa_u \text{ is offloaded} \end{cases} \tag{1}$$

$$\alpha_{u,\kappa}^2 = \begin{cases} 0 & \kappa_u \text{ is offloaded to a BS server} \\ 1 & \kappa_u \text{ is offloaded to a Wi-Fi server} \end{cases} \tag{2}$$

in which $\alpha_{u,\kappa}^1$ expresses the task κ_u computed locally or offloaded. In the second decision, $\alpha_{u,\kappa}^2 = 0$ or 1 means task κ_u is offloaded to a MEC server equipped with a cellular BS or Wi-Fi AP, which occurs only when $\alpha_{u,\kappa}^1 = 1$. These divergent task computing modes will enable the UAVs to efficiently implement tasks and obtain great service performance.

3.2. Task Computing Models

Considering the limitations on the energy and computational capabilities of UAVs, it is important to study the hybrid task computing modes to support the UAVs in this paper.

(1) Local computing model

After the UAV u perceives the task data, it may process the tasks locally. The local execution time can be computed as

$$l_{u,\kappa}^{loc} = \frac{c_{u,\kappa}}{\lambda_u} \tag{3}$$

where λ_u indicates the computational capability of the UAV u and $c_{u,\kappa}$ denotes the needed CPU amount to complete the task κ.

Let $e_{u,\kappa}^{loc}$ denote the energy consumed on local computing, which is represented by

$$e_{u,\kappa}^{loc} = \rho_u^e c_{u,\kappa} \tag{4}$$

where ρ_u^e means the local energy consumption coefficient per CPU cycle.

(2) MEC offloading model

In this model, we consider that there is more than one UAV that will offload tasks to the same MEC server in the same time period. In this case, if the UAV u performs the task κ by MEC, the achieved data transmission rates via the cellular and Wi-Fi networks are denoted by $\zeta_{u,\kappa}^c$ and $\zeta_{u,\kappa}^w$, which are, respectively, presented in Equations (5) and (6) [32].

$$\zeta_{u,\kappa}^c = B_u^c \cdot \log_2 \left(1 + \frac{P_{u,\kappa}^c G_{u,\kappa}^c}{(\sigma_{u,\kappa}^c)^2 + \sum_{u' \neq u} P_{u',\kappa'}^c G_{u',\kappa'}^c} \right) \tag{5}$$

$$\zeta_{u,\kappa}^w = B_u^w \cdot \log_2 \left(1 + \frac{P_{u,\kappa}^w G_{u,\kappa}^w}{(\sigma_{u,\kappa}^w)^2 + \sum_{u' \neq u} P_{u',\kappa'}^w G_{u',\kappa'}^w} \right) \tag{6}$$

In Equation (5), $P_{u,\kappa}^c$ means the transmission power of the UAV u for offloading the task to the MEC server m via cellular connectivity; $G_{u,\kappa}^c = d_{u,m}^{-\iota}$ is the channel gain because of the path loss effect and shadowing, where the path loss coefficient is denoted by ι, the distance between the UAV u and the server m is $d_{u,m}$, $d_{u,m} = \sqrt{dv_{u,m}^2 + dh_{u,m}^2}$, where $dv_{u,m}$ and $dh_{u,m}$, respectively, indicate the vertical and horizontal distances between the UAV u and the server m; $(\sigma_{u,\kappa}^c)^2$ denotes the noise power of the channel, u' defines the other UAVs that access the server m to process its task κ', and B_u^c expresses the allocated bandwidth from the cellular network. Additionally, the variables in Equation (6) have the same meanings as those in Equation (5).

Then the transmission time $l_{u,\kappa,m}^{tr}$ of the task data for the UAV u can be represented as

$$l_{u,\kappa,m}^{tr} = \begin{cases} l_{u,\kappa}^c = \frac{d_{u,\kappa}}{\zeta_{u,\kappa}^c} & \alpha_{u,\kappa}^1 = 1, \alpha_{u,\kappa}^2 = 0 \\ l_{u,\kappa}^w = \frac{d_{u,\kappa}}{\zeta_{u,\kappa}^w} & \alpha_{u,\kappa}^1 = 1, \alpha_{u,\kappa}^2 = 1 \end{cases} \tag{7}$$

Here, $l_{u,\kappa,m}^{tr}$ is a general variable; it defines the transmission time $l_{u,\kappa}^c$ and $l_{u,\kappa}^w$ occurs in the data transmission through the cellular or Wi-Fi network, respectively.

Accordingly, if the transmission power of the UAV u is indicated by P_u, and $P_u \in \{P_{u,\kappa}^c, P_{u,\kappa}^w\}$, the energy $e_{u,\kappa}^{mec}$ consumed by the UAV u during data transmission can be written as

$$e_{u,\kappa}^{mec} = P_u l_{u,\kappa,m}^{tr} \tag{8}$$

After the task data are transmitted to the server m, similar to the local computing model, the data processing time $l^{exe}_{u,\kappa,m}$ on the server m can be represented by

$$l^{exe}_{u,\kappa,m} = \frac{c_{u,\kappa}}{\lambda_m} \tag{9}$$

in which λ_m denotes the computational capability of the server m. Thereby, the total time consumed by the UAV u during offloading is expressed as

$$l^{mec}_{u,\kappa} = l^{tr}_{u,\kappa,m} + l^{exe}_{u,\kappa,m} \tag{10}$$

3.3. Utility Model in Task Computing

According to the utility theory, we designed a utility function to effectively evaluate the consumed time and energy during the processing tasks. The function designed is formulated as

$$\phi(z) = \begin{cases} 1 & z = 0 \\ \dfrac{1}{1+(\frac{z}{z^{mid}})^{\eta_z}} & 0 < z \le z^{mid} \\ \dfrac{(\frac{z^{max}-z}{z^{mid}})^{\eta_z}}{1+(\frac{z^{max}-z}{z^{mid}})^{\eta_z}} & z^{mid} < z < z^{max} \\ 0 & z \ge z^{max} \end{cases} \tag{11}$$

The value of $\phi(z)$ is mainly determined by the property of the designed function, such as twice differentiability, monotonicity, and convexity-concavity, as well as the tolerable upper bound z^{max} of the attribute z for an application task. Moreover, $z^{mid} = \frac{z^{max}}{2}, \eta_z \ge 2$ characterizes the sensitivity of the application task to a specific attribute and determines the steepness of the function.

To this end, when the UAV u adopts any task computing modes, the utility of the consumed time and energy can be measured by the above-designed utility function in Equation (11); we used $\phi(l)$ and $\phi(e)$ to express them, respectively. Then, on the basis of the multi-attribute utility principle, performing the tasks for the UAV u can be measured by the following utility F_u, which is the function of time and energy consumed when the UAV u chooses a certain task computing mode. Moreover, F_u is represented as follows:

$$F_u = w_{d,\kappa}\phi(l) + w_{e,\kappa}\phi(e) \tag{12}$$

in which $w_{d,\kappa}$ and $w_{e,\kappa}$ characterize the balance factors between the consumed time and energy; hence, $w_{d,\kappa} + w_{e,\kappa} = 1$.

3.4. Optimization Problem Formulation

Based on the system model constructed, our objective is to maximize the utility of UAVs by making optimal task computation decisions. Thus, the optimization problem under related constraints can be defined as follows:

$$P1 : \max_{\mathcal{A}_u} \sum_{u=1}^{U} F_u$$

$$s.t. \begin{cases} C_1 : \alpha^1_{u,\kappa} \in \{0,1\}, \alpha^2_{u,\kappa} \in \{0,1\}, \forall u \in \mathcal{U} \\ C_2 : C'_m(t) \le \hat{C}_m, \forall m \in \mathcal{M} \\ C_3 : e < \hat{e}_u \\ C_4 : l < \hat{l}_{u,\kappa} \\ C_5 : p < \hat{p}_{u,\kappa} \\ C_6 : b < \hat{b}_{u,\kappa} \end{cases} \tag{13}$$

The defined optimization problem is constrained by the binary offloading decision, computation capacity of the MEC servers, battery energy of UAVs, and the QoS demands of the tasks performed. In Equation (13), $\mathcal{A}_u$ denotes the decision set of each UAV. C_1 indicates the offloading decision constraint, and $C'_m(t) = \sum c_{u,\kappa}$ expresses the computation resources of the server m occupied by UAVs; therefore, C_2 denotes whether server $m \in \mathcal{M}$ is selected to provide service. The computation resources used by UAVs cannot exceed the computation capacity of the server m at a certain time slot t. C_3 indicates the battery energy constraint of an UAV $u \in \mathcal{U}$; C_4 denotes that the consumed time when executing task κ_u should be controlled within the allowable delay threshold $\hat{l}_{u,\kappa}$. C_5 and C_6 express whether the PLR and BER occurring in the data transmission should satisfy the tolerable upper bound values for a certain task κ_u if processed by the MEC.

The optimization problem is an integer-programming problem; the feasible decision number for task computation is $(M+1)^U$, which is commonly non-convex and NP-hard. Conventional mathematical-based optimization approaches can work out the optimal solution for the proposed problem theoretically but are unable to realize it in a short time. The DRL approach is applicable for settling the decision-making problems with high-dimensional solution spaces effectively, especially for the increased number of offloaded tasks in future wireless networks. In view of the above, we will develop a multi-agent A2C-based DRL scheme, which can find feasible offloading actions in polynomial time.

4. Offloading Assessment Based on Fuzzy Logic

In this section, by applying the fuzzy logic theory, we propose an assessment mechanism to let the UAV swarm adaptively screen the available offloading nodes to reduce the complexity and speed up the training progress of the MODUS proposed.

Fuzzy logic is quite distinct from binary logic, which is capable of making a decision based on multi-valued logic. This characteristic enables the fuzzy logic system to handle the input variables that have uncertain and incomplete data. The fuzzy logic-based approach can effectively rapidly respond to the dynamicity of the changing environment and adaptively produce crisp values. The proposed offloading assessment algorithm based on fuzzy logic involves processing the inputs containing the PLR, BER, and velocity parameters, then the input variables are processed through fuzzification, fuzzy inference, and defuzzification; finally, the fuzzy logic system outputs a crisp value, i.e., offloading probability, which signifies the fitness of a specific MEC server for the UAV task.

In particular, the fuzzy logic-based offloading assessment algorithm is deployed on the UAV in a decentralized manner, and can regard all MEC servers perceived as the candidate-offloading targets to be assessed. Algorithm 1 depicts the procedures of the proposed offloading assessment scheme.

Algorithm 1 Fuzzy logic-based offloading assessment.

Input: Set of candidate-offloading nodes $\mathcal{M}$.
Output: Available offloading node set $\tilde{\mathcal{M}}$ for the UAV u.
  1: **while** Obtaining sensor data in a time period **do**
  2:     **for** $m = 1 : M$ **do**
  3:         Velocity, PLR, BER, $\longleftarrow \mathcal{M}[m]$.velocity, .PLR, .BER according to Equations (14) and (15);
  4:         BER, PLR $\longleftarrow Normal$(BER), $Normal$(PLR);
  5:         $\chi_m \longleftarrow$ Fuzzy Logic (velocity, PLR, BER);
  6:         **if** $\chi_m < \hat{\chi}_m$ **then**
  7:             $\tilde{\mathcal{M}} \longleftarrow m$;
  8:         **end if**
  9:     **end for**
 10: **end while**

In step 3 of Algorithm 1, each UAV senses nearby offloading nodes, observes the flying velocity, and perceives the data of the PLR and BER with respect to candidate-offloading targets by assuming that the UAV task u will be offloaded to the node.

The PLR occurring in data transmission can be commonly evaluated by

$$p_{u,\kappa}^m = \varsigma \cdot d_{u,\kappa} \cdot \exp\left(-\vartheta \cdot \frac{P_{rss}}{\sigma^2 \cdot d_{u,m}^2}\right) \tag{14}$$

in which P_{rss} indicates the received signal strength and $\sigma^2 \in \{(\sigma_{u,\kappa}^c)^2, (\sigma_{u,\kappa}^w)^2\}$ denotes the noise power, ς and ϑ are two tunable parameters; both of them meet the condition that $0 < \varsigma < 1, 0 < \vartheta < 1$.

The BER for a binary phase-shift keying modulation in an additive Gaussian white noise environment is expressed as follows

$$b_{u,\kappa}^m = \frac{1}{2} erfc\left(\sqrt{\frac{P_{rss}}{\sigma^2}}\right) \tag{15}$$

where the $erfc(.)$ is a Gauss complementary error function, it can be written as

$$erfc(x) = \frac{2}{\sqrt{\pi}} \int_x^\infty e^{-\mu^2} d\mu \tag{16}$$

In particular, in step 4 of the algorithm, the obtained PLR and BER are normalized by $\frac{p_{u,\kappa}^m}{\hat{p}_{u,\kappa}}$ and $\frac{b_{u,\kappa}^m}{\hat{b}_{u,\kappa}}$ so as to eliminate the unit difference before inputting it into the fuzzy logic system. In step 5, the designed fuzzy logic system for offloading the assessment maps the sensed data, including velocity, PLR, and BER into fuzzy sets according to the membership functions (MFs) for each one; this process is called fuzzification. Afterward, the fuzzy inference procedure infers the fuzzified inputs and produces fuzzy output based on multiple IF-AND-THEN rules, which are designed by following empirically fuzzy rule sets. Furthermore, based on the triggered fuzzy rule, the fuzzy logic system proceeds to the defuzzification stage, which calculates and outputs a scalar value χ_m for the node $m \in \mathcal{M}$ by applying the centroid defuzzifier method [33]. Moreover, $\chi_m \in [0, 1]$ can characterize the fitness of the offloading node for the task κ_u of the UAV u; the higher the χ_m, the better the fitness. Finally, in steps 6 and 7, the obtained χ_m is compared with its permitted upper bound $\hat{\chi}_m$; if the condition is satisfied, the node $m \in \mathcal{M}$ will be selected by the UAV u as the available offloading target, and be included in $\tilde{\mathcal{M}}$.

5. Multi-Agent A2C-Based Decentralized Task Offloading

The optimization problem formulated in Equation (13) is a sequential decision-making problem in the dynamic environment. In this section, the problem of multi-UAV offloading is a time-varying multi-agent MDP; we propose a decentralized task-offloading scheme (DOMUS). We consider that the environmental dynamics of wireless networks are always unknown; thus, the proposed algorithm applies a model-free DRL framework based on the multi-agent A2C to enable each UAV agent to learn the optimal computing policy task via training in polynomial time.

5.1. Multi-Agent MDP Model in the A2C Framework

In the A2C framework, the optimization problem in Equation (13) for the task implementation of multi-UAVs can be defined as a multi-agent MDP $\langle \mathcal{U}, \mathcal{S}, \{\mathcal{A}_u\}_{u \in \mathcal{U}}, \mathcal{P}, \{\mathcal{R}_u\}_{u \in \mathcal{U}} \rangle$, which will be interpreted in detail as follows.

(1) State space $\mathcal{S}$. In a time slot t, each UAV agent observes the system state $s^t \in \mathcal{S}$, which involves the location of the UAV and relevant information of tasks and situations of the MEC environment; thus, the state s^t is constituted by a group of parameter metrics. (1) $loc(x, y, h)$: three-dimensional location of the UAV agent u; (2) $(d_{u,\kappa}, c_{u,\kappa})$: data size and required computation resources for the task κ; (3) $\hat{l}_{u,\kappa}$: maximum tolerable delay for the κ_u

task; (4) $SR = \{I_{u,1}, \ldots, I_{u,M}\}$: the signal-to-noise ratio vector between the UAV u and its available offloading nodes in $\tilde{\mathcal{M}}$; (5) $Dist = \{d_{u,1}, \ldots, d_{u,M}\}$: the distance vector between the UAV u and its available offloading nodes in $\tilde{\mathcal{M}}$.

(2) Action space $\mathcal{A}_u$. The action taken in the time slot t for each UAV u is to decide whether the task should be performed locally or offloaded to a MEC server, and if offloaded, which server will be selected. Thus, according to the definitions of the task computing decisions in Equations (1) and (2) in the system model, the action set for each UAV can be represented as $\mathcal{A}_u = \{a_u^1, a_u^2, a_u^3\}$, in which a_u^1 indicates $\alpha_{u,\kappa}^1 = 0$, a_u^2 denotes $\alpha_{u,\kappa}^1 = 1$ and $\alpha_{u,\kappa}^2 = 0$, and a_u^3 expresses $\alpha_{u,\kappa}^1 = 1$ and $\alpha_{u,\kappa}^2 = 1$.

(3) Reward function $\mathcal{R}_u$. At state s^t, each UAV chooses an action and receives an instant reward r_u^t from the environment. It is known that the purpose of each agent is to maximize its utility through improving the policy of task computing. For this reason, we define the reward r_u^t as the performance improvement between two utility values obtained by the UAV within two consecutive time slots; the r_u^t is written as

$$r_u^t = \begin{cases} \varepsilon_1 & F_u^t - F_u^{t-1} > \beta \\ \varepsilon_2 & F_u^t - F_u^{t-1} < -\beta \\ 0 & \text{otherwise,} \end{cases} \tag{17}$$

where F_u^t refers to the utility of the UAV u for processing tasks at time slot t, $\varepsilon_1 > 0$, and $\varepsilon_2 < 0$; both ε_1 and ε_2 denote the obtained instant rewards under two different situations, i.e., $F_u^t - F_u^{t-1} > \beta$ and $F_u^t - F_u^{t-1} < -\beta$, respectively. Moreover, $\beta > 0$ means the sensitivity to utility changes of the UAV in MDP. Therefore, the reward r_u^t can effectively characterize the change directions of two utilities corresponding to two consecutive time slots. Furthermore, the reward function $\mathcal{R}_u$ can be presented as $\mathcal{R}_u(s,a) = E[r_u^{t+1}|s_t, a_t]$, which is the expected value of the instant reward.

Therefore, in the multi-agent MDP model, at the current time slot t, if the state is $s^t \in \mathcal{S}$ and the joint actions of agents in the system can be denoted as $a^t = \{a_1, \ldots, a_U\} \in \mathcal{A}$, each agent $u \in \mathcal{U}$ can obtain a reward r_u^{t+1}. Then the state will be transformed into a new state $s^{t+1} \in \mathcal{S}$ according to the transition probability $P(s^{t+1}|s^t, a^t)$. Additionally, the policy of agent u is denoted as the probability that the agent selects the action at a given state, which can be expressed as $\pi_u(s, a_u)$. Then the joint policy of all agents can be formulated as $\pi(s,a) = \Pi_{u=1}^{U} \pi_u(s, a_u)$, and the $\pi(s,a)$ is written as π for simplicity.

5.2. Multi-Agent A2C Framework

As deep neural networks (DNNs) can offer accurate regression, A2C applies DNNs to the actor and the critic networks to approximate the policy and value function. The actor is a policy function $\pi_u(a_u|s; \theta_u)$, which allows agent u to yield a policy and select an action a_u under state s, where θ_u is a parameter of the DNN. We pack θ_u in a set $\theta = \theta_1, \ldots, \theta_u, \ldots, \theta_U$. The critic is a state value function $V(s)$ used to evaluate the state. Additionally, the advantage term expresses that there is a function $\zeta(s,a) = Q(s,a) - V(s)$ to indicate the advantage of the selected action under a given state, where $Q(s,a)$ represents the action-value function.

The learning objective of the agent in A2C is to find a policy π that can maximize the expected long-term system reward $J(\pi)$ over all possible trajectories. Accordingly, our optimization objective is to learn the optimal joint task computing policy $\pi^\theta = \pi(a|s; \theta) = \Pi_{u=1}^{U} \pi_u(a_u|s; \theta_u)$ so as to maximize the globally averaged expected reward $J(\pi)$ for agents, which are represented as

$$\pi^\theta = Arg \max_{\pi} J(\pi) \tag{18}$$

in which $J(\pi)$ is given as follows:

$$J(\pi) = \lim_{T} \frac{1}{T} E\left[\sum_{t=1}^{T} \frac{1}{U} \sum_{u \in \mathcal{U}} r_u^{t+1}\right]$$
$$= \sum_{s \in \mathcal{S}} \eta_\pi(s) \sum_{a \in \mathcal{A}} \pi(s,a) \frac{1}{U} \sum \mathcal{R}_u(s,a) \tag{19}$$

where $\eta_\pi(s) = \lim_{t \to \infty} \mathbf{Pr}(s_t = s|\pi)$ denotes the stationary probability distribution in the Markov chain when the policy π is given.

Furthermore, our optimization problem aims to work out

$$\max J(\theta) = \sum_{s \in \mathcal{S}} \eta_\theta(s) \sum_{a \in \mathcal{A}} \pi(a|s;\theta) \frac{1}{U} \sum \mathcal{R}_u(s,a) \tag{20}$$

where θ will be learned by the policy gradient method [34]. Moreover, based on the objective function, the gradients for θ can be calculated as

$$\nabla_{\theta_u} J(\theta) = E[\nabla_{\theta_u} \log \pi_u^{\theta_u} \zeta_u^{\pi^\theta}(s,a)] \tag{21}$$

in which $\zeta_u^{\pi^\theta}(s,a)$ is the advantage function, represented by

$$\zeta_u^{\pi^\theta}(s,a) = Q^{\pi^\theta}(s,a) - V_u^{\pi^\theta}(s,a_{-u}) \tag{22}$$

where we use a_{-u} to present the actions adopted by other agents, except for agent u; Q^{π^θ} indicates the action value function under the policy π^θ for a given state–action pair (s,a), while $V_u^{\pi^\theta}$ is the state value function. They are given as follows:

$$Q^{\pi^\theta}(s,a) = \sum_t E\left[\frac{1}{U} \sum_{u \in \mathcal{U}} r_u^{t+1} - J(\theta)|s^0 = s, a^0 = a, \pi^\theta)\right] \tag{23}$$

$$V_u^{\pi^\theta}(s,a_{-u}) = \sum_{a_u \in \mathcal{A}_u} \pi_u(a_u|s;\theta_u) Q^{\pi^\theta}(s,a_u,a_{-u})] \tag{24}$$

A2C takes the temporal difference (TD) error as an unbiased estimation to evaluate the advantage function, which reduces the complexity of the parameter update and improves the stability of the algorithm. In this case, the advantage is approximated as

$$\zeta(s_t,a_t) \approx \frac{1}{U} r^{t+1} + \gamma V(s^{t+1}|s^t,a^t) - V(s^t) = \delta(s^t) \tag{25}$$

in which γ indicates the discounted factor.

The critic network estimates $Q(s,a)$ with $Q^{\pi^\theta}(s,a)$ and generates a TD error to express whether the action taken by the agent is good or not, as well as updates the DNN parameter θ^c with the gradient descent method. Additionally, each UAV u can share estimations from the critic network with other UAVs nearby to effectively evaluate the actions. Then the output of the critic network is further used to update the parameter θ^a for the actor network of the UAV agent u, which aims at improving the probabilities of actions that perform relatively well. In particular, the update to θ^a and θ^c can be presented as

$$\theta^a \leftarrow \theta^a + \frac{\partial \log \pi(a^t|s^t;\theta^a)}{\partial \theta^a} \delta^t(s^t;\theta^c) \tag{26}$$

$$\theta^c \leftarrow \theta^c + \delta^t(s^t;\theta^c) \frac{\partial V(s^t;\theta^c)}{\partial \theta^c} \tag{27}$$

5.3. A2C-Based Decentralized Offloading Algorithm

In this section, based on the A2C model, we propose a distributed offloading algorithm for multi-UAVs; its implementation is summarized in Algorithm 2.

Algorithm 2 A2C-based decentralized offloading algorithm.

Input: UAV swarm $\mathcal{U}$, MEC server $\mathcal{M}$, the learning rates lr^a, lr^c of the actor and critic network, the maximum episodes Ep_{max}, the step size of one episode Ep_i, the update interval Δt, and the discount factor γ;
Output: $\pi^* = \{\pi_u^*, u \in \mathcal{U}\}$ for all UAVs.
  1: **for** UAV $u = 1 : U$ **do**
  2:    Initialize the parameters θ_u^a and θ_u^c with respect to the actor and critic network;
  3: **end for**
  4: **for** Episode $i = 1 : Ep_{max}$ **do**
  5:    Reset the state: $loc(x,y,h)$, $(d_{u,\kappa}, c_{u,\kappa})$, $\hat{l}_{u,\kappa}$, $SR = \{I_{u,1}, \ldots, I_{u,M}\}$ and $Dist = \{d_{u,1}, \ldots, d_{u,M}\}$;
  6:    **for** UAV $u = 1 : U$ **do**
  7:       Execute Algorithm 1 to obtain $\tilde{\mathcal{M}}$;
  8:       Obtain the state s^0;
  9:    **end for**
 10:    **for** Step $t = 1 : Ep_i$ **do**
 11:       **for** UAV $u = 1 : U$ **do**
 12:          Takes action a_u^t by actor $\pi_u(a_u^t|s^t; \theta_u^a)$;
 13:       **end for**
 14:       Perform computation offloading according to the joint actions $a^t = \{a_1, \ldots, a_U\}$;
 15:       Obtain the current reward $r^t = \{r_1, \ldots, r_U\}$ and calculate the new state s^{t+1} ;
 16:       **if** $mod(\Delta t, t) == 0$ **then**
 17:          Update θ^c for the critic networks based on Equation (26);
 18:          Compute θ^a for the actor networks using Equation (27);
 19:       **end if**
 20:    **end for**
 21: **end for**

At the initial stage, we give the related parameters including the set of UAVs and MEC servers, i.e., $\mathcal{U}$, $\mathcal{M}$, the learning rate lr^a, lr^c of the actor and critic network, maximum number of training episodes Ep_{max} and the step size Ep_i of one episode, update interval Δt, as well as the discount factor γ. For each UAV $u \in \mathcal{U}$, we initialize the actor parameter θ_u^a and critic parameter θ_u^c. Afterward, at the start of each episode in the training stage, the system state is randomly initialized, including the locations of UAVs and the relevant information of tasks and situations of the MEC environment; each UAV will execute Algorithm 1 to obtain the available offloading node set $\tilde{\mathcal{M}}$, then the initial state s^0 is obtained (from steps 5 to 8).

Without loss of generality, one training episode is divided into Ep_i time slots. At time slot t, each UAV adopts an action according to the policy $\pi_u(a_u^t|s^t; \theta_u^a)$ in the actor, then performs computation offloading according to the adopted action $a_u \in a_t$, and obtains the instant reward $r^u \in r^t$; next, the state is updated to s^{t+1} (from steps 11 to 15). Finally, once every Δt, the algorithm updates the parameters of the actor and critic network by only sampling the (s^{t+1}, a^t, s^t) (from steps 16 to 19). In order to enable the average reward to converge to a stable value and learn the optimal policy, the iterative training will last for Ep_{max} episodes. After convergence, the algorithm only needs to save the actor network to make offloading decisions for UAVs.

The computational complexity of Algorithm 1 is to explore the available offloading nodes by each UAV, and the complexity of the designed fuzzy logic module is a constant; thus, the complexity for Algorithm 1 is $O(M)$ in the worst case. In Algorithm 2, at the training stage, each UAV agent evaluates the Q-value with the critic network by inputting the joint actions of UAVs and the environment state; thus, the input and output sizes of the critic network in UAV are $U|\mathcal{S}|$ and 1, respectively. Moreover, each UAV makes an action by mapping the current state to the actor network; thereby, the input and output sizes of the actor network in UAV are $|\mathcal{S}|$ and 1, respectively. After training is finished, the action for each UAV can be obtained from its actor network only with the $|\mathcal{S}|$ input size and 1 output size. The computational complexity is proportional to the input and output sizes; thus, the overall complexity of our DOMUS proposed is $O(M + U|\mathcal{S}|)$.

6. Performance Evaluation

In this section, we perform a series of numerical simulations and evaluate the proposed task-offloading scheme for the UAV swarm in MEC-assisted heterogeneous networks.

6.1. Parameter Settings

We consider a MEC-assisted heterogeneous network scenario where MEC servers are randomly deployed in the 1000×1000 m area, UAVs are randomly distributed, and each UAV has a task to be processed. Concerning the communication parameters, the maximum communication ranges for cellular and Wi-Fi networks are 400 and 200 m, respectively [35,36]. The bandwidths of cellular and Wi-Fi networks are $B^c = 4$ MHz, $B^w = 5$ MHz, the transmission power P_u of UAV is set at 10 W [37], and the Gaussian noise power $(\sigma_{u,\kappa}^c)^2$ and $(\sigma_{u,\kappa}^w)^2$ are set as the same value, i.e., -100 dBm. The path loss follows a distance-dependent model with a path loss coefficient of $\iota \geq 1$ [38,39]. Additionally, the computational capability of MEC servers and UAVs are characterized by uniformly distributed variables λ_m and λ_u, which are uniformly distributed in $[5,8]$ and $[0.7,1]$ Gcycles/s [40], respectively. The computational capacity $\hat{C}_m$ of the server is uniformly distributed in $[9,11]$ Gcycles. The energy consumed per CPU cycle denoted by ρ_u^e is 5×10^{-10} J/cycle. For the computation tasks to be completed by UAVs, the data size $d_{u,\kappa}$ and the needed computation resources $c_{u,\kappa}$ are uniformly distributed in $[4,5]$ MB and $[1.6,2]$ Gcycles. The weighting factors $w_{d,\kappa}$ and $w_{e,\kappa}$ for delay and energy are commonly set to be the same, i.e., 0.5. Furthermore, for the learning parameters, we set the learning rate to $lr^a = 0.001$ and $lr^c = 0.004$, and the discount factor γ to be equal to 0.99. Finally, we summarize the above key parameters in Table 2.

Table 2. Parameter settings.

Symbol	Value	Symbol	Value
B^c(MHz)	4	B^w (MHz)	5
σ^2 (dBm)	-100	P_u (W)	10
λ_m (Gcycles/s)	$[0.7,1]$	λ_u (Gcycles/s)	$[5,8]$
$\hat{C}_m$(Gcycles)	$[9,11]$	ι	≥ 1
$d_{u,\kappa}$(MB)	$[4,5]$	$c_{u,\kappa}$(Gcycles)	$[1.6,2]$
ρ_u^e (J/cycles)	5×10^{-10}	γ	0.99
$w_{d,\kappa}$	0.5	$w_{e,\kappa}$	0.5
lr^a	0.001	lr^c	0.004

6.2. Fitness Demonstration of Offloading Targets

As shown in Figure 2, by executing Algorithm 1, we depict the relationship between the fitness of servers, the velocity of UAVs, and the PLR and BER that occur during task offloading. The fitness is characterized by the offloading probability in Algorithm 1. It can be observed that the offloading probability is negatively correlated with the above mentioned indicators, i.e., velocity, PLR, and BER. More specifically, Figure 2a,b shows that the offloading probability experiences a rapid decline as the three indicators increase, which validates the effectiveness and validity of the devised fuzzy logic-based offloading assessment mechanism in adaptively evaluating offloading targets at the UAV side.

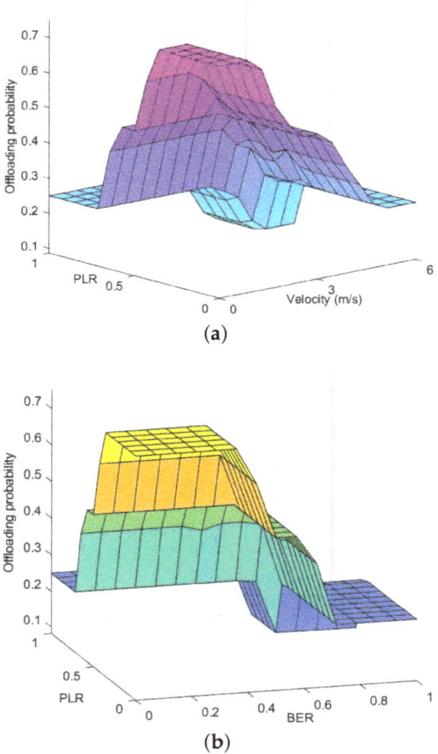

(a)

(b)

Figure 2. Offloading probability versus velocity, PLR, and BER (**a**,**b**).

6.3. Convergence Performance

In order to demonstrate the convergence performance of the proposed DOMUS scheme, we use the same parameter settings listed in Table 2 and plot the variation of the average rewards for four and six UAVs in Figure 3. As shown in Figure 3, the reward curves for different numbers of agents can rapidly converge and fluctuate within a small range. This phenomenon can be attributed to the multi-agent A2C-based distributed offloading mechanism, in which the critic networks assess and guide the actor networks to output better offloading policies for multi-UAVs at each learning episode.

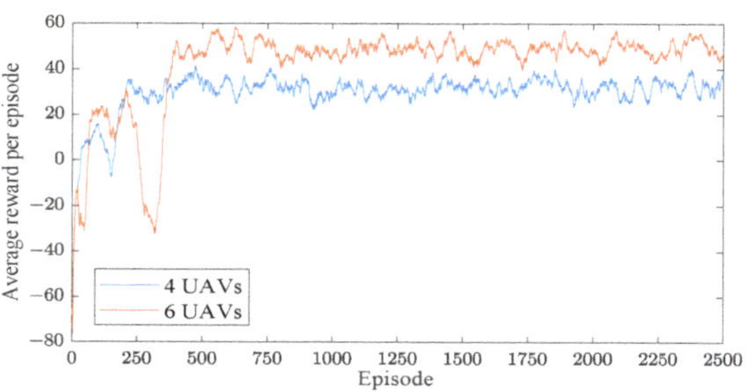

Figure 3. Convergence of the proposed DOMUS.

178

6.4. Impact of Weighting Factors

Figure 4a and b, respectfully, demonstrate the impact of energy consumption and delay weighting factors on two performance metrics when UAVs perform tasks using the proposed DOMUS scheme. Moreover, the number of UAVs is fixed at 6, and the average data size of generated tasks is varied from 4 to 16 MB. In Figure 4a, the curves of energy consumption increase with the average data size of tasks. However, a higher energy consumption weighting factor results in lower energy consumption needed to complete the tasks under the same data size.

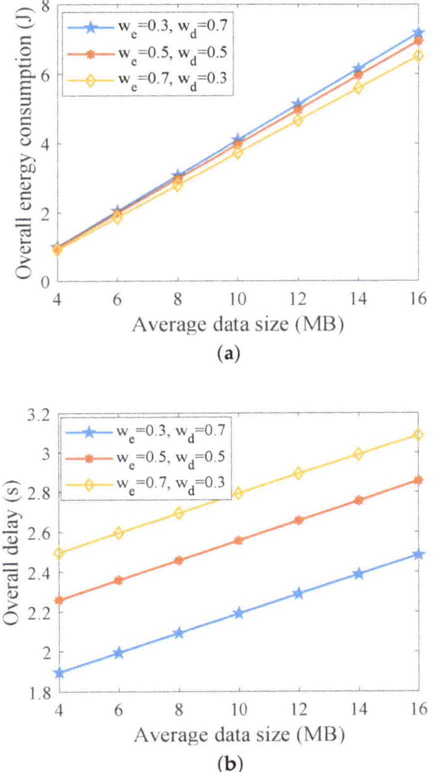

Figure 4. Energy consumption and delay comparison in DOMUS under different weighting factors (**a**,**b**).

From Figure 4b, we can see that the delay incurred by performing tasks shows a linearly increasing trend as the delay weighting factor increases. However, as the delay weighting factor grows larger, less delay is required to complete the UAV tasks under the same average data size. Therefore, the comparisons depicted in Figure 4a,b are consistent with the theoretical data that both energy consumption and delay show noticeable differences under different weighting factors when processing UAV tasks.

6.5. Performance Comparison

In the following section, we compare the proposed DOMUS with four task-offloading schemes under different parameter settings. The comparative algorithms considered are: (1) Greedy-based sequential tuning computation-offloading scheme (STCO) [41]; (2) Weight improvement-based particle swarm optimization offloading algorithm (IWPSO) [42]; (3)

Distance-dependent offloading scheme (DDO); (4) Smart ant colony optimization task-offloading algorithm (SACO) [31].

(1) Impact of the number of UAVs. First, we set the weighting factors as $w_{d,\kappa} = w_{e,\kappa} = 0.5$ and evaluated the performance of the proposed DOMUS under different numbers of UAVs. Figure 5a shows a comparison of the overall energy consumption of all algorithms as the number of UAVs increases. As shown, the DOMUS achieves lower energy consumption compared to the STCO and SACO schemes. By relying on the multi-agent DRL model, the DOMUS can learn the distribution of computation tasks and enable multi-UAVs to almost always select proper offloading targets as the number of UAVs increases. The STCO ignores future offloading decisions, resulting in decisions that are suboptimal from a long-term perspective. The SACO algorithm may become trapped in local optimization due to the feedback of pheromones in suboptimal solutions obtained in early iterations. In Figure 5b, we plot the overall delay under different UAV numbers, which shows the same trend with Figure 5a; the delay curves of other offloading schemes increase significantly compared with DOMUS. Because the DOMUS can select actions without a non-minimum delay for the current task of the UAV, it optimizes long-term performance. To prove this, as shown in Figure 5c, we also recorded the average utility of UAVs as the number of UAVs increased. Combined with Figure 5a,b, our proposed DOMUS can achieve lower energy consumption and lower delay compared to the STCO offloading approaches, with minimum improvements of 8.29% and 7.75%, respectively. Accordingly, the average utility achieved by DOMUS is the highest among the different algorithms, with an improvement of up to 12.82%.

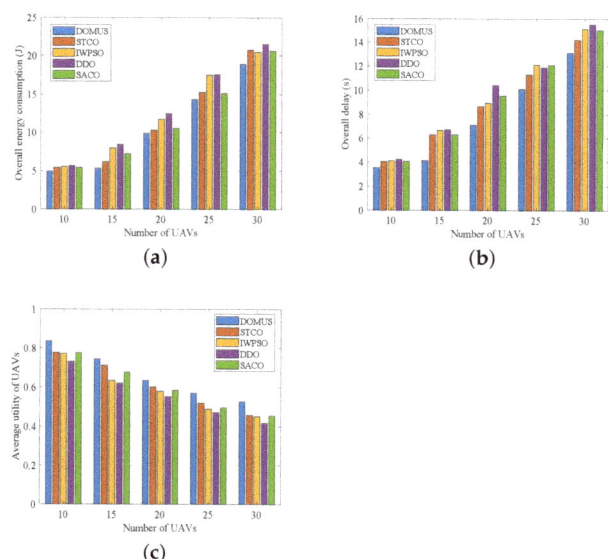

Figure 5. Energy consumption, delay, and utility comparison under different UAV numbers (**a**–**c**).

(2) Impact of data size. We set the number of UAVs to six and then investigated the energy consumption, delay, and average utility required to complete tasks for UAVs with different average task data sizes.

Figure 6 shows the impact of data transmission on energy consumption. Transmitting larger amounts of data requires more communication resources, leading to increased communication delays. In this case, as the data size increases, UAVs will consume more energy for data transmission. Moreover, SACO's energy consumption performance deteriorates as the data size increases, primarily due to the gradually increasing tabu lists in the SACO algorithm that restrict the UAV selection. Our proposed DOMUS can explore policy learning with great effectiveness, resulting in reduced energy consumption. In general, DOMUS reduces energy consumption by up to 13.13% compared to other schemes.

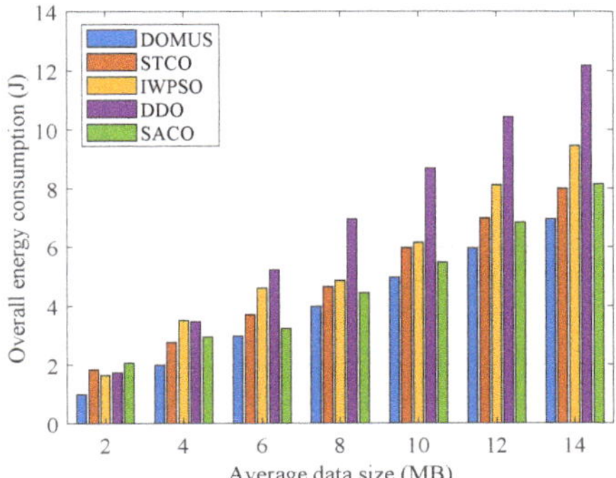

Figure 6. Energy consumption comparison under different average data sizes.

Figure 7 compares the impact of average data size on task completion delay for UAVs. As the data size increases, the delay to complete tasks also increases; when the data size of the task is big, computing the task locally is necessary, which can reduce the transmission delay but correspondingly increase the execution delay. Since the DNN can offer accurate regression, in the proposed DOMUS, DNN is used both in the actor and the critic to approximate the offloading policy and value function interactively, which enables each UAV agent to select appropriate task-processing strategies. Nonetheless, when adopting the STCO scheme, optimal task-processing strategies cannot be extensively derived. This is because the STCO may not effectively take into account the optimization of the subsequent execution of tasks. In summary, the DOMUS optimization results in up to a 6.77% delay compared to other schemes.

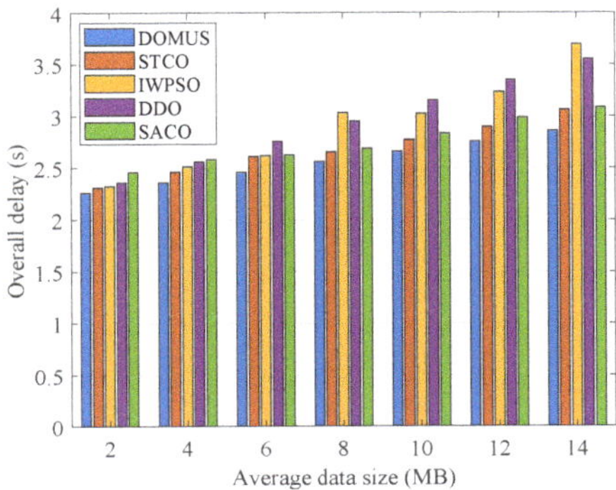

Figure 7. Delay comparison under different average data sizes.

Figure 8 shows the average utility of UAVs for task processing with varying data sizes. The figure shows that our DOMUS mechanism outperforms other schemes in terms of utility, especially for large data sizes. This is because our proposed DOMUS maximizes the utility of task processing for each UAV agent by learning the offloading policy over long training episodes. STCO and SACO obtain similar utility, while the IWPSO method only optimizes the utility from a single UAV perspective, resulting in poor performance, and the DDO method presents the worst utility. Finally, the average utility of UAVs is improved by at least 11.39% compared to the comparative schemes.

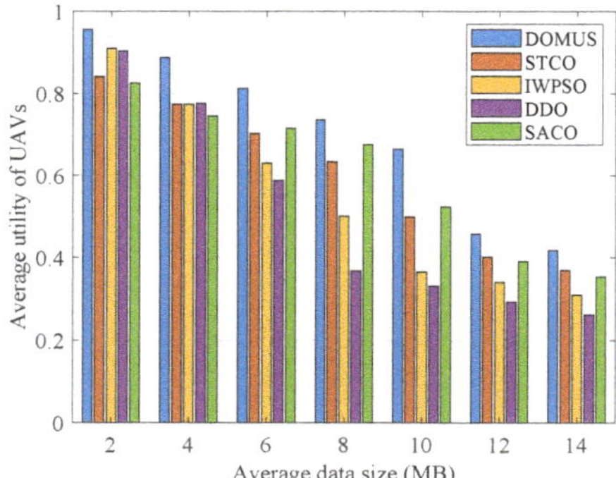

Figure 8. Average utility comparison under different average data sizes.

(3) Joint impact of computational capability and network bandwidth. We conducted a performance comparison between our proposed DOMUS and other benchmarks by increasing the computational capability of MEC servers and network bandwidth. The computational capability was changed from 2 to 5 Gcycles/s and the bandwidth (indicated by B) increased from 1 to 4 MHz.

Figure 9 shows the joint impact of computational capability and bandwidth on the transmission energy consumption required to complete the UAV tasks. As the computational capability and bandwidth increase, energy consumption decreases. This is because most tasks are offloaded rather than processed locally, so as to reduce the task completion time. The energy consumption in local computing is reduced for UAVs and the transmission energy also reduces with the increased bandwidth. Therefore, whether the computation resources and bandwidth resources are abundant will greatly affect the selection of the computing mode and optimal offloading node. In addition, in our proposed DOMUS scheme, the energy consumption curve rises moderately, and the energy consumption performance for UAVs is significantly better than the DDO method, with an improvement of at least 15.69% over other offloading methods.

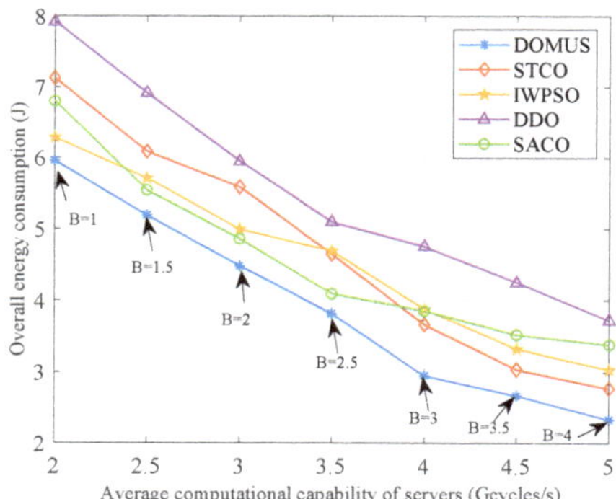

Figure 9. Energy consumption comparison under different computational capabilities and network bandwidths.

Figure 10 shows the task completion delay for UAVs with the variation in computational capability and bandwidth. It is observed that the delay performance changes similarly to the consumed energy consumption. When more computational resources and bandwidths are allocated to UAVs, they are attracted to offload tasks, resulting in degraded delay not only in transmission links but also on servers. However, there still exists a performance gap between our proposed scheme and the other four benchmarks. This is because our proposed DOMUS evaluates the quality of the communication link and offloading nodes for each UAV using the fuzzy logic-based offloading assessment mechanism, and further uses the A2C model to make offloading decisions by continuously updating the parameters of DNNs to enhance the prediction ability of the critic network for actions. This efficiently facilitates the optimization of offloading decisions. In general, the overall delay in task processing was reduced by at least 4.89% compared to other offloading approaches.

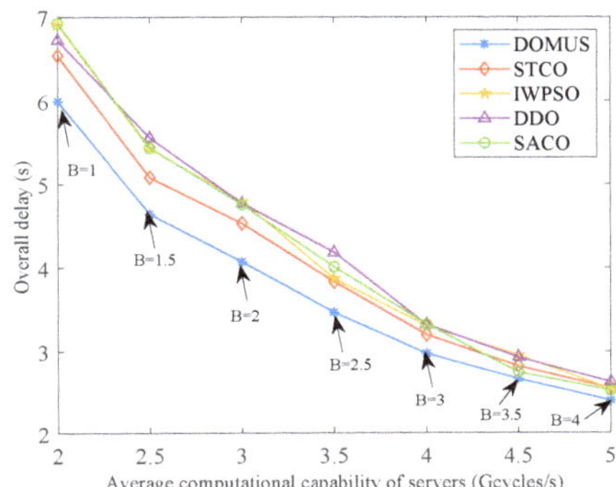

Figure 10. Delay comparison under different computational capabilities and network bandwidths.

Additionally, Figure 11 shows the variation of the average utility of UAVs as computational capability and network bandwidth vary. We can see from the figure that the utilities of different offloading schemes increase as the computational capability bandwidth increases. In particular, our proposed approach outperforms other approaches and improves the utility of UAVs by up to 8.14%. This phenomenon indicates that the DOMUS proposed can effectively enable multi-UAVs to explore the joint optimal task computing policy under the guidance of the devised offloading A2C-based DRL framework in a dynamic network environment.

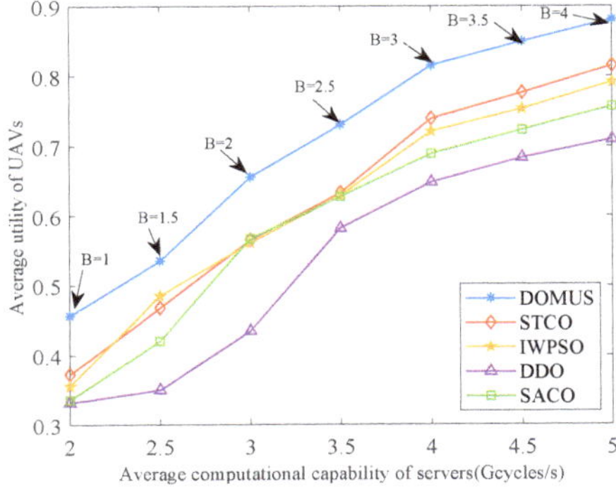

Figure 11. Average utility comparison under different computational capabilities and network bandwidths.

(4) Impact of transmission power. To further demonstrate the scalability of the proposed DOMUS scheme, we investigated the impact of varying transmission power of UAVs on the overall delay and energy consumption to complete tasks, as shown in Figure 12. The number of UAVs was set to 6. Generally, increasing the transmission power can result in a higher data transmission rate, which helps to reduce data transmission delay. Accordingly,

we observe from Figure 12a that all delay curves become smaller as the power gradually increases. Nonetheless, as depicted in Figure 12b, a higher transmission power can lead to a higher energy consumption when UAVs transmit task data due to the linear relationship between them. Moreover, the proposed DOMUS achieves the lowest delay and energy consumption among the five offloading approaches, which reduces the two metrics by at least 4.66% and 9.26%, respectively.

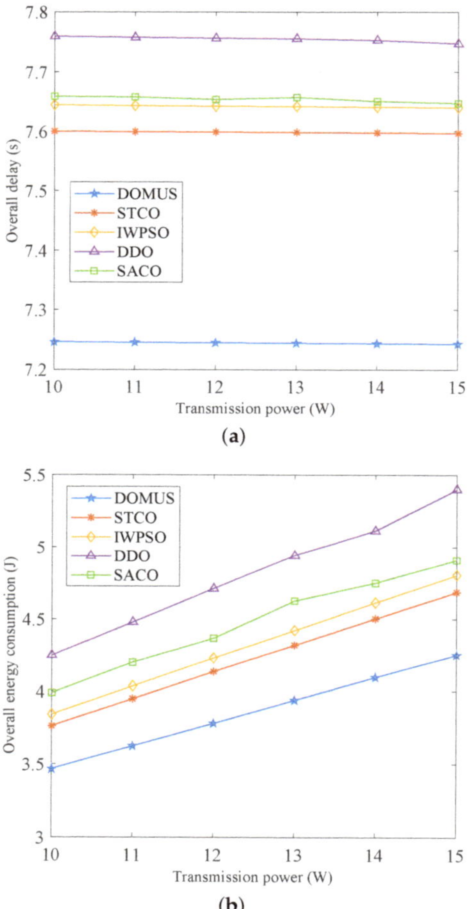

Figure 12. Delay and energy comparison under different transmission power (**a**,**b**).

7. Conclusions

This paper addresses the task offloading of an UAV swarm in MEC-assisted 5G heterogeneous networks. The objective is to optimize the utility of the multi-UAV system for task processing and prevent UAVs from offloading via easily disconnected wireless links and poorly-performing service nodes. We first devise an assessment mechanism to evaluate the candidate-offloading nodes by utilizing fuzzy logic theory. Afterward, considering the unknown environmental dynamics in heterogeneous networks, we model the optimization problem as a multi-agent MDP and propose a decentralized task-offloading scheme called DOMUS using the model-free DRL framework based on multi-agent A2C. In particular, the simulation results reveal that the proposed DOMUS can achieve effective convergence as well as reduce the delay and energy consumption under various settings for completing UAV tasks.

In future work, we will integrate the swarm intelligence approach into the proposed learning framework to enhance the services of drone base stations with multiple UAVs. By providing drone base station-enabled MEC architecture and realizing reasonable resource utilization with more advanced approaches, the much stricter requirements of next-generation Internet of Things applications on reliable and efficient service performances will be further satisfied.

Author Contributions: Conceptualization, M.M. and Z.W.; methodology and writing—original draft, M.M.; writing—review and editing, Z.W. All authors have read and agreed to the published version of the manuscript.

Funding: This research was funded by the National Key R&D Program of China No. 2020YFA0713504.

Institutional Review Board Statement: Not applicable.

Informed Consent Statement: Not applicable.

Data Availability Statement: Not applicable.

Conflicts of Interest: The authors declare no conflict of interest.

References

1. Dai, M.; Su, Z.; Xu, Q.; Zhang, N. Vehicle Assisted Computing Offloading for Unmanned Aerial Vehicles in Smart City. *IEEE Trans. Intell. Transp. Syst.* **2021**, *22*, 1932–1944. [CrossRef]
2. Liu, Z.; Wang, X.; Shen, L.; Zhao, S.; Cong, Y.; Li, J.; Yin, D.; Jia, S.; Xiang, X. Mission-Oriented Miniature Fixed-Wing UAV Swarms: A Multilayered and Distributed Architecture. *IEEE Trans. Syst. Man Cybern. Syst.* **2022**, *52*, 1588–1602. [CrossRef]
3. Sigala, A.; Langhals, B. Applications of Unmanned Aerial Systems (UAS): A Delphi Study projecting future UAS missions and relevant challenges. *Drones* **2020**, *4*, 8. [CrossRef]
4. Yan, S.; Hanly, S.V.; Collings, I.B. Optimal Transmit Power and Flying Location for UAV Covert Wireless Communications. *IEEE J. Sel. Areas Commun.* **2021**, *39*, 3321–3333. [CrossRef]
5. Hu, P.; Zhang, R.; Yang, J.; Chen, L. Development Status and Key Technologies of Plant Protection UAVs in China: A Review. *Drones* **2022**, *6*, 354. [CrossRef]
6. Yazid, Y.; Ez-Zazi, I.; Guerrero-González, A.; El Oualkadi, A.; Arioua, M. UAV-enabled mobile edge-computing for IoT based on AI: A comprehensive review. *Drones* **2021**, *5*, 148. [CrossRef]
7. Ma, M.; Zhu, A.; Guo, S.; Yang, Y. Intelligent Network Selection Algorithm for Multiservice Users in 5G Heterogeneous Network System: Nash Q-Learning Method. *IEEE Internet Things J.* **2021**, *8*, 11877–11890. [CrossRef]
8. Zhou, H.; Jiang, K.; Liu, X.; Li, X.; Leung, V.C.M. Deep Reinforcement Learning for Energy-Efficient Computation Offloading in Mobile-Edge Computing. *IEEE Internet Things J.* **2022**, *9*, 1517–1530. [CrossRef]
9. Chinchali, S.; Sharma, A.; Harrison, J.; Elhafsi, A.; Kang, D.; Pergament, E.; Cidon, E.; Katti, S.; Pavone, M. Network offloading policies for cloud robotics: A learning-based approach. *Auton. Robot.* **2021**, *45*, 997–1012. [CrossRef]
10. Zhu, A.; Ma, M.; Guo, S.; Yu, S.; Yi, L. Adaptive Multi-Access Algorithm for Multi-Service Edge Users in 5G Ultra-Dense Heterogeneous Networks. *IEEE Trans. Veh. Technol.* **2021**, *70*, 2807–2821. [CrossRef]
11. Zhang, X.; Cao, Y. Mobile data offloading efficiency: A stochastic analytical view. In Proceedings of the 2018 IEEE International Conference on Communications Workshops (ICC Workshops), Kansas City, MO, USA, 20–24 May 2018; pp. 1–6.
12. Li, H.; Wu, S.; Jiao, J.; Lin, X.H.; Zhang, N.; Zhang, Q. Energy-Efficient Task Offloading of Edge-Aided Maritime UAV Systems. *IEEE Trans. Veh. Technol.* **2023**, *72*, 1116–1126. [CrossRef]
13. Guo, M.; Huang, X.; Wang, W.; Liang, B.; Yang, Y.; Zhang, L.; Chen, L. Hagp: A heuristic algorithm based on greedy policy for task offloading with reliability of mds in mec of the industrial internet. *Sensors* **2021**, *21*, 3513. [CrossRef]
14. Zhang, D.; Li, X.; Zhang, J.; Zhang, T.; Gong, C. New Method of Task Offloading in Mobile Edge Computing for Vehicles Based on Simulated Annealing Mechanism. *J. Electron. Inf. Technol.* **2022**, *44*, 3220–3230.
15. Huang, J.; Wang, M.; Wu, Y.; Chen, Y.; Shen, X. Distributed Offloading in Overlapping Areas of Mobile-Edge Computing for Internet of Things. *IEEE Internet Things J.* **2022**, *9*, 13837–13847. [CrossRef]
16. Xia, S.; Yao, Z.; Li, Y.; Mao, S. Online Distributed Offloading and Computing Resource Management With Energy Harvesting for Heterogeneous MEC-Enabled IoT. *IEEE Trans. Wirel. Commun.* **2021**, *20*, 6743–6757. [CrossRef]
17. Zhou, H.; Wang, Z.; Cheng, N.; Zeng, D.; Fan, P. Stackelberg-Game-Based Computation Offloading Method in Cloud-Edge Computing Networks. *IEEE Internet Things J.* **2022**, *9*, 16510–16520. [CrossRef]
18. Gu, Q.; Shen, B. An Evolutionary Game Based Computation Offloading for an UAV Network in MEC. In *Wireless Algorithms, Systems, and Applications: Proceedings of the 17th International Conference, WASA 2022, Dalian, China, 24–26 November 2022*; Springer: Cham, Switzerland, 2022; pp. 586–597.

19. You, Q.; Tang, B. Efficient task offloading using particle swarm optimization algorithm in edge computing for industrial internet of things. *J. Cloud Comput.* **2021**, *10*, 41. [CrossRef]
20. Li, F.; He, S.; Liu, M.; Li, N.; Fang, C. Intelligent Computation Offloading Mechanism of UAV in Edge Computing. In Proceedings of the 2022 2nd International Conference on Frontiers of Electronics, Information and Computation Technologies (ICFEICT), Wuhan, China, 19–21 August 2022; pp. 451–456.
21. Asaamoning, G.; Mendes, P.; Rosário, D.; Cerqueira, E. Drone swarms as networked control systems by integration of networking and computing. *Sensors* **2021**, *21*, 2642. [CrossRef]
22. Pliatsios, D.; Goudos, S.K.; Lagkas, T.; Argyriou, V.; Boulogeorgos, A.A.A.; Sarigiannidis, P. Drone-base-station for next-generation internet-of-things: A comparison of swarm intelligence approaches. *IEEE Open J. Antennas Propag.* **2021**, *3*, 32–47. [CrossRef]
23. Amponis, G.; Lagkas, T.; Zevgara, M.; Katsikas, G.; Xirofotos, T.; Moscholios, I.; Sarigiannidis, P. Drones in B5G/6G networks as flying base stations. *Drones* **2022**, *6*, 39. [CrossRef]
24. Chen, M.; Wang, T.; Zhang, S.; Liu, A. Deep reinforcement learning for computation offloading in mobile edge computing environment. *Comput. Commun.* **2021**, *175*, 1–12. [CrossRef]
25. Zhang, D.; Cao, L.; Zhu, H.; Zhang, T.; Du, J.; Jiang, K. Task offloading method of edge computing in internet of vehicles based on deep reinforcement learning. *Clust. Comput.* **2022**, *25*, 1175–1187. [CrossRef]
26. Xu, J.; Li, D.; Gu, W.; Chen, Y. Uav-assisted task offloading for iot in smart buildings and environment via deep reinforcement learning. *Build. Environ.* **2022**, *222*, 109218. [CrossRef]
27. Vhora, F.; Gandhi, J.; Gandhi, A. Q-TOMEC: Q-Learning-Based Task Offloading in Mobile Edge Computing. In *Proceedings of the Futuristic Trends in Networks and Computing Technologies: Select Proceedings of Fourth International Conference on FTNCT 2021*; Springer: Singapore, 2022; pp. 39–53.
28. Zhu, D.; Li, T.; Tian, H.; Yang, Y.; Liu, Y.; Liu, H.; Geng, L.; Sun, J. Speed-aware and customized task offloading and resource allocation in mobile edge computing. *IEEE Commun. Lett.* **2021**, *25*, 2683–2687. [CrossRef]
29. Ma, L.; Wang, P.; Du, C.; Li, Y. Energy-Efficient Edge Caching and Task Deployment Algorithm Enabled by Deep Q-Learning for MEC. *Electronics* **2022**, *11*, 4121. [CrossRef]
30. Naouri, A.; Wu, H.; Nouri, N.A.; Dhelim, S.; Ning, H. A novel framework for mobile-edge computing by optimizing task offloading. *IEEE Internet Things J.* **2021**, *8*, 13065–13076. [CrossRef]
31. Kishor, A.; Chakarbarty, C. Task offloading in fog computing for using smart ant colony optimization. *Wirel. Pers. Commun.* **2022**, *127*, 1683–1704. [CrossRef]
32. Guo, H.; Liu, J. Collaborative computation offloading for multiaccess edge computing over fiber–wireless networks. *IEEE Trans. Veh. Technol.* **2018**, *67*, 4514–4526. [CrossRef]
33. Pekaslan, D.; Wagner, C.; Garibaldi, J.M. ADONiS-Adaptive Online Nonsingleton Fuzzy Logic Systems. *IEEE Trans. Fuzzy Syst.* **2020**, *28*, 2302–2312. [CrossRef]
34. Zhou, W.; Jiang, X.; Luo, Q.; Guo, B.; Sun, X.; Sun, F.; Meng, L. AQROM: A quality of service aware routing optimization mechanism based on asynchronous advantage actor-critic in software-defined networks. *Digit. Commun. Netw.* **2022**. [CrossRef]
35. Athanasiadou, G.E.; Fytampanis, P.; Zarbouti, D.A.; Tsoulos, G.V.; Gkonis, P.K.; Kaklamani, D.I. Radio network planning towards 5G mmWave standalone small-cell architectures. *Electronics* **2020**, *9*, 339. [CrossRef]
36. Garroppo, R.G.; Volpi, M.; Nencioni, G.; Wadatkar, P.V. Experimental Evaluation of Handover Strategies in 5G-MEC Scenario by using AdvantEDGE. In Proceedings of the 2022 IEEE International Mediterranean Conference on Communications and Networking (MeditCom), Athens, Greece, 5–8 September 2022; pp. 286–291.
37. Liu, Y.; Dai, H.N.; Wang, Q.; Imran, M.; Guizani, N. Wireless powering Internet of Things with UAVs: Challenges and opportunities. *IEEE Netw.* **2022**, *36*, 146–152. [CrossRef]
38. Feng, W.; Liu, H.; Yao, Y.; Cao, D.; Zhao, M. Latency-aware offloading for mobile edge computing networks. *IEEE Commun. Lett.* **2021**, *25*, 2673–2677. [CrossRef]
39. Zhou, H.; Wu, T.; Chen, X.; He, S.; Guo, D.; Wu, J. Reverse auction-based computation offloading and resource allocation in mobile cloud-edge computing. *IEEE Trans. Mob. Comput.* **2022**, 1–5. [CrossRef]
40. Huang, S.; Zhang, J.; Wu, Y. Altitude Optimization and Task Allocation of UAV-Assisted MEC Communication System. *Sensors* **2022**, *22*, 8061. [CrossRef]
41. Zhang, K.; Gui, X.; Ren, D.; Li, D. Energy-Latency Tradeoff for Computation Offloading in UAV-Assisted Multiaccess Edge Computing System. *IEEE Internet Things J.* **2021**, *8*, 6709–6719. [CrossRef]
42. Deng, X.; Sun, Z.; Li, D.; Luo, J.; Wan, S. User-centric computation offloading for edge computing. *IEEE Internet Things J.* **2021**, *8*, 12559–12568. [CrossRef]

Article

UAV Deployment Optimization for Secure Precise Wireless Transmission

Tong Shen [1,*]**, Guiyang Xia** [1]**, Jingjing Ye** [2]**, Lichuan Gu** [1]**, Xiaobo Zhou** [1] **and Feng Shu** [3]

[1]   School of Information and Computer, Anhui Agricultural University, Hefei 230036, China
[2]   Mingguang Meteorological Mureau, Chuzhou 239400, China
[3]   School of Information and Communication Engineering, Hainan University, Haikou 570228, China
*   Correspondence: shentong@ahau.edu.cn

Abstract: This paper develops an unmanned aerial vehicle (UAV) deployment scheme in the context of the directional modulation-based secure precise wireless transmissions (SPWTs) to achieve more secure and more energy efficiency transmission, where the optimal UAV position for the SPWT is derived to maximize the secrecy rate (SR) without frequency diverse array (FDA) and injecting any artificial noise (AN) signaling. To be specific, the proposed scheme reveals that the optimal position of UAV for maximizing the SR performance has to be placed at the null space of Eves channel, which impels the received energy of the confidential message at the unintended receiver deteriorating to zero, whilst benefits the one at the intended receiver by achieving its maximum value. Moreover, the highly cost FDA structure is eliminated and transmit power is all allocated for transmitting a useful message which shows its energy efficiency. Finally, simulation results verify the optimality of our proposed scheme in terms of the achievable SR performance.

Keywords: three-dimensional UAV deployment; precise wireless transmission; physical layer security; secrecy rate

Citation: Shen, T.; Xia, G.; Ye, J.; Gu, L.; Zhou, X.; Shu, F. UAV Deployment Optimization for Secure Precise Wireless Transmission. *Drones* **2023**, *7*, 224. https://doi.org/10.3390/drones7040224

Academic Editor: Petros Bithas

Received: 28 February 2023
Revised: 21 March 2023
Accepted: 21 March 2023
Published: 24 March 2023

1. Introduction

1.1. Background and Related Works

As a potential candidate for benefiting wireless transmission in physical layers [1–4], the directional modulation (DM) technique has attracted extensive attention from the fact that DM can transmit a confidential signal to a specific direction [5–7]. Technically, DM is generally employed in two fashions: one is implemented on the radio frequency (RF) frontend, employing the phased array (PA) to optimize the phase shift [5] while another one is implemented on the baseband by utilizing orthogonal vector [6] or beamforming operations [7]. Due to the broadcast nature of DM-based wireless communications, security issues are unavoidable since the distance dimension of DM is threatened by the unintended facility, although DM has an ability to enable the security in a directional aspect. To address this problem, the authors of [8] proposed a linear frequency diverse array (LFDA) method to achieve SPWT. However, the direction angle and distance achieved by LFDA may be coupled, which means that there may exist multiple directions and distances receiving the same confidential message as the desired users. As a further advancement, a number of DM schemes taking the advantage of random frequency diverse array (RFDA) [9–11] have been proposed to achieve secure precise wireless transmission (SPWT) for both the direction and distance domains in which the frequency is randomly allocated on each transmit antenna and the correlation direction and distance are decoupled.

Thereafter, SPWT develops repidly. In [12], the authors proposed two practical random subcarrier selection schemes which increase the randomness of subcarriers based on the random FDA, moreover it improves the stability of the SPWT system since a more random selected subcarriers lead to a better performance. In [13], a hybrid beamforming scheme with hybrid digital and analog SPWT structure was proposed, which reduces the circuit budget with low computational complexity and comparable secrecy performance, it significantly increases the practicability of SPWT system. Since the wireless propagation channel results in the security problem becoming more formidable due to the accessibility of diverse devices. For [14] as an example, the authors clarify that the energy of the main lobe is always formed around the desired receiver, but a number of the non-negligible side lobes remain having comparatively appreciable power. It is indicated that the position of the transmitter affects the distribution of those lobes, and then leading to a security risk. To elaborate a little, when the eavesdropper is located on the side-lobe peak, the achievable secrecy rate (SR) performance will be gravely degraded. Therefore, deploying the transmitter at an appropriate position has an important meaning for the secrecy performance.

As a matter of fact, most existing works regarding the SPWT focus on static scenarios, which severely limits its application. Considering that unmanned aerial vehicle (UAV) has been widely utilized in wireless communications due to bringing extensive benefits (e.g., high probability air-to-ground channel [15] and mobility controllable [16]). Moreover, UAV transmission technology has been researched wildly. In [17], the authors presented an UAV autonomous landing scheme with model predictive controlled moving platform. The authors in [18] proposed a novel approach for the Drone-BS in 5G communication systems, using the meta-heuristic algorithm. Ref. [19] considered the UAV-enabled networks under the probabilistic line-of-sight channel model in complex city environments and jointly optimize the communication connection, the three-dimensional (3D) UAV trajectory, and the transmit power of the UAV to increase the average secrecy rate. In [20], the authors consider UAV networks for collecting data securely and covertly from ground users, a full-duplex UAV intends to gather confidential information from a desired user through wireless communication and generate artificial noise (AN) with random transmit power in order to ensure a negligible probability of the desired user's transmission being detected by the undesired users. The authors in [21] proposed a detection strategy based on multiple antennas with beam sweeping to detect the potential transmission of UAV in wireless networks. In [22], a novel framework is established by jointly utilizing multiple measurements of received signal strength from multiple base stations and multiple points on the trajectory to improve the localization precision of UAV. In [23], the authors considered the region constraint and proposed a received-signal-strength-based optimal scheme for drones swarm passive location measurement. Thus, in this work we consider SPWT in the context of UAV networks, which not only extends the static scenarios to the dynamic situations but also matches the stringent requirement of SPWT for line-of-sight (LoS) communication link from the transmitter to receiver.

1.2. Motivation

Note that in the previous works (e.g., [9–11]), the authors consider the SPWT system model only in two-dimensional scenarios, which limits the practical applications scenarios. At the same time, the FDA technique is generally employed to determine a specific position, while this work casts off the high-cost FDA scheme but ingeniously achieves this goal by fully taking advantage of the angle information in the three-dimensional (3D) space. Against this background, this paper considers an SPWT scheme with the aid of a UAV to improve the system's security level. For a removable UAV transmitter, an analytical solution to the optimal UAV position is derived for reducing the computational complexity. Finally, simulation results show the efficiency of the proposed UAV deployment scheme in terms of the achievable SR performance.

1.3. Contributions

In this paper, the main contributions are as follows.

1. We propose a novel SPTW framework with UAV secure communications which improves transmission security by change UAV's position.

2. Proposed UAV SPWT scheme is based on DM but not FDA, which can reduce the radio frequency chains' cost. Meanwhile, the computational complexity will be significantly reduced.

3. Conventional SPWT improves security performance with aided AN, while our proposed scheme deploys the transmitter, e.g., UAV on the zero SINR space of Eve, and the power originally allocated to artificial noise can be used to transmit useful information, which greatly improves Bob's signal-to-interference-and-noise ratio (SINR) without affecting Eve's SINR, thus enhancing the security performance.

The remainder of this paper is organized as follows: in Section 2, our system model of proposed UAV SPWT is described, then the secrecy capacity performance based on UAV SPWT structure and proposed UAV deployment scheme is analyzed. Section 3 presents the simulation and the analysis. Finally, the conclusion is drawn in Section 4.

Notations: In this paper, scalar variables are denoted by italic symbols, vectors, and matrices are denoted by letters of bold upper case and bold lower case, respectively. Sign $(\cdot)^T$, $(\cdot)^*$ $tr(\cdot)$ and $(\cdot)^H$ denote transpose,conjugation, trace, and conjugate transpose, respectively. $\|\cdot\|$ and $|\cdot|$ denote the norm and modulus, respectively. $\mathbb{E}[\cdot]$ denotes expectation operation.

2. Method

2.1. System Model

As shown in Figure 1, our considered SPWT system is composed of an UAV, a desired user (Bob) and an eavesdropper (Eve). Herein, both Bob and Eve are equipped with a single antenna while the UAV is equipped with a $M \times N$ rectangular antenna array, namely the distance between any two adjacent antenna elements is identical. It should be noted that, the antennas array should be any planar array, such as rectangular array or circular array which can form both angle and distance depended 3D beams, in this paper, without loss generality we assume the transmit antenna forms a rectangular antenna array. For expression convenience, we set Bob as the origin and the ray formed from Bob to Eve is defined as the positive direction of the X-axis. Moreover, we considered that Bob and Eve are ground users, i.e., the Z-axis coordinates of both Bob and Eve are 0. We assume that the UAV flies at a predetermined height g and parallel to the ground. The channels between the users (Bob or Eve) are assumed as LoS channels which have been widely used in UAV communication scenarios [24]. What is more important, it is difficult for SPWT of applying in None-LoS (NLoS) channels, the reasons are as follows: Firstly, as NLOS channel is independent on θ and R, the designed beamforming vectors have the possibility of transmitting the confidential signal to any location. As a result, the security performance might be seriously degraded. Secondly, as for NLOS channels, it changes with time, the designed beamforming vectors can be only applied for a specific time. Therefore, with the designed beamforming vectors in NLOS channel, confidential message may be transmitted to any location as time goes. Lastly, in our proposed scheme, the invoked artificial noise (AN) has an ability of disturbing the signal received at Eve, but having a negligible effect on Bob. However, in NLOS channels, there exists an effect of gathering AN for Bob, thus resulting in a serious secrecy rate performance degradation at Bob. Thus, the assumption of the LoS channel in our proposed UAV and SPWT system is completely reasonable.

Due to the UAV serving the ground users, the relative positions of Bob and Eve is assumed to be known by the UAV. Rationality analysis of this assumption is as follows. For Bob, it is reasonable that the UAV knows his target user's location. For Eve, we consider a large number of users in our system, and all the users' location (including desired users, undesired users, and eavesdroppers) are known for Alice. The part of undesired users are legitimate users in other time periods and they will not eavesdrop the the confidential message. The other part of eavesdroppers are illegitimate users in all the time, however, the eavesdroppers hidden in all the users, who is an undesired user or who is an eavesdropper can not be determined. Thus, we can not determine which user is an eavesdropper, but can consider a user who is most likely to be a eavesdropper. This scheme is obviously more reasonable and practical, and thus the Eve in our proposed system model is the most likely eavesdropper, and we try to prevent his eavesdropping by our proposed scheme. In conclusion, the assumption that the relative positions of Bob and Eve is known to the UAV.

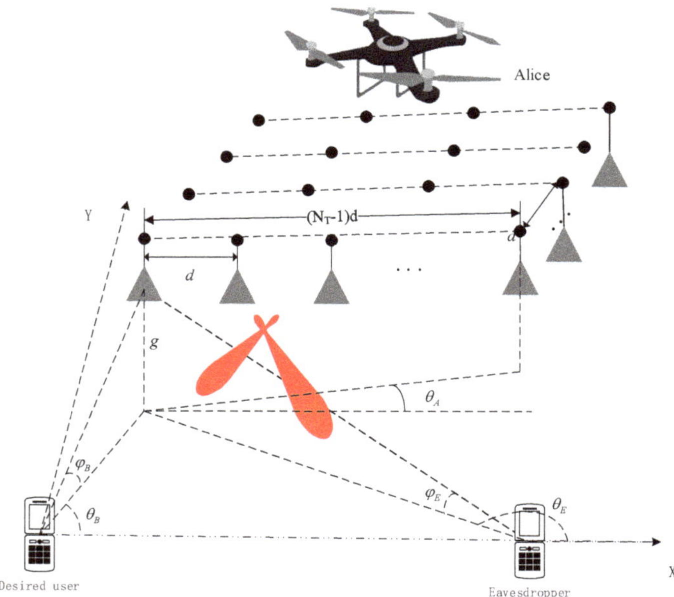

Figure 1. System model.

According to the definition, the 3D coordinates of Bob is $(0,0,0)$, of Eve is $(X_E,0,0)$ while of the UAV is (X_A, Y_A, g). Moreover, $\theta_A \in (0, 2\pi)$ is the yawing angle between a UAV-filed direction and X-axis. Then, based on the established 3D rectangular coordinate system, the angle relationship among UAV, Bob, and Eve can be derived. Specifically, the azimuth angle of Bob with respect to UAV follows $\sin \theta_B = \frac{Y_A}{\sqrt{X_A^2+Y_A^2}}$ and $\cos \theta_B = \frac{X_A}{\sqrt{X_A^2+Y_A^2}}$, while the azimuth angles of Eve with respect to UAV satisfies $\sin \theta_E = \frac{Y_A}{\sqrt{(X_A-X_E)^2+(Y_A)^2}}$ and $\cos \theta_E = \frac{X_A-X_E}{\sqrt{(X_A-X_E)^2+(Y_A)^2}}$. Owing to the existence of θ_A, the azimuth angles with respect to the antenna array of the UAV are given by $\theta'_B = \theta_B - \theta_A$ and $\theta'_E = \theta_E - \theta_A$, respectively. Similarly, the pitch angles of Bob and of Eve related to the UAV can be, respectively, derived as $\sin \varphi_B = \frac{g}{\sqrt{(X_A)^2+(Y_A)^2+g^2}}$ and $\sin \varphi_E = \frac{g}{\sqrt{(X_A-X_E)^2+(Y_A)^2+g^2}}$, where both φ_B and φ_E satisfy $\varphi_B, \varphi_E \in [0, \pi/2]$.

Due to UAV being a $M \times N$ rectangular antenna array, we define the array parallel to the flight direction as the row, whilst arrays perpendicular to the direction of flight as columns. As such, the subscripts m and n are used to denote the m-th row and n-th column of the rectangular antenna array. Then, the steering vector for the antenna array is expressed as

$$\mathbf{h}(\theta, \varphi) = \frac{1}{\sqrt{MN}}[e^{j2\pi\psi_{1,1}} \ldots e^{j2\pi\psi_{m,n}} \ldots e^{j2\pi\psi_{M,N}}], \tag{1}$$

where θ and φ respectively denote the receiver's azimuth angle and the pitch angle relative to the UAV, where $\psi_{m,n}$ is

$$\psi_{m,n} = -\frac{f_c}{c}[(m-1)d\cos\theta' + (n-1)d\sin\theta']\cos\varphi. \tag{2}$$

Herein, f_c is the central carrier frequency, $d = c/(2f_c)$ is the element spacing in the transmit antenna array and c is the speed of light. Substituting (θ'_B, φ_B) and (θ'_E, φ_E) into (1) and (2), respectively, we obtain the steering vector $\mathbf{h}(\theta'_B, \varphi_B)$ and $\mathbf{h}(\theta'_E, \varphi_E)$.

At the baseband, the transmit signal can be expressed as

$$s = \sqrt{\alpha P_s}\mathbf{v}x + \sqrt{(1-\alpha)P_s}\mathbf{w}, \tag{3}$$

where α, P_s and x refer to the power allocation factor, total transmit power and a complex symbol following $\mathbb{E}[|x|^2] = 1$. In addition, $\mathbf{v} \in \mathbb{C}^{MN \times 1}$ and $\mathbf{w} \in \mathbb{C}^{MN \times 1}$ denote the beamforming and AN vectors, respectively.

To achieve precise transmission, $\mathbf{v} = \mathbf{h}(\theta_B, \varphi_B)$ is set to maximize the received power of the confidential signal at Bob, while $\mathbf{w} = [\mathbf{I}_{MN} - \mathbf{h}(\theta'_B, \varphi_B)\mathbf{h}^H(\theta'_B, \varphi_B)]\mathbf{z}$ is to project the AN into the null space of Bob, where $\mathbf{z}$ is an AN vector consisting of MN complex Gaussian variables with normalized power, i.e., $\mathbf{z} \sim \mathcal{CN}(0, \mathbf{I}_{MN})$. Notably, UAV is usually an aerial platform serving for terrestrial nodes, hence all the communication channels follow the light of sight (LOS) model. Then, the received signal at Bob can be expressed as

$$y_B = \sqrt{\alpha P_s}\mathbf{h}_B^H\mathbf{v}x + \sqrt{(1-\alpha)P_s}\mathbf{h}_B^H\mathbf{w} + n_B = \sqrt{\alpha P_s}x + n_B, \tag{4}$$

while the received signal at Eve is

$$y_E = \sqrt{\alpha P_s}\mathbf{h}_E^H\mathbf{v}x + \sqrt{(1-\alpha)P_s}\mathbf{h}_E^H\mathbf{w} + n_E. \tag{5}$$

In (4) and (5), $\mathbf{h}_B$ and $\mathbf{h}_E$ are the abbreviations of $\mathbf{h}(\theta'_B, \varphi_B)$ and $\mathbf{h}(\theta'_E, \varphi_E)$, respectively. Moreover, n_B and n_E are the additive white Gaussian noises (AWGNs) at Bob and Eve satisfying $n_B \sim \mathcal{CN}(0, \sigma_B^2)$ and $n_E \sim \mathcal{CN}(0, \sigma_E^2)$.

In this subsection, we propose the system model and give the received signal expression, it is clear to find that the received signal at Bob is independent of the UAV position while the the received signal at Eve is related to $\mathbf{h}_E$ which is a function of UAV position information, e.g., (θ_E, φ_E). This shows that it is feasible to reduce the useful signal energy at Eve by moving the position of UAV without reducing the useful signal energy at Bob, so as to increase the security of UAV-based SPWT system.

2.2. Proposed UAV Deployment Schemes

In accordance with (4), the SPWT employs $\mathbf{v}$ to impel the power of the confidential message at the desired receiver achieving its maximum. However, such a conventional SPWT does not guarantee the power received by Eve at a minimum level, since the channel spanning from UAV to Eve (i.e., $\mathbf{h}_E$) is affected by the relative position of them. As a result, the security for the transmission of the confidential signal is unable to completely ensured. To further enhance the security level of the UAV-dominated moving network, we maximize the security rate (SR) performance by deploying the manoeuvrable UAV for shifting the channel attribute.

Traditionally, the SR performance can be characterized by $\text{SR} = \log(1 + \text{SINR}_B) - \log(1 + \text{SINR}_E)$, where SINR_B and SINR_E refer to the signal-to-interference-and-noise ratio of Bob and of Eve, respectively. In SPWT networks, $\mathbf{v}$ has been assigned as $\mathbf{h}_B$ for benefiting SINR_B to arrive its maximum value. Naturally, the optimization problem related to θ'_E and φ_E can be simplified as

$$\min_{\theta'_E, \varphi_E} \text{SINR}_E = \frac{\alpha P_s |\mathbf{h}_E^H \mathbf{h}_B|^2}{(1-\alpha) P_s \mathbf{h}_E^H \mathbf{w} + \sigma_E^2}$$

$$\text{s.t. } 0 \leq \varphi_E \leq \frac{\pi}{2}, \ 0 \leq \theta'_E \leq 2\pi. \tag{6}$$

Since φ_E is the angle between the line between the from UAV to Eve and the horizontal plane, thus in the constraint of Equation (6) it can not be larger than $\frac{\pi}{2}$. Considering the fact that SINR_E is certainly non-negative, we associate that SINR_E arrives its minimum when the numerator of the objective function of (6) is equal to 0. Pertinently, we expand the term $\mathbf{h}_E^H \mathbf{h}_B$ as

$$\mathbf{h}_E^H \mathbf{h}_B = \frac{1}{MN} \sum_{m,n}^{M,N} e^{j\pi[(m-1)\cos\theta'_E + (n-1)\sin\theta'_E]\cos\varphi_E} \times$$

$$e^{-j\pi[(m-1)\cos\theta'_B + (n-1)\sin\theta'_B]\cos\varphi_B}. \tag{7}$$

Ideally, SINR_E degrades to 0 as $\mathbf{h}_E^H \mathbf{h}_B = 0$, which indicates the maximum value of the SR is reached in this case. Toward this direction, we aware from (7) that the optimality condition of (6) achieves as the UAV deployment impels that $\mathbf{h}_E^H$ is orthogonal to $\mathbf{h}_B$.

However, it can be shown from (7) that θ'_E and φ_E are highly coupled, leading to a tricky task in terms of simultaneously obtaining their optimal solutions. To mitigate the difficulty of addressing the problem with respect to θ'_E or φ_E, we do not jointly optimize them but solve it by an ingenious way, which is also able to obtain the optimal solutions to both θ'_E or φ_E. To guarantee the ultimate result that $\mathbf{h}_E^H \mathbf{h}_B = 0$ and decouple the indivisible relationship between θ'_E and φ_E, without loss of generality, we can define $\varphi_B = \varphi_E$ or $\theta'_B = \theta'_E$ to ease the conceived problem. For $\varphi_B = \varphi_E$ as an example, the potential null points of array pattern along the azimuth angle dimension satisfy the following condition [25]

$$\mathbf{h}_E^H \mathbf{h}_B = \frac{1}{MN} \sum_{m}^{M} e^{j\pi(m-1)(\cos\theta'_E - \cos\theta'_B)\cos\varphi_E} \times \sum_{n}^{N} e^{j\pi(n-1)(\sin\theta'_E - \sin\theta'_B)\cos\varphi_E},$$

$$= \frac{1}{MN} \cdot \frac{e^{jM\pi(\cos\theta'_E - \cos\theta'_B)\cos\varphi_E} - 1}{e^{j\pi(\cos\theta'_E - \cos\theta'_B)\cos\varphi_E} - 1} \times \frac{e^{jN\pi(\sin\theta'_E - \sin\theta'_B)\cos\varphi_E} - 1}{e^{j\pi(\sin\theta'_E - \sin\theta'_B)\cos\varphi_E} - 1}. \tag{8}$$

To compel that $\mathbf{h}_E^H \mathbf{h}_B = 0$, the following condition has to be satisfied, given by

$$M\pi(\cos\theta'_E - \cos\theta'_B)\cos\varphi_E = \pm 2k\pi, \tag{9}$$

or

$$N\pi(\sin\theta'_E - \sin\theta'_B)\cos\varphi_E = \pm 2k\pi, \tag{10}$$

where k has to ensure that $k \neq k'M$ and $k \neq k'N$, herein k' is an integer (i.e., $k' \in \mathbb{Z}$). Then, the relationship between the optimal θ'_E and the optimal θ'_B can be obtained as

$$\cos\theta'_E - \cos\theta'_B = \frac{\pm 2k}{M\cos\varphi_E}, \tag{11}$$

or

$$\sin\theta'_E - \sin\theta'_B = \frac{\pm 2k}{N\cos\varphi_E}. \tag{12}$$

Taking $\varphi_B = \varphi_E$ and UAV flies at a constant height into account, we are aware that the X-coordinate of UAV satisfies $X_A = X_E/2$ and $\theta_B = \pi - \theta_E$, which can be readily verified according to our system model. Substituting $\theta_B = \pi - \theta_E$, $\theta'_B = \theta_B - \theta_A$ and $\theta'_E = \theta_E - \theta_A$ into (22), we obtain a correspondingly modified expression regarding θ_B, shown as

$$\cos \theta_B = \frac{\pm k}{M \cos \varphi_E \cos \theta_A}. \tag{13}$$

Alternatively, the other modified expression with respect to θ_B according to (23) is

$$\cos \theta_B = \frac{\pm k}{N \cos \varphi_E \sin \theta_A}. \tag{14}$$

Remark 1. *Considering that* $(\cos \theta'_E - \cos \theta'_B) \in [-2, 2]$ *and* $\cos \varphi_E \in [0, 1]$, *thus* $(\cos \theta'_E - \cos \theta'_B) \cos \varphi_E \in [-2, 2]$. *When* $(\cos \theta'_E - \cos \theta'_B) \neq 0$, *the denominator of the first term in* (8) *regarding* θ'_B *is unequal to 0. Naturally,* $(\sin \theta'_E - \sin \theta'_B) \neq 0$ *holds. In a nutshell, our analysis for the two denominators of* (8) *derives that* $\sin \theta'_E - \sin \theta'_B \neq 0$ *and* $\cos \theta'_E - \cos \theta'_B \neq 0$. *Taking into account the relationship among* $\theta_A, \theta'_B, \theta_B, \theta'_E$, *and* θ_E, *the flight angle of UAV follows* $\cos \theta_B \cos \theta_A \neq 0$ *and* $\cos \theta_B \sin \theta_A \neq 0$, *which further arrives* $\theta_A \neq \frac{p\pi}{2}$, $(p \in \mathbb{Z})$, *thus the denominators of* (8) *are unequal to 0. As a result, our analysis and derivation for the solution* θ_B *to expression* (8) *is of physical significance.*

Since $\varphi_E \in [0, \pi/2]$ increases as Y_A increases in the domain of $(-\infty, 0)$, hence $\cos \varphi_E$ decreases as $Y_A \in (-\infty, 0)$ increases. Similarly, $\cos \varphi_E$ increases as $Y_A \in (0, +\infty)$ increases. With those in mind, we then have $\frac{1}{N \cos \varphi_E} \in (1/N, 1/(N \cos \varphi_E^*)]$, where φ_E^* refers to the optimal pitch angle for maximizing $\frac{1}{N \cos \varphi_E}$. As a matter of fact, φ_E^* arrives in the case of $Y_A = 0$, which further derives that

$$\cos \varphi_E^* = \frac{X_E/2}{\sqrt{(X_E/2)^2 + g^2}}. \tag{15}$$

Furthermore, we note that the term $\frac{\pm k}{\sin \theta_A}$ of (14) follows $\left| \frac{\pm k}{\sin \theta_A} \right| \in [1, +\infty)$, herein the left extremum arrives as $k = 1$ and $\sin \theta_A = 1$. Hence, we conclude that θ_B has at least one solution when $|1/(N \cos \varphi_E^*)| \leq 1$.

With the above conclusion, we further derive the optimal solution θ_B^* to characterize Y_A, where Y_A can be determined once θ_B^* is optimized. Upon substituting the original definition $\cos \theta_B = \frac{X_A}{\sqrt{X_A^2 + Y_A^2}}$ and $\cos \varphi_E = \frac{\sqrt{(X_A - X_E)^2 + (Y_A)^2}}{\sqrt{(X_A - X_E)^2 + (Y_A)^2 + g^2}}$ into the derivation of (13), we have

$$\frac{X_A}{\sqrt{X_A^2 + Y_A^2}} \frac{\sqrt{(X_A - X_E)^2 + (Y_A)^2}}{\sqrt{(X_A - X_E)^2 + (Y_A)^2 + g^2}} = \frac{\pm k}{M \cos \theta_A}. \tag{16}$$

Taking into account that $X_A = X_E/2$, Y_A can be obtained as

$$Y_A = \pm \sqrt{\frac{M^2 \cos^2 \theta_A X_E^2 - k^2 X_E^2 - 4k^2 g^2}{4k^2}}. \tag{17}$$

Similarly, Y_A can also be obtained in accordance with (14) as

$$Y_A = \pm \sqrt{\frac{N^2 \sin^2 \theta_A X_E^2 - k^2 X_E^2 - 4k^2 g^2}{4k^2}}. \tag{18}$$

Until now, we have obtained an analytical expression of UAV's coordinate shown of (17) or of (18). It is worth noting that, in a potentially infancy stage, the right term $M^2 \cos^2 \theta_A X_E^2 - k^2 X_E^2 - 4k^2 g^2$ of (17) or the term $N^2 \sin^2 \theta_A X_E^2 - k^2 X_E^2 - 4k^2 g^2$ of (18) is not inherently greater than or equal to 0, there is no solution satisfying (17) or (18). However, the parameters θ_A (the yawing angle of UAV) and g (the height of UAV) can be strategically regulated by UAV's attitude, hence at least one feasible solution to (17) or (18) is able to be gained. Because of the analytical solution, the computational complexity is significantly reduced when compared to directly addressing problem (6). For the given X_A, once Y_A is optimized by (17) or (18), we finish the UAV deployment problem. At such an optimal coordinate (X_A^*, Y_A^*, g), SINR_E degrades to zero while SINR_B achieves its maximum value $\alpha P_S / \sigma_B^2$, thus ensuring that the maximum SR performance can be achieved.

Remark 2. *Upon optimizing UAV deployment, we promulgate that the maximum SR performance remains achievable, even though we have not split any precious power to the AN signaling. The benefits are threefold: (1) from UAV's perspective, the hardware structure becomes more simple owing to removing the module of AN generator, which cuts down the expenditure and favors lightening the weight of UAV; (2) for the perspective of energy efficiency, all the transmit power is used to convey the confidential message, which contributes to Bob receiving a high-quality signal; and (3) while for the computational complexity, our derived analytical solution to Y_A on the basis of the 3-D spatial relation conspicuously mitigates the computational burden in terms of the UAV deployment.*

For $\varphi_B = \varphi_E$ as our anther instance to decouple the indivisible relationship between $\theta_B = \theta_E$, the potential null points of array pattern along the pitch angle dimension satisfy the following condition,

$$
\begin{aligned}
\mathbf{h}_E^H \mathbf{h}_B &= \frac{1}{MN} \sum_m^M e^{j\pi(m-1)\cos\theta_E(\cos\varphi_E - \cos\varphi_B)} \cdot \sum_n^N e^{j\pi(n-1)\sin\theta_E(\cos\varphi_E - \cos\varphi_B)}, \\
&= \frac{1}{MN} \cdot \frac{e^{jM\pi\cos\theta_E(\cos\varphi_E - \cos\varphi_B)} - 1}{e^{j\pi\cos\theta_E(\cos\varphi_E - \cos\varphi_B)} - 1} \cdot \frac{e^{jN\pi\sin\theta_E(\cos\varphi_E - \cos\varphi_B)} - 1}{e^{j\pi\sin\theta_E(\cos\varphi_E - \cos\varphi_B)} - 1}.
\end{aligned}
\tag{19}
$$

Similarly, let $\mathbf{h}_E^H \mathbf{h}_B = 0$, we have,

$$
M\pi\cos\theta_E(\cos\varphi_E - \cos\varphi_B) = \pm 2l\pi,
$$
$$
l \neq l' M(l' = 1,2,3,\ldots).
\tag{20}
$$

or

$$
N\pi\sin\theta_E(\cos\varphi_E - \cos\varphi_B) = \pm 2l\pi,
$$
$$
l \neq l' N(l' = 1,2,3,\ldots).
\tag{21}
$$

Then, the relationship θ_E and θ_B can be obtained as,

$$
\cos\varphi_E - \cos\varphi_B = \frac{\pm 2l}{M\cos\theta_E},
$$
$$
l \neq l' M(l' = 1,2,3,\ldots).
\tag{22}
$$

or

$$
\cos\varphi_E - \cos\varphi_B = \frac{\pm 2l}{N\sin\theta_E},
$$
$$
l \neq l' N(l' = 1,2,3,\ldots).
\tag{23}
$$

Note that, the azimuth angles satisfy the constraint of $\theta_B = \theta_E$. According to the system model, the Y-coordinate of Alice must satisfy $Y_A = 0$. Since the azimuth angle can be adjusted with the flight angle of Alice, i.e., the flight direction which we defined as θ_A, its value range is $(0, 2\pi)$. The X-coordinate range of Alice is $(-\infty, 0) \cup (X_E, +\infty)$ while $\cos\varphi_E - \cos\varphi_B$ is a monotone decreasing function about X_A. Let $l = 1$, it is clear that when

$\frac{2}{M\cos\theta_E} < 0$ or $\frac{2}{N\sin\theta_E} < 0$, the X-coordinate $X_A \in (-\infty, 0)$. Conversely, when $\frac{2}{M\cos\theta_E} > 0$ or $\frac{2}{N\sin\theta_E} > 0$, the X-coordinate $X_A \in (X_E, +\infty)$. Thus, the value of X_A is existed and can be easily obtained similar to the pitch angles scheme or by dichotomy method.

In this subsection, we propose the UAV deployment strategies from azimuth angel dimension and pitch angel dimension, respectively. According to the derived equations, it is clear that, based on our proposed two schemes, the received SINR at Eve achieves zero. At the same time, the received SINR at Bob reaches its maximum value. Thus, the secrecy rate is improved.

3. Simulation Results and Analysis

In this section, we evaluate the achievable SR performance of our proposed scheme via numerical simulations, where the detailed system parameters are set as shown in Table 1. The SR is defined as $SR = log_2(SINR_B - SINR_E)$. Without loss of generality, we assume that the noise levels at Bob and Eve are identical, i.e., $\sigma_B^2 = \sigma_E^2$. Moreover, the variable argument k in both (13) and (14) are set to be 1.

Table 1. Simulation parameters setting.

Parameter	Value
The number of transmitter antennas ($M \times N$)	4×4
Total signal bandwidth (B)	5 MHz
Total transmit power (P)	1 W
The height of UAV (g)	200 m
The eavesdropper's position (X_E, Y_E)	(500 m, 0)
The flight angle of UAV (θ_A)	$\pi/4$
Central carrier frequency (f_c)	3 GHz

Figure 2 shows the attainable SR performance of our proposed azimuth angles scheme versus the signal-to-noise ratio (SNR), where SNR $= 10\log(P_s/\sigma^2)$. For comparison, we consider the beamforming scheme proposed in [14], then three conventionally random deployments are invoked to validate the efficiency of our proposed UAV deployment. Firstly, it can be obviously noted that our proposed UAV deployment scheme is superior to the other three random schemes in terms of the SR performance, albeit a negligible computational complexity is increased. Moreover, the SR performance gap between the proposed scheme and any other scheme becomes distinct as the SNR increases. Therefore, the main benefits of our considered UAV deployment scheme stem from not only assuring the precise transmission but also improving the security performance. On the other hand, we note from Figure 2 that the theoretical SR performance, i.e., the maximum performance, is coincident with that of our proposed scheme at any SNR. The result further verifies the validness of our derived solution to the UAV deployment in the context of SPWT.

Figure 3 shows the attainable SR performance of our proposed pitch angles scheme versus the SNR. Similarly, the three conventionally random deployments are invoked for comparison. The results show that this proposed scheme is also superior to the conventionally random deployments. This also proves that the basic ideas of the two methods we proposed are completely correct. From Figures 2 and 3, it is clear to find, no matter the azimuth angles scheme or pitch angles scheme we adopt, if only the deployed UAV satisfies the constraint of $\mathbf{h}_E^H \mathbf{h}_B = 0$, the optimal secrecy rates can achieve by the optimal beamforming of $\mathbf{v} = \mathbf{h}(\theta_B, \varphi_B)$.

To illustrate the efficiency of our proposed UAV-assisted SPWT scheme in gaining the security, Figure 4 shows that the achievable SR performance varies as α increases. It can be noted that the SR performances of the three conventional comparison schemes are constantly unable to achieve the optimal SR performance even with the aid of power allocation. In fact, the SR performance of the conventional scheme seriously degrades when the UAV is randomly deployed. Hence, properly arranging UAV has a momentous impact on the SPWT. While for our proposed UAV deployment eliminating the AN signaling, it overleaps the power split operation but remains gaining the maximum SR performance, which further corroborates the potential value for the SPWT. Another interesting conclusion is that the SR of our proposed scheme remains the maximum value and unchanged, this means our proposed scheme do not need the artificial noise, this is because $\mathbf{h}_E^H \mathbf{h}_B = 0$, regardless of the power of artificial noise, the received SNR at Eve is always 0. Thus, artificial noise is unnecessary, and this brings the benefit of a budget reduction.

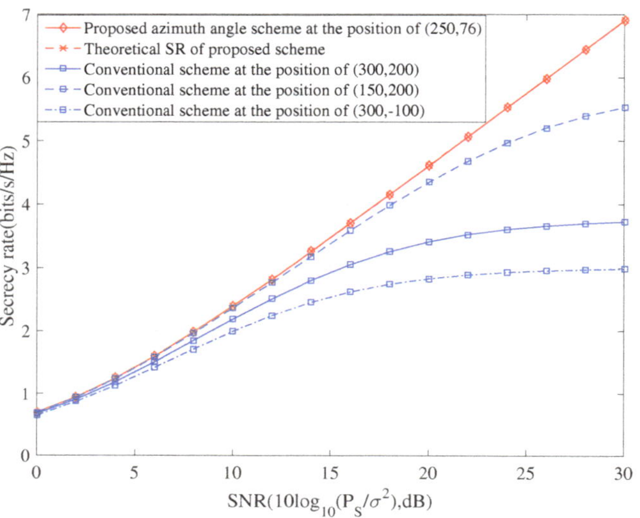

Figure 2. The achievable SR performance versus SNR for our proposed azimuth angles scheme, where three typically random deployments are invoked as the baselines.

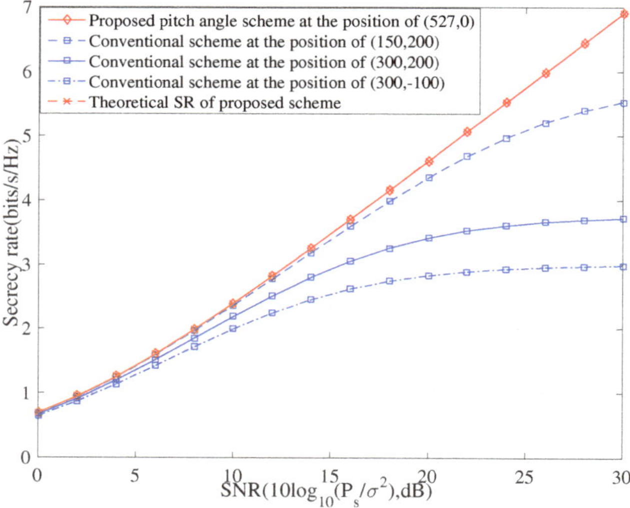

Figure 3. The achievable SR performance versus SNR for our proposed pitch angle scheme, where three typically random deployments are invoked as the baselines.

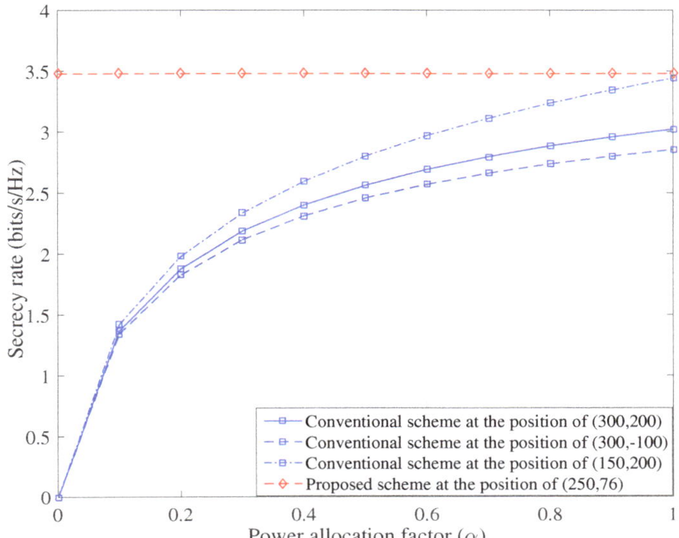

Figure 4. The achievable SR performance versus parameter α for the different scheme when SNR is 15 dB.

In summary, simulation results show that our proposed UAV deployment schemes achieve the SR enhanced SPWT compared with an UAV randomly distributed scheme. Moreover, the UAV transmit power is concentrated on transmitting confidential message but not part-allocated to AN. Thus, our proposed UAV deployment SPWT schemes are more secure and more energy efficiency.

4. Conclusions

In this letter, we proposed an UAV deployment scheme in the context of SPWT from the perspective of azimuth angle and pitch angle. In this scheme, we abandoned FDA for the first time and adopted the method of combining DM with 3D scenario, which reduces the system budget significantly. Compared to the conventional method, the proposed scheme is more superior in terms of the attainable SR performance. Moreover, our proposed UAV deployment algorithm gives the analytical solutions which has almost no complexity, this is also an important benefit of our proposed scheme. Interestingly, although we introduced AN in this hybrid SPWT system, the mathematic analysis shows that when the allocated power in AN is zero, the performance achieves the optimal, this means our proposed scheme do not need AN assistance to achieve the optimal SR performance, which has a powerful ability in economizing the precious energy resource.

Author Contributions: Conceptualization, T.S.; methodology, T.S.; software, L.G.; validation, X.Z., G.X. and F.S.; formal analysis, T.S.; investigation, T.S.; resources, G.X.; data curation, G.X.; writing—original draft preparation, T.S.; writing—review and editing, G.X. and J.Y.; visualization, F.S.; supervision, X.Z.; project administration, F.S.; funding acquisition, F.S. All authors have read and agreed to the published version of the manuscript

Funding: This work was supported in part by the Scientific Research Fund Project of Anhui Agricultural University under Grant rc482106, Grant rc482103 and Grant rc482108, in part by National Natural Science Foundation of China under Grant 61901121, Grant 62001225, Grant 62071234, Grant 62071289, and Grant 61972093, in part by the Natural Science Research Project of Education Department of Anhui Province of China under Grant KJ2021A0183 and K2248004, in part by the Key Research and Development Project of Anhui Province in 2022(202204c06020022), in part by the Natural Science Foundation of Jiangsu Province under Grant BK20190454, in party by the open research fund of National Mobile Communications Research Laboratory, Southeast University under Grant 2022D07, in part by the Scientific Research Fund Project of Hainan University under Grant KYQD(ZR)-21007 and Grant KYQD(ZR)-21008, in part by the Hainan Major Projects under Grant ZDKJ20211022, and in part by the National Key R and D Program of China under Grant 2018YFB1801102.

Institutional Review Board Statement: Not applicable.

Informed Consent Statement: Not applicable.

Data Availability Statement: Not applicable.

Conflicts of Interest: The authors declare no conflict of interest.

Abbreviations

The following abbreviations are used in this manuscript:

UAV	Unmanned Aerial Vehicle
SPWT	Secure Precise Wireless Transmissions
SR	Secrecy Rate
AN	Artificial Noise
DM	Directional Modulation
RF	Radio Frequency
FDA	Frequency Diverse Array
LoS	Line-of-Sight
3D	Three-Dimensional

References

1. Wu, Y.; Khisti, A.; Xiao, C.; Caire, G.; Wong, K.-K.; Gao, X. A survey of physical layer security techniques for 5G wireless networks and challenges ahead. *IEEE J. Sel. Areas Commun.* **2018**, *36*, 679–695. [CrossRef]
2. Chen, X.; Ng, D.W.K.; Gerstacker, W.H.; Chen, H.-H. A survey on multiple-antenna techniques for physical layer security. *IEEE Commun. Surv. Tutor.* **2017**, *19*, 1027–1053. [CrossRef]
3. Bai, J.; Wang, H.-M.; Liu, P. Robust irs-aided secrecy transmission with location optimization. *IEEE Trans. Commun.* **2022**, *70*, 6149–6163. [CrossRef]
4. Rao, H.; Xiao, S.; Yan, S.; Wang, J.; Tang, W. Optimal geometric solutions to uav-enabled covert communications in line-of-sight scenarios. *IEEE Trans. Wirel. Commun.* **2022**, *21*, 10633–1064. [CrossRef]
5. Daly, M.P.; Bernhard, J.T. Directional modulation technique for phased arrays. *IEEE Trans. Antennas Propag.* **2009**, *57*, 2633–2640. [CrossRef]
6. Yuan, D.; Fusco, V.F. Orthogonal vector approach for synthesis of multi-beam directional modulation transmitters. *IEEE Trans. Antennas Wirel. Propagat. Lett.* **2015**, *14*, 1330–1333.
7. Jinsong, H.; Feng, S.; Jun, L. Robust synthesis method for secure directional modulation with imperfect direction angle. *IEEE Commun. Lett.* **2016**, *20*, 1084–1087.
8. Sammartino, P.F.; Baker, C.J.; Griffiths, H.D. Frequency diverse MIMO techniques for radar. *IEEE Trans. Aerosp. Electron. Syst.* **2013**, *49*, 201–222. [CrossRef]
9. Nusenu, S.Y.; Huaizong, S.; Ye, P. Power allocation and equivalent transmit fda beamspace for 5G mmwave noma networks: Meta-heuristic optimization approach. *IEEE Trans. Veh. Technol.* **2022**, *71*, 9635–9646. [CrossRef]
10. Hu, Y.Q.; Chen, H.; Ji, S.-L.; Wang, W.-Q.; Chen, H. Adaptive detector for fda-based ambient backscatter communications. *IEEE Trans. Wirel. Commun.* **2022**, *21*, 10381–10392. [CrossRef]
11. Wang, L.; Wang, W.-Q.; So, H.C. Covariance matrix estimation for fda-mimo adaptive transmit power allocation. *IEEE Trans. Signal Process.* **2022**, *70*, 3386–3399. [CrossRef]
12. Shen, T.; Zhang, S.; Chen, R.; Wang, J.; Hu, J.; Shu, F.; Wang, J. Two Practical Random-Subcarrier-Selection Methods for Secure Precise Wireless Transmissions. *IEEE Trans. Veh. Technol.* **2019**, *68*, 9018–9028. [CrossRef]
13. Shen, T.; Lin, Y.; Zou, J.; Wu, Y.; Shu, F.; Wang, J. Low-Complexity Leakage-Based Secure Precise Wireless Transmission with Hybrid Beamforming. *IEEE Wirel. Commun. Lett.* **2020**, *9*, 1687–1691. [CrossRef]
14. Shu, F.; Wu, X.; Hu, J.; Li, J.; Chen, R.; Wang, J. Secure and precise wireless transmission for random-subcarrier-selection-based directional modulation transmit antenna array. *IEEE J. Sel. Areas Commun.* **2018**, *36*, 890–904. [CrossRef]
15. Zhou, X.; Yan, S.; Hu, J.; Sun, J.; Li, J.; Shu, F. Joint optimization of a uav's trajectory and transmit power for covert communications. *IEEE Trans. Signal Process.* **2019**, *67*, 4276–4290. [CrossRef]
16. Wu, Q.; Zeng, Y.; Zhang, R. Joint trajectory and communication design for multi-uav enabled wireless networks. *IEEE Trans. Wirel. Commun.* **2018**, *17*, 2109–2121. [CrossRef]

17. Yi, F.; Zhang, C.; Baek, S.; Rawashdeh, S.; Mohammadi, A. Autonomous Landing of a UAV on a Moving Platform Using Model Predictive Control. *Drones* **2018**, *2*, 34. [CrossRef]
18. Ouamri, M.A.; Oteşteanu, M.-E.; Barb, G.; Gueguen, C. Coverage Analysis and Efficient Placement of Drone-BSs in 5G Networks. *Eng. Proc.* **2022**, *14*, 18. [CrossRef]
19. Shen, A.; Luo, J.; Ning, J.; Li, Y.; Wang, Z.; Duo, B. Safeguarding UAV Networks against Active Eavesdropping: An Elevation Angle-Distance Trade-Off for Secrecy Enhancement. *Drones* **2023**, *7*, 109. [CrossRef]
20. Zhou, X.; Yan, S.; Shu, F.; Chen, R.; Li, J. UAV-Enabled Covert Wireless Data Collection. *IEEE J. Sel. Areas Commun.* **2021**, *39*, 3348–3362. [CrossRef]
21. Hu, J.; Wu, Y.; Chen, R.; Shu, F.; Wang, J. Optimal Detection of UAV's Transmission with Beam Sweeping in Covert Wireless Networks. *IEEE Trans. Veh. Technol.* **2020**, *69*, 1080–1085. [CrossRef]
22. Li, Y.; Shu, F.; Shi, B.; Cheng, X.; Song, Y.; Wang, J. Enhanced RSS-based UAV localization via trajectory and multi-base stations. *IEEE Commun. Lett.* **2021**, *25*, 1881–1885. [CrossRef]
23. Cheng, X.; Shu, F.; Li, Y.; Zhuang, Z.; Wu, D.; Wang, J. Optimal Measurement of Drone Swarm in RSS-Based Passive Localization with Region Constraints. *IEEE Open J. Veh. Technol.* **2023**, *4*, 1–11. [CrossRef]
24. Zhou, X.; Li, J.; Shu, F.; Wu, Q.; Wu, Y.; Chen, W.; Hanzo, L. Secure SWIPT for Directional Modulation-Aided AF Relaying Networks. *IEEE J. Sel. Areas Commun.* **2021**, *37*, 253–268. [CrossRef]
25. Kraus, J.D.; Marhefka, R.J. *Antennas For All Applications*; McGraw-Hill, Inc.: New York, NY, USA, 2002.

Article

Joint Trajectory Planning, Time and Power Allocation to Maximize Throughput in UAV Network

Kehao Wang [1,*], Jiangwei Xu [1], Xiaobai Li [2], Pei Liu [1,3], Hui Cao [1] and Kezhong Liu [4]

[1] School of Information Engineering, Wuhan University of Technology, Wuhan 430070, China
[2] Air Force Early Warning Academy , Wuhan 430019, China
[3] Integrated Computing and Chip Security Sichuan Collaborative Innovation Center, Chengdu University of Information Technology, Chengdu 610059, China
[4] School of Navigation, Wuhan University of Technology, Wuhan 430070, China
* Correspondence: kehao.wang@whut.edu.cn

Abstract: Due to the advantages of strong mobility, flexible deployment, and low cost, unmanned aerial vehicles (UAVs) are widely used in various industries. As a flying relay, UAVs can establish line-of-sight (LOS) links for different scenarios, effectively improving communication quality. In this paper, considering the limited energy budget of UAVs and the existence of multiple jammers, we introduce a simultaneous wireless information and power transfer (SWIPT) technology and study the problems of joint-trajectory planning, time, and power allocation to increase communication performance. Specifically, the network includes multiple UAVs, source nodes (SNs), destination nodes (DNs), and jammers. We assume that the UAVs need to communicate with DNs, the SNs use the SWIPT technology to transmit wireless energy and information to UAVs, and the jammers can interfere with the channel from UAVs to DNs. In this network, our target was to maximize the throughput of DNs by optimizing the UAV's trajectory, time, and power allocation under the constraints of jammers and the actual motion of UAVs (including UAV energy budget, maximum speed, and anti-collision constraints). Since the formulated problem was non-convex and difficult to solve directly, we first decomposed the original problem into three subproblems. We then solved the subproblems by applying a successive convex optimization technology and a slack variables method. Finally, an efficient joint optimization algorithm was proposed to obtain a sub-optimal solution by using a block coordinate descent method. Simulation results indicated that the proposed algorithm has better performance than the four baseline schemes.

Keywords: UAV network; trajectory planning; power allocation; time allocation

Citation: Wang, K.; Xu, J.; Li, X.; Liu, P.; Cao, H.; Liu, K.; Cong, Y. Joint Trajectory Planning, Time and Power Allocation to Maximize Throughput in UAV Network. *Drones* **2023**, *7*, 68. https://doi.org/10.3390/drones7020068

Academic Editor: Andrey V. Savkin

Received: 28 December 2022
Revised: 14 January 2023
Accepted: 16 January 2023
Published: 18 January 2023

1. Introduction

In recent years, due to high mobility, flexible deployment, and low cost, UAVs have been widely used in many scenarios, such as intelligent transportation systems [1–3], disaster relief, military activities, emergency communications, and so on [4–6]. In particular, with the development of 5G technology, non-terrestrial networks have become the next hot spot [7]. For example, [8] studied the extreme performance of a cognitive uplink fixed satellite service and a fixed terrestrial service in the Ka-band 27.5–29.5 GHz frequency range. Ref. [9] studied the physical layer security problem of a satellite network.

Compared with traditional communication methods, most of the wireless communication channels in UAV networks are dominated by line-of-sight (LOS) links [10], which can reduce the obstruction of mountains and buildings so as to obtain better data transmission effects. For example, for natural disasters scenes and ground network communication failures, Ref. [11] used UAVs to provide support for ground base stations and proposed an adaptive UAV deployment scheme to solve the communication network coverage problem. Ref. [12] proposed a task-driven routing strategy for emergency UAVs network to enhance rescue efficiency. Aiming at the post-disaster areas where infrastructure has been destroyed,

Ref. [13] proposed an efficient data transmission scheme based on a particle swarm algorithm to ensure communication quality. Ref. [14] studied a UAV deployment strategy for disaster-affected areas to maximize the number of communication-coverage nodes.

Despite the widespread adoption of UAV-assisted communication technologies, there are still some challenges in UAV networks. First, due to the openness of a UAV network interface and the broadcast nature of electromagnetic waves, a UAV network is susceptible to radio interference. For instance, the existence of malicious jammers could reduce the communication quality between the nodes in a UAV network. Second, the battery capacity of the UAV is limited, which greatly shortens total mission time [15–17].

In view of the problem that a UAV network is susceptible to radio interference, the high mobility of UAVs could be used to improve a system's performance through the reasonable optimization of network resources [18–21]. Whereas, in a UAV network, the resources are limited, coupled with each other, and the established problems are usually non-convex and difficult to solve. Therefore, it is necessary to design a feasible optimization algorithm to obtain a solution to the original problem. Based on this, a successive convex approximation (SCA) [22] can be used as an effective method to solve the original non-convex problem.

To address the problem that the battery capacity of the UAV is limited, traditional energy harvesting mechanisms represented by solar and wind energy have been extensively studied [23,24]. For example, Ref. [23] studied the energy efficiency problem in solar-powered UAV systems by optimizing the speed, acceleration, heading angle, and transmission power of a UAV. Ref. [24] proposed an energy harvesting model based on hybrid solar and wind power for a UAV system and obtained a solution to the signal-to-noise ratio (SNR) outage minimization problem. Unfortunately, due to the limitations of hardware technology, the traditional energy harvesting scheme will significantly increase the take-off weight and lead to the degradation of the system's performance. Based on that, radio frequency (RF)-based SWIPT technology combined with UAV network resource allocation optimization could provide an effective solution [25].

To be specific, SWIPT is a technology that integrates wireless power transfer (WPT) and wireless information transfer (WIT). The power and information could be transferred at the same time, as an RF signal carries both power and information [26]. Typically, time switching (TS) and power splitting (PS) protocols are two common methods to implement SWIPT [27]. The former depends on time-slot allocation, where part of the time slot is used for energy transmission, and the other is used for information transmission and processing. The latter depends on power allocation, where part of the power is used for information transmission and processing, and the other is used for energy harvesting (EH).

1.1. Related Work

For UAV-assisted communication networks, due to the broadcast characteristics of electromagnetic waves, the UAV network is vulnerable to malicious interference. Generally speaking, an attack on the jammers on a UAV communication channel usually focuses on the physical layer and is completed by transmitting high-power interference signals. Therefore, many optimization schemes have been proposed to improve the system's performance.

Based on the high mobility of UAVs, Refs. [28–31] studied the UAV communication system in jamming environments from the perspective of trajectory planning. In [28], the authors studied a dual-UAV communication system with malicious interference and minimized the flight time by optimizing the UAV trajectory with steering angle constraints under the premise of meeting the information transmission requirements. In the scenario with a dynamic jammer, Ref. [29] proposed an offline algorithm based on deep reinforcement learning to optimize the trajectory and minimize the mission completion time. To ensure the reliability of the communication link, Ref. [30] maximized the average SINR by planning a trajectory for UAVs under energy constraints. In [31], the authors studied a UAV communication network with multiple jammers and optimized the UAV's trajectory by using the Dinklebach method and a non-convex optimization method to maximize the energy efficiency of the system.

However, the research on UAV network channel optimization in the above-mentioned interference environment mainly focused on trajectory planning in the space domain. Considering more degrees of freedom for a UAV network, Refs. [32–34] studied the multi-domain optimization problem based on trajectory planning. In [32], the authors mitigated the influence of jammers by jointly optimizing the UAV trajectory and signal transmission power and maximizing the average-secrecy rate of an uplink and a downlink. For the probabilistic LOS channel system, Ref. [33] studied a single UAV-assisted dual-user anti-jamming network and maximized the system throughput by optimizing the UAV's three-dimensional position and the ground nodes' signal transmission power, wherein the UAV provides communication services for two users. In the scenario where jammers have imperfect locations, considering the different requirements for service quality, Ref. [34] maximized the minimum throughput, the average throughput, and the minimum throughput with delay constraints on all nodes by optimizing the UAV trajectory, the task scheduling, and the ground nodes' signal transmission power, based on a successive convex approximation method and a block coordinate descent method. In addition, for the application scenarios of drones in package delivery, Ref. [35] maximized delivery and sensing utility under energy constraints by jointly optimizing the route selection, sensing time, and delivery weight allocation.

Moreover, on account of the fact that an RF signal carries both power and information, many resource allocation schemes were proposed to make the best of finite energy based on RF energy harvesting technology.

Aiming at the vulnerability of the UAV network, some existing works optimized UAV channel quality in terms of security [25,36–38]. In [36], the authors proposed a scheme based on power division and time switching to improve the security rate in a single UAV relay system with eavesdroppers. To analyze the security of the physical layer, Ref. [37] considered three attack scenarios: (1) a free-space optical eavesdropping attack, (2) a radio frequency attack, and (3) both a free-space optical and a radio frequency attack. The authors further studied the effects of SWIPT parameters and power amplifier efficiency on the security of the system. In [25], the authors investigated two cooperative UAV-assisted SWIPT networks. Specifically, they aimed to maximize the average-secrecy rate by jointly optimizing the trajectory and power of the UAVs, as well as the PS ratio. In the scenario where multiple eavesdroppers have imperfect locations, Ref. [38] improved the security of the network by jointly optimizing the UAV's position, noise power, PS, and TS ratios.

In addition, outage probability (OP) describes the probability of link failure and is often used to evaluate the performance of SWIPT communication systems. In [39], the authors obtained the closed-form expressions of outage probability and the bit-error rate in a UAV-relay-assisted decode and forward (DF) network based on TS and PS protocols and analyzed the transmission rate and delay limitation state. Moreover, a fast convergence algorithm based on bandwidth and time allocation was proposed by [40] to optimize the outage probability. In [41], the authors derived closed expressions of OP and throughput over Nakagami-m fading channels in a DF two-way relay system. They also analyzed the effects of the PS factor, threshold, fading severity, and parameters on the network's performance, wherein the two source nodes could communicate with each other with the help of a relay. Also exploiting SWIPT, Ref. [42] proposed a novel UAV-relay-assisted amplifiers and forwards (AF) network and derived connection- and secrecy-outage probabilities based on a PS scheme.

Furthermore, considering the energy-constrained system, some existing works studied the energy efficiency (EE) of networks. In [43], the energy efficiency was maximized by optimizing the UAV position. However, the authors did not use the detailed energy consumption expression of the rotary-wing UAV, but simply adopted the constant power. Ref. [44] investigated a UAV-assisted non-orthogonal multiple access (NOMA) networks, where the Dinkelbach method and successive convex optimization techniques were used to maximize the energy efficiency of the system by designing a UAV's location, beam pattern, power, and time schedule. For a multi-user distributed antenna system, Ref. [45]

maximized the energy efficiency by jointly-optimizing energy allocation and a PS ratio based on the Lagrangian method. Ref. [46] explored SWIPT techniques for device-to-device (D2D) communications based on a nonlinear energy system, optimized power, and the PS ratio maximizing energy efficiency.

Meanwhile, as a promising technology, intelligent reflecting surface (IRS)-aided SWIPT in UAV networks has drawn much attention recently [47,48]. In [47], the authors studied a UAV network supported by IRS and SWIPT and proposed an alternating optimization (AO) algorithm to minimize all users' energy consumption. Moreover, the authors in [48] investigated a UAV network equipped with IRS-aided SWIPT and developed an efficient iterative algorithm based on successive convex approximation, block coordinate descent, and time division multiple access (TDMA) protocols, to maximize the average-transmission rate. Aiming at the security problem in an aerial intelligent reflecting surface (AIRS) network, Ref. [49] proposed an AO algorithm based on the Riemannian manifold optimization method, the SCA technique, and the element-wise BCD method to jointly design the AIRS's deployment and phase shift, so as to maximize the system's secrecy rate.

In addition to the joint UAV trajectory planning and resource allocation optimization methods mentioned above, some existing works proposed new optimization methods to support a UAV communication network [50–52]. For example, considering the trajectory planning problem of multi-UAV assisted networks in a post-disaster scenario, Ref. [50] studied two heuristic algorithms to effectively utilize the UAVs' energy, to improve the communication coverage performance. In the RIS-based UAV-assisted IoT communication scenario, Ref. [51] proposed a multi-UAV path planning/transmission scheduling algorithm based on model predictive control (MPC) to improve system performance and reduce the total energy consumption of UAVs. For the UAV-assisted mobile edge computing (MEC) network communication scenario, Ref. [52] proposed a multi-agent deep reinforcement learning-based UAV trajectory control algorithm to jointly optimize the geographical fairness among all the user equipment, the fairness of every user-equipment load and the users' energy.

1.2. Motivation

In spite of the fact that the related works above have made great progress, there are still several problems needing to be resolved. To be specific, most existing studies did not consider the existence of multiple jammers, even though jammers have a significant impact on the legitimate communication of the system. In addition, most of the existing works considered a single UAV or a single ground node. That is because a multi-UAV system needs to meet a series of requirements, such as an anti-collision constraint and mission planning. These will increase the design difficulty and further increase the complexity of the algorithm. Furthermore, most existing works based on a SWIPT network focused on optimizing power or trajectory instead of multi-domains, including time, power, and trajectory. Most importantly, due to the development of onboard batteries, the flight time of UAVs is limited. The energy constraint problem greatly restricts further applications of UAVs. Therefore, how to improve the communication quality of a UAV network has always been a difficult and hot issue.

Inspired by the discussion above, we study a multi-UAV-assisted multi-user network system. Specifically, where the SN can send information and energy to the power-limited UAV, and the UAV uses the collected energy to communicate with the DN. It should be noted that there are multiple jammers in the network blocking legitimate communication. Different from the existing network, we introduce multiple UAVs based on SWIPT technology and fully consider the existence of jammers and the energy consumption of UAVs. In addition, due to the complexity of the network, solving this joint optimization problem was a considerable challenge. Thus, we introduced multiple slack variables and used the SCA method to make the original problem satisfy the disciplined convex program (DCP) rules so that the reformulated problem could be solved based on the solver CVX.

1.3. Contributions

For the sake of solving the problems given above, a multi-domain optimization algorithm based on the PS protocol that combines trajectory planning, time allocation, and power splitting is proposed by us. We aim to maximize the throughput by considering all constraints. The main contributions of this paper are as follows:

- We investigate a multi-UAV-assisted multi-user relay network in which the SNs use SWIPT technology to transmit wireless energy and information to UAVs. The UAVs use the collected energy to transmit information to the DNs, with the jammers interfering with legitimate channel communications.
- Our goal was to jointly optimize UAV trajectories, time allocation, and power-splitting factors, to mitigate interference and maximize the system throughput. Given that the original problem is non-convex and difficult to solve directly, we decomposed the original problem into three subproblems based on successive convex approximation, block coordinate descent, and a slack variables method presenting an efficient joint optimization algorithm to obtain a suboptimal solution.
- Simulation results indicate that the proposed scheme had better performance than the four benchmark schemes. In addition, we discuss the impact of the number of jammers and energy budgets on system performance and illustrate the effectiveness of joint trajectory planning, time, and power allocation to mitigate interference.

The rest part of the paper is organized as follows. In Section 2, the system model is introduced. In Section 3, we propose a joint optimization algorithm to solve the original problem. In Section 4, we provide simulation results and some necessary discussions. Finally, Section 5 concludes this paper.

2. System Model

Considering a multi-UAV enabled wireless communication network as shown in Figure 1, which includes K_1 quadcopter UAVs, K_2 source nodes (SNs) and destination nodes (DNs), and multiple jammers. Since each pair of ground nodes (SN and DN) is equipped with a fixed UAV to provide communication services for them, then $K_1 = K_2 = K$. In this system, we assume that the UAVs need to communicate with DNs, and the SNs use SWIPT technology to transmit wireless energy and information to UAVs. Specifically, K SNs stored information and energy. First, all SNs simultaneously send information and energy to the UAV relays, and then, the UAV relays use the collected energy to forward the information to the DNs in DF mode. It is assumed that the SNs, the UAVs, and the DNs are each equipped with a single antenna, the jammers are equipped with K antennas, and the jammers' antennas are aimed at the signal transmission direction of the UAVs [53]. Thus, The jammers which are far away from SNs and closer to UAVs interfere with the channel from UAVs to DNs.

In order to describe the model in mathematical terms, we introduce a 3D Cartesian coordinate system. Suppose the locations of SN k and DN k are $\mathbf{w}_{S_k} = (x_{S_k}, y_{S_k}, 0)$ and $\mathbf{w}_{D_k} = (x_{D_k}, y_{D_k}, 0)$ respectively, $k \in \mathcal{K} = \{1, 2, ..., K\}$. The system contains multiple jammers, denoted as $j \in \mathcal{J} = \{1, 2, ..., J\}$, and the location of the j-th jammer is $\mathbf{w}_j = (x_j, y_j, 0)$. At the same time, we discretize the UAVs mission period T into N time slots with equal length δ, i.e., $\delta = \frac{T}{N}$. Therefore, the position of UAV k flying at a height Z in any time slot $n \in \mathcal{N} = \{1, 2, ..., N\}$ is denoted as $\mathbf{q}_k[n] = (x_k[n], y_k[n])$. Moreover, we assume that the maximum flight speed of the UAV k is $V_{\max}$. Thus, we have the following:

$$\|\mathbf{q}_k[n] - \mathbf{q}_k[n-1]\| \leq V_{\max}\delta, \forall k, n = 2, ..., N. \tag{1}$$

which means that the UAV's speed between two adjacent time slots cannot exceed the maximum speed, where $\|\bullet\|$ represents the Euclidean norm. In addition, the distance between any two UAVs needs to be greater than a minimum safe distance of $D_{\min}$ to avoid collision and ensure safety. Thus,

$$\left\| \mathbf{q}_k[n] - \mathbf{q}_l[n] \right\|^2 \geq D_{\min}^2, \forall k, l, k \neq l. \tag{2}$$

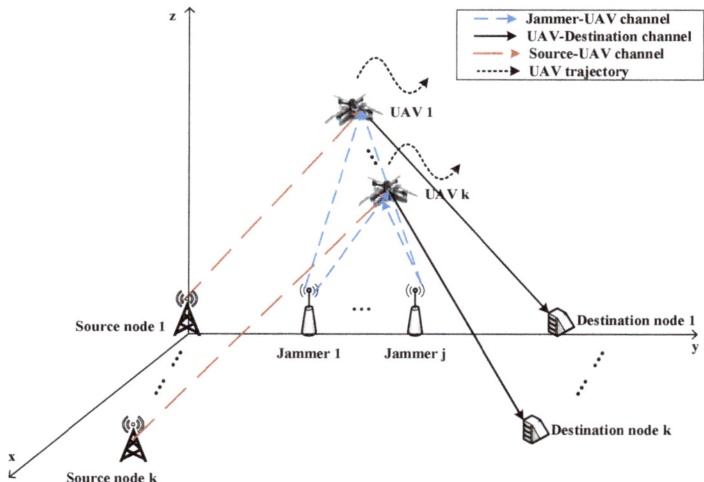

Figure 1. UAV communication network.

Generally speaking, most of the wireless communication channels in UAV networks are dominated by line-of-sight (LOS) links. Thus, following the free-space path loss model, we can get the channel-power gain between SN k and UAV k as follows:

$$h_{S_k U_k}^2 = \beta_{su} \left\| \mathbf{q}_k[n] - \mathbf{w}_{S_k} \right\|^{-2} \kappa[n], \forall n. \tag{3}$$

where β_{su} represents SN to UAV channel gain at the reference distance 1 m; $\kappa[n]$ is a small-scale fading with unit mean that is modeled by Rayleigh fading, i.e., $E[|\kappa[n]|^2] = 1$. Similarly, following the free-space path loss model, the channel-power gains from the jammer j to UAV k, and from UAV k to DN k are:

$$h_{j U_k}^2 = \beta_{Ju} \left\| \mathbf{q}_k[n] - \mathbf{w}_j \right\|^{-2}, \forall n. \tag{4}$$

$$h_{U_k D_k}^2 = \beta_{ud} \left\| \mathbf{q}_k[n] - \mathbf{w}_{D_k} \right\|^{-2}, \forall n. \tag{5}$$

where β_{Ju} and β_{ud} represent jammer j to UAV k and UAV k to DN k channel gain at the reference distance of 1 m, respectively.

2.1. Energy and Information Transmission Model

Inspired by [39], we use the PS protocol to describe the transmission process between the nodes of the network. Specifically, the PS protocol is divided into two steps, as shown in Figure 2, where τ is the time-allocation factor, α is the power-splitting factor, and P_k is the transmit power from SN k to UAV. We first divide each time slot into $\tau\delta$ and $(1 - \tau)\delta$. During the first $\tau\delta$ process, the SN sends a signal to the UAV. According to the PS protocol, the $1 - \alpha$ portion of the signal power is used by the UAV to receive and decode a specific signal from the SN, and the remaining portion is used for energy harvesting. Note that the energy collected by UAV from the SN is temporarily stored in the supercapacitor [54]. During the second $(1 - \tau)\delta$ process, the UAV uses the energy collected in the previous stage to transmit the decoded data to the DN. It should be noted that due to the limitation of energy harvesting efficiency and signal fading, we consider the harvested energy to be only used for signal processing and transmission. The energy consumed by the motor is determined by its own battery capacity, which we will discuss later.

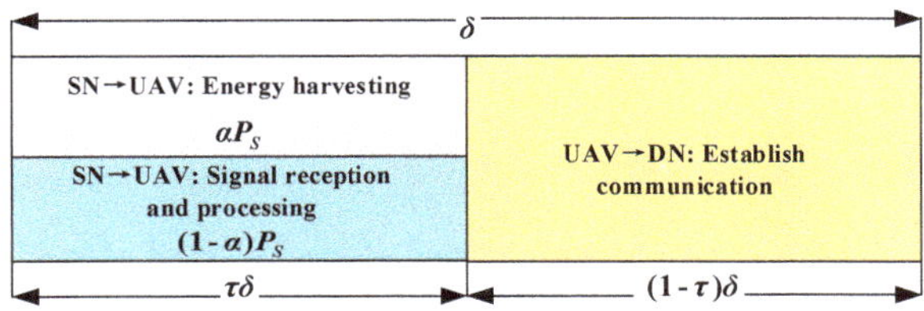

Figure 2. PS protocol.

Based on the above analysis, the UAV-received signal used for information processing during a time slot can be expressed as follows:

$$y_k^{IN} = (1-\alpha)\sqrt{P_k}h_{S_k U_k}x + n_{su}. \tag{6}$$

where x denotes a signal sent by SN, and n_{su} is additive white Gaussian noise (AWGN) with mean 0 and variance σ_{su}^2, i.e., $n_{su} \sim CN(0, \sigma_{su}^2)$. Therefore, the signal-to-noise ratio (SNR) of the received signal is as follows:

$$SNR_{k_{IN}} = \frac{(1-\alpha)P_k h_{S_k U_k}^2}{\sigma_{su}^2}. \tag{7}$$

Thus, the achievable rate from SN to UAV is as follows:

$$R_{SU_k} = \tau \log_2\left(1 + SNR_{IN_k}\right), \forall k. \tag{8}$$

It should be noted that in the actual network, the SNR needs to be greater than a threshold γ_{th1}; otherwise, the information transmission will be interrupted. Thus, we have the following:

$$SNR_{IN_k} \geq \gamma_{th1}, \forall k, \forall n. \tag{9}$$

According to the PS protocol, the energy harvesting time during each time slot is $\tau\delta$. Therefore, the energy collected by the UAV in a slot can be expressed as follows:

$$E_k^{EH} = \eta\alpha P_k h_{S_k U_k}^2 \tau\delta. \tag{10}$$

where η is energy collection efficiency. It should be noted that the collected energy cannot exceed the maximum capacity of the supercapacitor. Thus, we have the following:

$$E_k^{EH} \leq E_k^{cap} \tag{11}$$

where E_k^{cap} is the maximum capacity of the supercapacitor. Thus, the UAV's transmission power during the $(1-\tau)\delta$ can be expressed as follows:

$$P_{U_k} = \frac{E_k^{EH}}{(1-\tau)\delta} = \frac{\eta\alpha P_k h_{S_k U_k}^2 \tau}{(1-\tau)}. \tag{12}$$

Due to the existence of the jammers, the received signal-to-interference-plus-noise-ratio (SINR) at the DN needs to be greater than a threshold γ_{th2} to ensure that the signal transmission will not be interrupted. Thus, we have the following:

$$SINR_{D_k} = \frac{P_{U_k}[n]h_{U_k D_k}^2[n]}{\sum_{j=1}^{J} P_j h_{jU_k}^2[n] + \sigma_{ud}^2} \geq \gamma_{th2}, \forall n, \forall k. \tag{13}$$

where σ_{ud}^2 is the noise power, and P_j is the interference power. Thus, the achievable rate from the UAV to the DN is as follows:

$$R_{UD_k} = (1 - \tau)\log_2(1 + SINR_D), \forall k. \tag{14}$$

In addition, the information-causality constraints of the system can be written as

$$\sum_{t=1}^{n} (\tau)\log_2(1 + SNR_{IN_k}) \geq \sum_{t=1}^{n} (1 - \tau)\log_2(1 + SINR_D), n = 1, ..., N, \forall k. \tag{15}$$

2.2. Energy Consumption Model

Compared to energy consumption related to information transmission and signal decoding, UAVs need to consume more energy to maintain level flight. According to existing achievements in [17], the propulsion power of a rotary-wing UAV in 2D horizontal flight could be modeled as

$$P_{uav}(v) = \underbrace{P_B\left(1 + \frac{3v^2}{v_{tip}^2}\right)}_{blade\ profile} + \underbrace{P_I\left(\sqrt{1 + \frac{v^4}{4v_0^4} - \frac{v^2}{2v_0^4}}\right)^{\frac{1}{2}}}_{induced} + \underbrace{\frac{1}{2}d_0\rho s A_0 v^3}_{parasite} \tag{16}$$

where v means the UAV's speed, P_B and P_I are the blade profile and induced powers, respectively, when the UAV is hovering. v_{tip} represents the tip speed of the rotor blade, and v_0 is the mean rotor-induced velocity. In addition, d_0 is the fuselage drag ratio, ρ is the air density, s is rotor solidity, and A_0 is the rotor disc area. Therefore, we can get the sum of the energy consumption of the UAV in a mission period T by

$$E_{UAV}(v) = \int_0^T P_I\left(\sqrt{1 + \frac{v^4}{4v_0^4} - \frac{v^2}{2v_0^4}}\right)^{\frac{1}{2}} dt + \int_0^T P_B\left(1 + \frac{3v^2}{v_{tip}^2}\right) + \frac{1}{2}d_0\rho s A_0 v^3 dt \tag{17}$$

From the definition of time slot δ, we define the UAV's speed as $v = \|\mathbf{q}_k[n] - \mathbf{q}_k[n-1]\|/\delta$. Thus, we can rewrite E_{UAV} as

$$E_{UAV}(\Delta q) = \sum_{n=2}^{N} P_I\left(\sqrt{\delta^4 + \frac{\Delta q^4}{4v_0^4} - \frac{\Delta q^2}{2v_0^4}}\right)^{\frac{1}{2}} + \sum_{n=2}^{N} P_B\left(\delta + \frac{3\Delta q^2}{\delta v_{tip}^2}\right) + \frac{1}{2}d_0\rho s A_0 \frac{\Delta q^3}{\delta^2} \tag{18}$$

where $\Delta q = \|\mathbf{q}_k[n] - \mathbf{q}_k[n-1]\|$. In summary, we get the energy consumption expression of the rotary-wing UAV.

2.3. Problem Formulation

In this paper, our purpose is to maximize the throughput of UAV to the DN by optimizing the trajectory $\mathbf{q}[n]$, time-allocation factor τ, and power-splitting factor α. Thus, the throughput of the DN k over N time slots can be expressed as follows:

$$R_{D_k} = \sum_{n=1}^{N} (1 - \tau)\log_2(1 + SINR_D). \tag{19}$$

Let μ denote the the minimum throughput of DNs, i.e., $\mu = \min_{k \in \mathcal{K}} R_{D_k}$, and define $\mathbf{Q} = \{\mathbf{q}_k[n], \forall k, \forall n\}$, $\tau = \{\tau_k[n], \forall k, \forall n\}$, $\alpha = \{\alpha_k[n], \forall k, \forall n\}$, E_{th} as the UAV's energy budget, and $\mathbf{q}_{start}$ and $\mathbf{q}_{end}$ as the start and endpoints of the UAV. The joint trajectory planning, time, and power allocation optimization problem can be formulated as

$$P1: \max_{\{\mathbf{Q},\tau,\alpha,\mu\}} \mu \tag{20a}$$

$$s.t. \quad (1),(2),(9),(11),(13),(15) \tag{20b}$$

$$E_{UAV}(\Delta q) \leq E_{th}, \forall k \tag{20c}$$

$$\mu \leq R_{D_k}, \forall k \tag{20d}$$

$$0 \leq \tau \leq 1 \tag{20e}$$

$$0 \leq \alpha \leq 1 \tag{20f}$$

$$\mathbf{q}_k[1] = \mathbf{q}_{start} \tag{20g}$$

$$\mathbf{q}_k[N] = \mathbf{q}_{end} \tag{20h}$$

Note that problem P1 is difficult to solve directly since (2), (11), (13), (15), (20c), and (20d) are non-convex. In the next section, we propose an efficient iterative algorithm to obtain a feasible solution to original problem.

3. Joint Optimization

Since P1 is a non-convex problem, it is difficult to solve directly. In this section, we divide P1 into three subproblems and obtain suboptimal solutions by applying a successive convex approximation and a slack variables method. Then, we develop an overall iterative algorithm based on the block coordinate descent technique to get a locally optimal solution. The specific flow chart is shown in Figure 3.

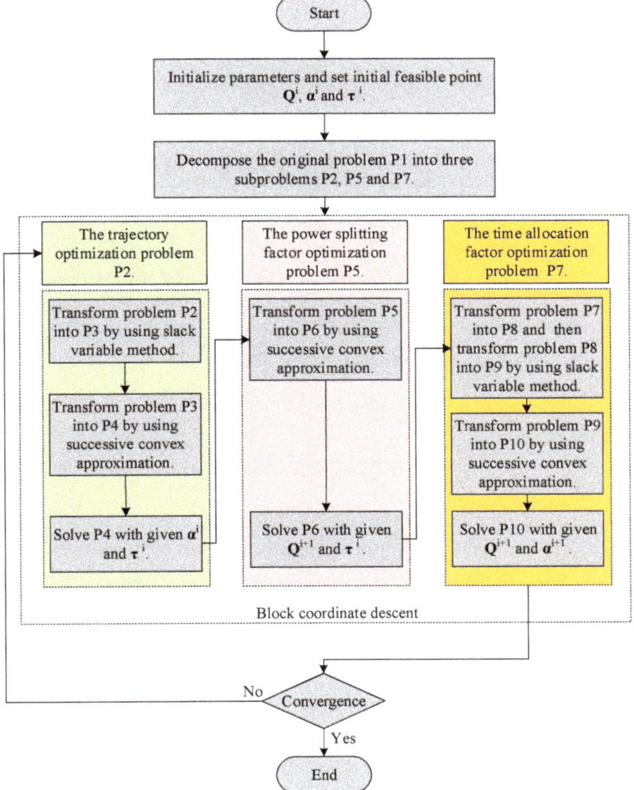

Figure 3. The flow chart of solution to problem P1.

3.1. Optimization of Trajectory

For any fixed power-splitting and time-allocation factors $\{\alpha, \tau\}$, the trajectory optimization problem can be expressed as follows:

$$P2 : \max_{\{\mathbf{Q}, \mu\}} \mu \tag{21a}$$

$$s.t. \quad (1), (2), (9), (11), (13), (15), (20c), (20d), (20g), (20h) \tag{21b}$$

Note that the problem P2 is intractable due to the non-convexity of (2), (11), (13), (15), (20c), and (20d). To tackle this issue, we first introduce slack variables $\{\mathbf{A}[n], \mathbf{B}[n], \mathbf{C}[n]\}$.

Theorem 1. *By introducing slack variables, problem P3 could be equivalently written as*

$$P3 : \max_{\{\mathbf{q}[n], \mu, \mathbf{A}[n], \cdots, \mathbf{D}[n]\}} \mu \tag{22a}$$

$$s.t. \quad (1), (2), (9), (11), (15), (20c), (20g), (20h) \tag{22b}$$

$$\mu \leq \sum_{n=2}^{N} (1 - \tau_k[n]) \log_2 \left(1 + \frac{1}{A_k[n] B_k[n] C_k[n]} \right) \tag{22c}$$

$$A_k[n] \geq \left(\frac{\eta \alpha P_k \beta_{su} \tau_k[n]}{1 - \tau_k[n]} \right)^{-1} \left\| \mathbf{q}_k[n] - \mathbf{w}_{S_k} \right\|^2, \forall k, \forall n \tag{22d}$$

$$B_k[n] \geq \beta_{ud}^{-1} \left\| \mathbf{q}_k[n] - \mathbf{w}_{D_k} \right\|^2, \forall k, \forall n \tag{22e}$$

$$C_k[n] \geq \sum_{j=1}^{J} P_j \beta_{Ju} \left\| \mathbf{q}_k[n] - \mathbf{w}_j \right\|^{-2} + \sigma_{ud}^2, \forall k, \forall n \tag{22f}$$

$$\frac{1}{A_k[n] B_k[n] C_k[n]} \geq \gamma_{th2}, \forall k, \forall n \tag{22g}$$

Proof. The theorem can be proved by the method of contradiction. Specifically, if (22d)–(22f) are strict equality constraints, problem P3 is equal to P2. Otherwise, by adjusting the slack variables, the value of the objective can always be further optimized. $\square$

However, P3 is still difficult to solve because (15), (20c), and (22f) are non-convex constraints, and the left-hand-side (LHS) of (2) and (22g), the right-hand-side (RHS) of (22c) is convex. Consider that any convex function is globally lower-bounded by its first-order Taylor expansion at any point [55]. Therefore, taking Taylor expansion approximately as lower bound, we can obtain the following:

$$\log_2 \left(1 + \frac{1}{\mathbf{A}[n]\mathbf{B}[n]\mathbf{C}[n]} \right) \geq \log_2 \left(1 + \frac{1}{\mathbf{A}^i \mathbf{B}^i \mathbf{C}^i} \right) - \frac{\mathbf{A}[n] - \mathbf{A}^i}{\mathbf{A}^i (1 + \mathbf{A}^i \mathbf{B}^i \mathbf{C}) \ln 2}$$

$$- \frac{\mathbf{B}[n] - \mathbf{B}^i}{\mathbf{B}^i (1 + \mathbf{A}^i \mathbf{B}^i \mathbf{C}^i) \ln 2} - \frac{\mathbf{C}[n] - \mathbf{C}^i}{\mathbf{C}^i (1 + \mathbf{A}^i \mathbf{B}^i \mathbf{C}^i) \ln 2} \tag{23}$$

where $(\mathbf{A}^i, \mathbf{B}^i, \mathbf{C}^i)$ is a given local point in the i-th iteration.

Ignoring the terms that are independent of the slack variables $(\mathbf{A}[n], \mathbf{B}[n], \mathbf{C}[n])$, we replace the RHS of (22c) with $\mathbf{S}[n]$:

$$\mathbf{S}[n] = -\frac{\mathbf{A}[n]/\mathbf{A}^i + \mathbf{B}[n]/\mathbf{B}^i + \mathbf{C}[n]/\mathbf{C}^i}{(1 + \mathbf{A}^i \mathbf{B}^i \mathbf{C}^i) \ln 2} \tag{24}$$

$$\mu \leq \sum_{n=2}^{N} (1 - \tau_k[n]) \mathbf{S}[n] \tag{25}$$

Similar to Theorem 1, by introducing slack variables $\mathbf{D}[n]$, (22f) could be equivalently written as follows:

$$C_k[n] \geq \sum_{j=1}^{J} P_j \beta_{Ju} D_{k,j}(n)^{-1} + \sigma_{ud}^2, \forall k, \forall n \tag{26a}$$

$$0 \leq D_{k,j}(n) \leq \left\| \mathbf{q}_k[n] - \mathbf{w}_j \right\|^2, \forall k, \forall n, \forall j \tag{26b}$$

However, we notice that the RHS of (26b) is convex with respect to trajectory $\mathbf{Q}$, thus, (26b) is still a non-convex constraint. Relying on the first-order Taylor expansion, we have the lower bound as

$$\left\| \mathbf{q}_k[n] - \mathbf{w}_j \right\|^2 \geq \left\| \mathbf{q}_k^i[n] - \mathbf{w}_j \right\|^2 + 2(\mathbf{q}_k^i[n] - \mathbf{w}_j)^T \times (\mathbf{q}_k[n] - \mathbf{q}_k^i[n]) = E_{k,j}[n], \forall k, \forall n, \forall j \tag{27}$$

Thus, we reformulate (26b) as

$$0 \leq D_{k,j}(n) \leq E_{k,j}(n), \forall k, \forall n, \forall j \tag{28}$$

For the non-convex constraint (22g). Since the LHS of (22g) is convex, we apply the first-order Taylor expansion to get the lower bound at the i-th iteration point

$$\frac{1}{\mathbf{A}[n]\mathbf{B}[n]\mathbf{C}[n]} \geq \frac{1}{\mathbf{A}^i\mathbf{B}^i\mathbf{C}^i} - \frac{(\mathbf{A}[n] - \mathbf{A}^i)}{(\mathbf{A}^i)^2\mathbf{B}^i\mathbf{C}^i} - \frac{(\mathbf{B}[n] - \mathbf{B}^i)}{\mathbf{A}^i(\mathbf{B}^i)^2\mathbf{C}^i} - \frac{(\mathbf{C}[n] - \mathbf{C}^i)}{\mathbf{A}^i\mathbf{B}^i(\mathbf{C}^i)^2} \geq \gamma_{th2} \tag{29}$$

For the LHS of (2), we can obtain the lower bound according to the first-order Taylor expansion as

$$\left\| \mathbf{q}_k[n] - \mathbf{q}_l[n] \right\|^2 \geq -\left\| \mathbf{q}_k^i[n] - \mathbf{q}_l^i[n] \right\|^2 + 2(\mathbf{q}_k^i[n] - \mathbf{q}_l^i[n])^T \times (\mathbf{q}_k[n] - \mathbf{q}_l[n]) \tag{30}$$

Therefore, the non-convex constraint (2) can be rewritten as a convex constraint:

$$\left\| \mathbf{q}_k^i[n] - \mathbf{q}_l^i[n] \right\|^2 + 2(\mathbf{q}_k^i[n] - \mathbf{q}_l^i[n])^T \times (\mathbf{q}_k[n] - \mathbf{q}_l[n]) \geq D_{\min}^2, \forall k, l, k \neq l. \tag{31}$$

For the information-causality constraint (15), by introducing slack variables $\{\mathbf{F}, \mathbf{G}, \mathbf{H}, \mathbf{I}\}$, we have the following:

$$\sum_{t=1}^{n} (\tau)\log_2\left(1 + \frac{\eta \alpha P_k \tau}{(1-\tau)F_k[t]G_k[t]H_k[t]}\right)$$
$$\leq \sum_{t=1}^{n} (1-\tau)\log_2\left(1 + \frac{(1-\alpha)P_k}{\sigma^2 I_k[t]}\right), n = 1, ..., N, \forall k. \tag{32a}$$

$$F_k[n] \leq \sum_{j=1}^{J} P_j h_{jU_k}^2[n] + \sigma_{ud}^2, n = 1, ..., N. \tag{32b}$$

$$G_k[n] \leq \beta_{su}^{-1} \left\| \mathbf{q}_k[n] - \mathbf{w}_{S_k} \right\|^2, n = 1, ..., N. \tag{32c}$$

$$H_k[n] \leq \beta_{ud}^{-1} \left\| \mathbf{q}_k[n] - \mathbf{w}_{D_k} \right\|^2, n = 1, ..., N. \tag{32d}$$

$$I_k[n] \geq \beta_{su}^{-1} \left\| \mathbf{q}_k[n] - \mathbf{w}_{S_k} \right\|^2, n = 1, ..., N. \tag{32e}$$

Since the RHS of (32a) is convex with respect to $\mathbf{I}$, relying on the first-order Taylor expansion, we have the following:

$$\log_2\left(1 + \frac{(1-\alpha)P_k}{\sigma^2 I_k}\right) \geq \log_2\left(1 + \frac{(1-\alpha)P_k}{\sigma^2 I_k^i}\right) - \frac{(1-\alpha)P_k(I_k - I_k^i)}{\left(\sigma^2(I_k^i)^2 + (1-\alpha)P_k I_k^i\right)\ln 2} = L \tag{33}$$

Thus, (32a) can be written as

$$\sum_{t=1}^{n} (\tau)\log_2\left(1 + \frac{\eta\alpha P_k\tau}{(1-\tau)F_k[t]G_k[t]H_k[t]}\right) \leq \sum_{t=1}^{n}(1-\tau)L_k[t], n = 1,\ldots,N, \forall k. \tag{34}$$

For the non-convex constraint (32b), by introducing slack variables $\mathbf{M}$, we have:

$$F_k[n] \leq \sum_{j=1}^{J} P_j M_{k,j}[n] + \sigma_{ud}^2, \forall k, \forall n. \tag{35a}$$

$$\frac{1}{M_{k,j}[n]} \geq \beta_{Ju}^{-1}\left\|\mathbf{q}_k[n] - \mathbf{w}_j\right\|^2, \forall k, \forall n, \forall j. \tag{35b}$$

Since the LHS of (35b) is convex, we have the following:

$$\frac{1}{M_{k,j}[n]} \geq \frac{1}{M_{k,j}^i[n]} - \frac{M_{k,j}[n] - M_{k,j}[n]}{M_{k,j}^i[n]^2} \geq \beta_{Ju}^{-1}\left\|\mathbf{q}_k[n] - \mathbf{w}_j\right\|^2 \tag{36}$$

Similar to (27), for the non-convex constraints (32c) and (32d), we have:

$$\beta_{su}^{-1}\left\|\mathbf{q}_k^i[n] - \mathbf{w}_{S_k}\right\|^2 + 2\beta_{su}^{-1}(\mathbf{q}_k^i[n] - \mathbf{w}_{S_k})^T \times (\mathbf{q}_k[n] - \mathbf{q}_k^i[n]) \geq G_k[n], \forall k, \forall n. \tag{37a}$$

$$\beta_{ud}^{-1}\left\|\mathbf{q}_k^i[n] - \mathbf{w}_{D_k}\right\|^2 + 2\beta_{ud}^{-1}(\mathbf{q}_k^i[n] - \mathbf{w}_{D_k})^T \times (\mathbf{q}_k[n] - \mathbf{q}_k^i[n]) \geq H_k[n], \forall k, \forall n. \tag{37b}$$

Similarly, for the non-convex constraint (11), we have the following:

$$\frac{\eta\alpha P_k\beta_{su}\tau\delta}{E_k^{cap}} \leq \left\|\mathbf{q}_k^i[n] - \mathbf{w}_{S_k}\right\|^2 + 2(\mathbf{q}_k^i[n] - \mathbf{w}_{S_k})^T \times (\mathbf{q}_k[n] - \mathbf{q}_k^i[n]), \forall k, \forall n. \tag{38}$$

Since the energy constraint expression (20c) is very complex and difficult to solve directly, in order to facilitate the analysis, we introduce a slack variable $\mathbf{O}$ as follows:

$$O_k[n] \geq \left(\sqrt{\delta^4 + \frac{\Delta q_k^4}{4v_0^4}} - \frac{\Delta q_k^2}{2v_0^4}\right)^{\frac{1}{2}} \tag{39}$$

We can further obtain:

$$O_k^2[n] + \frac{\Delta q_k^2}{v_0^2} \geq \frac{\delta^4}{O_k^2[n]}, n = 2,\ldots,N, \forall k. \tag{40}$$

Therefore, the energy consumption of the UAV k can be equivalently expressed as follows:

$$E_{th} \geq E_k(\Delta q_k, O_k[n]) = \sum_{n=2}^{N} P_B\left(\delta + \frac{3\Delta q^2}{\delta v_{tip}^2}\right) + \frac{1}{2}d_0\rho s A_0\frac{\Delta q^3}{\delta^2} + \sum_{n=2}^{N} P_I O_k[n] \geq E_{UAV}(\Delta q_k), \forall k \tag{41}$$

Finally, for non-convex constraint (40), we have the following:

$$O_k^2[n] + \frac{\Delta q_k^2}{v_0^2} \geq O_k^i[n]^2 + \frac{\Delta q_k^{i\,2}}{v_0^2} + 2O_k^i[n](O_k[n] - O_k^i[n]) + \frac{2\Delta q_k^i}{v_0^2}(\Delta q_k - \Delta q_k^i) \geq \frac{\delta^4}{O_k^2[n]}, \forall k. \tag{42}$$

As a result, the lower bound problem of P3 can be approximated as

$$P4 : \max_{\{\mathbf{Q},\mathbf{A},\dots,\mathbf{O},\mu\}} \mu \tag{43a}$$

$$s.t. \quad (1),(9),(20g),(20h),(22d),(22e),(25),(26a),(28),(29),$$
$$(31),(32e),(34),(35a),(36),(37a),(37b),(38),(41),(42) \tag{43b}$$

Obviously, P4 is a convex optimization problem that could be solved efficiently by some classical optimization techniques, such as the interior point method. In addition, it is worth noting that the optimal objective value obtained from P4 usually serves as a lower bound of P3.

3.2. Optimization of Power Splitting Factor

For any fixed UAV trajectory and time-allocation factor $\{\mathbf{Q}, \tau\}$, the power-splitting factor optimization problem can be expressed as follows:

$$P5 : \max_{\{\boldsymbol{\alpha},\mu\}} \mu \tag{44a}$$

$$s.t. \quad (9),(11),(13),(15),(20d),(20f) \tag{44b}$$

It should be noted that problem P5 is not a standard convex optimization problem because the LHS and the RHS of (15) are concave. Thus, consider that any concave function is globally upper-bounded by its first-order Taylor expansion. Therefore, we can obtain the following:

$$\log_2(1 + \alpha P) \le \log_2\left(1 + \alpha^i P\right) + \frac{(\alpha - \alpha^i)P}{(1 + \alpha^i P)\ln 2} \tag{45}$$

where

$$P = \frac{\eta P_k h^2_{S_k U_k} \tau h^2_{U_k D_k}}{(1 - \tau)\left(\sum\limits_{j=1}^{J} P_j h^2_{jU_k} + \sigma^2_{ud}\right)} \tag{46}$$

Thus, the lower bound problem of P5 can be approximated as

$$P6 : \max_{\{\boldsymbol{\alpha},\mu\}} \mu \tag{47a}$$

$$s.t. \quad (9),(11),(13),(20d),(20f) \tag{47b}$$

$$\sum_{t=1}^{n} (1 - \tau)\left(\log_2\left(1 + \alpha_k^i[t]P[t]\right) + \frac{(\alpha_k[t] - \alpha_k^i[t])P[t]}{(1 + \alpha_k^i[t]P[t])\ln 2}\right)$$
$$\le \sum_{t=1}^{n} (\tau)\log_2(1 + SNR_{IN_k}), n = 1,\dots,N, \forall k. \tag{47c}$$

P6 is also a convex optimization problem that can be solved like P4. Additionally, the optimal objective value obtained from P6 usually serves as a lower bound of P5.

3.3. Optimization of Time-Allocation Factor

For any given power-splitting factor and trajectory $\{\boldsymbol{\alpha}, \mathbf{Q}\}$, we consider the subproblem of optimizing the time-allocation factor as follows:

$$P7 : \max_{\{\tau,\mu\}} \mu \tag{48a}$$

$$s.t. \quad (11),(13),(15),(20d),(20e) \tag{48b}$$

Further, we can transform the problem P7 into P8:

$$P8 : \max_{\{\tau,\mu\}} \mu \tag{49a}$$

$$s.t. \quad (11), (13), (15), (20e) \tag{49b}$$

$$\mu \le \sum_{n=1}^{N} (1 - \tau_k[n])\log_2\left(1 + \frac{\tau_k[n]}{1 - \tau_k[n]}R_k[n]\right), \forall k \tag{49c}$$

where

$$R_k[n] = \frac{\alpha_k[n]\eta P_k h_{S_k U_k}^2[n]h_{U_k D_k}^2[n]}{\sum\limits_{j=1}^{J} P_j h_{jU_k}^2[n] + \sigma_{ud}^2} \tag{50}$$

Since (15) and (49c) are non-convex, P8 is difficult to solve directly. To this end, we introduce slack variables to solve this problem.

Theorem 2. *By introducing slack variables* $\{\mathbf{U}, \mathbf{V}, \mathbf{W}, \mathbf{X}\}$, *problem P8 could be equivalently written as follows:*

$$P9 : \max_{\{\tau,\mu,\mathbf{U},...,\mathbf{X}\}} \mu \tag{51a}$$

$$s.t. \quad (11), (13), (20e) \tag{51b}$$

$$\mu \le \sum_{n=1}^{N} U_k[n]V_k[n], \forall k \tag{51c}$$

$$0 \le U_k[n] \le 1 - \tau_k[n], \forall k, \forall n \tag{51d}$$

$$0 \le V_k[n] \le \log_2\left(1 + \frac{\tau_k[n]}{1 - \tau_k[n]}R_k[n]\right), \forall k, \forall n \tag{51e}$$

$$W_k[n] \ge 1 - \tau_k[n], \forall k, \forall n \tag{51f}$$

$$X_k[n] \ge \log_2\left(1 + \frac{\tau_k[n]}{1 - \tau_k[n]}R_k[n]\right), \forall k, \forall n \tag{51g}$$

$$\sum_{t=1}^{n} W_k[t]X_k[t] \le \sum_{t=1}^{n} (\tau)\log_2(1 + SNR_{IN_k}), n = 1, ..., N, \forall k. \tag{51h}$$

Proof. According to (51d) and (51e), we have the following:

$$\sum_{n=1}^{N} U_k[n]V_k[n] \le \sum_{n=1}^{N} (1 - \tau_k[n])\log_2\left(1 + \frac{\tau_k[n]}{1 - \tau_k[n]}R_k[n]\right) \tag{52}$$

According to (51f) and (51g), we have the following:

$$\sum_{t=1}^{n} (1 - \tau_k[t])\log_2\left(1 + \frac{\tau_k[t]}{1 - \tau_k[t]}R_k[t]\right) \le \sum_{t=1}^{n} W_k[t]X_k[t] \tag{53}$$

Therefore, we prove the theorem by the method of contradiction. Specifically, if (52) and (53) are strict equality constraints, combined with (51c) and (51h), we can know that P9 is equal to P8. Otherwise, by adjusting the slack variables, the value of the objective function can always be further optimized. $\square$

However, P9 is still a non-convex optimization problem that is difficult to solve directly. Considering that (51c) has a product of functions (PF) structure, we can rewrite (51c) as a function with the difference of convex (DC) structure, that is,

$$\mathbf{UV} = \frac{1}{2}(\mathbf{U} + \mathbf{V})^2 - \frac{1}{2}(\mathbf{U}^2 + \mathbf{V}^2) \tag{54}$$

Since the first term in the RHS of (54) is convex, we can obtain a lower bound for (54) by using first-order Taylor expansion, that is

$$\frac{1}{2}(\mathbf{U} + \mathbf{V})^2 - \frac{1}{2}(\mathbf{U}^2 + \mathbf{V}^2) \geq (\mathbf{U}^i + \mathbf{V}^i)(\mathbf{U} + \mathbf{V}) - \frac{1}{2}(\mathbf{U}^i + \mathbf{V}^i)^2 - \frac{1}{2}(\mathbf{U}^2 + \mathbf{V}^2) = \mathbf{Y} \quad (55)$$

Thus, (51c) can be rewritten as

$$\mu \leq \sum_{n=2}^{N} \mathbf{Y} \quad (56)$$

For the non-convex constraint (51e), by introducing slack variables $\mathbf{Z}$, we have:

$$0 \leq V_k[n] \leq \log_2(1 + Z_k[n]R_k[n]) \quad (57a)$$

$$0 \leq Z_k[n] \leq \frac{\tau_k[n]}{1 - \tau_k[n]} \quad (57b)$$

Furthermore, the RHS of the (57b) is convex on the domain ($\tau \in [0, 1]$). Thus, we have the following:

$$\frac{\tau_k[n]}{1 - \tau_k[n]} \geq \left(\frac{1}{1 - \tau_k^i[n]} - 1\right) + \frac{\tau_k[n] - \tau_k^i[n]}{(1 - \tau_k^i[n])^2} = \Gamma \quad (58)$$

Thus, (57b) can be rewritten as

$$0 \leq \mathbf{Z} \leq \Gamma \quad (59)$$

Similar to the procedure of handling (51e), for the non-convex constraint (51g), by introducing the slack variable Λ, we have

$$X_k[n] \geq \log_2\left(1 + \frac{R_k[n]}{\Lambda_k[n]}\right) \quad (60a)$$

$$0 \leq \Lambda_k[n] \leq \frac{1 - \tau_k[n]}{\tau_k[n]} \quad (60b)$$

For the (60b), we have

$$\frac{1 - \tau_k[n]}{\tau_k[n]} \geq \frac{1}{\tau_k^i[n]} - \frac{(\tau_k[n] - \tau_k^i[n])}{(\tau_k^i[n])^2} - 1 \geq \Lambda_k[n] \geq 0 \quad (61)$$

According to (55), for the (51h), we have

$$\sum_{t=1}^{n} \mathbf{W}\mathbf{X} \leq \sum_{t=1}^{n} \frac{1}{2}(\mathbf{W} + \mathbf{X})^2 \leq \sum_{t=1}^{n} (\tau)\log_2(1 + SNR_{IN_k}), n = 1, ..., N, \forall k. \quad (62)$$

As a result, the lower bound problem of P9 can be rewritten as

$$P10: \max_{\{\tau, \mu, \mathbf{U}, ..., \Lambda\}} \mu \quad (63a)$$

$$s.t. \ (11), (13), (20e), (51d), (51f), (56), (57a), (59), (60a), (61), (62) \quad (63b)$$

P10 is also a convex optimization problem that can be solved like P6. In addition, the optimal objective value obtained from P10 usually serves as a lower bound of P9.

3.4. Algorithmic Architecture

According to the above analysis, we obtain the suboptimal solution of the original problem P1 based on the block coordinate descent (BCD) method. As shown in Algorithm 1, the algorithm alternately optimizes $\mathbf{Q}$, α, and τ until convergence. Note that the initial point is the Taylor expansion point within the feasible region. Then, the convergence is proved as follows:

Algorithm 1 Overall algorithm

1: **Initialize** $i = 1$. Set initial feasible point $\{\mathbf{Q}^i, \boldsymbol{\alpha}^i, \boldsymbol{\tau}^i\}$ and other slack variables.
2: **Do**
3:     Solve problem P4 with given $\{\mathbf{Q}^i, \boldsymbol{\alpha}^i, \mathbf{A}^i, ..., \mathbf{O}^i\}$ to obtain the optimal solution $\{\mathbf{Q}^{i+1}, \mathbf{A}^{i+1}, ..., \mathbf{O}^{i+1}\}$,
4:     Solve problem P6 with given $\{\boldsymbol{\alpha}^i, \mathbf{Q}^{i+1}, \boldsymbol{\tau}^i\}$ to obtain the optimal solution $\{\boldsymbol{\alpha}^{i+1}\}$,
5:     Solve problem P10 with given $\{\boldsymbol{\alpha}^{i+1}, \mathbf{Q}^{i+1}, ..., \boldsymbol{\Lambda}^i\}$ to obtain the optimal solution $\{\boldsymbol{\tau}^{i+1}, ..., \boldsymbol{\Lambda}^{i+1}\}$.
6:     **Update** $i = i + 1$.
7: **Until** the objective value converges.
8: **Output** $\boldsymbol{\alpha}^* \leftarrow \boldsymbol{\alpha}^i, \mathbf{Q}^* \leftarrow \mathbf{Q}^i, \boldsymbol{\tau}^* \leftarrow \boldsymbol{\tau}^i$.

We define $\mu(\mathbf{Q}^i, \boldsymbol{\alpha}^i, \boldsymbol{\tau}^i)$, $\mu_1(\mathbf{Q}^i, \boldsymbol{\alpha}^i, \boldsymbol{\tau}^i)$, $\mu_2(\mathbf{Q}^i, \boldsymbol{\alpha}^i, \boldsymbol{\tau}^i)$, and $\mu_3(\mathbf{Q}^i, \boldsymbol{\alpha}^i, \boldsymbol{\tau}^i)$ as the objective values of problem P1, P4, P6 and P10 based on $\mathbf{Q}^i$, $\boldsymbol{\alpha}^i$, and $\boldsymbol{\tau}^i$ over i-th iteration. Thus, we have

$$
\mu(\mathbf{Q}^i, \boldsymbol{\alpha}^i, \boldsymbol{\tau}^i) \overset{(a)}{\leq} \mu_1(\mathbf{Q}^{i+1}, \boldsymbol{\alpha}^i, \boldsymbol{\tau}^i) \overset{(b)}{\leq} \mu_2(\mathbf{Q}^{i+1}, \boldsymbol{\alpha}^{i+1}, \boldsymbol{\tau}^i)
$$
$$
\overset{(c)}{\leq} \mu_3(\mathbf{Q}^{i+1}, \boldsymbol{\alpha}^{i+1}, \boldsymbol{\tau}^{i+1}) \overset{(d)}{\leq} \mu(\mathbf{Q}^{i+1}, \boldsymbol{\alpha}^{i+1}, \boldsymbol{\tau}^{i+1}) \tag{64}
$$

where (a) holds because in Algorithm 1, problem P4 is solved to obtain the optimal solution $\mathbf{Q}^{i+1}$ with given $\boldsymbol{\alpha}^i$ and $\boldsymbol{\tau}^i$ at step 3; (b) holds because problem P6 is solved to obtain the optimal solution $\boldsymbol{\alpha}^{i+1}$ with given $\mathbf{Q}^{i+1}$ and $\boldsymbol{\tau}^i$ at step 4; (c) holds because problem P10 is solved to obtain the optimal solution $\boldsymbol{\tau}^{i+1}$ with given $\mathbf{Q}^{i+1}$ and $\boldsymbol{\alpha}^{i+1}$ at step 5; (d) holds because the optimal objective values of P4, P6 and P10 are upper bounded by original problem P1, then the convergence can be guaranteed.

Finally, we briefly analyze the overall complexity of the algorithm. According to Algorithm 1, the complexity of the algorithm is mainly dominated by steps 3, 4, and 5, and the number of optimization variables increases with the multiples of K, J, and N. Hence, the total computational complexity is $O((KJN)^{3.5} \log \frac{1}{\varepsilon})$, where K is the number of UAVs, J is the number of jammers, N is the number of time slots, and ε is the convergence accuracy. In addition, it should be noted that the proposed scheme is an offline algorithm, which requires path planning and resource allocation through a specific ground station (such as QGroundControl in LINUX) before the mission is executed and does not need to run on UAVs.

4. Simulation Results

In this section, simulation results and some detailed discussions are provided. We first present the simulation settings and then analyze the effect of different energy budgets and the number of jammers on the experimental results. Finally, we compare the proposed algorithm with four baseline schemes to further illustrate the superiority of the joint trajectory planning, time, and power allocation scheme.

4.1. Simulation Settings

In the simulation, we considered a communication system with four UAV nodes, i.e., $k = 4$. In addition, it was assumed that the initial and end positions of the UAV 1–4 were $(200, 50)$, $(200, 70)$, $(200, 90)$, $(200, 110)$, and $(50, 50)$, $(50, 70)$, $(50, 90)$, $(50, 110)$, respectively. The rest of the parameter settings are shown in Table 1.

Table 1. Simulation Parameters Setting.

Parameter	Notation	Value
time slots	N	50
minimum safe distance	D_{min}	10 m
Bandwidth	B	10 MHz
SN to UAV channel gain	β_{su}	-30 dB
Jammer to UAV channel gain	β_{Ju}	-60 dB
transmit power of SN	P_k	30 W
transmit power of jammer	P_j	5 W
energy collection efficiency	η	0.8
additive white Gaussian noise	σ^2	-169 dBm
UAV maximum speed	V_{max}	10 m/s
the blade profile power	P_B	79.86 W
the induced power	P_I	88.63 W
the tip speed of the rotor blade	v_{tip}	120 m/s
the mean rotor induced velocity	v_0	4.03 m/s
the fuselage drag ratio	d_0	0.6
the air density	ρ	1.225 kg/m^3
the rotor solidity	s	0.05

4.2. Effect of Energy Budgets

Figure 4 shows the 2D trajectories of four UAVs with different energy budgets. We plotted the trajectories for $E_{th} = 10{,}000$ J, $E_{th} = 15{,}000$ J, and $E_{th} = 20{,}000$ J. It can be seen from Figure 4 that from four initial points, UAVs $1-4$ needed to approach the SNs according to the arc trajectory away from the jammers to ensure that more energy was collected to maximize the throughput of the DNs, and then fly back to the endpoints we set. For a different since the initial point was far from the source node, and in order to satisfy the minimum distance constraint between UAVs, it needed to fly a greater distance. Since UAV 4 was closest to the jammers, in order to ensure the communication quality, UAV 4 needed to be far away from the jammers under the constraint of the minimum safe distance and closer to the corresponding source node so as to collect more energy to communicate with the DN. Note that the UAVs cannot be infinitely close to the SNs, because while being closer to the SNs could guarantee enough energy to be collected, it would make the UAVs far away from the DNs, which would lead to the deterioration of the throughput. In addition, we noticed that the flying distance of the UAV increased with the energy budget, because a sufficient energy budget would ensure that the UAV was farther away from the jammers when planning a more reasonable path to maximize the throughput of the DN.

Figure 5 demonstrates the speed of four UAVs with different energy budgets. We observed that the UAVs' speed could be divided into two stages. In the first stage, the UAVs moved at high speed, and the speed decreased with time, but in the second stage, the UAVs accelerated to the endpoints. This is because in the first stage, the jammers were closer to the UAVs, and the UAVs needed to move away from the jammers at high speed to ensure communication quality. As the UAVs kept getting closer to the optimal positions, the speed needed to be reduced to collect more energy. However, due to the limited time, and energy budgets, the UAVs could not fly at low speed for a long time; thus, the UAVs needed to accelerate to the endpoints in the second stage. In addition, we can see that UAV 1 flew the fastest, while UAV 3 was the slowest. This is because UAV 1 was the farthest from its corresponding SN, and it needed to fly farther to collect more energy, while UAV 3 was the closest to its corresponding SN, so the budget was sufficient to allow it to collect the required energy with a lower speed. Finally, we observed that for the first 30 time slots and the last 10 time slots of the total mission, the UAVs' speed increased with energy budgets. That was because larger energy budgets could keep the UAVs away from the jammers and back to the endpoints at higher speed. These were as expected.

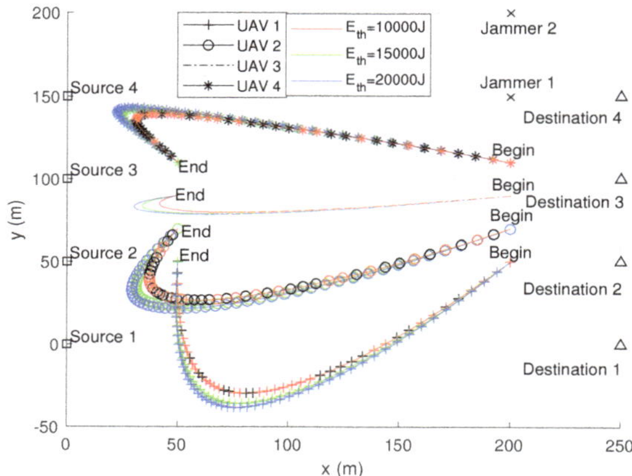

Figure 4. Optimal trajectories in different energy budgets.

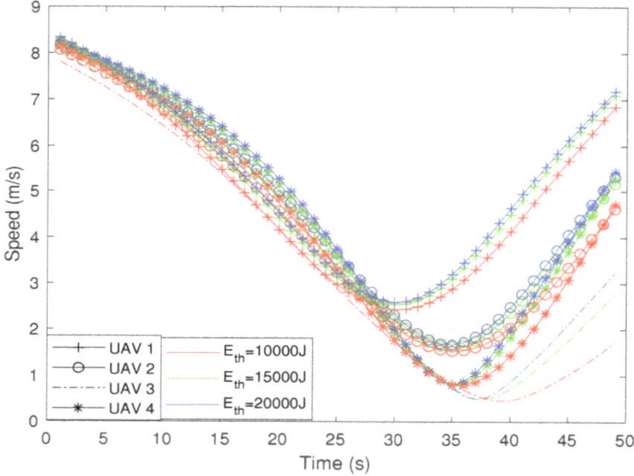

Figure 5. UAV's speed for different energy budgets.

Figure 6 and Figure 7 illustrate the variation of the power-splitting factor α and the time-allocation factor τ. It can be seen that α increased with time, and τ first increased and then decreased with time. This is because the UAVs moved away from the jammers and approached optimal points over time, at which point the SNs needed to consume less energy to ensure the SNR threshold constraint. As for τ, in the process of approaching the optimal points, τ first increased to ensure that enough energy was collected. When returning to the endpoints, in order to ensure the communication quality of the DNs, τ decreased to improve the throughput of the DNs.

Figure 8 presents the achievable throughput over every time slot. It is shown that the throughput increased first and then decreased. This was because the UAVs were initially far away from the jammers and closer to the optimal points, thereby collecting enough energy to increase the throughput. When returning to the endpoints, the throughput dropped as the UAVs moved away from the optimal points and closer to the jammers. Moreover, we noticed that the larger the UAV's energy budget, the greater the achievable throughput, which was in line with expectations.

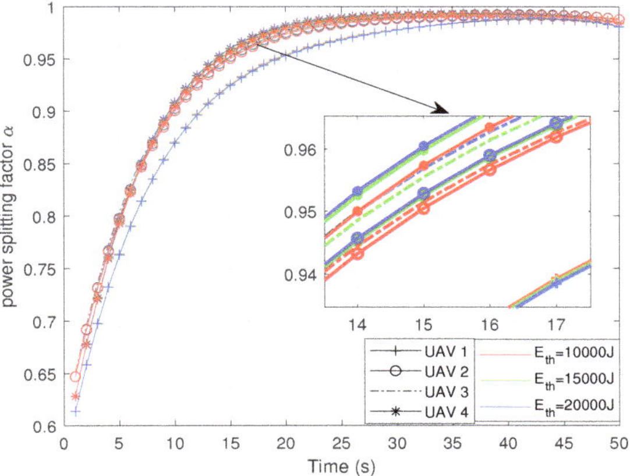

Figure 6. Power splitting factor in each time slot.

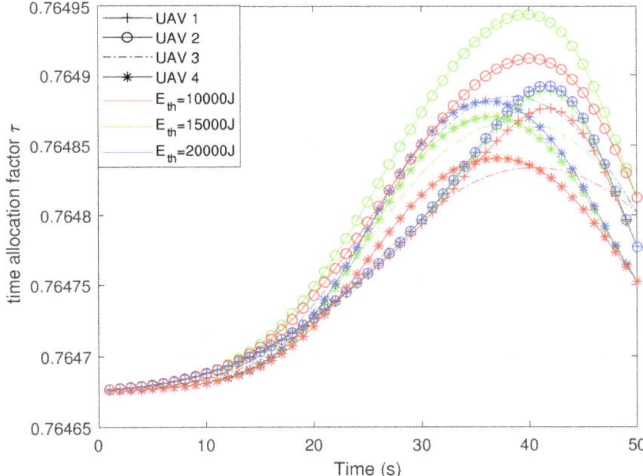

Figure 7. Time allocation factor in each time slot.

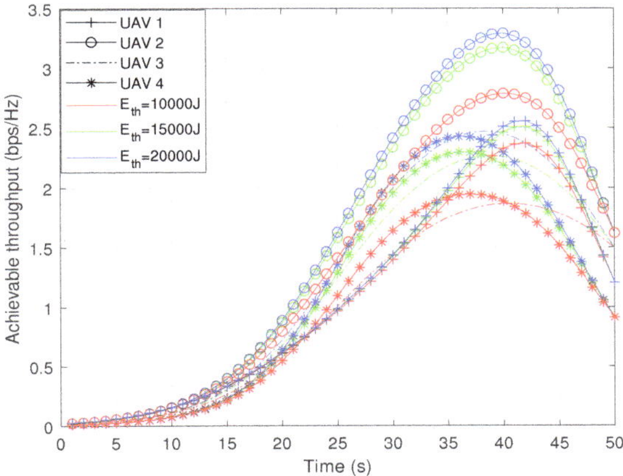

Figure 8. Achievable throughput over every time slot.

Furthermore, we increased the number of UAVs from 4 to 6, then 8 to further illustrate the performance of the proposed scheme. We set the initial and end positions of UAV 5–8 as (200, 240), (200, 260), (200, 280), and (200, 300); and (50, 240), (50, 260), (50, 280), and (50, 300), respectively. The achievable system throughput versus time is shown in Figure 9. It can be seen from Figure 9 that the throughput of the system increased with the number of UAVs. Given the energy budget, when $K = 8$, the throughput of the system was at the maximum; when $K = 2$, the throughput of the system was at the minimum. Moreover, it can be seen that given the number of UAVs, the throughput of the system increased with the energy budget. This was as expected.

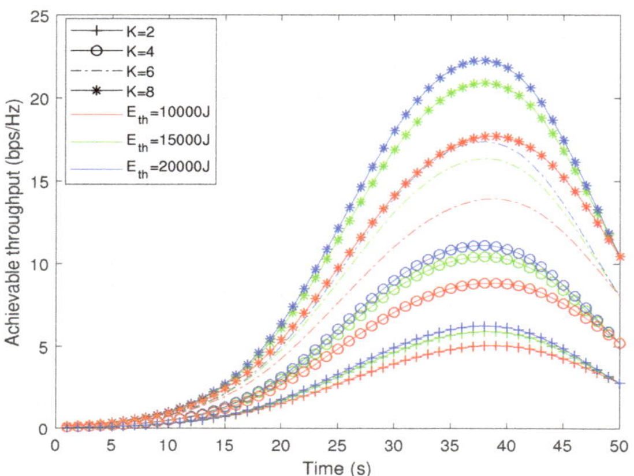

Figure 9. Achievable throughput over every time slot.

4.3. Effect of Jammers

Figure 10 shows the 2D trajectories of four UAVs with differing numbers of jammers. We plotted the trajectories for $J = 2$, $J = 3$, and $J = 4$. The basic trajectories of the four UAVs were consistent with those from Figure 4, and we will not repeat them here. It is worth noting that the flight distance of the UAVs increased with the number of jammers. Specifically, the more the number of jammers, the more obvious the interference effect, so the UAVs needed to be farther away from the jammers to ensure the channel throughput.

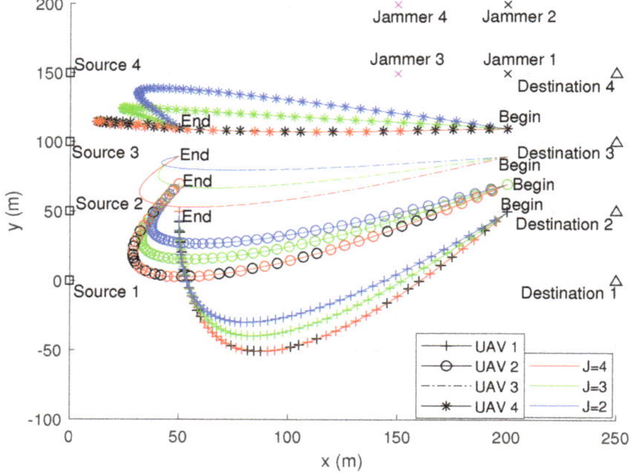

Figure 10. Optimal trajectories with different number of jammers.

Figure 11 displays the achievable throughput with different numbers of jammers. As can be seen from the figure, UAV 2 had the highest throughput. This is because UAV 2 was closer to the corresponding SN than UAV 1 and farther away from the jammers than UAV 3 and UAV 4. Finally, we observed that the throughput of all four UAVs decreased with the number of jammers. This was in line with our expectations and showed the significance of our study.

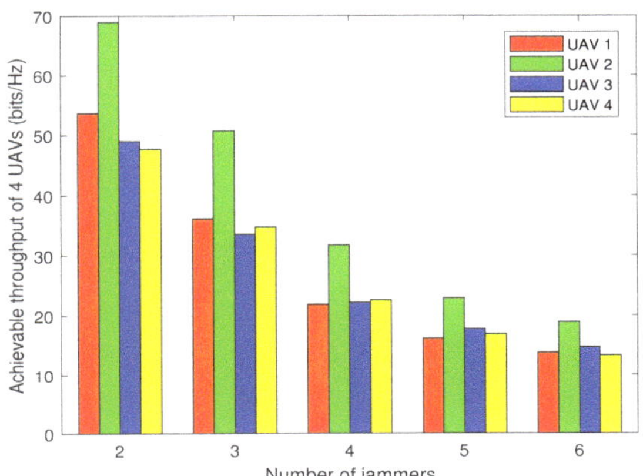

Figure 11. Achievable throughput with differing numbers of jammers.

Further, we increased the number of UAVs from 4 to 6, then 8 to further illustrate the impact of jammers on system performance. The achievable system throughput versus jammers is shown in Figure 12. As can be seen from Figure 12, the system throughput increased with the number of UAVs. Given the same number of jammers, when $K = 8$, the system throughput was at the maximum; when $K = 2$, the system throughput was at the minimum. Additionally, it can be seen that given the same number of UAVs, the throughput of the system decreased with the jammers. This was as expected.

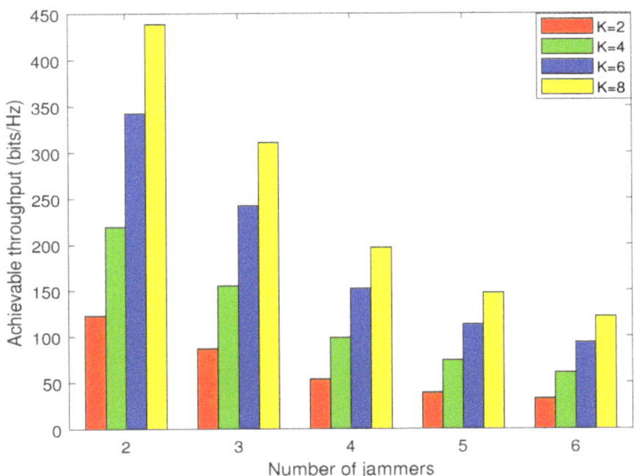

Figure 12. Achievable throughput with differing numbers of jammers.

4.4. Performance Comparison

In order to further illustrate the superiority of the proposed algorithm, in this subsection, we will compare our scheme with four baseline schemes:

- Scheme 1: Our proposed joint trajectory planning, time, and power allocation scheme.
- Scheme 2: Optimizing the power-splitting factor α and UAV's trajectory $\mathbf{Q}$ under the fixed time-allocation factor τ.
- Scheme 3: Optimizing the time-allocation factor τ and UAV's trajectory $\mathbf{Q}$ under the fixed power-splitting factor α.
- Scheme 4: Optimizing the UAV's trajectory $\mathbf{Q}$ under the fixed time-allocation factor τ and power-splitting factor α.
- Scheme 5: Optimizing the power-splitting factor α and time-allocation factor τ under circular trajectory.

We evaluated the average throughput of the two UAVs, as shown in Figure 13. For dynamic schemes 1–5, the throughput of the system first increased over time and then decreased as the UAVs moved away from the optimal positions and returned to the endpoints. Also, we noticed that scheme 5 had the worst performance since the circular trajectory had been set in advance. Moreover, at the best time slot, the average throughput of the proposed scheme 1 was two times higher than schemes 2 and 3.

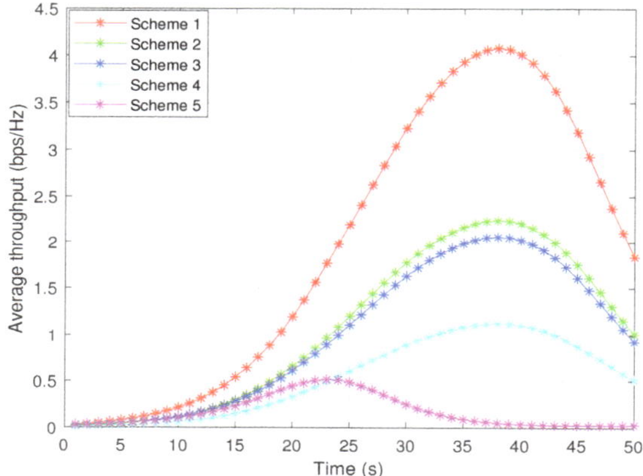

Figure 13. Average throughput in different schemes.

Figure 14 shows the average throughput with different energy budgets. It can be seen from Figure 14 that the average throughput of schemes 1–4 increased with the energy budget. For scheme 5, since the flight trajectory had been set in advance, increasing the energy budget did not bring about an improvement in average throughput. In addition, we observed that the proposed scheme 1 had the best performance, and the average throughput was increased by 40%, 50%, 150%, and 550% compared with schemes 2–5, respectively.

Figure 15 shows the average throughput of the system with differing number of jammers. Consistent with our expectations, the average throughput of all schemes decreased as the number of jammers increased. However, in comparison, the proposed scheme 1 had the best performance. Even in the extreme case with 6 jammers, the throughput of scheme 1 was still improved by 26%, 33%, 160%, and 500% compared with schemes 2–5.

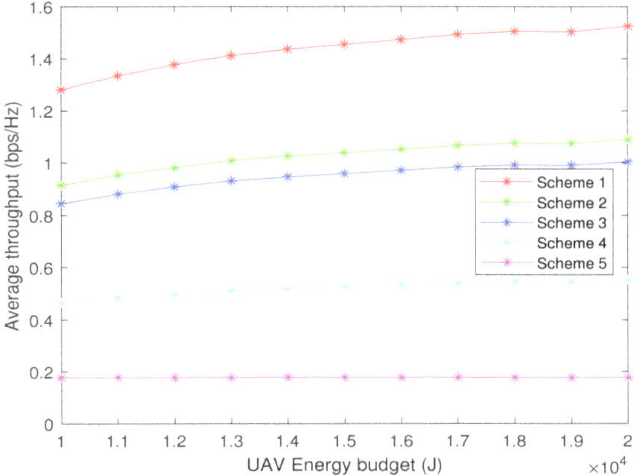

Figure 14. Average throughput with different energy budgets.

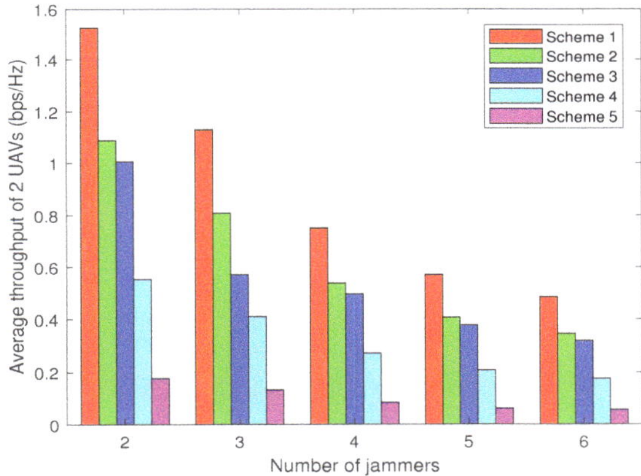

Figure 15. Average throughput with differing numbers of jammers.

5. Conclusions

This paper investigated joint trajectory planning, time, and power resource allocation to maximize the throughput in UAV networks. Considering the limited energy budget of UAVs and the existence of multiple jammers, we introduced SWIPT technology to improve channel quality. Our goal was to maximize the throughput of the DNs. Since the original problem is non-convex, taking into account the actual flight constraints of the UAVs, we proposed an efficient joint optimization algorithm based on successive convex approximations, a block coordinate descent, and the slack variables method to obtain a suboptimal solution. Simulation results corroborated that the proposed scheme can significantly improve the channel throughput and illustrated the effectiveness of joint trajectory planning, time, and power allocation in mitigating interference. Finally, we compared the proposed scheme with four benchmark schemes to highlight the superiority of our study. In future work, we will consider the UAVs scenario with mobile nodes, more complex channel models, and/or scheduling schemes such as multi-UAV coordination and multi-point access.

Author Contributions: Conceptualisation, K.W.; methodology, J.X.; formal analysis, P.L.; investigation, H.C.; writing—original draft preparation, J.X.; writing—review and editing, K.W.; validation, X.L.; supervision, K.L. All authors have read and agreed to the published version of the manuscript.

Funding: This work was supported in part by the National Natural Science Foundation of China under Grant 62172313, in part by the Natural Science Foundation of Hunan Province under Grant 2021JJ20054, in part by the National Key Research and Development Program of China under Grant 2021YFB3901503, in part by the National Natural Science Foundation of China under Grant 62001336, and in part by the open research fund of Integrated Computing and Chip Security Sichuan Collaborative Innovation Center of Chengdu University of Information Technology under Grant CXPAQ202204.

Data Availability Statement: Not Applicable.

Conflicts of Interest: The authors declare no conflict of interest.

Notations

The following notations are used in this manuscript:

Notation	Definition
$\mathbf{w}_{S_k}$	Location of the SN
$\mathbf{w}_{D_k}$	Location of the DN
$\mathbf{w}_j$	Locations of the jammer
Z	Height of the UAV
$\mathbf{q}_k$	Locations of the UAV
T	Total task time
N	Number of time slots
δ	Duration of each time slot
$V_{\max}$	The maximum speed of the UAV
$D_{\min}$	The minimum safe distance
$h_{S_k U_k}$	The channel-power gain between the SN and the UAV
$h_{j U_k}$	The channel-power gain between a jammer and a UAV
$h_{U_k D_k}$	The channel-power gain between a UAV and the DN
P_k	The transmit power of the SN
n_{su}	Additive white Gaussian noise
α	Power-splitting factor
τ	Time-allocation factor
η	Energy collection efficiency
P_B	The blade profile power
P_I	The induced power
v_{tip}	The tip speed of the rotor blade
v_0	The mean rotor induced velocity
d_0	The fuselage drag ratio
ρ	The air density
s	The rotor solidity
A_0	The rotor disc area

References

1. Xiang, C.; Li, Y.; Zhou, Y.; He, S.; Qu, Y.; Li, Z.; Gong, L.; Chen, C. A Comparative Approach to Resurrecting the Market of MOD Vehicular Crowdsensing. In Proceedings of the IEEE INFOCOM 2022—IEEE Conference on Computer Communications, London, UK, 2–5 May 2022; pp. 1479–1488.
2. Xiang, C.; Zhang, Z.; Qu, Y.; Lu, D.; Fan, X.; Yang, P.; Wu, F. Edge computing-empowered large-scale traffic data recovery leveraging low-rank theory. *IEEE Trans. Netw. Sci. Eng.* **2020**, *7*, 2205–2218. [CrossRef]
3. Ma, B.; Ren, Z.; Cheng, W. Traffic Routing-Based Computation Offloading in Cybertwin-Driven Internet of Vehicles for V2X Applications. *IEEE Trans. Veh. Technol.* **2021**, *71*, 4551–4560. [CrossRef]
4. Zhao, N.; Lu, W.; Sheng, M.; Chen, Y.; Tang, J.; Yu, F.R.; Wong, K.K. UAV-assisted emergency networks in disasters. *IEEE Wirel. Commun.* **2019**, *26*, 45–51. [CrossRef]
5. Jiang, X.; Sheng, M.; Zhao, N.; Xing, C.; Lu, W.; Wang, X. Green UAV communications for 6G: A survey. *Chin. J. Aeronaut.* **2022**, *35*, 19–34. [CrossRef]
6. Tran, D.H.; Vu, T.X.; Chatzinotas, S.; ShahbazPanahi, S.; Ottersten, B. Coarse trajectory design for energy minimization in UAV-enabled. *IEEE Trans. Veh. Technol.* **2020**, *69*, 9483–9496. [CrossRef]

7. Miao, J.; Wang, P. Power Control for Multi-UAV Location-aware Wireless Powered Communication Networks. In Proceedings of the 2020 IEEE/CIC International Conference on Communications in China (ICCC), Xiamen, China, 28–30 July 2020; pp. 225–230.

8. An K.; Liang, T.; Zheng, G.; Yan, X.; Li, Y.; Chatzinotas, S. Performance limits of cognitive-uplink FSS and terrestrial FS for Ka-band. *IEEE Trans. Aerosp. Electron. Syst.* **2018**, *55*, 2604–2611. [CrossRef]

9. An, K.; Lin, M.; Ouyang, J.; Zhu, W.-P. Secure Transmission in Cognitive Satellite Terrestrial Networks. *IEEE J. Sel. Areas Commun.* **2016**, *34*, 3025–3037. [CrossRef]

10. Zeng, Y.; Zhang, R. Energy-Efficient UAV Communication with Trajectory Optimization. *IEEE Trans. Wirel. Commun.* **2017**, *16*, 3747–3760. [CrossRef]

11. Lin, N.; Liu, Y.; Zhao, L.; Wu, D.O.; Wang, Y. An Adaptive UAV Deployment Scheme for Emergency Networking. *IEEE Trans. Wirel. Commun.* **2022**, *21*, 2383–2398. [CrossRef]

12. Ma, B.; Ren, Z.; Cheng, W. Credibility Computation Offloading Based Task-Driven Routing Strategy for Emergency UAVs Network. In Proceedings of the 2021 IEEE Global Communications Conference (GLOBECOM), Rio de Janeiro, Brazil, 4–8 December 2021; pp. 1–6.

13. Fu, Y.; Li, D.; Tang, Q.; Zhou, S. Joint Speed and Bandwidth Optimized Strategy of UAV-Assisted Data Collection in Post-Disaster Areas. In Proceedings of the 2022 20th Mediterranean Communication and Computer Networking Conference (MedComNet), Pafos, Cyprus, 1–3 June 2022; pp. 39–42.

14. Peer, M.; Bohara, V.A.; Srivastava, A. Multi-UAV Placement Strategy for Disaster-Resilient Communication Network. In Proceedings of the 2020 IEEE 92nd Vehicular Technology Conference (VTC2020-Fall), Virtual, 18 November–16 December 2020; pp. 1–7.

15. Kim, Y.H.; Chowdhury, I.A.; Song, I. Design and Analysis of UAV-Assisted Relaying with Simultaneous Wireless Information and Power Transfer. *IEEE Access* **2020**, *8*, 27874–27886. [CrossRef]

16. Zhan, C.; Hu, H.; Wang, Z.; Fan, R.; Niyato, D. Unmanned Aircraft System Aided Adaptive Video Streaming: A Joint Optimization Approach. *IEEE Transactions on Multimedia* **2020**, *22*, 795–807. [CrossRef]

17. Sun, Z.; Yang, D.; Xiao, L.; Cuthbert, L.; Wu, F.; Zhu, Y. Joint Energy and Trajectory Optimization for UAV-Enabled Relaying Network with Multi-Pair Users. *IEEE Trans. Cogn. Commun. Netw.* **2021**, *7*, 939–954. [CrossRef]

18. Gao, Y.; Wu, Y.; Cui, Z.; Yang, W.; Hu, G.; Xu, S. Robust trajectory and communication design for angle-constrained multi-UAV communications in the presence of jammers. *China Commun.* **2022**, *19*, 131–147. [CrossRef]

19. Feng, Z.; Ren, G.; Chen, J.; Zhang, X.; Luo, Y.; Wang, M.; Xu, Y. Power control in relay-assisted anti-jamming systems: A Bayesian three-layer Stackelberg game approach. *IEEE Access* **2019**, *7*, 14623–14636. [CrossRef]

20. Xu, Y.; Ren, G.; Chen, J.; Zhang, X.; Jia, L.; Feng, Z.; Xu, Y. Joint Power and Trajectory Optimization in UAV Anti-Jamming Communication Networks. In Proceedings of the ICC 2019—2019 IEEE International Conference on Communications (ICC), Shanghai, China, 20–24 May 2019; pp. 1–5.

21. Wu, Y.; Yang, W.; Guan, X.; Wu, Q. UAV-Enabled Relay Communication Under Malicious Jamming: Joint Trajectory and Transmit Power Optimization. *IEEE Trans. Veh. Technol.* **2021**, *70*, 8275–8279. [CrossRef]

22. Lin, Z.; An K.; Niu, H.; Hu, Y.; Chatzinotas, S.; Zheng, G.; Wang, J. SLNR-based Secure Energy Efficient Beamforming in Multibeam Satellite Systems. *IEEE Trans. Aerosp. Electron. Syst.* 2022. [CrossRef]

23. Song, X.; Chang, Z.; Guo, X.; Wu, P.; Hämäläinen, T. Energy Efficient Optimization for Solar-Powered UAV Communications System. In Proceedings of the 2021 IEEE International Conference on Communications Workshops (ICC Workshops), Montreal, QC, Canada, 14–23 June 2021; pp. 1–6.

24. Sekander, S.; Tabassum, H.; Hossain, E. Statistical Performance Modeling of Solar and Wind-Powered UAV Communications. *IEEE Trans. Mob. Comput.* **2021**, *20*, 2686–2700. [CrossRef]

25. Mamaghani, M.T.; Hong, Y. Improving PHY-Security of UAV-Enabled Transmission with Wireless Energy Harvesting: Robust Trajectory Design and Communications Resource Allocation. *IEEE Trans. Veh. Technol.* **2020**, *69*, 8586–8600. [CrossRef]

26. Lu, W.; Fang, S.; Gong, Y.; Qian, L.; Liu, X.; Hua, J. Resource Allocation for OFDM Relaying Wireless Power Transfer Based Energy-Constrained UAV Communication Network. In Proceedings of the 2018 IEEE International Conference on Communications Workshops (ICC), Kansas City, MO, USA, 20–24 May 2018; pp. 1–6.

27. Ramzan, M.R.; Naeem, M.; Altaf, M.; Ejaz, W. Multi-Criterion Resource Management in Energy Harvested Cooperative UAV-enabled IoT Networks. *IEEE Internet Things J.* **2021**, *9*, 2944–2959. [CrossRef]

28. Wu, Y.; Yang, W.; Guan, X. UAV-UAV Communication Under Malicious Jamming: Trajectory Optimization with Turning Angle Constraint. In Proceedings of the 2020 International Conference on Wireless Communications and Signal Processing (WCSP), Nanjing, China, 21–23 October 2020; pp. 26–31.

29. Wang, X.; Gursoy, M.C.; Erpek, T.; Sagduyu, Y.E. Jamming-Resilient Path Planning for Multiple UAVs via Deep Reinforcement Learning. In Proceedings of the 2021 IEEE International Conference on Communications Workshops (ICC Workshops), Montreal, QC, Canada, 14–23 June 2021; pp. 1–6.

30. Zhou, L.; Zhao, X.; Guan, X.; Song, E.; Zeng, X.; Shi, Q. Robust trajectory planning for UAV communication systems in the presence of jammers. *Chin. J. Aeronaut.* **2022**, *35*, 265–274. [CrossRef]

31. Wu, Y.; Yang, W.; Guan, X.; Wu, Q. Energy-Efficient Trajectory Design for UAV-Enabled Communication Under Malicious Jamming. *IEEE Wirel. Commun. Lett.* **2021**, *10*, 206–210. [CrossRef]

32. Duo, B.; Luo, J.; Li, Y.; Hu, H.; Wang, Z. Joint trajectory and power optimization for securing UAV communications against active eavesdropping. *China Commun.* **2021**, *18*, 88–99. [CrossRef]
33. Li, X.; Xu, J. Positioning Optimization for Sum-Rate Maximization in UAV-Enabled Interference Channel. *IEEE Signal Process. Lett.* **2019**, *26*, 1466–1470. [CrossRef]
34. Wu, Y.; Fan, W.; Yang, W.; Sun, X.; Guan, X. Robust Trajectory and Communication Design for Multi-UAV Enabled Wireless Networks in the Presence of Jammers. *IEEE Access* **2020**, *8*, 2893–2905. [CrossRef]
35. Xiang, C.; Zhou, Y.; Dai, H.; Qu, Y.; He, S.; Chen, C.; Yang, P. Reusing delivery drones for urban crowdsensing. *IEEE Trans. Mob. Comput.* **2021**. [CrossRef]
36. Li, J.; Tian, Y.; Zhang, Y. Destination-Based Cooperative Jamming in Security UAV Relay System with SWIPT. In Proceedings of the 2021 13th International Conference on Communication Software and Networks (ICCSN), Chongqing, China, 4–7 June 2021; pp. 160–167.
37. Singh, R.; Rawat, M.; Jaiswal, A. On the Physical Layer Security of Mixed FSO-RF SWIPT System with Non-Ideal Power Amplifier. *IEEE Photonics J.* **2021**, *13*, 1–17. [CrossRef]
38. Wang, W.; Li, X.; Zhang, M.; Cumanan, K.; Ng, D.W.K.; Zhang, G.; Tang, J.; Dobre, O.A. Energy-constrained UAV-assisted secure communications with position optimization and cooperative jamming. *IEEE Trans. Commun.* **2020**, *68*, 4476–4489. [CrossRef]
39. Ji, B.; Li, Y.; Zhou, B.; Li, C.; Song, K.; Wen, H. Performance Analysis of UAV Relay Assisted IoT Communication Network Enhanced with Energy Harvesting. *IEEE Access* **2019**, *7*, 38738–38747. [CrossRef]
40. Park, J.C.; Kang, K.-M.; Choi, J. Low-Complexity Algorithm for Outage Optimal Resource Allocation in Energy Harvesting-Based UAV Identification Networks. *IEEE Commun. Lett.* **2021**, *25*, 3639–3643. [CrossRef]
41. Kumar, D.; Singya, P.K.; Bhatia, V. Performance Analysis of SWIPT Enabled Decode-and-Forward based Cooperative Network. In Proceedings of the 2022 IEEE 11th International Conference on Communication Systems and Network Technologies (CSNT), Indore, India, 23–24 April 2022; pp. 476–481.
42. Hu, T.; Ma, F.; Shang, Y.; Cheng, Y. Physical Layer Security of Untrusted UAV-enabled Relaying NOMA Network Using SWIPT and the Cooperative Jamming. In Proceedings of the 2021 IEEE 94th Vehicular Technology Conference (VTC2021-Fall), Virtual, 27 September–28 October 2021; pp. 1–6.
43. Najmeddin, S.; Bayat, A.; Aïssa, S.; Tahar, S. Energy-Efficient Resource Allocation for UAV-Enabled Wireless Powered Communications. In Proceedings of the 2019 IEEE Wireless Communications and Networking Conference (WCNC), Marrakesh, Morocco, 15–18 April 2019; pp. 1–6.
44. Su, Z.; Tang, J.; Feng, W.; Chen, Z.; Fu, Y.; Wong, K.-K. Energy Efficiency Optimization for D2D communications in UAV-assisted Networks with SWIPT. In Proceedings of the 2021 IEEE Global Communications Conference (GLOBECOM), Madrid, Spain, 7–11 December 2021; pp. 1–7.
45. Yang, Y.; Xiao, K. Energy Efficiency Optimization of Multi-user Distributed Antenna Systems with SWIPT Technique. In Proceedings of the 2021 International Conference on Communications, Information System and Computer Engineering (CISCE), Beijing, China, 14–16 May 2021; pp. 118–122.
46. Yang, H.; Ye, Y.; Chu, X.; Dong, M. Resource and Power Allocation in SWIPT-Enabled Device-to-Device Communications Based on a Nonlinear Energy Harvesting Model. *IEEE Internet Things J.* **2020**, *7*, 10813–10825. [CrossRef]
47. Zargari, S.; Hakimi, A.; Tellambura, C.; Herath, S. User Scheduling and Trajectory Optimization for Energy-Efficient IRS-UAV Networks with SWIPT. *IEEE Trans. Veh. Technol.* **2022**. [CrossRef]
48. Liu, Y.; Han, F.; Zhao, S. Flexible and Reliable Multiuser SWIPT IoT Network Enhanced by UAV-Mounted Intelligent Reflecting Surface. *IEEE Trans. Reliab.* **2022**, *71*, 1092–1103. [CrossRef]
49. Niu, H.; Chu, Z.; Zhu, Z.; Zhou, F. Aerial intelligent reflecting surface for secure wireless networks: Secrecy capacity and optimal trajectory strategy. *Intell. Converg. Netw.* **2020**, *3*, 119–133. [CrossRef]
50. Lin, Y.; Wang, T.; Wang, S. Trajectory Planning for Multi-UAV Assisted Wireless Networks in Post-Disaster Scenario. In Proceedings of the 2019 IEEE Global Communications Conference (GLOBECOM), Waikoloa, HI, USA, 9–13 December 2019; pp. 1–6.
51. Savkin, A.V.; Huang, C.; Ni, W. Joint Multi-UAV Path Planning and LoS Communication for Mobile Edge Computing in IoT Networks with RISs. *IEEE Internet Things J.* **2022**. [CrossRef]
52. Wang, L.; Wang, K.; Pan, C.; Xu, W.; Aslam, N.; Hanzo, L. Multi-Agent Deep Reinforcement Learning-Based Trajectory Planning for Multi-UAV Assisted Mobile Edge Computing. *IEEE Trans. Cogn. Commun. Netw.* **2021**, *7*, 73–84. [CrossRef]
53. Li, S.; He, C.; Liu, M.; Wan, Y.; Gu, Y.; Xie, J.; Fu, S.; Lu, K. Design and implementation of aerial communication using directional antennas: Learning control in unknown communication environments. *IET Control. Theory Appl.* **2019**, *13*, 2906–2916. [CrossRef]
54. Jayakody, D.N.K.; Perera, T.D.P.; Nathan, M.C.; Hasna, M. Self-energized Full-Duplex UAV-assisted Cooperative Communication Systems. In Proceedings of the 2019 IEEE International Black Sea Conference on Communications and Networking (BlackSeaCom), Sochi, Russia, 3–6 June 2019; pp. 1–6.
55. Boyd, S.; Boyd, S.P.; Vandenberghe, L. *Convex Optimization*; Cambridge University Press: Cambridge, UK, 2004.

 *drones*

Article

Research on the Cooperative Passive Location of Moving Targets Based on Improved Particle Swarm Optimization

Li Hao [1], Fan Xiangyu [2,*] and Shi Manhong [3]

[1] Department of Intelligence, Air Force Early Warning Academy, Wuhan 430010, China
[2] Department of Bomber and Transport Aircraft Pilots Conversion, Air Force Harbin Flying College, Harbin 150088, China
[3] Department of Information Countermeasures, Air Force Early Warning Academy, Wuhan 430010, China
* Correspondence: panda0077@163.com; Tel.: +86-177-0360-5050

Abstract: Aiming at the cooperative passive location of moving targets by UAV swarm, this paper constructs a passive location and tracking algorithm for a moving target based on the A optimization criterion and the improved particle swarm optimization (PSO) algorithm. Firstly, the localization method of cluster cooperative passive localization is selected and the measurement model is constructed. Then, the problem of improving passive location accuracy is transformed into the problem of obtaining more target information. From the perspective of information theory, using the A criterion as the optimization target, the passive localization process for static targets is further deduced. The Recursive Neural Network (RNN) is used to predict the probability distribution of the target's location in the next moment so as to improve the localization method and make it suitable for the localization of moving targets. The particle swarm algorithm is improved by using grouping and time period strategy, and the algorithm flow of moving target location is constructed. Finally, through the simulation verification and algorithm comparison, the advantages of the algorithm in this paper are presented.

Keywords: passive location; UAV swarm; moving target; A optimization criterion; particle swarm optimization; recursive neural network

Citation: Hao, L.; Xiangyu, F.; Manhong, S. Research on the Cooperative Passive Location of Moving Targets Based on Improved Particle Swarm Optimization. *Drones* **2023**, *7*, 264. https://doi.org/10.3390/drones7040264

Academic Editors: Zhihong Liu, Shihao Yan, Yirui Cong and Kehao Wang

Received: 26 February 2023
Revised: 3 April 2023
Accepted: 10 April 2023
Published: 12 April 2023

1. Introduction

As electromagnetic space has become the fifth-dimensional battlefield after "land, sea, air, and sky", the importance and research efforts of various countries in electromagnetic space have increased considerably. When using and radiating electromagnetic waves, the position of electromagnetic space is exposed, and passive location emerges as the times require [1–5]. However, the location accuracy of passive location decreases significantly with the increase in the distance from the target, and the location efficiency is highly related to the spatial position distribution of the location points. With the rapid development of UAV technology, UAV has gradually become a new type of combat force in the future battlefield with its unique advantages. Utilizing the distributed characteristics of UAV swarms to optimize their spatial distribution and trajectory has become a new way to improve the ability to passively locate targets.

The current research on passive location can be divided into two main directions. The first is to study and improve the location accuracy algorithm, such as improving the time of arrival (TOA) [6], time difference of arrival (TDOA) [7], received signal strength (RSS) [8], and angle of arrival (AOA) [9,10]. Since this article does not involve the improvement of the location algorithm, it will not be considerably discussed here.

The other major direction is to optimize the spatial location of passive location points to improve location performance. It mainly includes two research contents: optimizing the time-series spatial position of a single station and the spatial distribution position of multiple stations. For a single-station location, [11] deduced the factors affecting the location error based on the AOA-based airborne platform location method and constructs a method to reduce the single-station error. The authors of [12,13] extend the passive motion location of a single station to a multi-station, and optimized the corresponding location mode and designed a new objective function.

In studying the optimal configuration of a multi-station location, the general paradigm is to first select or design a certain location index as the objective function. Then, through theoretical derivation or numerical calculation, the aircraft coordinate parameters under the optimal objective function are obtained, which is the optimal configuration of passive location.

In [14,15], geometric dilution of precision (GDOP) is used as the objective function of location, and the corresponding optimization function is designed to further improve the accuracy of a passive location. The authors of [16] took the AOA location system as the research object and deduced the conditions of the optimal passive location configuration with the minimum circular error probable (CEP) as the criterion. In [17,18], the Fisher information matrix (FIM) was considered as the objective function to study the optimal multi-aircraft passive location configuration when FIM is the largest. In [19,20], the value of the Cramer–Rao lower bound (CRLB) determinant was used as the objective function to study the optimal location configuration of multiple stations under the TDOA location system.

Table 1 shows a comparison of the main work and related research of this article and the selection of the articles from the above-mentioned literature that conducted in-depth research into this field of study.

Table 1. Comparison of main work.

Functions Implemented	Algorithm in This Paper	Article [8]	Article [13]	Article [20]
Improved location algorithm	Yes	Yes	Yes	No
Location using multiple stations	Yes	No	No	Yes
Real-time optimization trajectory	Yes	No	No	No
Positioning by target's motion characteristics	Yes	No	No	No

It can be seen from the above-mentioned literature and Table 1 that research on passive location at this stage has mainly focused on the improvement of the passive location method and static station deployment. That is, by designing various criteria to improve the accuracy of passive location algorithms or based on different location systems, research has been conducted on optimizing the station layout. However, there is little research on the cooperative passive location of moving targets. At the same time, the method of static station placement cannot be directly applied to the problem of the cooperative passive location of moving objects because the passive location of stationary targets has no constraints on the target point. The location of moving targets is a sequential decision-making problem. That is, the optimization result in the next moment is subject to the constraints of the position in the present moment and the performance parameters of the platform. The subsequent location performance is also affected by the location accuracy of the previous sequence. Although the localization of stationary targets cannot be directly used to solve the problem of localization of dynamic targets, the two are not completely unrelated. It can learn from the research ideas and methods of stationary target location, combined with the characteristics of the moving target location. Thus, we aim to improve the location method and promote its scope of application.

The results of the above-mentioned literature also focus on obtaining the optimal spatial configuration. For the static layout of the site, the above-mentioned research has a strong practical significance. However, for a spatial motion platform such as an unmanned aerial vehicle cluster, the optimal configuration can be obtained directly, while ignoring the process of forming the optimal configuration, which requires a lot of time and computing resources. Therefore, it is necessary to optimize the space location of UAVs in real time and to achieve global optimization gradually.

Based on the perspective of information theory, this paper optimizes the spatial trajectory of each UAV in the UAV swarm to improve the location efficiency. The main contributions are as follows:

1. The real-time trajectory planning for the passive location of the UAV cluster is implemented based on the RSS model.
2. Using the improved deep learning network to correct the target location probability parameters in the positioning algorithm, a more accurate positioning of the moving target is achieved.
3. The depth network can identify the target movement trend in complex mixed noise, which provides a method to solve the problem of recognition in complex noise.
4. Designing particle grouping and time period to improve the particle swarm optimization algorithm, the algorithm effect is improved.

The article is organized as follows. The passive location principle of the cluster and the corresponding measurement model are constructed in Section 2. The optimization process of static target and dynamic target location is analyzed, and the optimization target function for the passive location of a moving target is constructed and derived in Section 3. To address the shortcomings of particle swarm optimization, the grouping and time period strategies are used to improve it in Section 4. The optimization function and corresponding constraints for moving target localization are constructed, and the passive location optimization process based on improved particle swarm optimization algorithm are presented in Section 5. Simulation verification and algorithm comparison are performed to highlight the advantages of the method in Section 6. The discussion and final conclusion are presented in Sections 7 and 8, respectively.

2. Location Model and Optimization Criteria

2.1. Principles of RSS

Due to the attenuation of electromagnetic signals as they propagate in space, the attenuation model of the electromagnetic signal is first constructed and the corresponding parameters are determined. Then, according to the strength of the signal received by the platforms in different positions, the position of the target can be calculated. This is the principle of received signal strength (RSS) [21–24].

Therefore, it is only necessary to obtain the strength of the signal received by each platform and the location parameters of each platform, and then passively locate the target by using RSS, as shown in Figure 1.

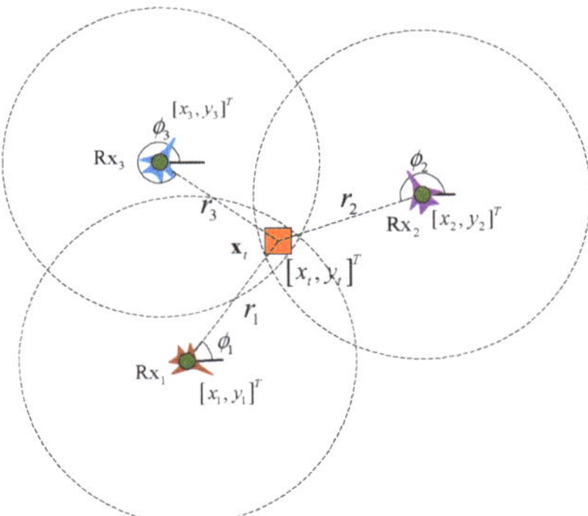

Figure 1. Schematic diagram of RSS location.

In Figure 1, the target radiates electromagnetic signals and its coordinates are $R_t = [x_t, y_t]^T$. The three platforms Rx_1, Rx_2, and Rx_3 receive radiation signals. Combined with the constructed signal attenuation model, the distance r_i between the target to be located and each detection platform can be obtained. The RSS location equation is:

$$\begin{cases} \sqrt{(x_t - x_1)^2 + (y_t - y_1)^2} = r_1 \\ \sqrt{(x_t - x_2)^2 + (y_t - y_2)^2} = r_2 \\ \sqrt{(x_t - x_3)^2 + (y_t - y_3)^2} = r_3 \end{cases} \tag{1}$$

By solving Formula (1), the RSS envelope of each receiving platform in Figure 1 can be obtained. The place where the three circles overlap each other in Figure 1 is the area where the target is located.

2.2. Measurement Model

This section builds a measurement model for the passive location of targets by UAV swarms. The positioning target studied in this paper was located on the ground or sea, and the height was set to zero. It was also assumed that the UAV flies on the same altitude plane. Therefore, the positioning of this article did not consider the issue of height.

Assuming that there are M UAVs in the UAV swarm, the positional parameters and spatial relationship between the UAV swarm and the target are shown in Figure 2.

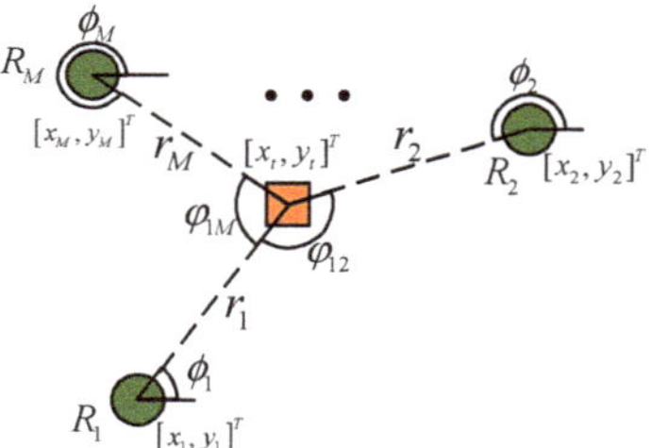

Figure 2. Schematic diagram of the passive location of the UAV swarm.

The location of the target is $R_t = [x_t, y_t]^T$. The position and velocity of the i-th UAV are $R_i = [x_i, y_i]^T$ and $Rv_i = [vx_i, vy_i]^T$, $I = 1, 2, \ldots, M$, respectively. The connecting line between the drone and the target has an included angle ϕ_i with the x-axis. The distance from the target is $r_i = ||R_i - R_t||_2$, and the angle between any two UAVs and the target is φ_{ij}, $j = 1, 2, \ldots, M$.

The attenuation model of the signal in the atmosphere is:

$$p_s = p_o - 10\gamma_i \log_{10} d_i \tag{2}$$

where p_o is the equivalent radiated power of the target-radiated signal. That is, the product of the target-radiated power and the antenna gain. As these two parameters are not the concern of the research in this paper, they are not introduced in detail here. γ_i is the attenuation factor of the electromagnetic wave, and di is the length of the signal propagation path. This paper assumed that the signal is not refracted. That is, di is the distance r_i between the UAV and the target [25]. Then, the signal strength p_s of the signal reaching the UAV receiving end can be calculated by Formula (2).

Due to the existence of electromagnetic interference and clutter in the atmosphere and the thermal noise of the system in the signal receiver, the actual signal $p_{ir}(k)$ received by the receiver of the i-th UAV at time k can be expressed as:

$$p_{ir}(k) = p_{is}(k) + n(k) \tag{3}$$

Among them, $n(k)$ represents the measurement error that obeys the Gaussian distribution, that is, $n(k) \sim N(0, \sigma_i^2(d_i))$. The error is related to the distance d_i between the targets, satisfying:

$$\sigma_i^2(d_i) = d_i^\alpha \sigma_0^2 \tag{4}$$

where σ_0^2 is a constant and is the basic unit of measure for variance. α is the path attenuation factor. According to Formulas (2)–(4) and the signal $P_{ir}(k)$ received by each UAV at time k, the matrix of the received signal strength distribution of the UAV swarm can be obtained as $P_r(k)$. The $P_r(k)$ covariance matrix is $\sigma_p = \mathrm{diag}(\sigma_1^2(k), \sigma_2^2(k), \ldots \sigma_M^2(k))$. Then, the signal received by the UAV swarm can be denoted as $P_r(k) \sim N(P_s(k), \sigma_p)$, where $P_s(k)$ represents the estimated target position using the pure signal that reaches the UAV.

After acquiring the signal energy of each point, the distance r from the target to the sensor can be estimated according to the signal attenuation model. Since the positions of the UAVs themselves are known, the multiple circles shown in Figure 1 can then be obtained using Formula (1). The overlapping areas of the different circles are the target position.

It can be seen that positioning accuracy is related to the accuracy of the signal attenuation model. The attenuation characteristics and corresponding parameters of the signal attenuation model are accurate, and the distance between the UAV and the target can be estimated well. Otherwise, the error is large. Scholars have conducted in-depth research on this and constructed a variety of attenuation models to further ensure the accuracy of distance estimation.

2.3. A Optimization Criterion

CRLB represents the theoretical limit of the error estimation performance when making unbiased estimates. In practice, CRLB can be obtained by calculating the inverse matrix of the FIM.

Evidently, the performance of CRLB is highly correlated with the accuracy of the measured parameters. The more precise the measurement, the lower the error. As shown in Figure 2, the measurement parameters obtained by the UAV about the target are highly correlated with the spatial distribution of the UAV swarm. That is, a different spatial distribution corresponds to a different CRLB. Therefore, based on CRLB, this paper optimized the trajectory of the UAV swarm to achieve the efficiency of the passive location of moving targets.

Since CRLB is in matrix form, it is not easy to use in conventional applications. Scholars have proposed the A optimization criterion for CRLB whose physical meaning is to minimize the mean square error (MSE).

The A optimization criterion can be expressed as:

$$F_A = \mathrm{argmin}[tr(\mathbf{CRLB})] = \mathrm{argmin}\, tr\left(\mathbf{J}^{-1}\right) \tag{5}$$

where $\mathbf{J}$ represents the FIM of the measurement matrix, and -1 represents the inverse of this matrix. Then, $\mathbf{J}^{-1}$ is CRLB.

3. Configuration Optimization Method for the Passive Location of Moving Target

3.1. Passive Location Methods for Static Objects

The passive location of stationary targets using UAV swarms includes three processes. First, the relationship between CRLB and UAV swarm coordinates is constructed. Then, using the A criterion, the space configuration of the UAV swarm corresponding to the optimal CRLB matrix is obtained. By optimizing each subsequent moment in turn, the trajectory of each UAV in the cluster can be obtained.

Assuming that the position of the target is $R_t = [x_t, y_t]^T$ and the measurement set of M UAVs at a certain moment is P_r, FIM can be expressed as:

$$\mathbf{J} = \begin{bmatrix} \mathbf{J}_{xx} & \mathbf{J}_{xy} \\ \mathbf{J}_{yx} & \mathbf{J}_{yy} \end{bmatrix} \tag{6}$$

The elements of the i-th row and the j-th column of the four matrices in Formula (6) can be expressed as:

$$\begin{cases} \mathbf{J}_{xx}^{i,j} = E\left[\frac{\partial}{\partial x_{ti}}\ln(f(P_r;R_t))\frac{\partial}{\partial x_{tj}}\ln(f(P_r;R_t))\right] \\ \mathbf{J}_{xy}^{i,j} = E\left[\frac{\partial}{\partial x_{ti}}\ln(f(P_r;R_t))\frac{\partial}{\partial y_{tj}}\ln(f(P_r;R_t))\right] \\ \mathbf{J}_{xy}^{i,j} = E\left[\frac{\partial}{\partial y_{ti}}\ln(f(P_r;R_t))\frac{\partial}{\partial x_{tj}}\ln(f(P_r;R_t))\right] \\ \mathbf{J}_{yy}^{i,j} = E\left[\frac{\partial}{\partial y_{ti}}\ln(f(P_r;R_t))\frac{\partial}{\partial y_{tj}}\ln(f(P_r;R_t))\right] \end{cases} \tag{7}$$

where $f(P_r;R_t)$ is the probability density distribution function of P_r, namely:

$$f(P_r;R_t) = \frac{1}{(2\pi)^{M/2}\sqrt{\det(\sigma_p)}}\exp\left[-\frac{1}{2}(P_r-R_t)^T\sigma_p^{-1}(P_r-R_t)\right] \tag{8}$$

According to the definition of FIM and as shown in Formula (7), it is necessary to obtain $\mathbf{J}_{xx}$ by continuously calculating the derivative twice. x_{ti} and x_{tj} are related, that is, the second derivative is not zero.

The horizontal axis position x_t and the vertical axis position y_t of the target coordinates are independent of each other. Being independent of each other means that both $\mathbf{J}_{yx}$ and $\mathbf{J}_{xy}$ are 0. Then, Formula (6) can be rewritten as:

$$\mathbf{J} = \begin{bmatrix} \mathbf{J}_{xx} & \mathbf{0} \\ \mathbf{0} & \mathbf{J}_{yy} \end{bmatrix} \tag{9}$$

Similarly, since the horizontal and vertical coordinates of the target are relatively independent, the processes of obtaining $\mathbf{J}_{xx}$ and $\mathbf{J}_{yy}$ are independent of each other, and the calculation process is similar. This section analyzes $\mathbf{J}_{xx}$.

Substituting Formula (8) into Formula (7), we obtained [26]:

$$\mathbf{J}_{xx}^{i,j} = \frac{1}{\sigma_p^2}\frac{\partial P_r}{\partial x_{ti}}\frac{\partial P_r}{\partial x_{tj}} + \frac{1}{2}\frac{1}{\sigma_p^2}\frac{\partial \sigma_p}{\partial x_{ti}}\frac{\partial \sigma_p}{\partial x_{tj}} \tag{10}$$

The right side of the equal sign of Formula (10) can be regarded as the sum of two parts, which can be expressed as:

$$\begin{cases} \mathbf{J}_{xx}^{i,j} = \mathbf{J}_{xx1}^{i,j} + \mathbf{J}_{xx2}^{i,j} \\ \mathbf{J}_{xx1}^{i,j} = \nabla_{R_{tx}}P_r{}^T\sigma_p^{-1}\nabla_{R_{tx}}P_t \\ \mathbf{J}_{xx2}^{i,j} = \frac{1}{2}Tr\left(\sigma_p^{-1}\frac{\partial\sigma_p}{\partial x_i}\sigma_p^{-1}\frac{\partial\sigma_p}{\partial x_j}\right) \end{cases} \tag{11}$$

Among them, $\nabla_{R_{tx}}P_r{}^T$ is the Jacobian matrix obtained after the derivation of the target abscissa R_{tx} using the measured value $P_r{}^T$, which is expressed as:

$$J_{1,1} = \frac{50}{(\ln(10))^2\sigma_0^2}\sum_{i=1}^{M}\frac{\gamma_i^2(1+\cos 2\phi_i)}{d_i^{a+2}} \tag{12}$$

$$J_{1,2} = J_{2,1} = \frac{50}{(\ln(10))^2\sigma_0^2}\sum_{i=1}^{M}\frac{\gamma_i^2\cos 2\phi_i}{d_i^{a+2}} \tag{13}$$

$$J_{2,2} = \frac{50}{(\ln(10))^2\sigma_0^2}\sum_{i=1}^{M}\frac{\gamma_i^2(1-\cos 2\phi_i)}{d_i^{a+2}} \tag{14}$$

The meanings of the parameters in Formulas (11)–(13) are the same as those in Formulas (2)–(4), which are not repeated here.

In Formula (11), *Tr* represents the trace of the matrix. Then, the two partial derivatives are:

$$\frac{\partial \sigma_p}{\partial x_i} = \sigma_0^2 \begin{bmatrix} d_1^{\alpha-1}\cos\phi_1 & 0 & \cdots & 0 \\ 0 & d_2^{\alpha-1}\cos\phi_2 & \cdots & 0 \\ \vdots & \vdots & \ddots & \vdots \\ 0 & 0 & \cdots & d_M^{\alpha-1}\cos\phi_M \end{bmatrix} \tag{15}$$

$$\frac{\partial \sigma_p}{\partial x_j} = \sigma_0^2 \begin{bmatrix} d_1^{\alpha-1}\sin\phi_1 & 0 & \cdots & 0 \\ 0 & d_2^{\alpha-1}\sin\phi_2 & \cdots & 0 \\ \vdots & \vdots & \ddots & \vdots \\ 0 & 0 & \cdots & d_M^{\alpha-1}\sin\phi_M \end{bmatrix} \tag{16}$$

To further simplify Formula (11), let:

$$\beta_i = \frac{\alpha^2}{d_i^2} + \frac{25\gamma_i^2}{(\ln(10))^2 \sigma_0^2 d_i^{a+2}} \tag{17}$$

Then, $\mathbf{J}_{xx}$ can be expressed as:

$$\mathbf{J}_{xx} = \begin{bmatrix} \sum\limits_{i=1}^{M}\beta_i + \sum\limits_{i=1}^{M}\beta_i\cos 2\phi_i & \sum\limits_{i=1}^{M}\beta_i\sin 2\phi_i \\ \sum\limits_{i=1}^{M}\beta_i\sin 2\phi_i & \sum\limits_{i=1}^{M}\beta_i - \sum\limits_{i=1}^{M}\beta_i\cos 2\phi_i \end{bmatrix} \tag{18}$$

Then, the corresponding CRLB can be expressed as:

$$\mathbf{J}_{xx}^{-1} = \frac{1}{\det(\mathbf{J}_{xx})} \begin{bmatrix} \sum\limits_{i=1}^{M}\beta_i - \sum\limits_{i=1}^{M}\beta_i\cos 2\phi_i & -\sum\limits_{i=1}^{M}\beta_i\sin 2\phi_i \\ -\sum\limits_{i=1}^{M}\beta_i\sin 2\phi_i & \sum\limits_{i=1}^{M}\beta_i + \sum\limits_{i=1}^{M}\beta_i\cos 2\phi_i \end{bmatrix} \tag{19}$$

The value of the $\mathbf{J}_{xx}$ determinant can be expressed as:

$$\begin{aligned} \det(\mathbf{J}_{xx}) &= \frac{1}{4}\left[\left(\sum\limits_{i=1}^{M}\beta_i\right)^2 - \left(\sum\limits_{i=1}^{M}\beta_i\cos(2\phi_i)\right)^2 - \left(\sum\limits_{i=1}^{M}\beta_i\sin(2\phi_i)\right)^2 \right] \\ &= \frac{1}{2}\sum\limits_{i=1}^{M}\sum\limits_{j=i+1}^{M}\beta_i\beta_j\left[1 - \cos(2\phi_i - 2\phi_j)\right] \end{aligned} \tag{20}$$

Then, according to the A optimization criterion, the objective function can be expressed as:

$$F_x^{opt} = \operatorname{argmin} tr(\mathbf{J}_{xx}^{-1}) = \operatorname{argmin} \frac{8\sum\limits_{i=1}^{M}\beta_i}{\sum\limits_{i=1}^{M}\sum\limits_{j=i+1}^{M}\beta_i\beta_j\left[1 - \cos(2\phi_i - 2\phi_j)\right]} \tag{21}$$

Combining Formulas (21) and (17), it can be seen that the location accuracy of the target abscissa x_t is related to the distance di between each UAV and the target. It is also related to the angle difference $\phi_i - \phi_j$ between any two drones.

Formula (21) only involves the estimation of the target abscissa x_t. The estimation of the target ordinate y_t is the same as x_t; thus, Formula (10) is modified as:

$$\mathbf{J}_{yy}^{i,j} = \frac{1}{\sigma_p^2}\frac{\partial P_r}{\partial y_{ti}}\frac{\partial P_r}{\partial y_{tj}} + \frac{1}{2}\frac{1}{\sigma_p^2}\frac{\partial \sigma_p}{\partial y_{ti}}\frac{\partial \sigma_p}{\partial y_{tj}} \tag{22}$$

The subsequent operation process is completely similar to $\mathbf{J}_{xx}$ in the previously mentioned article and is not repeated in this article.

Since the horizontal and vertical coordinates of the targets are independent of each other, the effects of directly calculating $\mathbf{J}$ as well as $\mathbf{J}_{xx}$ and $\mathbf{J}_{yy}$ are equivalent. Therefore, the optimization objective function for the passive location of stationary targets is:

$$F^{opt} = \mathrm{argmin}\left(F_x^{opt} + F_y^{opt} \right) \tag{23}$$

3.2. The Main Difference between the Location of Moving Objects and Stationary Objects

The key difference between the location of moving targets and stationary targets is $f(P_r; R_t)$ in Formula (8), that is, the probability density distribution function of the target position changes in different trends with the location of the target.

In the process of locating a stationary target, since there is no prior information as a support, the target obeys a uniform distribution on the x-axis and y-axis. That is, $f(P_r; R_t)$ obeys an equal probability distribution on the abscissa and ordinate axes. Then, as the location progresses, it obeys the Gaussian distribution.

In the process of locating the moving target, as the location continues, the coordinates of the target in the next moment does not obey a uniform distribution on the entire coordinate axis. Instead, the $f(P_r; R_t)$ of the target position in the next moment should be derived by combining the existing multiple location results and the target movement trend.

That is, the main difference between the location of moving objects and stationary objects is that, in the process of location moving objects, the probability density $f(P_r; R_t)$ of the spatial distribution of the objects should be adjusted in real time.

3.3. Probability Distribution Determination Method Based on Deep Combinatorial Network

With the continuous location, the probability distribution characteristics and parameters of $f(P_r; R_t)$ continue to change. However, due to differences in target characteristics and intent, it is impossible to obtain a common or unambiguous expression. Therefore, this section adopts an approach based on deep combinatorial networks. By training the deep combinatorial network, a large number of iterations predicts the position of the target in the next moment. Thus, probability is replaced by frequency, and the probability density function of the spatial distribution of the target is quantified.

In order to improve the accuracy of target location prediction, the motion state of the target must be identified first. The discrimination of the motion state is essentially a classification problem. Because the types of target motion patterns are fixed, that is, the total number of categories for classification is determined, this paper utilized convolutional neural networks (CNN) to determine the motion state. The target trajectory prediction is actually the prediction of the time series. A recursive neural network (RNN) has good processing ability for time-series data.

Therefore, this section uses CNN and RNN to build a combined network architecture to achieve target intent recognition and trajectory prediction. The network is divided into two parts: offline training and online application; offline training is shown in Figure 3.

The specific process of offline training in Figure 3 can be described as:

Step 1: Set the target motion state and generate trajectory parameters in combination with performance indicators. Then, data corresponding to different motion states are generated. It is assumed that the target motion state includes three types: Constant Velocity (CV), Constant Acceleration (CA), and Constant Turn Rate (CT).

Step 2: Combine the characteristics of the environment and noise to generate the corresponding noise.

Step 3: Train the CNN for recognizing motion states.

Step 4: Train the RNN network parameters for predicting the trajectories of different motion states.

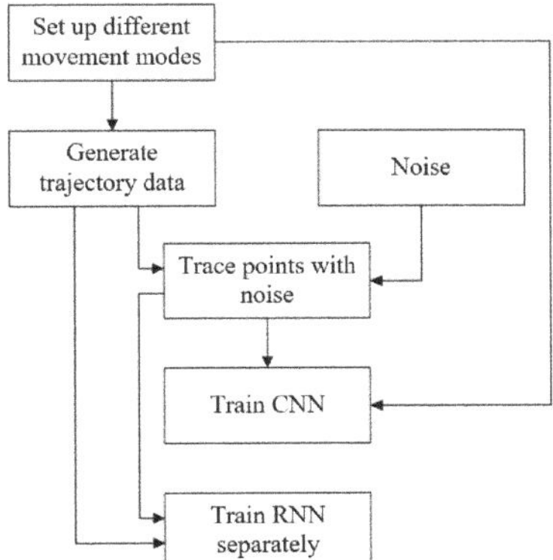

Figure 3. The process of offline training.

Through the above process, the CNN and RNN network training can be achieved.

Among them, Step 4 trains the corresponding network parameters according to the different motion states of the target, which can improve the applicability of the network and further improve the prediction accuracy.

The specific process of online application in Figure 4 can be described as:

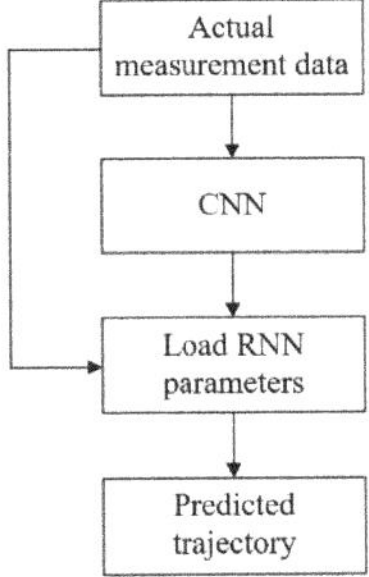

Figure 4. The process of online application.

Step 1: Use the RSS passive location method to obtain the trajectory parameters of the target. Input it into the CNN to identify the motion state of the target.

Step 2: According to the identified motion state, select and load the corresponding RNN network parameters.

Step 3: Input the trajectory parameters of the target into the RNN to obtain the predicted trajectory points of the target.

To date, the single prediction of the target trajectory using the deep combination network has been achieved.

The core purpose of constructing a combined network is not to accurately predict the position of the target, but to obtain $f(P_r; R_t)$ in Formula (8). When used online, step 3 is repeatedly executed to obtain the predicted values of the multiple sets of target positions. Frequency is used instead of probability, as $f(P_r; R_t)$ of the target in the next moment.

This way of obtaining $f(P_r; R_t)$ is not limited by the probability density distribution function and corresponding parameters. At the same time, it does not require sufficient professional knowledge and mathematical skills to obtain the probability density distribution function of the target in the next moment. This method is easy to operate and the results are more accurate.

At the same time, this strategy has another advantage. In practical situations, environmental noise is generally a mixture of multiple different parameters and distribution types of noise, and has time-varying characteristics. However, it is impossible to obtain the type and corresponding parameters of each noise in this mixed noise. This also leads to actual noise being much more complex than theoretical noise and inability to build a theoretical model of environmental noise. Furthermore, subsequent quantitative analysis and formula derivation cannot be carried out. The CNN network in this paper can construct noise distribution based on actual measured parameters. The CNN network can be trained using the previously measured target and noise measurements. This research can greatly improve the accuracy of trajectory recognition in complex noise backgrounds.

Although deep learning can be used to predict the position of the target, it is still necessary to combine the FIM to optimize the spatial position of the UAV and improve the passive location accuracy. Therefore, its essence is still an NP-hard problem, and it is difficult to obtain an analytical solution.

Therefore, this paper improved the particle swarm algorithm and optimized the spatial position and trajectory of the UAV to improve the accuracy of the passive location of moving targets. There are two main reasons for using the PSO algorithm in this article.

The first reason is that it is difficult to obtain the expression of the parameter $f(P_r; R_t)$ through theoretical derivation. Due to such constraints, even if $f(P_r; R_t)$ is set, deriving an analytical solution is extremely difficult and not universal. Therefore, this article used intelligent optimization algorithms to solve it.

The second reason is that, compared to many other intelligent optimization algorithms, the PSO algorithm is recognized as being the fastest. The in-depth research that has been conducted on PSO is sufficient to ensure the effectiveness of PSO and, also due to the extensive research on PSO, its algorithm has good stability.

4. Improved Particle Swarm Optimization Algorithm

4.1. Particle Swarm Optimization Algorithm and Its Shortcomings

Particle swarm optimization (PSO) [27,28] was established by observing the predation characteristics of birds. The algorithm is simple to operate, efficient in searches, and has been widely used in many fields.

Assume that the dimension of the search space to be optimized is D, the total number of particles is N, and the total number of search iterations is T. Then, the updated iterative formula for optimization is:

$$v_{id}^{t+1} = \omega v_{id}^t + c_1 r_1 \left(p_{ibest}^t - x_{id}^t \right) + c_2 r_2 \left(p_{gbest}^t - x_{id}^t \right) \tag{24}$$

$$x_{id}^{t+1} = x_{id}^t + v_{id}^{t+1} \tag{25}$$

where $v_i^t = \left(v_{i1}^t, v_{i2}^t, \cdots, v_{iD}^t \right)$ represents the set of velocities of the i-th particle in each dimension during the t-th iteration; x_i^t represents the set of particle position, $i = 1, 2, \ldots, N, d = 1, 2, \ldots, D, t = 1, 2, \ldots, T$; ω is the inertia coefficient; c_1 and c_2 are learning factors; and r_1 and r_2 are random numbers uniformly distributed between $[0, 1]$. p_{ibest}^t and p_{gbest}^t are the best positions in individual history and population history, respectively.

Then, the fitness function corresponding to the particle position is calculated. The better the fitness, the better the position of the particle. All particles adjust their speed direction and move towards a better position by comparing their fitness functions with that of other particles.

The above is the core formula and basic principle of the PSO algorithm. It can be seen that the PSO algorithm only needs to adjust the flying speed of the particles to achieve optimization.

Although PSO can easily achieve the local optimal solution, especially for typical multimodal functions, its search efficiency is limited. This is due to the fact that particles are easily influenced by other particles. Some particles are affected by other better particles when they do not search a certain area completely. All move towards the position of the optimal particle at this stage, resulting in premature maturity.

If it is possible to conduct a complete and thorough search of each area, a global comparison can be established. Or during the movement, a detailed search for the area within the movement track can be performed. This can reduce the possibility of falling into a local optimum. Therefore, this paper constructs a time-period-based hierarchical PSO improvement strategy to improve the search performance of PSO.

4.2. Time-Period-Based Hierarchical PSO

The architecture of the time-period-based hierarchical PSO is shown in Figure 5.

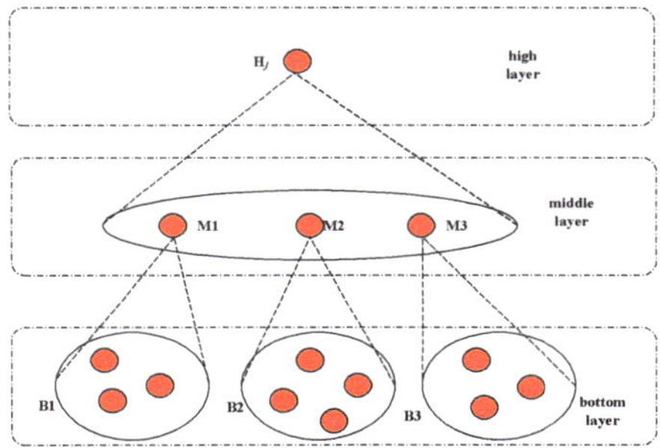

Figure 5. Schematic diagram of the layered architecture.

The core idea of layering is to construct three groups according to the distribution of particles: bottom layer, middle layer, and high layer. The bottom layer is explored in real time, and after interaction, the fitness function is compared to obtain the middle and high layers. The bottom layer of each group only interacts with the group, which ensures that an area is fully searched. At the same time, the best bottom layer data in this group are used as the middle layer. Then, the middle layer interacts occasionally, which balances the contradiction between the global and local searches. Afterwards, the middle and high layers guide the work of the lower layers, and the upper layers of different ethnic groups occasionally interact, thereby changing the movement pattern.

In the above discussion, how the particles are grouped and how often the particles between the middle and high layers exchange information seriously affect the algorithm performance. For this reason, it is introduced in detail later.

4.3. Particle Grouping Strategy

First, the initial population is randomly generated, and the initial position of each particle is obtained. Particles are grouped using the hierarchical clustering method. Hierarchical clustering method combines particles with similar distances into a group. In this way, particles that are close to each other can be clustered, and the result is shown in Figure 6.

The relationship between particles can be directly seen from Figure 6. Then, the number of groups is set, and the particles in the group are obtained. As shown in Figure 6, using the hierarchical clustering method, the result obtained is a typical binary tree structure. This structure is more intuitive. The red and blue lines are the grouped lines. If set to four groups, the particles below the red line become a group according to the cross-linking relationship of the grouping. The above is the process of grouping particles.

The hierarchical clustering method is a mature algorithm, and as there is a corresponding code in MATLAB, it is not repeated in this article.

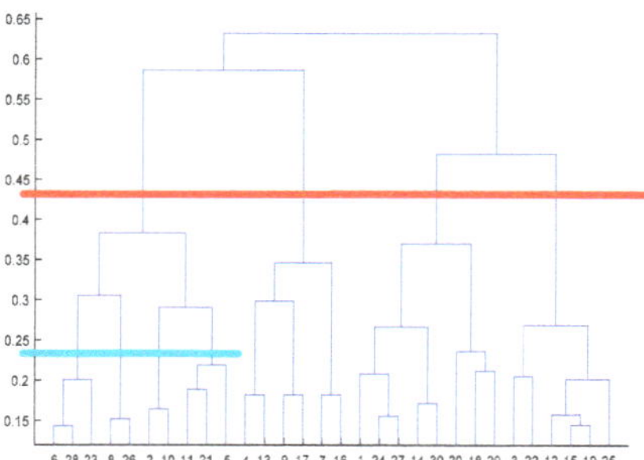

Figure 6. Schematic diagram of the hierarchical clustering method results.

After that, the particles start to be optimized. In the initial stage, the fitness function corresponding to each particle position is calculated. Then, a comparison within the group is performed to obtain the optimal particle within the group. That is, p^t_{Mbest} is the best particle of the bottom layer, and it also becomes the particle of the middle layer. Afterwards, each middle-layer particle is compared to obtain the position p^t_{Hbest} of the optimal particle of the group, that is, the high-level particles in Figure 5. Then, Formula (24) can be modified as:

$$v^{t+1}_{id} = \omega v^t_{id} + c_1 r_1 \left(p^t_{ibest} - x^t_{id} \right) + c_2 r_2 \left(p^t_{Mbest} - x^t_{id} \right) + c_3 r_3 \left(p^t_{Hbest} - x^t_{id} \right) \tag{26}$$

The parameter definitions in Formula (26) are the same as those in Formula (24), and are therefore not repeated here.

It can be seen from Formula (26) that the improved PSO is less affected by the global optimal solution. At the same time, each ethnic group searches for the optimal solution within its own territory as much as possible. This enables the adequate exploration of multiple regions. Occasional high-level interactions between groups can ensure that each group moves toward the optimal solution within the group. Ultimately, the possibility of the premature maturity of the PSO algorithm is reduced.

4.4. Time Period

The frequency of interaction between particles in the middle layer affects the direction of particle optimization. Therefore, this paper constructs a pattern of time periods to optimize the interaction frequency of the middle layer.

Assuming that the update times of the middle and high particles are t_M and t_H, respectively, that is, the middle layer optimal is updated only after every t_M iterations, Formula (26) is further modified as:

$$\begin{aligned} v^{t+1}_{id} &= \omega v^t_{id} + c_1 r_1 \left(p^t_{ibest} - x^t_{id} \right) + \frac{\mathrm{mod}(t, t_M)}{t_M} r_2 \left(p^t_{Mbest} - x^t_{id} \right) \\ &+ \frac{\mathrm{mod}(t, t_H)}{t_H} r_3 \left(p^t_{Hbest} - x^t_{id} \right) \end{aligned} \tag{27}$$

where $\mathrm{mod}(a,b)$ is the remainder operation, that is, the remainder obtained after dividing a by b.

In Formula (27), when the mod operation result is small, it means that the corresponding optimal value has just been updated. At this time, it is more focused on letting the particles search in their respective areas to obtain a better p^t_{ibest} for subsequent updates. As the search progresses, the mod results gradually increase, and the particles move closer to the local optimum. It is ensured that, before the next update of the local optimal value, the particle performed a more comprehensive search for the region where it is located, thereby reducing the possibility of falling into the local optimal value.

However, as the search progresses, particles within a group do not always belong to the same group. Instead, they regroup after multiple searches. This ensures a comprehensive search of the area. Therefore, in this paper, after every t_G search, all particles were regrouped according to the hierarchical clustering method in the previous section to improve the search efficiency.

The idea of the time period is borrowed from the clock model. That is, important parameters, such as the hour hand, should be updated slowly. Exploratory particles, such as the minute and second hands, should be updated faster. In this way, the effective search for the full dimension is better achieved, and the possibility of falling into a local optimum is reduced.

To sum up, this section improves the PSO algorithm by designing the particle grouping architecture and building the time period.

4.5. Algorithm Complexity Analysis

Due to the few parameters involved, the PSO algorithm has a significantly better optimization speed than other intelligent algorithms. Therefore, the improvement of the PSO algorithm should not affect its algorithm speed as much as possible. Therefore, this section analyzes the computational complexity of the improved algorithm to ensure that the speed of the algorithm does not drop significantly.

In terms of iterative update strategy, comparing Formulas (24) and (27), it can be seen that the original PSO considers the influence of individual historical optimal and global optimal on particle velocity. The improved PSO increases the impact of local optima on the speed. The amount of calculation becomes 1.5 times the original, but only the subtraction operation is performed without changing the complexity of the algorithm.

In terms of coefficients, the improved algorithm adjusts the learning factor c1 to mod operation. This operation is linear, and only needs to be performed once per iteration and the result recorded. That is, in each iteration, only one operation is performed. At the same time, subsequent particles directly use this result without repeating the calculation. The added computation has little impact compared to iterative operations.

At the level of algorithm architecture, in each iteration process, the original PSO algorithm compares the fitness functions of all particles. Thereby, the maximum value among the N fitness functions is obtained. In the improved PSO algorithm, due to the design of the time period, although the comparison is also required, the comparison within the group is mainly performed. Compared with the global comparison of the original PSO algorithm, the computational complexity of the improved PSO algorithm is significantly reduced. Although the improved algorithm also involves global comparison, due to the hierarchical structure and time period, the fitness function and the number of comparisons involved in this comparison are significantly lower than the global comparison of the original algorithm.

In the improved algorithm, the hierarchical clustering method is used to group the particles. This method needs to calculate the distance matrix between particles and then classify them according to the distance. However, due to the design of the time period, regrouping is performed only once after t_G searches. Compared with the original algorithm, each particle needs to update the fitness function corresponding to the calculation, and the increased calculation amount of the improved algorithm is very small.

To sum up, the time-period-based hierarchical PSO improvement strategy constructed in this paper only approximately increased the amount of computation to 1.5 times that of the original PSO, without changing the algorithm's complexity. Therefore, the improved algorithm still retains the efficient characteristics of PSO.

5. Passive Location Algorithm Flow of the Moving Target Based on Improved PSO

5.1. Objective Function

Using UAV swarms to locate moving targets is an asymptotically optimal process. Therefore, not only the location effect in the present moment, but also the subsequent impact of the decision in the present moment, should be considered. In this way, the best location effect can be achieved at a faster speed.

Assuming a time k, the subsequent motion state of the UAV swarm and the target is shown in Figure 7.

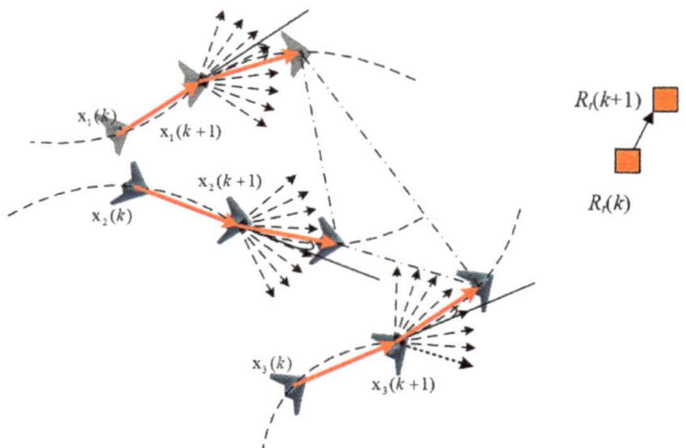

Figure 7. Movement situation diagram.

The coordinate of the i-th UAV in our UAV swarm is $x_i(k)$ and the target coordinate is $R_t(k)$.

At this time, the model predictive control (MPC) method was adopted. That is, the optimization method of predicting H steps and executing one step was adopted. On the basis of Formula (23), the objective function is adjusted as:

$$FH^{opt} = \operatorname{argmin} \sum_{i=0}^{H-1} \gamma^i F^{opt}(k+i) \tag{28}$$

where γ is the decay factor. The MPC method used in Formula (28) is relatively mature, and is not repeated in this paper.

5.2. Constraints

Constraints mainly include individual motion constraints and obstacle avoidance constraints, as well as cluster communication constraints and collision avoidance constraints.

It was assumed that the motion state of the UAV is at k and the next moment, that is, the motion state at the moment $k + \Delta k$, as shown in Figure 8.

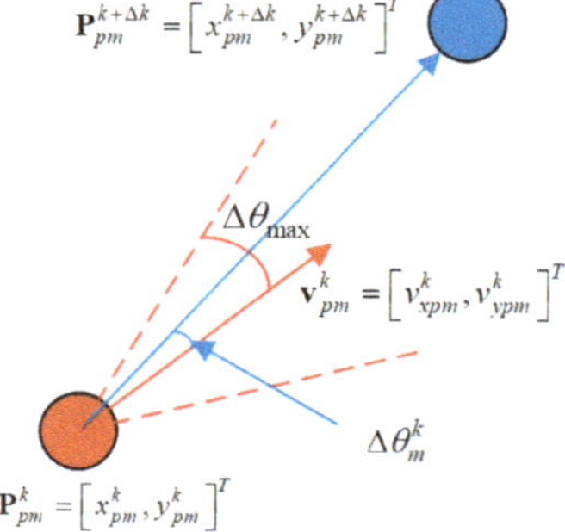

Figure 8. Schematic diagram of motion constraints.

The position and speed of m-th UAV at time k are $\mathbf{P}_m^k = [x_m^k, y_m^k]$ and $\mathbf{v}_m^k = [v_{xm}^k, v_{ym}^k]$. Taking it as the initial condition, it was optimized to obtain the position and velocity in the next moment as $\mathbf{P}_m^{k+\Delta k} = [x_m^{k+\Delta k}, y_m^{k+\Delta k}]$ and $\mathbf{v}_m^{k+\Delta k} = [v_m^{k+\Delta k}, v_m^{k+\Delta k}]$, respectively. The corresponding relationship is shown in Figure 8. Then, the motion constraints should be satisfied, namely:

$$\begin{cases} \mathbf{P}_m^{k+\Delta k} = \mathbf{P}_m^k + \mathbf{v}_m^k \Delta k \\ \|\mathbf{P}_m^{k+\Delta k} - \mathbf{P}_m^k\|_2 \leq \|\mathbf{v}_m^k\|_2 \Delta k \end{cases} \tag{29}$$

where $\| \|_2$ means to take the 2-norm. The relationship between the speeds is:

$$\mathbf{v}_m^{k+\Delta k} = \mathbf{v}_m^k + \Delta \mathbf{v}_m^k \tag{30}$$

where $\Delta \mathbf{v}_m^k$ is the value of the velocity change, which should satisfy:

$$\begin{cases} \|\Delta \mathbf{v}_m^k\|_2 \leq \Delta v_{\max} \\ v_{\min} \leq \|\mathbf{v}_m^{k+\Delta k}\|_2 \leq v_{\max} \end{cases} \tag{31}$$

That is, the speed and the amount of speed change cannot exceed their allowable limit.

Similarly, the change amount $\Delta \theta_m^k$ of the UAV direction can be calculated according to the velocity vector at two moments, which should satisfy:

$$\Delta \theta_m^k \leq \Delta \theta_{\max} \tag{32}$$

where $| \ |$ represents the absolute value.

The above are the motion constraints that the UAV should meet.

The remaining three constraints are mainly reflected in the spatial distance. Among them, the individual obstacle avoidance constraints are mainly that the minimum distance between the UAV and the obstacle during the entire flight process cannot be lower than the set safe distance.

The communication constraints of the swarm require that, for any UAV, there is at least one UAV whose distance to the UAV is less than the set communication distance.

The collision avoidance constraint is the opposite, requiring that the distance between any two UAVs is not lower than the set collision avoidance distance.

The above three constraints are relatively simple and are not described in detail in this article.

5.3. Algorithm Optimization Process

In order to achieve the passive location of moving targets, the construction algorithm flow is shown in Figure 9.

The algorithm flow of Figure 9 can be described as:

Step 1: Obtain the position of each UAV in swarm at time k and the signals received by each platform.

Step 2: Construct the objective function shown in Formula (28) and construct the corresponding constraints. Use the improved PSO algorithm and MPC for optimization.

Step 3: Judge whether the result satisfies the constraint conditions; if not, return to Step 2; if it is satisfied, execute Step 4.

Step 4: Construct a time series of the location points obtained at this time and the previous five moments. It is fed into the combined network to predict the target trajectory. The predicted result is used to correct $f(P_r; R_t)$ in Formula (8).

Step 5: Obtain the optimal coordinates of each drone in the next moment and then update the position of the drone. Determine whether the final optimization time k is reached. If it is not reached, return to Step 1; otherwise, the optimization ends.

The above is the algorithm flow of using UAV swarm to locate the moving target.

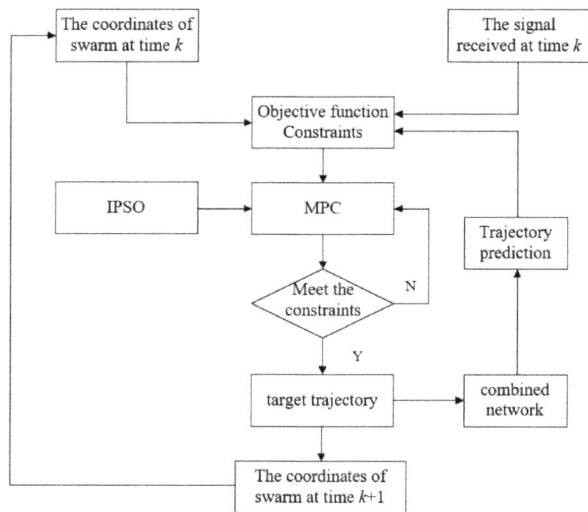

Figure 9. Flowchart of the passive location algorithm for the moving target by UAV swarm.

6. Simulation and Verification

To verify the performance of the algorithm in this paper, it was assumed that the target run for 60 min, 1–20 min for CV, 21–40 min for CT, and 41–60 min for CA.

Five UAVs took off near (0,0) with a speed limit of 200 m/min and performed the cooperative passive location of the target.

The simulation environment was I7-10750H, with 2.60 GHz dominant frequency and 16 G memory, and the simulation experiment was made on a platform based on MATLAB 2020b.

6.1. Performance Verification of Deep Networks

To verify the performance of the network built in Section 3.3, this section conducts simulation experiments on the network.

The data used to train the CNN network were the track data added with standard Gaussian white noise. At the same time, it was necessary to identify the target's motion state in this minute; so, the training data were 60 s, which means to generate a data sequence with a length of 60 based on the above motion state and corresponding time. The output of the training was the motion state of the trajectory, namely, the three motion states of CV, CA, and CT. This article set 60 points per minute. The target movement lasted for a total of 60 min. To ensure sufficient training data, 360,000 sets of data were generated for training, and an additional 3600 sets of data were generated for testing. The test was passed when the test error was set to not less than 95%.

Since the length of the data used for training was only 60, the data were not long. Thus, the number of network layers was set to 7, that is, 5 of them were hidden layers, the learning rate was 0.3, eight neurons per layer, and the number of iterations was 2000. Comparing the algorithm with IMM-EKF [29], the result is shown in Figure 10.

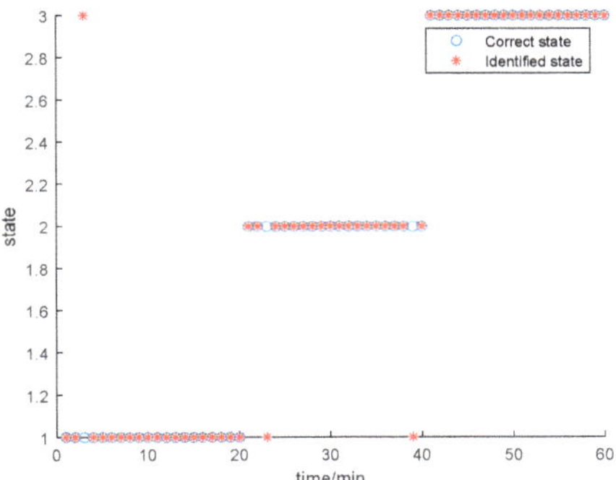

Figure 10. Comparison of the recognition results.

From Figure 10, it can be seen that CNN has three errors and IMM-EKF has six. CNN is more accurate than IMM-EKF. This is because the basic function of CNN is recognition, and the recognition effect will increase with the increase in training data. However, the recognition effect of IMM-EKF is affected by noise, and the performance does not change with the amount of training data. Therefore, CNN is more suitable for target motion state recognition.

The data used to train the CNN-RNN composite network were input for 60 points, that is, the position of the target and the motion state identified by the CNN per second. The output was 60 track points that predict the target for the next minute. A total of 360,000 sets of data were used for training and an additional 3600 sets for testing. The training was completed after the number of iterations was reached.

Because the length of the data used for training was only 60, the number of RNN network layers was set to 8 layers, that is, 6 layers were hidden layers, the learning rate was 0.3, 8 neurons per layer, and the number of iterations was 50,000. Comparing the algorithm with the classical RNN, the result is shown in Figure 11.

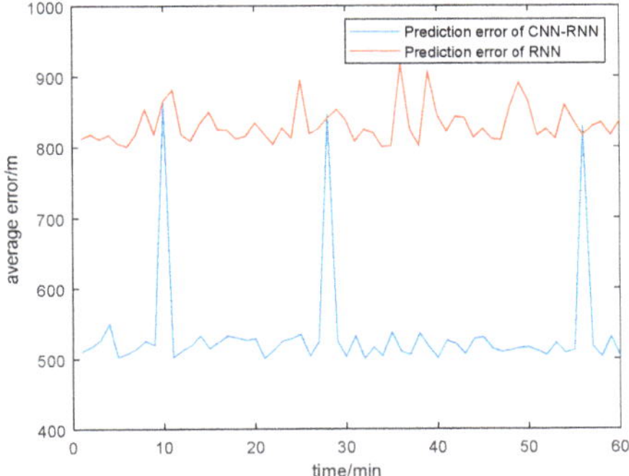

Figure 11. Comparison of the average error.

As can be seen from Figure 11, the prediction results of the CNN-RNN network were generally better than that of RNN. This is because there are actually three sets of RNN networks with different parameters in CNN-RNN. That is, after the CNN identifies the target motion state, the RNN loads the corresponding parameters to perform the prediction. With more targeted networks, the results will certainly be more accurate. However, once the CNN recognizes an error, the error spikes, as shown in Figure 11.

6.2. Passive Location Performance Verification and Algorithm Comparison

The algorithm in this paper, the IMM-EFK in [29], and the location method in [30] were compared, and the results are shown in Figure 12.

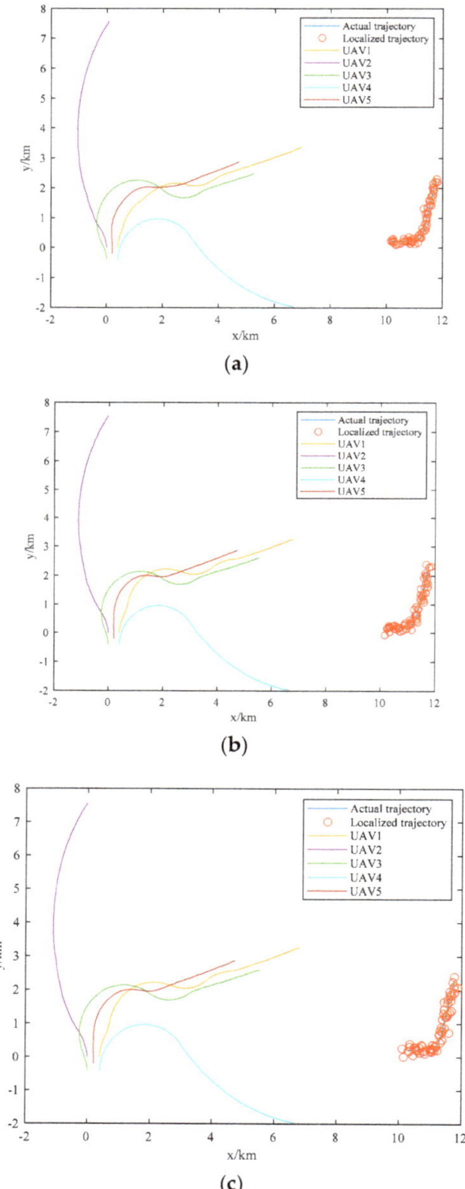

Figure 12. Comparison of algorithm optimization results. (**a**) Optimization results of the algorithm in this paper; (**b**) IMM−EKF optimization results; and (**c**) the results of the location method in [30].

By comparing the three sets of results in Figure 12, it can be seen that the location points of the algorithm in this paper are more coincident with the target trajectory.

It can be seen from Figure 12b that the method of IMM-EKF has better localization accuracy. However, when the motion state of the target is converted, the IMM-EKF cannot quickly identify the change of the motion state of the target. At the same time, after identification, it is difficult to quickly establish a new tracking equation, resulting in a significant decrease in the location efficiency at this time.

Literature [30] uses Doppler rate to improve the accuracy of moving target positioning based on time delay and Doppler shift. Meanwhile, literature [30] establishes a pseudolinear set of equations by introducing some additional variables. The analytic solution for moving target positioning is given. The positioning CRLB is derived. However, by comparing Figure 12a,c, it can be seen that the positioning method in literature [30] differs from that in this paper in positioning accuracy. There are two main reasons.

The first is that, as can be seen from Figure 12c, the method in literature [30] has always had a large error. This is because the method in literature [30] does not consider the sequential nature of target motion, treating each localization as an independent localization. As a result, its positioning performance will not improve with the progress of positioning. The second reason is that the method in literature [30] does not achieve real-time planning for the trajectory of unmanned aerial vehicles, but rather provides the ultimate ideal location point distribution method. The real-time optimization is not achieved, and motion conditions such as platform motion are not considered. This results in poor performance during the positioning process.

The core reason why the algorithm in this paper is superior to other algorithms is that this paper constructs a model of cooperative passive location from the perspective of clusters. This article does not disassemble the five UAV into a "2 + 3" model, but optimizes the five UAVs as a whole. It can be seen from Figure 12a that among the 5 UAVs, 3 UAVs are flying towards the target, which is pulling in the relative distance between the cluster and the target. The 2 UAVs flew towards a wide area, increasing the observation angles of different drones. This also conforms to Formula (21), that is, the UAV swarm adjusts the distance and angle factors that affect the location accuracy.

The algorithm in this paper obtains better location performance by adjusting the distance between the cluster and the target and forming different observation angles at the same time.

In order to further quantify and compare the location performance. Under the condition that the simulation conditions remain unchanged, 30 Monte Carlo experimental simulations are carried out for each algorithm. Take the average value of the errors at each moment to obtain a comparison chart, as shown in Figure 13.

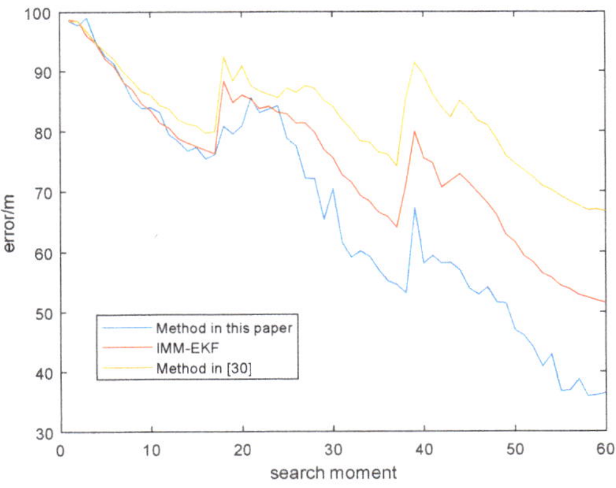

Figure 13. Comparison of average errors [30].

As can be seen from Figure 13, the algorithm in this paper has two obvious advantages over other algorithms. One is that the MPC is involved in the algorithm in this paper, so its error decreases significantly faster than other algorithms.

The other is that the stability of the algorithm in this paper is stronger. When the motion state of the target changes, it is difficult for each algorithm to judge the change in the state at the first time, so there is a sudden change in error in Figure 13. By comparison, it can be seen that, because the algorithm in this paper uses a combined network, the error is less affected. At the same time, the algorithm also stabilizes faster.

To further compare the effectiveness of the positioning methods, this section counts the positioning time of 30 Monte Carlo experiments of the above three methods. The results are shown in Table 2.

Table 2. Comparison of time consumption of the three positioning algorithms.

Time Consumption	Algorithm in This Paper	IMM-EKF	Method in [30]
Average total time	163.26	205.81	732.42
Average time for each point	2.71	3.43	12.21

As can be seen from Table 2, the algorithm in this paper is superior to the other two algorithms in terms of efficiency. This is because, when using IMM-EKF to determine the motion state of a target, it is necessary to calculate the probability of the target's motion state in the next moment based on its previous motion trajectory. The algorithm in this paper only needs to input the trajectory into the trained network, and can directly predict the position of the target in the next moment, which is faster.

The method in [30] provides an analytical solution, which can intuitively see the relationship between factors affecting the target's positioning accuracy and quantification. However, in the solution process of [30], it involves performing inverse operations on a large number of matrices, Which seriously affects the speed of the algorithm. Therefore, it takes a long time.

6.3. Optimization Algorithm Performance Comparison

In order to further measure and compare the performance of the improved PSO algorithm, the improved PSO in this paper was compared with the PSO in [31] and the Holonic-PSO in [32]. The simulation conditions were the same, and 30 Monte Carlo simulations were performed to obtain a comparison chart of the mean error value, as shown in Figure 14.

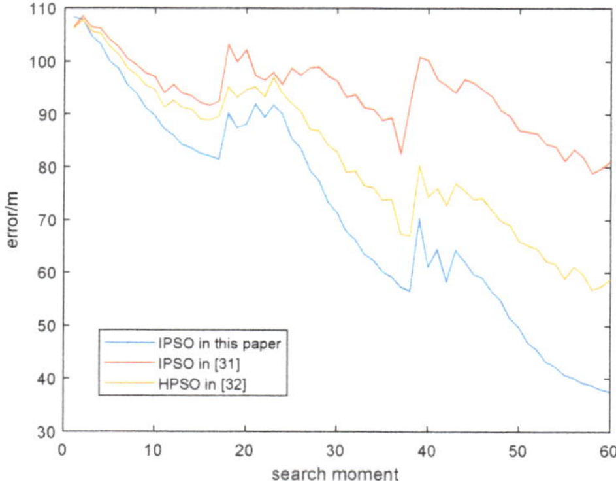

Figure 14. Error comparison chart [31,32].

It can be seen from Figure 14 that the performance of the algorithm in this paper is more stable, because the algorithm in this paper can perform a more global search and improve the algorithm efficiency.

The method in [31] is more focused on enabling PSO particles to jump out of the local optimization with maximum probability, thereby achieving global search. To achieve this goal, Formulas (5)–(7) in [31] set a method for generating approximately random search directions. This setting can reduce the possibility of falling into a local optimum, but this near-random approach has no significant effect on improving search performance.

The improvement idea of this article was inspired by [32] to group particles for search. One disadvantage of [32] is that its particle search strategy, i.e., the updated equation of particle state, is artificially adjusted. In the iterative process of [32], the first 80% of searches and the last 20% of searches use different update equations. However, in [32], a simple comparative experiment shows that the ratio of 80% to 20% is better, without indicating whether it is optimal. Obviously, this ratio may vary depending on the issue.

At the same time, there is another reason why the method in this article is superior to the above two methods. What this article aimed to solve is a sequential decision-making problem. The optimization results of the previous moment affect the next moment. The positioning accuracy of the previous moment is good, providing a good initial condition for the next moment, and the positioning accuracy of the next moment will not be poor. If the positioning effect at the previous moment is poor, it will also affect the positioning at the next moment. Therefore, over time, compared to the other two methods, the effect of this article becomes better and better.

In order to further compare the performance of the algorithms, the time of the three optimization algorithms is also counted, and the results are shown in Table 3.

Table 3. Comparison of the time consumption of the three optimization algorithms.

Time Consumption	Algorithm in This Paper	IPSO in [31]	HPSO in [32]
Average total time	163.26	116.46	164.47
Average time for each point	2.71	1.94	2.71

Through comparison, it can be seen that the algorithm speed in this article is weaker than IPSO [31], but better than HPSO [32].

The improvement of the IPSO algorithm on the search direction of the particles is still based on a random mode. Compared with the PSO algorithm, this search mode has almost no significant change in the additional computation amount generated by the PSO algorithm. Therefore, IPSO still maintains its high-speed solution efficiency. The algorithm in this paper involves further information interaction between groups and particles, with a significant increase in computational complexity. Therefore, the performance is weaker than IPSO.

Both this algorithm and HPSO [32] involve particle grouping and information interaction. However, in each iteration of the HPSO algorithm, the particle parameters at each level are updated. In this article, by designing a clock cycle, particles at different levels were updated according to the cycle, which reduces the amount of computation. This can also allow different particle populations to conduct more detailed searches of their regions.

Under the main premise of ensuring positioning accuracy, the effectiveness of the algorithm in this paper is even higher.

7. Discussion

This section mainly discusses the main contributions, application scenarios, algorithm deficiencies, and follow-up work of this article.

This paper built a passive location method for moving targets based on RSS for UAV clusters. The target probability distribution network was designed to predict the subsequent location of the target more clearly and easily. Thus, the mature static target positioning method was extended to the target positioning. At the same time, the PSO algorithm was improved in this paper. From the simulation comparison, the improved method had a good performance.

The research results can be applied in many ways, mainly using a UAV cluster to locate a target and achieve navigation without a GPS signal. UAV clusters can also be used to search and rescue people with mobile phones. Sound and electromagnetic information can be collected to build digital maps. It can also locate ships on the sea, or discover and locate concealed radar.

Although this paper has conducted some research work, there are still some limitations. Firstly, the positioning model does not take the altitude direction into account, so in practice, this study is still far from achieving more accurate applications. Secondly, although the network can suppress complex noise, its effect is limited. Finally, the real-time performance of the algorithm needs further design. The PSO algorithm cannot increase speed further, but as UAV clusters are multiple platforms, parallel computing can be considered. It is feasible to exchange computing resources for optimization time.

To overcome these shortcomings, a passive positioning model of UAV in a three-dimensional scene will be built in future research to improve the network to improve its ability of target state recognition under strong noise background. Additionally, a framework of parallel computing will be designed to test and improve the algorithm.

8. Conclusions

In this paper, the problem of improving passive location accuracy will be transformed into the problem of obtaining more target information. Based on RSS and the A criterion, a passive location method for moving objects was constructed. Firstly, the measurement model of cluster passive location was constructed. After that, the relationship between the UAV spatial position and the static target localization effectiveness was derived and constructed. Then, the difference between stationary target and moving target location was analyzed. In order to expand the scope of the application of the algorithm, the prediction of the target position was realized by designing a deep combined network. Thereby, the probability density distribution function required in the passive location process of the moving target was obtained. Considering that trajectory optimization is an NP-Hard problem and addressing the problem that the PSO algorithm easily falls into the local optimum, a layered improvement strategy based on time period was designed to improve PSO performance. Then, a passive location algorithm flow based on the improved PSO was constructed. Through simulation verification and algorithm comparison, the feasibility and performance advantages of the algorithm in this paper were highlighted.

Author Contributions: Conceptualization, L.H. and F.X.; methodology, F.X.; software, S.M.; validation, L.H. and F.X.; formal analysis, F.X.; writing—original draft preparation, F.X.; writing—review and editing, L.H. and F.X. All authors have read and agreed to the published version of the manuscript.

Funding: This research was funded by [National Natural Science Foundation of China] grant number [61502522 and 61502523].

Data Availability Statement: The data can be found for https://pan.baidu.com/s/1lf1zE2eMLW7mCsQi42NrkQ, and the extract code is nu82.

Conflicts of Interest: The authors declare no conflict of interest.

References

1. Zou, Y.; Wan, Q. Asynchronous Time-of-Arrival-Based Source Localization with Sensor Position Uncertainties. *IEEE Commun. Lett.* **2016**, *20*, 1860–1863. [CrossRef]
2. Dang, L.; Yang, H.; Teng, B. Application of Time-Difference-of-Arrival Localization Method in Impulse System Radar and the Prospect of Application of Impulse System Radar in the Internet of Things. *IEEE Access* **2018**, *6*, 44846–44857. [CrossRef]
3. Chitte, S.D.; Dasgupta, S.; Ding, Z. Distance Estimation from Received Signal Strength under Log-Normal Shadowing: Bias and Variance. *IEEE Signal Process. Lett.* **2009**, *16*, 216–218. [CrossRef]

4. Cheng, Y.; Lin, Y. A new received signal strength based location estimation scheme for wireless sensor network. *IEEE Trans. Consum. Electron.* **2009**, *55*, 1295–1299. [CrossRef]

5. Zhao, Y.; Hu, D.; Liu, Z.; Zhao, Y. Calibrating the Transmitter and Receiver Location Errors for Moving Target Localization in Multistatic Passive Radar. *IEEE Access* **2019**, *7*, 118173–118187. [CrossRef]

6. Li, B.; Zhao, K.; Shen, X. Dilution of Precision in Location Systems Using both Angle of Arrival and Time of Arrival Measurements. *IEEE Access* **2020**, *8*, 192506–192516. [CrossRef]

7. Wei, H.; Ye, S. Comments on "A Linear Closed-Form Algorithm for Source Localization from Time-Differences of Arrival". *IEEE Signal Process. Lett.* **2008**, *15*, 895. [CrossRef]

8. Abeywickrama, S.; Samarasinghe, T.; Ho, C.K.; Yuen, C. Wireless Energy Beamforming Using Received Signal Strength Indicator Feedback. *IEEE Trans. Signal Process.* **2018**, *66*, 224–235. [CrossRef]

9. Maric, A.; Kaljic, E.; Njemcevic, P.; Lipovac, V. Projective Approach in Determining Homogeneous Hyperspherical Geometrically-Based Stochastic Channel Model's Statistics: Angle of Departure, Angle of Arrival and Time of Arrival. *IEEE Trans. Wirel. Commun.* **2020**, *19*, 7864–7880. [CrossRef]

10. Xu, H.; Zhang, Y.; Ba, B.; Wang, D.; Li, X. Fast Joint Estimation of Time of Arrival and Angle of Arrival in Complex Multipath Environment Using OFDM. *IEEE Access* **2018**, *6*, 60613–60621. [CrossRef]

11. Schau, H.; Robinson, A. Passive source localization employing intersecting spherical surfaces from time-of-arrival differences. *IEEE Trans. Acoust. Speech Signal Process.* **1987**, *35*, 1223–1225. [CrossRef]

12. Wang, X.; Huang, Z.; Zhou, Y. Underdetermined DOA estimation and blind separation of non-disjoint sources in time-frequency domain based on sparse representation method. *J. Syst. Eng. Electron.* **2014**, *25*, 17–25. [CrossRef]

13. Zhang, M.; Guo, F.; Zhou, Y.; Yao, S. A Single Moving Observer Direct Position Determination Method Using Interferometer Phase Difference. *Acta Aeronaut. Astronaut. Sin.* **2013**, *34*, 2185–2193.

14. Liu, Y.; Guo, F.; Yang, L.; Jiang, W. Source localization using a moving receiver and noisy TOA measurements. *Signal Process.* **2016**, *119*, 185–189. [CrossRef]

15. Xi, L.; Guo, F.; Le, Y.; Min, Z. Improved solution for geolocating a known altitude source using TDOA and FDOA under random sensor location errors. *Electron. Lett.* **2018**, *54*, 597–599.

16. Wang, G.H.; Bai, J.; He, Y.; Xiu, J.J. Optimal deployment of multiple passive sensors in the sense of minimum concentration ellipse. *IET Radar Sonar Navig.* **2009**, *3*, 8–17. [CrossRef]

17. Bishop, A.N.; Fidan, B.; Anderson, B.D.O. Optimality Analysis of Sensor-Target Geometries in Passive Location: Part 2—Time-of-Arrival Based Localization. In Proceedings of the 3rd International Conference on Intelligent Sensors Sensor Networks and Information Processing, Melbourne, VIC, Australia, 3–6 December 2007.

18. Bishop, A.N.; Fidan, B.; Anderson, B.D.O. Optimality Analysis of Sensor-Target Geometries in Passive Location: Part 1—Bearing-Only Localization. In Proceedings of the 3rd International Conference on Intelligent Sensors Sensor Networks and Information Processing, Melbourne, VIC, Australia, 3–6 December 2007; pp. 7–12.

19. Lui, K.W.K.; So, H.C. A Study of Two-Dimensional Sensor Placement Using Time-Difference-of-Arrival Measurements. *Digit. Signal Process.* **2009**, *19*, 650–659. [CrossRef]

20. Yang, B.; Scheuing, J. *A Theoretical Analysis of 2D Sensor Arrays for TDOA Based Localization*; ICASSP: Toulouse, France, 2006; pp. 901–904.

21. Wu, Z.; Fu, K.; Jedari, E.; Shuvra, S.R.; Rashidzadeh, R.; Saif, M. A Fast and Resource Efficient Method for Indoor Location Using Received Signal Strength. *IEEE Trans. Veh. Technol.* **2016**, *65*, 9747–9758. [CrossRef]

22. Xu, S.; Ou, Y.; Zheng, W. Optimal Sensor-Target Geometries for 3-D Static Target Localization Using Received-Signal-Strength Measurements. *IEEE Signal Process. Lett.* **2019**, *26*, 966–970. [CrossRef]

23. Vaghefi, R.M.; Gholami, M.R.; Buehrer, R.M.; Strom, E.G. Cooperative Received Signal Strength-Based Sensor Localization with Unknown Transmit Powers. *IEEE Trans. Signal Process.* **2013**, *61*, 1389–1403. [CrossRef]

24. Hemavathi, N.; Meenalochani, M.; Sudha, S. Influence of Received Signal Strength on Prediction of Cluster Head and Number of Rounds. *IEEE Trans. Instrum. Meas.* **2020**, *69*, 3739–3749. [CrossRef]

25. So, H.C.; Lin, L. Linear Least Squares Approach for Accurate Received Signal Strength Based Source Localization. *IEEE Trans. Signal Process.* **2011**, *59*, 4035–4040. [CrossRef]

26. Kay, S.M. *Fundamentals of Statistical Signal Processing: Estimation Theory*; Prentice Hall PTR: Upper Saddle River, NJ, USA, 1993; pp. 47, 73–76.

27. Zhang, B.; Zhuang, L.; Gao, L.; Luo, W.; Ran, Q.; Du, Q. PSO-EM: A Hyperspectral Unmixing Algorithm Based on Normal Compositional Model. *IEEE Trans. Geosci. Remote Sens.* **2014**, *52*, 7782–7792. [CrossRef]

28. Sun, T.; Tang, C.; Tien, F. Post-Slicing Inspection of Silicon Wafers Using the HJ-PSO Algorithm under Machine Vision. *IEEE Trans. Semicond. Manuf.* **2011**, *24*, 80–88. [CrossRef]

29. Hernandez, M.; Farina, A. PCRB and IMM for Target Tracking in the Presence of Specular Multipath. *IEEE Trans. Aerosp. Electron. Syst.* **2020**, *56*, 2437–2449. [CrossRef]

30. Yongsheng, Z.; Dexiu, H.; Yongjun, Z.; Zhixin, L. Moving target localization for multistatic passive radar using delay, Doppler and Doppler rate measurements. *J. Syst. Eng. Electron.* **2020**, *31*, 939–949. [CrossRef]

31. Mistry, K.; Zhang, L.; Neoh, S.C.; Lim, C.P.; Fielding, B. A Micro-GA Embedded PSO Feature Selection Approach to Intelligent Facial Emotion Recognition. *IEEE Trans. Cybern.* **2017**, *47*, 1496–1509. [CrossRef]
32. Roshanzamir, M.; Balafar, M.A.; Razavi, S.N. Empowering particle swarm optimization algorithm using multi agents' capability: A holonic approach. *Knowl. Based Syst.* **2017**, *136*, 58–74. [CrossRef]

Article

Adjustable Fully Adaptive Cross-Entropy Algorithms for Task Assignment of Multi-UAVs

Kehao Wang [1], Xun Zhang [1], Xuyang Qiao [1], Xiaobai Li [2,*], Wei Cheng [2], Yirui Cong [3] and Kezhong Liu [4]

[1] School of Information Engineering, Wuhan University of Technology, Wuhan 430070, China
[2] Department of Early Warning Intelligence, Air Force Early Warning Academy, Wuhan 430070, China
[3] College of Intelligence Science and Technology, National University of Defense Technology, Changsha 410073, China
[4] School of Navigation, Wuhan University of Technology, Wuhan 430070, China
* Correspondence: lxb2cici@163.com

Abstract: This paper investigates the multiple unmanned aerial vehicle (multi-UAV) cooperative task assignment problem. Specifically, we assign different types of UAVs to accomplish the classification, attack, and verification tasks of targets under resource, precedence, and timing constraints. Due to complex coupling among these tasks, we decompose the considered problem into two subproblems: one with continuous and independent tasks and another with continuous and correlative tasks. To solve them, we first present an adjustable, fully adaptive cross-entropy (AFACE) algorithm based on the cross-entropy (CE) method, which serves as a stepping stone for developing other algorithms. Secondly, to overcome task precedence in the first subproblem, we propose a mutually independent AFACE (MIAFACE) algorithm, which converges faster than the CE method when obtaining the optimal scheme vectors of these continuous and independent tasks. Thirdly, to deal with task coupling in the second subproblem, we present a mutually correlative AFACE (MCAFACE) algorithm to find the optimal scheme vectors of these continuous and correlative tasks, while its computational complexity is inferior to that of the MIAFACE algorithm. Finally, numerical simulations demonstrate that the proposed MIAFACE (MCAFACE, respectively) algorithm consumes less time than the existing algorithms for the continuous and independent (correlative, respectively) task assignment problem.

Keywords: multi-UAVs; task assignment; AFACE algorithm; MIAFACE algorithm; MCAFACE algorithm

Citation: Wang, K.; Zhang, X.; Qiao, X.; Li, X.; Cheng, W.; Cong, Y.; Liu, K. Adjustable Fully Adaptive Cross-Entropy Algorithms for Task Assignment of Multi-UAVs. *Drones* **2023**, 7, 204. https://doi.org/10.3390/drones7030204

Academic Editors: Andrey V. Savkin and Oleg Yakimenko

Received: 13 February 2023
Revised: 5 March 2023
Accepted: 13 March 2023
Published: 16 March 2023

1. Introduction

Due to its rapid deployment and nearly unlimited mobility, an unmanned aerial vehicle (UAV) has great potential in both military and civilian applications, including modern warfare, disaster search and rescue, traffic control, celestial exploration, and a variety of other fields [1–4]. UAVs for these applications have limited capabilities and require sufficient resources to perform tasks autonomously. As a result, multi-UAVs can be regarded as a promising method by which to handle complex tasks. As more attention is focused on them, two problems in multi-UAV collaboration, such as multi-UAV cooperative path planning and cooperative task assignment, are becoming more widely recognized. The main consideration of this paper is the multi-UAV cooperative task assignment problem.

In recent years, many scholars have paid attention to the multi-UAV cooperative task assignment problem, while the related research of this problem is as follows. Chen et al. [5] utilized mixed integer linear programming (MILP) to address the problem of multi-UAV cooperative task assignment and path planning for moving targets on the ground, but it had low scalability while maintaining global optimality. References [6,7] used a heuristic approach to produce near-optimal results in real time, which has been widely considered for large-scale problems and dynamic scenarios. For swarm intelligence algorithms, e.g., particle swarm optimization (PSO) [8], ant colony optimization (ACO) [9], and genetic algorithm (GA) [10], when solving the task assignment problem, they had a fast convergence speed and could effectively obtain optimal assignment schemes, but there is a possibility of falling into local optimum. Moreover, the auction algorithm, game theory, and reinforcement learning have also been applied to the multi-UAV task assignment problem. Duan et al. [11] presented a novel hybrid "two-stage" auction algorithm that combines the structural advantages of the centralized and distributed auction algorithms, which greatly facilitates the performance of UAVs in dynamic task assignments. Chen et al. [12] studied the cooperative reconnaissance and spectrum access (CRSA) problem for task-driven heterogeneous coalition-based UAV networks, and proposed a joint bandwidth allocation and coalition formation (JBACF) algorithm to solve the task assignment and bandwidth allocation. Qie et al. [13] proposed an artificial intelligence method called simultaneous target assignment and path planning (STAPP) to solve the multi-UAV target assignment and path planning problem, and the effectiveness of the algorithm was experimentally verified. In addition, references [14–21] provide a variety of alternative algorithms for the solution of analogous problems.

Similarly, some novel works on task assignment, e.g., UAV-assisted task assignment, have been presented. Liu et al. [22] studied a UAV-assisted IoT system while presenting a nonconvex age-of-information (AoI) minimization problem, which was solved by jointly optimizing task assignment, interaction point selection (IPT), and UAV trajectories. Zhu et al. [23] considered the problem of task loss rate (TLR) fairness among IoTs and equal energy consumption (EC) fairness among UAVs, and proposed a multiagent deep deterministic policy gradient (MA-DDPG) method by which to assign UAVs to accomplish tasks and guarantee the balance between IoT TLR and UAV EC. Seid et al. [24] considered the assignment of UAVs to perform aerial base station tasks based on a multi-UAV-assisted IoT network framework, while presenting a joint optimization problem for computational offloading with energy harvesting (EH) and resource price, and the resource demands and pricing strategies between IoT devices and UAVs were continuously adjusted by the Stackelberg game. Hu et al. [25] considered the aging of cache refreshing, computation offloading, and state updates in UAV-assisted vehicle task awareness, and formulated a task-assignment energy-minimization problem that was solved by a deep deterministic policy gradient (DDPG) method. Zhou et al. [26] studied UAV-assisted mobile crowd sensing (MCS) scenarios and proposed a UAV-assisted multitasking assignment (UMA) method, while demonstrating the effectiveness of UMA. In addition, compared the UAV-assisted task assignment with the UAV task assignment, the difference is that UAVs play a secondary role in the former while serving as the primary reconnaissance and attack objects in the latter. Furthermore, the simulation scenarios in the paper are not consistent with the existing works (e.g., references [22–26]).

In the complex stochastic network, the cross-entropy (CE) method [27], a relatively new technique for dealing with combinatorial optimization problems, was initially utilized to estimate rare event probabilities. Then, references [28,29] discussed and analysed its convergence. Additionally, the cross-entropy (CE) method was proved by the authors in [30] to be particularly meaningful for handling combinatorial optimization problems. Since then, it has also been proven by many scholars to be a simple and effective tool for different fields, e.g., vehicle routing [31], buffer allocation [31], and machine learning [32]. In addition, researchers have also considered applying the cross-entropy (CE) method to the UAV task assignment [33–35]. However, the authors of these papers did not consider the specific precedence and timing constraints among these tasks.

When it comes to task-assignment schemes in the field of UAVs, some researchers usually assume that each UAV is assigned to only one target, and they rarely consider the execution sequence and the time constraints among tasks. On the other hand, multi-UAVs are sometimes needed to perform some complex combinatorial tasks, such as classifying the target, attacking it, and then verifying the target's damage level in a reasonable time on the battlefield. In addition, such deterministic approaches may not be able to find the optimal solution in a reasonable time for large-scale task assignment problems. Under these circumstances, we present an adjustable fully adaptive cross-entropy (AFACE) algorithm based on CE method.

Therefore, the purpose of this paper is to study the AFACE algorithm for the multi-UAV cooperative task assignment problem under resource, precedence, and timing constraints. The main contributions are summed up as follows.

- We consider the multi-UAV cooperative task assignment problem in which different types of UAVs are assigned to perform classification, attack and verification tasks of targets under resource, and precedence and timing constraints. Considering complex coupling among these tasks, we decompose the considered problem into two subproblems: one with continuous and independent tasks and another with continuous and correlative tasks.
- We propose an AFACE algorithm, which changes the random sample and the quantile at each iteration and adds a parameter to adjust the maximum sample based on the CE method. Meanwhile, the algorithm serves as a stepping stone for developing other algorithms.
- To overcome task precedence and task coupling existing in these two problems, respectively, we present a mutually independent AFACE (MIAFACE) algorithm and a mutually correlative AFACE (MCAFACE) algorithm with polynomial time complexity. The former algorithm converges faster than the CE method, while the computational complexity of the latter algorithm is inferior to that of the former algorithm.
- Simulation results demonstrate that both MIAFACE and MCAFACE algorithms consume less time than other existing optimization algorithms for solving the corresponding problem.

The rest of this paper is organized as follows. In Section 2, we introduce the related works of the CE method and other algorithms for the UAV task assignment. Section 3 depicts the multi-UAV cooperative task assignment problem with its mathematical formulation. In Section 4, we decompose the considered problem into two subproblems, and propose an AFACE algorithm, a MIAFACE algorithm, and a MCAFACE algorithm, and apply the latter two algorithms to solving the corresponding problem. Section 5 conducts several simulations and comparisons to verify the feasibility and effectiveness of the proposed algorithms. This paper is concluded in Section 6.

2. Related Work

This section reviews the related works on CE method and other algorithms used for UAV task assignment.

2.1. CE Method Used for UAV Task Assignment

Due to CE's merits, the authors of [33] first proposed using the CE method for tackling the multi-UAV task assignment problem to tackle the large traveling salesman problem (TSP), the vehicle routing problem (VRP), and Markov decision process (MDP). In particular, compared to other algorithms, CE could solve optimization problems efficiently because of its ability to deal with these problems with nonlinear objective functions. Three separate multi-UAV task assignment problems were then formulated, including a nonlinear objective function with distance penalty, a nonlinear objective function with no distance penalty, and nonlinear constraints. In these problems, the authors considered the distance penalty and required that each task must be assigned to at least one vehicle. Then, the task scores were considered as nonlinear functions, and the CE method was used to determine the optimal schemes for the functions of these problems. Finally, simulation results verified that the performance of the CE method was superior to other algorithms.

The authors in [34] considered the multi-UAV task assignment problem. Then, the score function of this problem was determined with the constraint that each UAV was used for only one task. Subsequently, the CE method was used to find the optimal scheme of this problem. Finally, simulation results showed that the CE method outperformed the Branch and Bound algorithm in solving the above problem, especially on a large scale.

Referring to [33,34], the authors in [35] described the multitype UAV task assignment problem. In this problem, different types of UAVs, or the same type of UAV as well as resource constraints, were considered. The authors then formulated the problem and provided a score function under resource constraints. Then, the CE method was used to determine the optimal scheme of this problem by assigning multitype UAVs to complete tasks. Finally, numerical simulations of the CE method for task assignment, as well as comparisons with the exhaust search method, were conducted to verify its merits in solving the considered problem.

In [36], the authors first analyzed the CE method, then redefined its construct and applied it to UAV swarms. Subsequently, due to the robustness of this method, it could be used as an effective measure to control UAV swarms in the face of obstacles and unforeseen problems. Finally, it was validated to support UAV swarms in achieving mission objectives.

The authors of [37] considered the multi-UAV task assignment problem under resource constraint and precedence constraint. The fully adaptive cross-entropy (FACE) algorithm based on the CE method was then applied to solve the considered problem. Then, simulation results verified that the FACE algorithm was better than the CE method and PSO algorithm in terms of convergence speed.

2.2. Other Algorithms Used for UAV Task Assignment

The authors in [8] improved the PSO algorithm with an inertia weight factor and applied it to handle the multi-UAV task assignment problem, then conducted several simulations and comparisons. Then, it was verified that the improved algorithm has a faster convergence speed and global optimization capability compared with the standard PSO algorithm.

In [38], the authors presented a novel hierarchical task assignment method to solve the multi-UAV task assignment problem, and the method was decomposed into two phases, including the hierarchical decomposition phase and the task assignment phase. The former phase reduced the computational complexity by using the balance cluster method to simplify the large-scale UAV model; the latter phase maintained the diversity of the population by an improved firefly algorithm. Then, simulations showed that compared with other algorithms, the proposed hierarchical method becomes more efficient in terms of search ability and convergence speed.

The authors in [39] defined the task assignment problem for cooperative multi-UAV road network reconnaissance and formulated a multi-UAV road network reconnaissance traveling salesman problem (MRRTSP) model. Furthermore, a customized genetic algorithm for road network reconnaissance (CGA-RNR) was proposed and used to solve the considered problem. Then, simulations showed that the algorithm can quickly obtain feasible solutions and converge to the optimal solution.

3. Problem Description and Formulation

The main parameters of this paper is shown in Table 1.

Problem Description

On the battlefield, multi-UAVs are deployed to perform different tasks, for example, to classify targets before attacking them, and then to verify them to check whether these tasks have been accomplished. The problem considered in this paper is the selection of a mix of the same type of UAV or different types of UAVs from their bases to perform the classification, attack and verification tasks of targets. As shown in Figure 1, there are N_b types of UAVs with the same speed, and the related components of this problem can be defined as a 5-tuple $\{\mathcal{A}, \mathcal{B}, \mathcal{G}, \mathcal{K}, \mathcal{T}\}$. In the 5-tuple, $\mathcal{A} := \{1, 2, \ldots, N_m\}$ denotes the set of task index of targets, $\mathcal{B} := \{1, 2, \ldots, N_b\}$ represents the set of N_b bases, $\mathcal{G} := \{1, 2, \ldots, N_t\}$ denotes the set of N_t targets with known positions, $\mathcal{K} := \{K_1, K_2, \ldots, K_{N_m}\}$ represents the set of N_m tasks of targets, and $\mathcal{T} := \{T_1, T_2, \ldots, T_{N_\mathrm{m}}\}$ denotes the set of the execution time of N_m tasks of targets. Note that the time required to allocate tasks is ignored.

Table 1. Simulation parameter settings.

Variables	Explanation
N_b	The number of bases
N_t	The number of targets
j	The target index
$\mathcal{K}$	The set of tasks of targets
N_m	The number of tasks of targets
m	The task index of targets
$\mathcal{X}$	The set of all possible UAV deployment schemes
L	The number of $\mathcal{X}$ for each task
z	The maximum number of UAVs in each scheme of $\mathcal{X}$
$\mathcal{Z}$	The set of all possible UAV deployment scheme indexes
k	A UAV deployment scheme index or a UAV formation index
$x(m)$	A feasible UAV deployment scheme vector of task m
$x_j(m)$	A feasible UAV deployment scheme or a UAV formation of task m of target j
$g(x_j(m); k)$	A 0–1 decision variable
$\mathcal{Y}$	The set of all feasible $x(m)$
$\Omega(x(m))$	The performance of task m
Ω	The performance vector of $\Omega(x(m))$
ρ	The total objective function
$\psi(x_j(2))$	The reward benefit of the attack task of target j
$\varphi(x_j(m))$	The cost of assigning UAV formation $x_j(m)$ to accomplish task m of target j
p_k^j	The probability of killing target j
p_s^j	The UAV survival probability of accomplishing tasks of target j
w_1, w_2 and w_3	Weight coefficients
P_c and V	The target identification certainty and the constant velocity of each UAV
y_j and s_j	The value and the threat level of target j
$d_j(m)$	The farthest distance from the base corresponding to UAV formation $x_j(m)$ to target j
$D_{\max}$	The maximum flying distance
T_m	The execution time of task m

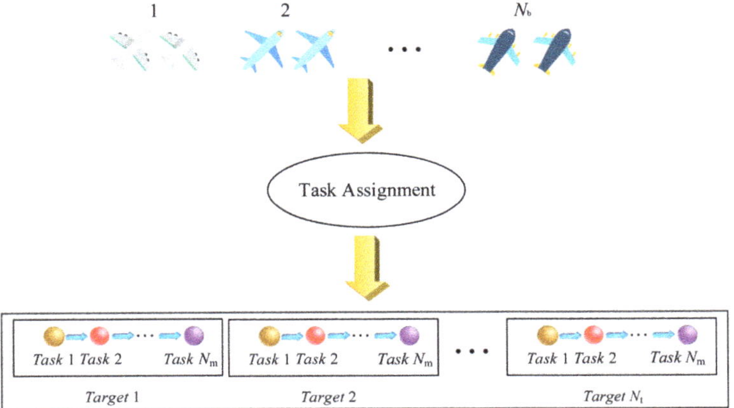

Figure 1. Task assignment diagram.

Moreover, let $x(m) = [x_1(m), x_2(m), \ldots, x_{N_t}(m)]^T$ denote a feasible UAV deployment scheme vector, and define $\mathcal{Y}$ as the set of all feasible $x(m)$. Let $\mathcal{Z} := \{1, 2, \ldots, L\}$ be the set of all possible UAV deployment scheme indices. Thus, $x(m)$ satisfies

$$g(x_j(m); k) = \begin{cases} 1, if\ \phi(x_j(m)) = k, j \in \mathcal{G}, m \in \mathcal{A}, k \in \mathcal{Z} \\ 0, if\ \phi(x_j(m)) \neq k, x_j(m) \in \mathcal{X} \end{cases}, \tag{1}$$

where j denotes the target index, m designates the task index, $x_j(m)$ is a feasible UAV deployment scheme or a UAV formation of task m of target j, k represents a UAV deployment scheme index or a UAV formation index, $\phi(x_j(m))$ is an index function that serves to output the subscript corresponding to $x_j(m)$ in $\mathcal{X}$, and $g(x_j(m); k)$ is a 0–1 decision variable, i.e., the kth UAV formation is assigned to accomplish task m of target j.

Then, the total objective function ρ based on $x(m)$ is defined as

$$\begin{aligned} \rho &= \sum_{m=1}^{N_m} \Omega(x(m)) \\ &= \sum_{j=1}^{N_t} \psi(x_j(2)) - \sum_{m=1}^{N_m} \sum_{j=1}^{N_t} \varphi(x_j(m)), \end{aligned} \tag{2}$$

where $\Omega(x(m))$ is the subobjective function of task m, $\psi(x_j(2))$ and $\varphi(x_j(m))$ denote the reward benefit of the attack task and the cost of assigning $x_j(m)$ to accomplish tasks of target j, respectively, which are

$$\psi(x_j(2)) = w_1 \times P_c \times p_k^j \times y_j \tag{3}$$

$$\varphi(x_j(m)) = w_2 \times p_s^j \times s_j + w_3 \times (V \times T_m + d_j(m)), \tag{4}$$

where P_c is the target identification certainty, y_j represents the value of target j, s_j denotes the threat level of target j, V is the constant velocity of each UAV, T_m represents the execution time of task m, $d_j(m)$ denotes the farthest distance from the bases corresponding to UAV formation $x_j(m)$ to target j, w_1, w_2 and w_3 represent weight coefficients, indicating the information about the relative importance of each subobjective, p_k^j denotes the probability of killing target j, and p_s^j is the UAV survival probability of accomplishing task m of target j.

In addition, p_k^j and p_s^j are defined as

$$p_k^j = \prod_{a \in x_j(2)} p_{aj} \tag{5}$$

$$p_s^j = 1 - \prod_{b \in x_j(m)} p_{bj}, \tag{6}$$

where a stands for a UAV in UAV formation $x_j(2)$; p_{aj} is the probability of killing target j with UAV a, b represents a UAV in UAV formation $x_j(m)$, and p_{bj} is the UAV survival probability of accomplishing task m of target j with UAV b.

According to Equations (2)–(6), ρ is rewritten as

$$\begin{aligned} \rho = \sum_{j=1}^{N_t} w_1 \times P_c \times p_k^j \times y_j - \sum_{m=1}^{N_m} \sum_{j=1}^{N_t} [w_2 \times p_s^j \times s_j \\ + w_3 \times (V \times T_m + d_j(m))]. \end{aligned} \tag{7}$$

Then, our objective is to maximize ρ, and the considered problem can be formulated as

$$\mathcal{P}: \max_{x(m) \in \mathcal{Y}} \rho = \sum_{m=1}^{N_m} \Omega(x(m)) \tag{8}$$

$$\text{s.t.} \quad w_1 + w_2 + w_3 = 1, 0 \leq w_1, w_2, w_3 \leq 1 \tag{9}$$

$$d_j(m) + V \times T_m \leq D_{\max} \quad \forall j, m \tag{10}$$

$$K_1^j \prec K_2^j \prec K_3^j \quad \forall j. \tag{11}$$

Constraint (9) represents the range of w_1, w_2, and w_3. Constraint (10) is that, for target j, the sum of $d_j(m)$ and the farthest flying distance performed by the UAV formation $x_j(m)$ does not exceed the maximum flying distance $D_{\max}$. Constraint (11) means that K_1^j, K_2^j, and K_3^j are the classification, attack, and verification tasks of the target j, which are executed in a specific order, and $\prec$ denotes the preceding symbol.

According to Equation (11), the specific precedence and timing constraints are equal to

$$\begin{cases} t_{s1}^j \geq s_1, e_1 \geq t_{s1}^j + T_1 \\ t_{s2}^j \geq s_2, e_2 \geq t_{s2}^j + T_2 \\ t_{s3}^j \geq s_3, e_3 \geq t_{s3}^j + T_3 \end{cases}, \tag{12}$$

where $[s_1, e_1]$, $[s_2, e_2]$, and $[s_3, e_3]$ represent the classification, attack, and verification time windows and t_{s1}^j, t_{s2}^j and t_{s3}^j denote the start time of classification, attack, and verification tasks of the target j, respectively.

Moreover, we set a certain value γ, which ensures that the optimal scheme vector $x^*(m)$ conforms to $\Omega(x^*(m)) \geq \gamma$. After that, the maximum ρ^* is written as

$$\rho^* = \sum_{m=1}^{N_m} \Omega(x^*(m)) \geq N_m \gamma. \tag{13}$$

Therefore, to obtain $x^*(m)$, we present an AFACE algorithm.

4. Algorithm Analysis

In this section, an AFACE algorithm will be introduced for the considered problem, and the differences between the algorithm and cross-entropy (CE) method are that the former changes the random sample N_d^t and the quantile θ_t at each iteration t, and adds a parameter to adjust the maximum sample $N^{\max}$. For details, please refer to the analysis of the algorithm below.

4.1. Adjustable Fully Adaptive Cross-Entropy Algorithm

Referring to the principle of CE method in references [30,35] and maximizing the subobjective function $\Omega(x(m))$ of the considered problem, we have

$$\gamma^* = \Omega(x^*(m)) = \max_{x(m) \in \mathcal{Y}} \Omega(x(m)), \tag{14}$$

where γ^* is the maximum of $\Omega(x(m))$ on $\mathcal{Y}$; that is, the optimal scheme vector is $x^*(m)$.

After that, transform this problem into a probability estimator problem, which can be explained by the probability density function (PDF) $f(\cdot; u)$ with respect to u, and the problem can be written as

$$\begin{aligned} \ell(\gamma) &= \mathbb{P}_u(\Omega(x(m)) \geq \gamma) \\ &= \sum_{x(m)} I_{\{\Omega(x(m)) \geq \gamma\}} f(x(m); u) \\ &= \mathbb{E}_u I_{\{\Omega(x(m)) \geq \gamma\}}, \end{aligned} \tag{15}$$

where γ denotes a value close to γ^*, $\mathbb{P}_u$ represents the probability measure under which the random vector $x(m)$ has the PDF $f(\cdot; u)$, $\mathbb{E}_u$ is the corresponding expectation operator, and $I(x(m); \gamma)$, i.e., $I_{\{\Omega(x(m)) \geq \gamma\}}$, denotes the indicator function, which is

$$I(\cdot; \gamma) = \begin{cases} 1, & if \ \Omega(x(m)) \geq \gamma \\ 0, & if \ \Omega(x(m)) < \gamma \end{cases}. \tag{16}$$

Then, at the tth iteration of AFACE algorithm, we obtain

$$\Omega_{t,1} \leq \cdots \Omega_{t,i} \leq \cdots \leq \Omega_{t,N_d^t}, \tag{17}$$

where $\Omega_{t,i}$ ($i = 1, 2, \ldots, N_d^t$) denotes the ith sample performance, and $\Omega(x_i(m)))$ and Ω_{t,N_d^t} are defined by $\Omega_{t,i}$ and Ω_t^* for convenience. Meanwhile, AFACE algorithm parameters N_d^t and θ_t satisfy

$$\begin{cases} N^{\min} \leq N_d^t \leq N^{\max} \\ \theta_t = \beta_m / N_d^t \end{cases}, \tag{18}$$

where N_d^t denotes the random sample of the tth iteration, varying between $N^{\min}$ and $N^{\max}$ ($N^{\min} = N, N^{\max} = hN, h \in \{2, 3, 4, 5\}$) and θ_t represents the quantile of the tth iteration. The reason for presenting h is that by adjusting the size of $N^{\max}$, we can obtain optimal $N^{\max}$ that matches the combat scenario, which can be conducted by the following simulations in Section 5.

For the AFACE algorithm, the main idea is to update N_d^t and θ_t based on the elite sample β_m ($\beta_m = c_m N$), where c_m and N are the elite sample influence coefficient of task m (usually $0.01 \leq c_m \leq 0.1$) and the fixed random sample, respectively. Therefore, the set of elite samples ε_t ($\varepsilon_t \in \mathcal{Y}$) are comprised of such β_m samples in $\{x_1(m), x_2(m), \ldots, x_{N_d^t}(m)\}$ with the highest performances $\Omega_{t,1}, \Omega_{t,2}, \ldots, \Omega_{t,N_d^t}$.

Next, referring to the formulas for solving $\widehat{\gamma}_t$ and $\widehat{v}_t$ of CE method [30], they are modified as

$$\widehat{\gamma}_t = \Omega_{(\lceil (1-\theta_t) N_d^t \rceil)} \tag{19}$$

$$\widehat{v}_t = \arg\max_v \sum_{x_i(m) \in \varepsilon_t} \ln f(x_i(m); v), \tag{20}$$

where $x_i(m)$ is generated from $f(\cdot; u)$, $f(\cdot; v)$ denotes another PDF with respect to v on $\mathcal{Y}$ via minimizing the Kullback–Leibler distance, $\widehat{\gamma}_t$ is equal to the worst sample performance among the elite performances, while Ω_t^* is the best sample performance among the elite performances, and $\widehat{v}_t$ converges to the probability density when Ω_t^* occurs.

Then, we devise a sampling scheme for each iteration t, ensuring high probability that

$$\begin{cases} \Omega_t^* > \Omega_{t-1}^* \\ \widehat{\gamma}_t > \widehat{\gamma}_{t-1} \end{cases}. \tag{21}$$

Moreover, we simultaneously generate two sequences to validate the correctness of AFACE algorithm. One is the levels $\widehat{\gamma}_1, \widehat{\gamma}_2, \ldots, \widehat{\gamma}_t$, and the other is the parameters $\widehat{v}_1, \widehat{v}_2, \ldots, \widehat{v}_t$. After that, the initialization process is set to $\widehat{v}_0 = u$, and the quantile $(1 - \theta_t)$ is calculated at the tth iteration according to Equation (18), followed by the next two steps of Algorithm 1.

In addition, the main steps of AFACE algorithm applied to solving the subobjective function $\Omega(x(m))$ of the considered problem are given by Algorithm 2.

Algorithm 1 Adaptive updating of $\widehat{\gamma}_t$ and $\widehat{v}_t$.

Adaptive updating of $\widehat{\gamma}_t$:
1: Given a fixed $\widehat{v}_{t-1}$ at the tth iteration;
2: Let γ_t be a $(1 - \theta_t)$-quantile of $\Omega(x(m))$ under $\widehat{v}_{t-1}$, then γ_t satisfies $\mathbb{P}_{\widehat{v}_{t-1}}(\Omega(x(m)) \leq \gamma_t) \geq 1 - \theta_t$, where $x(m) \sim f(\cdot; \widehat{v}_{t-1})$;
3: Obtain a simple estimator $\widehat{\gamma}_t$ of γ_t by drawing N_{d}^t random samples $x_1(m), x_2(m), \ldots, x_{N_{\mathrm{d}}^t}(m)$ from $f(\cdot; \widehat{v}_{t-1})$;
4: Calculate and order all performances of $\Omega(x(m))$ from smallest to biggest: $\Omega_{t,1} \leq \cdots \leq \Omega_{t,N_{\mathrm{d}}^t}$;
5: Compute $\widehat{\gamma}_t$ according to Equation (19);
Adaptive updating of $\widehat{v}_t$:
6: Given a fixed $\widehat{\gamma}_t$ and $\widehat{v}_{t-1}$ at the tth iteration, then derive $\widehat{v}_t$ according to Equation (20).

Algorithm 2 AFACE algorithm.

Input: $\widehat{v}_0$, h, N.
Output: Ω_t^*.

1: Set $t = 1$, $N^{\min} = N$ and $N^{\max} = hN$;
2: **while** at the tth iteration ($t \geq 1$) **do**
3:    **if** $t = 1$ **then**
4:       Generate N_{d}^t ($N_{\mathrm{d}}^t = N^{\min}$) random samples $x_1(m), x_2(m), \ldots, x_{N_{\mathrm{d}}^t}(m)$ from $f(\cdot; \widehat{v}_0)$;
5:       Calculate $\widehat{\gamma}_t$ and $\widehat{v}_t$ according to Equations (19) and (20);
6:    **else**
7:       Draw N_{d}^t ($N^{\min} \leq N_{\mathrm{d}}^t \leq N^{\max}$) random samples $x_1(m), x_2(m), \ldots, x_{N_{\mathrm{d}}^t}(m)$ from $f(\cdot; \widehat{v}_{t-1})$;
8:    **end if**
9:    Update $\widehat{\gamma}_t$ and $\widehat{v}_t$ according to Algorithm 1, then calculate Ω_t^*;
10:   **if** Equation (21) occurs **then**
11:      Set $t = t + 1$ and go to step 2;
12:   **else**
13:      Check whether or not $\Omega_t^* = \cdots = \Omega_{t-d}^*$ for some $t \geq d$, e.g., $d = 5$;
14:      **if** $\Omega_t^* = \cdots = \Omega_{t-d}^*$ **then**
15:       Stop, obtain Ω_t^* and **return** Ω_t^*;
16:      **else**
17:       Set $t = t + 1$, take random integer N_{d}^t in $[N^{\min}, N^{\max}]$ and go to step 2;
18:      **end if**
19:   **end if**
20: **end while**

4.2. Adjustable Fully Adaptive Cross-Entropy Algorithm for Solving Problem

Considering complex coupling among the three tasks, we decompose the considered problem $\mathcal{P}$ into two subproblems: the problem $\mathcal{P}1$ with continuous and independent tasks and the problem $\mathcal{P}2$ with continuous and correlative tasks.

Before discussing the algorithm for solving problem $\mathcal{P}$, we have to determine the number of the available schemes for each task. Please refer to Theorem 1 for the specific derivation process.

Theorem 1. *Assume that $z \geq 1$ and $N_{\mathrm{b}} = 3$, the number of the available schemes for each task of targets is L. Then, according to the mathematical formulas of permutation and combination, we can obtain*

$$L = 3z + \frac{z(z-1)(z+7)}{6}, z \geq 1. \tag{22}$$

Proof. Please see Appendix A. $\square$

4.2.1. Mutually Independent AFACE Algorithm for Solving Problem $\mathcal{P}1$

In problem $\mathcal{P}1$, assume that there are N_{m} continuous and mutually independent tasks for each target. Time continuity among these tasks then needs to be considered. Assume that there are L available schemes for each task, i.e., the scheme chosen by the previous task has no effect on the choice of the scheme for the next task, indicating that the available schemes among these tasks are independent. Thus, the problem $\mathcal{P}1$ is rewritten as

$$\mathcal{P}1: \max_{x(m) \in \mathcal{Y}} \rho = \sum_{m=1}^{N_m} \Omega(x(m))$$
$$\text{s.t.} \quad (9) - (12)$$
$$l_m^1 = L \quad \forall m \tag{23}$$

where l_m^1 is the available schemes when performing the mth task.

Considering time sequence and independence of the available schemes among these tasks, we present a MIAFACE algorithm, which is a combination of N_m AFACE algorithms. For MIAFACE algorithm, we first introduce the probability matrix vector $\boldsymbol{P} = [\boldsymbol{P}(1), \boldsymbol{P}(2), \dots, \boldsymbol{P}(N_m)]^T$ and the performance vector $\Omega = [\Omega(x(1)), \Omega(x(2)), \dots, \Omega(x(N_m))]^T$, where $\boldsymbol{P}(m)$ and $\Omega(x(m))$ are the probability matrix and the performance of task m, respectively. Then, $\boldsymbol{P}(m)$ is defined as

$$\boldsymbol{P}(m) = \begin{pmatrix} p(1|1,m) & p(2|1,m) & \cdots & p(l_m^1|1,m) \\ p(1|2,m) & p(2|2,m) & \cdots & p(l_m^1|2,m) \\ \vdots & \vdots & \ddots & \vdots \\ p(1|N_t,m) & p(2|N_t,m) & \cdots & p(l_m^1|N_t,m) \end{pmatrix}_{N_t \times l_m^1} ,$$

where $p(k|j,m)$ represents the probability of assigning the kth UAV formation to accomplish task m of target j and $\boldsymbol{P}(m)$ is subjected to $\sum_{k=1}^{l_m^1} p(k|j,m) = 1$.

Then, for the mth task, we initialize $\boldsymbol{P}_0(m) = (p_0(k|j,m))_{N_t \times l_m^1}$ with a uniform distribution. Let n_{jm}^1 be the number of the feasible schemes of target j, and define $p_0(k|j,m) := \frac{1}{n_{jm}^1}$ as the element of $\boldsymbol{P}_0(m)$. After that, we set $\widehat{v}_0 = \boldsymbol{P}_0(m)$.

At the tth iteration, we assume that the samples $x_1(m), x_2(m), \dots, x_{N_{d1}^t}(m)$ are drawn from $f(x(m); \widehat{v}_{t-1}(m))$. In addition, we calculate the performances $\Omega_{t,i}$ $(i = 1, 2, \dots, N_{d1}^t)$, and order them from smallest to largest: $\Omega_{t,1} \leq \Omega_{t,2} \leq \cdots \leq \Omega_{t,N_{d1}^t}$. It is noted that β_m^1 is calculated by $\beta_m^1 = c_m^1 N$, and $\widehat{\gamma}_t$ is updated by Equation (20). After that, we compare $\Omega_{t,i}$ with $\widehat{\gamma}_t$, and obtain all eligible performances greater than $\widehat{\gamma}_t$ and merge them into a set $\mathcal{S}_1 := \{\Omega_{(t,\lceil(1-\theta_t)N_{d1}^t\rceil)}, \Omega_{(t,\lceil(1-\theta_t)N_{d1}^t\rceil+1)}, \dots, \Omega_{t,N_{d1}^t}\}$, where β_m^1 is the number of the element of $\mathcal{S}_1$, and Ω_t^* is the maximum element of $\mathcal{S}_1$. Then, $p_t(k|j,m)$ is calculated, and the specific derivation process can be seen in Theorem 2. Thus, $\boldsymbol{P}_t(m)$ is the probability matrix composed of $p_t(k|j,m)$, and $\widehat{v}_t$ is equal to $\boldsymbol{P}_t(m)$.

Theorem 2. *Assume that there are N_m continuous and mutually independent tasks for each target. After that, N_m tasks correspond to N_m AFACE algorithms, which has an elite sample of $\beta_m^1 = c_m^1 N$. In the MIAFACE algorithm, $\boldsymbol{c}_1$ is a combined vector of c_m^1, e.g., $\boldsymbol{c}_1 = [c_1^1, c_2^1, \dots, c_{N_m}^1]^T$. Thus, when performing the mth task, we can then obtain the updating formula of $\boldsymbol{P}(m)$ as follows:*

$$\begin{cases} p(k|j,m) = \dfrac{\sum\limits_{n=1}^{c_m^1 N} g(x_j^n(m); k)}{c_m^1 N} \\ k \in \{1, \dots, L\}, n \in \{1, \dots, c_m^1 N\}, c_m^1 \in \boldsymbol{c}_1 \end{cases} . \tag{24}$$

Proof. Please see Appendix B. $\square$

Through the iterative updating of $\boldsymbol{P}(m)$, the optimal probability matrix vector $\boldsymbol{P}^*$ and the maximum performance vector Ω^* are obtained. Then, the main steps of the MIAFACE algorithm applied to solving problem $\mathcal{P}1$ are described in Algorithm 3, and the convergence of the MIAFACE algorithm is similar to that of the CE method in [40].

Algorithm 3 MIAFACE algorithm.

Input: N_{m}, l_m^1, N, h.
Output: P^*, Ω^*.

1: Set $N^{\min} = N$ and $N^{\max} = hN$;
2: **for** $m = 1$; $m < N_{\mathrm{m}}$; $m++$ **do**
3:     Initialize $P_0(m)$ with a uniform distribution and define $\widehat{v}_0 = P_0(m)$, then set $t = 1$;
4:     **while** at the tth iteration ($t \geq 1$) **do**
5:         **if** $t = 1$ **then**
6:             Generate N_{d1}^t ($N_{\mathrm{d1}}^t = N^{\min}$) random samples $x_1(m), x_2(m), \ldots, x_{N_{\mathrm{d1}}^t}(m)$ from $f(\cdot; \widehat{v}_0)$;
7:         **else**
8:             Draw N_{d1}^t ($N^{\min} \leq N_{\mathrm{d1}}^t \leq N^{\max}$) random samples $x_1(m), x_2(m), \ldots, x_{N_{\mathrm{d1}}^t}(m)$ from $f(\cdot; \widehat{v}_{t-1})$;
9:         **end if**
10:         Update $\widehat{\gamma}_t$ according to Equation (19) and calculate Ω_t^*;
11:         Calculate $p_t(k|j, m)$ by Equation (A14) in Appendix B;
12:         **if** $\sum_{k=1}^{l_m^1} p_t(k|j, m) = 1$ and $p_t(k|j, m) \in \{0, 1\}$ **then**
13:             Stop, obtain $P_t^*(m)$ and Ω_t^*, then $P^*(m) \leftarrow P_t^*(m)$ and $\Omega(x^*(m)) \leftarrow \Omega_t^*$;
14:         **else**
15:             Calculate $P_t(m)$ and update $\widehat{v}_t$ by $\widehat{v}_t = P_t(m)$;
16:             Set $t = t + 1$, take random integer N_{d1}^t in $[N^{\min}, N^{\max}]$, then go to step 4;
17:         **end if**
18:     **end while**
19: **end for**
20: **Return** P^* and Ω^*.

4.2.2. Mutually Correlative AFACE Algorithm for Solving Problem $\mathcal{P}2$

In problem $\mathcal{P}2$, assume that there are N_{m} continuous and mutually correlative tasks for each target. Then, time continuity among these tasks also needs to be considered. Assume that when performing the mth task, there are only $L - m + 1$ available schemes since $m - 1$ schemes have been deleted before performing the mth task. It means that the available schemes among these tasks are correlative. Thus, the problem $\mathcal{P}2$ is rewritten as

$$\mathcal{P}2: \max_{x(m) \in \mathcal{Y}} \rho = \sum_{m=1}^{N_{\mathrm{m}}} \Omega(x(m))$$
$$\text{s.t. } (9) - (12) \tag{25}$$
$$l_m^2 = L - m + 1 \quad \forall m$$

where l_m^2 is the remaining schemes when performing the mth task.

Considering time sequence and relevance of the available schemes among these tasks, we present a MCAFACE algorithm, which is also combined by N_{m} AFACE algorithms. For the MCAFACE algorithm, we first introduce the probability matrix vector $Q = [Q(1), Q(2), \ldots, Q(N_{\mathrm{m}})]^T$ and the performance vector $\Omega = [\Omega(x(1)), \Omega(x(2)), \ldots, \Omega(x(N_{\mathrm{m}}))]^T$, where $Q(m)$ and $\Omega(x(m))$ are the probability matrix and the performance of task m, respectively. Then, $Q(m)$ is defined as

$$Q(m) = \begin{pmatrix} q(1|1, m) & q(2|1, m) & \cdots & q(l_m^2|1, m) \\ q(1|2, m) & q(2|2, m) & \cdots & q(l_m^2|2, m) \\ \vdots & \vdots & \ddots & \vdots \\ q(1|N_{\mathrm{t}}, m) & q(2|N_{\mathrm{t}}, m) & \cdots & q(l_m^2|N_{\mathrm{t}}, m) \end{pmatrix}_{N_{\mathrm{t}} \times l_m^2},$$

where $q(k|j,m)$ represents the probability of assigning the kth UAV formation to accomplish task m of target j and $Q(m)$ is subjected to $\sum_{k=1}^{l_m^2} q(k|j,m) = 1$.

Then, for the mth task, we initialize $Q_0(m) = (q_0(k|j,m))_{N_t \times l_m^2}$ with a uniform distribution. Let n_{jm}^2 be the number of the feasible schemes of target j and define $q_0(k|j,m) := \frac{1}{n_{jm}^2}$ as the element of $Q_0(m)$. After that, we set $\widehat{v}_0 = Q_0(m)$.

At the tth iteration, we assume that the samples $x_1(m), x_2(m), \ldots, x_{N_{d2}^t}(m)$ are drawn from $f(x(m); \widehat{v}_{t-1}(m))$. In addition, we calculate the performances $\Omega_{t,i}$ $(i = 1, 2, \ldots, N_{d2}^t)$, and order them from smallest to largest: $\Omega_{t,1} \leq \Omega_{t,2} \leq \cdots \leq \Omega_{t,N_{d2}^t}$. It is noted that β_m^2 is calculated by $\beta_m^2 = c_m^2 N$, and $\widehat{\gamma}_t$ is updated by (20). After that, we compare $\Omega_{t,i}$ with $\widehat{\gamma}_t$, and obtain all eligible performances greater than $\widehat{\gamma}_t$ and merge them into a set $S_2 := \{\Omega_{(t,\lceil(1-\theta_t)N_{d2}^t\rceil)}, \Omega_{(t,\lceil(1-\theta_t)N_{d2}^t\rceil+1)}, \ldots, \Omega_{t,N_{d2}^t}\}$, where β_m^2 is the number of the element of S_2 and Ω_t^* is the maximum element of S_2. Then, $q_t(k|j,m)$ is calculated and the specific derivation process can be found in Theorem 3. Thus, $Q_t(m)$ is the probability matrix composed of $q_t(k|j,m)$, and $\widehat{v}_t$ is equivalent to $Q_t(m)$.

Theorem 3. *Assume that there are N_m continuous and mutually correlative tasks for each target. After that, the selected scheme is required to be deleted after each task is accomplished. The other settings are the same as Theorem 2. Thus, when performing the mth task, we can obtain the updating formula of $Q(m)$, as follows:*

$$\begin{cases} q(k|j,m) = \dfrac{\sum\limits_{n=1}^{c_m^2 N} g(x_j^n(m);k)}{c_m^2 N} \\ k \in \{1, \ldots, L-m+1\}, n \in \{1, \ldots, c_m^2 N\}, c_m^2 \in c_2 \end{cases} \tag{26}$$

Proof. Please see Appendix C. $\square$

Through the iterative updating of $Q(m)$, the optimal probability matrix vector Q^* and the maximum performance vector Ω^* are obtained. Then, the main steps of the MCAFACE algorithm for dealing with problem $\mathcal{P}2$ are explained in Algorithm 4, and the convergence of the MCAFACE algorithm is also close to that of CE method in [40].

Algorithm 4 MCAFACE algorithm.

Input: $N_{\mathrm{m}}, l^2_m, N, h$.

Output: Q^*, Ω^*.

1: Set $N^{\min} = N$ and $N^{\max} = hN$;
2: **for** $m = 1; m < N_{\mathrm{m}}; m++$ **do**
3: Initialize $Q_0(m)$ with a uniform distribution and define $\widehat{v}_0 = Q_0(m)$, then set $t = 1$;
4: **while** at the t-th iteration $(t \geq 1)$ **do**
5: **if** $t = 1$ **then**
6:   Generate N^t_{d2} $(N^t_{\mathrm{d2}} = N^{\min})$ random samples $x_1(m), x_2(m), \ldots, x_{N^t_{\mathrm{d2}}}(m)$ from $f(\cdot; \widehat{v}_0)$;
7: **else**
8:   Draw N^t_{d2} $(N^{\min} \leq N^t_{\mathrm{d2}} \leq N^{\max})$ random samples $x_1(m), x_2(m), \ldots, x_{N^t_{\mathrm{d2}}}(m)$ from $f(\cdot; \widehat{v}_{t-1})$;
9: **end if**
10: Update $\widehat{\gamma}_t$ according to Equation (19) and calculate Ω^*_t;
11: Calculate $q_t(k|j, m)$ by Equation (A15) in Appendix C;
12: **if** $\sum_{k=1}^{l^2_m} q_t(k|j, m) = 1$ and $q_t(k|j, m) \in \{0, 1\}$ **then**
13:   Stop, obtain $Q^*_t(m)$ and Ω^*_t, then $Q^*(m) \leftarrow Q^*_t(m)$ and $\Omega(x^*(m)) \leftarrow \Omega^*_t$;
14: **else**
15:   Calculate $Q_t(m)$ and update $\widehat{v}_t$ by $\widehat{v}_t = Q_t(m)$;
16:   Set $t = t + 1$, take random integer N^t_{d2} in $[N^{\min}, N^{\max}]$, then go to step 4;
17: **end if**
18: **end while**
19: **end for**
20: **Return** Q^* and Ω^*.

4.3. Complexity Analysis of the MIAFACE Algorithm and the MCAFACE Algorithm

Let N_{m} represent the number of tasks, n_{d} denote the random sample to perform each task, n_{f} represent the iteration number of AFACE algorithm to perform each task, N_{e} denote the elite sample, N_{t} represent the number of targets, and L denote the number of all possible UAV deployment schemes. The computational complexity of AFACE algorithm is divided into four parts: initialization C_1, sample C_2, sort C_3, and update C_4. Meanwhile, these parts can be defined as

$$C_1 = N_{\mathrm{t}} \times L \tag{27}$$

$$C_2 = n_{\mathrm{f}} \times n_{\mathrm{d}} \tag{28}$$

$$C_3 = n_{\mathrm{f}} \times n_{\mathrm{d}} \log n_{\mathrm{d}} \tag{29}$$

$$C_4 = n_{\mathrm{f}} \times (n_{\mathrm{d}} - N_{\mathrm{e}}). \tag{30}$$

Specifically, the computational complexity of AFACE algorithm can be written as

$$\begin{aligned} C_{\mathrm{f}} &= C_1 + C_2 + C_3 + C_4 \\ &= N_{\mathrm{t}} \times L + n_{\mathrm{f}} \times (n_{\mathrm{d}} + n_{\mathrm{d}} \log n_{\mathrm{d}} + n_{\mathrm{d}} - N_{\mathrm{e}}) \end{aligned} \tag{31}$$

Obviously, n_{f} increases with the increment of $(N_{\mathrm{t}} \times L)$, i.e., $n_{\mathrm{f}} \propto (N_{\mathrm{t}} \times L)$. Then, Equation (31) is rewritten as

$$\begin{aligned} C_{\mathrm{f}} &= N_{\mathrm{t}} \times L \times (1 + n_{\mathrm{d}} + n_{\mathrm{d}} \log n_{\mathrm{d}} + n_{\mathrm{d}} - N_{\mathrm{e}}) \\ &= N_{\mathrm{t}} \times L \times (n_{\mathrm{d}} \log n_{\mathrm{d}} + 2n_{\mathrm{d}} - N_{\mathrm{e}} + 1) \end{aligned} \tag{32}$$

where $n_{\mathrm{d}} \log n_{\mathrm{d}}$ is greater than the other terms in the bracket on the right side of the equation. Thus, the time complexity of AFACE algorithm can be computed as $\mathcal{O}(N_{\mathrm{t}} \times L \times n_{\mathrm{d}} \log n_{\mathrm{d}})$.

When the proposed algorithms are applied to accomplishing N_m tasks of targets in problems $\mathcal{P}1$ and $\mathcal{P}2$, respectively, according to Algorithm 3 and Equation (32), the computational complexity of MIAFACE algorithm is

$$C_{mi} = N_m \times C_f. \tag{33}$$

Thus, its time complexity is written as $\mathcal{O}(N_m \times N_t \times L \times n_d \log n_d)$. However, based on Algorithm 4 and Equation (32), the computational complexity of the MCAFACE algorithm is

$$
\begin{aligned}
C_{mc} &= N_t \times (L + L - 1 + \cdots + L - N_m + 1) \times (n_d \log n_d + 2n_d - N_e + 1) \\
&= N_t \times (N_m \times L - (1 + 2 + \cdots + N_m - 1)) \times (n_d \log n_d + 2n_d - N_e + 1) \\
&= N_t \times N_m \times (L - \frac{(N_m - 1)}{2}) \times (n_d \log n_d + 2n_d - N_e + 1).
\end{aligned}
\tag{34}
$$

Since $N_m \geq 3$, its time complexity is approximately equal to $\mathcal{O}(N_m \times N_t \times (L-1) \times n_d \log n_d)$.

5. Simulation and Analysis

In order to verify the effectiveness of the proposed algorithms, we compared these proposed algorithms with the CE method and other intelligent algorithms by applying them to the multi-UAV cooperative task assignment problem. The simulations were implemented in Pycharm Community's 2019.1.1 x64 version of the programming environment on an Intel Core PC with 8 GB memory. The total cumulative reward that the UAV formations earn by successfully completing three tasks from all targets are used to measure the system performance.

On the basis of the above algorithms, various simulations were performed by assigning three types of UAVs located in the corresponding bases to accomplish three tasks of 20 targets in a 200 m $\times$ 200 m combat scenario. The position of each base and these targets are shown in Figure 2. Bases B1, B2, and B3 are located in (0,0), (0,200), and (200,0), respectively. The information of three types of UAVs and 20 target are given in Tables 2 and 3, respectively, where a and b represent two types of resources, for example, the number of resources a and b needed for different types of UAVs or to accomplish different tasks, and also they have no units.

Table 2. Information of three types of UAVs.

UAV	Base	$U_{resource}$ (Units)		p_k	p_s
		a	b		
Type A	B1	1	2	0.9	0.7
Type B	B2	2	2	0.8	0.8
Type C	B3	3	3	0.7	0.9

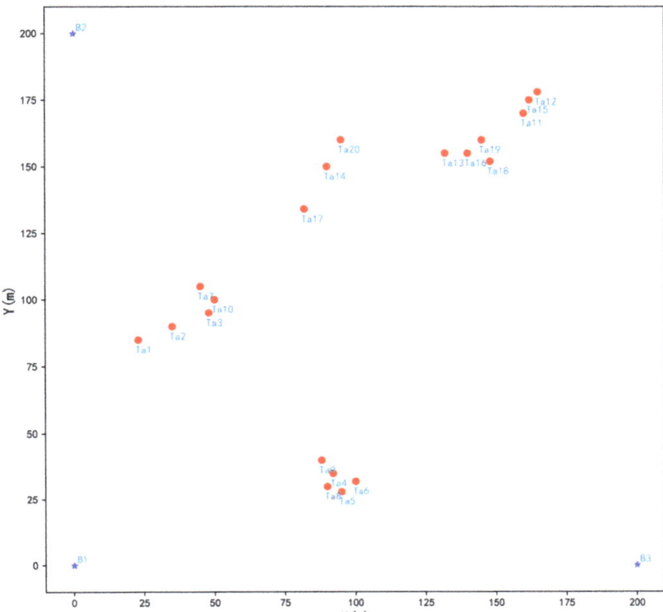

Figure 2. Initial bases and targets state.

Table 3. Information of 20 targets.

Target	Position	$T_{resource}$			y	s
		K_1 ([a,b])	K_2 ([a,b])	K_3 ([a,b])		
Target 1	(23,85)	[2,3]	[2,3]	[2,3]	30	2
Target 2	(35,90)	[3,3]	[3,3]	[3,3]	70	6
Target 3	(48,95)	[2,3]	[2,3]	[2,3]	50	4
Target 4	(92,35)	[3,2]	[3,2]	[3,2]	100	10
Target 5	(95,28)	[2,2]	[2,2]	[2,2]	120	8
Target 6	(100,32)	[3,2]	[3,2]	[3,2]	40	5
Target 7	(45,105)	[3,3]	[3,3]	[3,3]	65	3
Target 8	(90,30)	[2,2]	[2,2]	[2,2]	78	7
Target 9	(88,40)	[2,2]	[2,2]	[2,2]	35	9
Target 10	(50,100)	[3,2]	[3,2]	[3,2]	63	5
Target 11	(160,170)	[2,3]	[5,3]	[4,3]	30	2
Target 12	(165,178)	[3,3]	[5,3]	[5,3]	70	6
Target 13	(132,155)	[3,3]	[5,3]	[6,3]	50	4
Target 14	(90,150)	[3,3]	[3,5]	[4,4]	100	10
Target 15	(162,175)	[2,2]	[4,5]	[4,4]	120	8
Target 16	(140,155)	[2,3]	[6,3]	[5,3]	40	5
Target 17	(82,134)	[3,3]	[4,3]	[4,3]	65	3
Target 18	(148,152)	[2,3]	[4,2]	[5,2]	78	7
Target 19	(145,160)	[3,2]	[3,4]	[4,3]	35	9
Target 20	(95,160)	[2,2]	[3,4]	[4,4]	63	5

Referring to Theorem 1, we note that when z exceeds 3, these simulations are complicated. Thus, z is set to be 3, i.e., no more than 3 UAVs are needed to accomplish three tasks of targets in a specific order, and then the total number of each type of UAV is unrestricted. Then, each target in the following cases has 19 possible schemes, i.e., *A*, *B*, *C*, *AA*, *AB*, *AC*, *BB*, *BC*, *CC*, *AAA*, *AAB*, *AAC*, *ABB*, *ACC*, *BBB*, *BBC*, *BCC*, *CCC*, and

ABC, respectively, and these schemes correspond to numbers from 1 to 19. After that, we can use a matching approach to quickly find the feasible schemes. The resources needed to accomplish three tasks of targets are randomly generated and satisfy the maximum cooperative number of UAVs.

In the following simulations, the notations used in the tables and the figures are displayed as

- $U_{resource}$ represents the initial resources consumed by three types of UAVs;
- $T_{resource}$ represents the resources consumed by three tasks; and
- Time is CPU time in seconds for each case, and the time of each case is the average consumption time of running 100 times of each algorithm.

The parameters of the CE method, MIAFACE algorithm, MCAFACE algorithm, PSO algorithm, ACO algorithm, and GA algorithm are assumed to be set in Table 4, where the settings of the speed and maximum flying distance of the UAV are referred to [35] and they have no effect on the simulation results. For more detailed theory and parameter settings of CE, PSO, ACO, and GA (see [8–10,30,35,41]). For the targets in Table 3, there are two scenarios in the multi-UAV cooperative task assignment problem.

(1) In scenario 1, we consider the first 10 targets or more similar targets. When performing the three tasks of each target, we obtain the identical optimal scheme vector of each task. Therefore, the situation in which each target has different tasks but each task has the same optimal scheme is called the problem with continuous and independent tasks.

(2) In scenario 2, the last 10 targets or more similar targets are considered. When performing the three tasks of each target, we obtain the different optimal scheme vector of each task. Thus, the situation in which each target has different tasks and each task does not have the same optimal scheme is called the problem with continuous and correlative tasks.

Table 4. Simulation parameter settings.

Parameter	Value
The target identification	$P_c = 1$
Weight coefficients	$w_1 = 0.8, w_2 = 0.18, w_3 = 0.02$
The UAV's speed	$V = 40 \, \text{m/s}$
The maximum flying distance	$D_{max} = 1000 \, \text{m}$
Time window of task K_1 (s)	$[e_1, s_1] = [3, 10]$
Time window of task K_2 (s)	$[e_2, s_2] = [8, 20]$
Time window of task K_3 (s)	$[e_3, s_3] = [18, 26]$
Consumption time of task K_1	$T_1 = 5 \, \text{s}$
Consumption time of task K_2	$T_2 = 10 \, \text{s}$
Consumption time of task K_3	$T_3 = 5 \, \text{s}$
The number of targets	$N_t \in [5, 20]$
The fixed random samples	$N = 1000$
The quantile in CE	$\theta = 0.1$
Inertial weight in PSO	$w = 0.75$
Learning factors in PSO	$\eta_1 = \eta_2 = 0.5$
The number of ants in ACO	$N_a = 200$
Pheromone evaporation coefficient in ACO	$\varepsilon = 0.9$
Transfer probability in ACO	$P_a = 0.2$
Mating probability in GA	$P_1 = 0.8$
Mutation probability in GA	$P_2 = 0.01$

5.1. Scenario 1

In case 1, we used the first 10 targets in Table 3 to perform continuous and independent tasks of problem $\mathcal{P}1$, and the results are shown in Table 5.

According to Table 5, we note that the optimal scheme vector and the total result of CE and MIAFACE are identical, while that of MCAFACE is suboptimal to the other two algorithms. Moreover, we can obtain some observations. (i) For CE, the number of iterations and the optimal scheme vector are both 4 and [3,3,3,2,2,2,3,2,2,3], respectively, and the results of each task are -79.50, 274.90, and -79.50, and the sum of the results of each task is 115.9. The situations of MIAFACE are similar to CE, except that the number of iterations is 3. (ii) For MCAFACE, the numbers of iterations and the optimal scheme vectors are 3, 2, 1 and [3,3,3,2,2,2,3,2,2,3], [9,9,9,7,7,7,9,7,7,9], [18,18,18,15,15,15,18,15,15,18], respectively, and the results of each task are -79.50, 179.0, and -82.14, and the sum of the results of each task is 17.36. (iii) The total times of using CE, MIAFACE and MCAFACE are 3.36, 3.29, and 2.17, respectively.

In case 2, we tested the MIAFACE algorithm and MCAFACE algorithm under h and c_1, and their times change with N_t in Figures 3a–c and 4a–c, respectively.

From Figures 3 and 4, the curves of MIAFACE and MCAFACE both show an increasing trend as N_t grows, and their times increase with the increment of c_1 and h. Meanwhile, the time differences between the curves gradually increase with the growth of N_t in each figure. In Figure 3a, the curve with $h = 2$ is at the lowest of the four curves, while the curve with $h = 5$ is at the highest of the four curves. The remaining two curves are in the middle, and the curve with $h = 4$ is at the top and the other one is at the bottom. Moreover, the time ranges of the four curves are both approximately in [1,12]. In Figure 3b,c, their situations are described similarly to Figure 3a, and their time ranges are in [1,14] and [1,15], respectively. From Figure 4a, the order of the four curves is similar to Figure 3a. Moreover, their time ranges are both roughly in [0.3,10]. In Figure 4b,c, their situations are analogous to Figure 4a, and their time ranges are in [0.3,10] and [0.3,12], respectively.

In case 3, the CE method, PSO algorithm, ACO algorithm, and GA algorithm are both used three times for three tasks continuously. We compared them with MIAFACE algorithm by obtaining the same optimal score under $h = 2$ and c_1, and their times change with N_t in Figure 5a–c. Since MCAFACE algorithm obtains suboptimal results in scenario 1, it is not compared to other algorithms.

Table 5. Iterative results of three algorithms in case 1.

Algorithm	CE	MIAFACE	MCAFACE
Task	K_1 K_2 K_3	K_1 K_2 K_3	K_1 K_2 K_3
Iterations	4	3	3 2 1
Optimal scheme vector	[3,3,3,2,2,2,3,2,2,3]	[3,3,3,2,2,2,3,2,2,3]	[3,3,3,2,2,2,3,2,2,3] [9,9,9,7,7,7,9,7,7,9] [18,18,18,15,15,15,18,15,15,18]
Result of each task	-79.50 274.90 -79.50	-79.50 274.90 -79.50	-79.50 179.0 -82.14
Sum of each task's result	115.9	115.9	17.36
Total time(s)	3.36	3.29	2.17

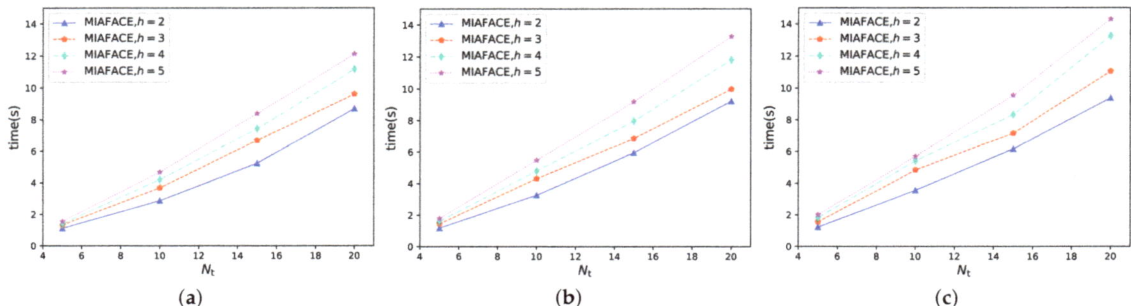

Figure 3. Time changing with N_t under MIAFACE algorithm in scenario 1. (**a**) $c_1 = [0.01, 0.02, 0.03]$. (**b**) $c_1 = [0.02, 0.03, 0.04]$. (**c**) $c_1 = [0.03, 0.04, 0.05]$.

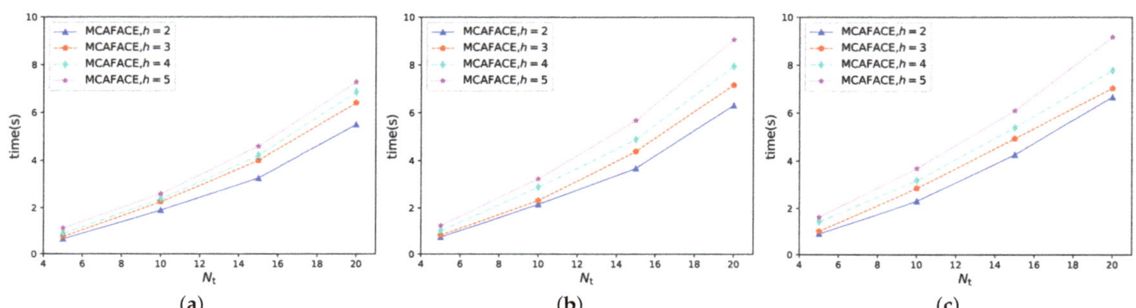

Figure 4. Time changing with N_t under MCAFACE algorithm in scenario 1. (**a**) $c_1 = [0.01, 0.02, 0.03]$. (**b**) $c_1 = [0.02, 0.03, 0.04]$. (**c**) $c_1 = [0.03, 0.04, 0.05]$.

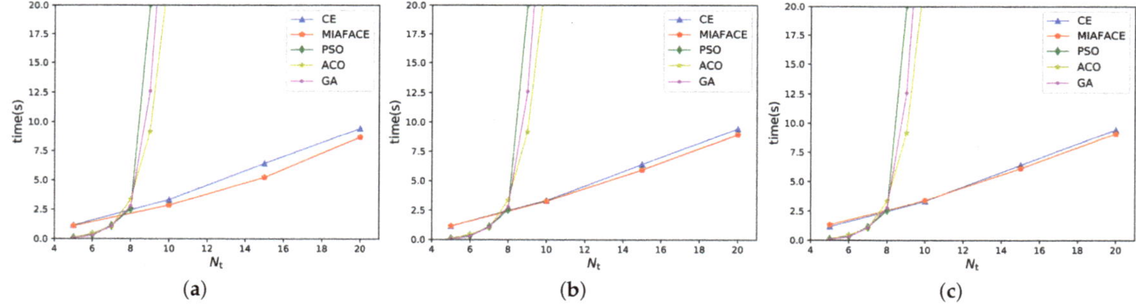

Figure 5. Time changing with N_t under $h = 2$ and different algorithms in scenario 1. (**a**) $c_1 = [0.01, 0.02, 0.03]$. (**b**) $c_1 = [0.02, 0.03, 0.04]$. (**c**) $c_1 = [0.03, 0.04, 0.05]$.

From Figure 5, we note that the curves of CE and MIAFACE grow linearly, while the curves of PSO, ACO, and GA increase exponentially. In addition, their times increase gradually with the increment of c_1 and N_t. In Figure 5a, when N_t is in [5,20], the time of MIAFACE is less than that of CE, and the time difference between the two algorithms grows as N_t increases. Meanwhile, when N_t is below 8, the times of PSO, ACO, and GA are lower than that of CE and MIAFACE, but when N_t is more than 8, the situation is reversed. In addition, the time ranges of CE and MIAFACE are both approximately in [1,10], while the times of other algorithms are over 20 when N_t is larger than 10. From Figure 5b,c, their situations are similar to Figure 5a, except that the time difference between CE and

MIAFACE in Figure 5b is lower than that in Figure 5a, and the time difference in Figure 5c first decreases gradually to intersect at a point where N_t is 10, then increases slowly with the increment of N_t.

5.2. Scenario 2

In case 4, we utilized the last 10 targets in Table 3 to perform continuous and correlative tasks of problem $\mathcal{P}2$, and the results are shown in Table 6.

According to Table 6, we note that the optimal scheme vectors and the total results of CE, MIAFACE, and MCAFACE are the same. The reason for this phenomenon is that for three tasks of the same 10 targets, the optimal scheme vectors are eventually obtained and identical by using the three algorithms, which leads to the same score of the total objective function; however, the consumption time by the different algorithms varies. Moreover, some observations are available. First, for CE, the number of iterations and the sum of each task's result are 5 and 218.72, and the optimal solution vectors are [3,3,3,3,3,3,3,3,3,3], [9,9,9,7,7,7,9,7,7,9], [18,18,18,15,15,15,18,15,15,18], and the results of each task are -298.65, 819.85, and -302.48, respectively. Secondly, the situations using MIAFACE and MCAFACE are similar to that of CE, apart from the fact that the number of iterations in MIAFACE is 4 and the numbers of iterations in MCAFACE are 4, 4, and 3. Finally, the total times using CE, MIAFACE, and MCAFACE are 7.33, 7.11, and 6.9, respectively.

In case 5, we tested the MIAFACE algorithm and MCAFACE algorithm under h and c_2, and their times change with N_t in Figures 6a–c and 7a–c, respectively.

From Figures 6 and 7, the variations of the curves, the times and the time differences are both similar to Figures 3 and 4, while in Figures 6a and 7a, the time grows rapidly when N_t is over 10. The reason is that the results of these two figures are suboptimal to others. In Figure 6a, the order of the curves is the same as that of each figure in Figures 3 and 4. In addition, the time ranges of these four curves are both approximately in [1,50]. From Figure 6b,c, the situations are described similarly to that of Figure 6a and their time ranges are in [2,30] and [2,32], except that their results are the optimal results. In Figure 7, the situation of each figure is roughly similar to that of the corresponding figure in Figure 6, apart from the fact that the time range is lower than that in Figure 6.

Table 6. Iterative results of three algorithms in case 4.

Algorithm	CE	MIAFACE	MCAFACE
Task	K_1 K_2 K_3	K_1 K_2 K_3	K_1 K_2 K_3
Iterations	5	4	4 4 3
Optimal scheme vector	[3,3,3,2,2,2,3,2,2,3] [9,9,9,7,7,7,9,7,7,9] [18,18,18,15,15,15,18,15,15,18]	[3,3,3,2,2,2,3,2,2,3] [9,9,9,7,7,7,9,7,7,9] [18,18,18,15,15,15,18,15,15,18]	[3,3,3,2,2,2,3,2,2,3] [9,9,9,7,7,7,9,7,7,9] [18,18,18,15,15,15,18,15,15,18]
Result of each task	-298.65 819.85 -302.48	-298.65 819.85 -302.48	-298.65 819.85 -302.48
Sum of each task's result	218.72	218.72	218.72
Total time(s)	7.33	7.11	6.9

In case 6, we compared the CE method, MIAFACE algorithm, MCAFACE algorithm, PSO algorithm, ACO algorithm, and GA algorithm by obtaining the same optimal score under $h = 2$ and c_2, and their times change with N_t in Figure 8a–c. The CE method, PSO algorithm, ACO algorithm, and GA algorithm are also used three times for three tasks continuously.

From Figure 8, we note that for CE, MIAFACE, MCAFACE, PSO, ACO, and GA, the variations of the curves and the times are similar to the case in Figure 5. In Figure 8a, when N_t is below 11, the times of MIAFACE and MCAFACE are relatively close and less than that of CE; however, when N_t is over 11, the times of MIAFACE and MCAFACE grow quickly and more than that of CE due to obtaining the suboptimal results. Moreover, the time ranges of CE, MIAFACE, and MCAFACE are both approximately in [1,30]. Meanwhile, the times of PSO, ACO, and GA are much higher than that of CE, MIAFACE, and MCAFACE, and their time ranges are over 30 when N_t is more than 6. From Figure 8b,c, the situations of PSO, ACO, and GA are similar to Figure 8a. In Figure 8b, the time differences between CE, MIAFACE, and MCAFACE grow as N_t increases. In addition, the time of MCAFACE is lower than that of CE and MIAFACE, and the curves of CE and MIAFACE intersect at $N_t = 8$ and the time of CE is also lower than that of MIAFACE when N_t is below 8, then the situation is reversed after N_t exceeds 8. Moreover, the time ranges of CE, MIAFACE, and MCAFACE are both in [2,22]. In Figure 8c, the time differences between CE, MIAFACE, and MCAFACE decrease, and then increase as N_t grows. Furthermore, the curves of CE, MIAFACE, and MCAFACE intersect at $N_t = 9$ and the time of CE is lower than that of MIAFACE and MCAFACE when N_t is below 9, then the situation is reversed after N_t exceeds 9.

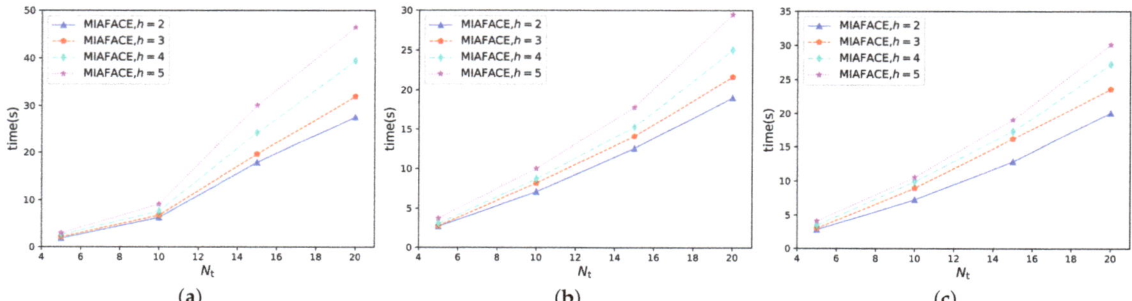

Figure 6. Time changing with N_t under MIAFACE algorithm in scenario 2. (**a**) $c_2 = [0.01,0.02,0.03]$. (**b**) $c_2 = [0.02,0.03,0.04]$. (**c**) $c_2 = [0.03,0.04,0.05]$.

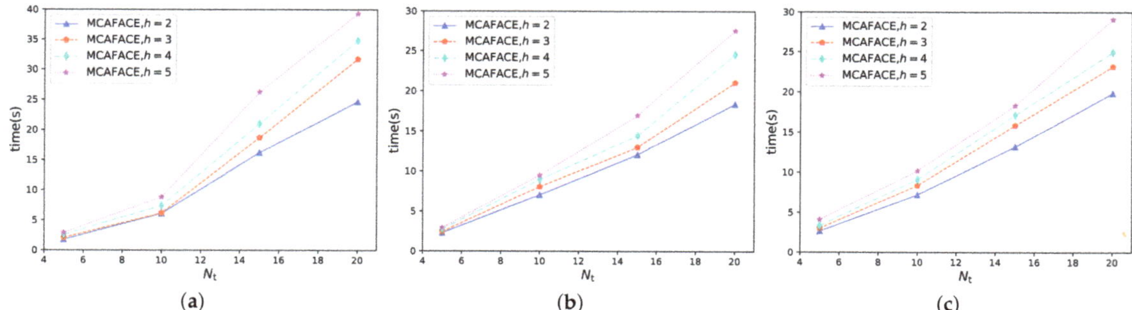

Figure 7. Time changing with N_t under MCAFACE algorithm in scenario 2. (**a**) $c_2 = [0.01,0.02,0.03]$. (**b**) $c_2 = [0.02,0.03,0.04]$. (**c**) $c_2 = [0.03,0.04,0.05]$.

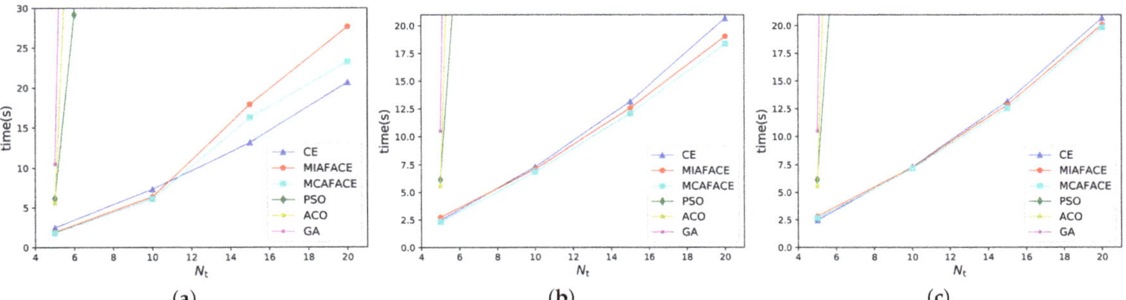

Figure 8. Time changing with N_t under $h = 2$ and different algorithms in scenario 2. (**a**) $c_2 = [0.01, 0.02, 0.03]$. (**b**) $c_2 = [0.02, 0.03, 0.04]$. (**c**) $c_2 = [0.03, 0.04, 0.05]$.

5.3. Analysis

Analysing the results of case 1 and case 4, we note that the optimal scheme vectors of using MIAFACE and MCAFACE algorithms in problems $\mathcal{P}1$ and $\mathcal{P}2$, respectively, are obtained by initializing and updating the probability matrices $\boldsymbol{P}(m)$ and $\boldsymbol{Q}(m)$, which conforms to Algorithms 3 and 4 described in Section 4.2. In addition, the result of MCAFACE in case 1 is suboptimal to that of other algorithms due to deleting the corresponding optimal solution after the end of each task.

Comprehensively considering the situations of case 2 and case 5, we note that the times of CE, MIAFACE, and MCAFACE increase with the increment of N_t, h, as well as $\boldsymbol{c}$ and the time complexity of MCAFACE is lower than that of MIAFACE, and these phenomena comply with the complexity analysis of MIAFACE and MCAFACE in Section 4.3. In addition, the time of case 5 is superior to that of case 2 because there are more available solutions for each target in case 5 than in case 2 after each iteration. Meanwhile, in case 5, using MIAFACE and MCAFACE for solving this problem is easy to fall into local optimum when $\boldsymbol{c}$ is inferior to a certain vector, e.g., $\boldsymbol{c} = [0.01, 0.02, 0.03]$. The reason behind this phenomenon is that when all elements in $\boldsymbol{c}$ are small and more solutions exist after each iteration, the optimal scheme may not be selected during one of the iterations of MIAFACE and MCAFACE, leading to a suboptimal result.

Comparing the situations of case 3 and case 6, we note that the times of PSO, ACO, and GA are only related to the growth of N_t. Meanwhile, CE, MIAFACE, and MCAFACE are superior to PSO, ACO, and GA for large-scale allocation problems, e.g., more than 8 targets of case 3 and 5 targets of case 6. Moreover, CE is inferior to MIAFACE in scenario 1, e.g., Figure 5, when N_t is over 10. Moreover, e.g., Figure 8c in scenario 2, MCAFACE is superior to MIAFACE and CE when N_t is over 9.

6. Conclusions

In this paper, the multi-UAV cooperative task assignment problem was described and formulated, and three types of UAVs were considered, cooperatively accomplishing the classification, attack, and verification tasks of targets under resource, precedence, and timing constraints. After that, considering complex coupling among these three tasks, we decomposed the considered problem into two subproblems. In order to solve them, we proposed an AFACE algorithm, a MIAFACE algorithm, and a MCAFACE algorithm. Finally, simulation results verified that both MIAFACE and MCAFACE consume less time than other intelligent algorithms for solving the corresponding problem.

Nevertheless, there still exist challenges when applying the MIAFACE algorithm and MCAFACE algorithm to processing optimization problems, e.g., appropriate parameter settings, falling into local optimum when using lower elements in $\boldsymbol{c}$, etc. In future work, it will be meaningful to concentrate on promoting these two algorithms on problems where it is vulnerable to local optimum when the number of samples is limited and on task assignment problems in complex dynamic scenarios.

Author Contributions: Conceptualization, K.W., X.Z. and X.L.; methodology, X.Z.; software, X.Z.; validation, K.W., X.Z., X.L. and W.C.; formal analysis, K.W. and X.Z.; investigation, X.Z.; resources, X.Q. and Y.C.; data curation, X.Q., Y.C. and K.L.; writing—original draft preparation, X.Z.; writing—review and editing, X.Z.; visualization, X.Z.; supervision, K.W.; project administration, Y.C. and K.L. All authors have read and agreed to the published version of the manuscript.

Funding: This research was funded in part by the National Natural Science Foundation of China under Grant 62172313 and 52031009, in part by the Natural Science Foundation of Hunan Province under Grant 2021JJ20054.

Data Availability Statement: Data sharing is not applied.

Conflicts of Interest: The authors declare no conflict of interest.

Appendix A

If one of each type of UAVs is selected, i.e., $z = 1$, the possible schemes are written as

$$L = C_{N_b}^1 C_z^1 = 3. \tag{A1}$$

When $z = 2$, we can choose no more than three types of UAVs, and then

$$L = C_{N_b}^1 C_z^1 + C_{N_b}^2 C_z^2 = 9. \tag{A2}$$

Once $z \geq 3$, we can choose no more than three types of UAVs; thus

$$\begin{aligned}
L &= \underbrace{C_{N_b}^1 C_z^1}_{1\ type} + \underbrace{C_{N_b}^2 C_z^2}_{2\ types} + \underbrace{C_{N_b}^3 C_z^3}_{3\ types} \\
&= 3z + \frac{3z(z-1)}{2} + \frac{z(z-1)(z-2)}{6}. \\
&= 3z + \frac{z(z-1)(z+7)}{6}
\end{aligned} \tag{A3}$$

As a conclusion, the number of the possible schemes for each task of targets can be defined as

$$L = 3z + \frac{z(z-1)(z+7)}{6}, z \geq 1. \tag{A4}$$

Thus, we have successfully proven Theorem 1.

Appendix B

Inserting $P(m)$ and Equation (1) into $f(x(m); u)$, we define the problem $\mathcal{P}1$ as

$$\begin{aligned}
f(x(m); P(m)) &= \prod_{j=1}^{N_t} p(x_j(m)|j, m) \\
&= \prod_{j=1}^{N_t} \prod_{k=1}^{l_m^1} p(k|j, m)^{g(x_j(m);k)}
\end{aligned} \tag{A5}$$

where $p(k|j, m)$ represents the coefficient in the column k and the row j of $P(m)$, $g(x_j(m); k)$ is 1 if $\phi(x_j(m))$ equals k and 0 otherwise according to Equation (1).

After that, at the tth iteration, we assume that the samples $x_1(m), x_2(m), \ldots, x_{N_{d1}^t}(m)$ are drawn from $f(x(m); \widehat{v}_{t-1}(m))$. In addition, we calculate the performances $\Omega_{t,i}$, and order them from smallest to largest: $\Omega_{t,1} \leq \Omega_{t,2} \leq \cdots \leq \Omega_{t,N_{d1}^t}$, and then define $\widehat{\gamma}_t(m) = \Omega_{(N_{d1}^t - \beta_m^1)}$.

Thus, Equation (A5) can be rewritten as follows:

$$\arg\max_{P(m)} \frac{1}{N_{\mathrm{d1}}^t} \sum_{i=1}^{N_{\mathrm{d1}}^t} I_{\{\Omega(x_i(m)) \geq \widehat{\gamma}_t(m)\}} \times \ln f(x_i(m); P(m)) \tag{A6}$$

In Equation (A6), $I(x_i(m); \widehat{\gamma}_t(m))$ is recognized, and when $N_{\mathrm{d1}}^t \to \infty$, the problem is equal to

$$\max_{P(m)} \sum_{n=1}^{\beta_m^1} \ln f(x(m); P(m)). \tag{A7}$$

Putting Equation (A5) into Equation (A7), we have

$$\begin{aligned}
&\max_{P(m)} \sum_{n=1}^{\beta_m^1} \ln f(x(m); P(m)) \\
&= \max_{p(k|j,m)} \sum_{n=1}^{\beta_m^1} \ln \left(\prod_{j=1}^{N_t} \prod_{k=1}^{l_m^1} p(k|j,m)^{g(x_j^n(m);k)} \right) \\
&= \max_{p(k|j,m)} \sum_{n=1}^{\beta_m^1} \sum_{j=1}^{N_t} \sum_{k=1}^{l_m^1} g(x_j^n(m);k) \ln(p(k|j,m)).
\end{aligned} \tag{A8}$$

Then, we assume that $r_{kj}(m) = p(k|j,m)$, $a_{kj}^n(m) = g(x_j^n(m);k)$, and Equation (A8) is modeled as

$$\begin{aligned}
\mathcal{P}11: \min_{r_{kj}(m)} \Big(- \sum_{n=1}^{\beta_m^1} \sum_{j=1}^{N_t} \sum_{k=1}^{l_m^1} a_{kj}^n(m) \ln\big(r_{kj}(m) \big) \Big) & \\
\text{s.t. } \sum_{k=1}^{l_m^1} r_{kj}(m) = 1 \quad \forall j, m & \\
r_{kj}(m) \geq 0 \quad \forall j, k, m & \\
l_m^1 = L \quad \forall m. &
\end{aligned} \tag{A9}$$

Considering $\mathcal{P}11$ as a convex problem and denoting the convex function by $f(r_{kj}(m))$, we can obtain the Lagrangian function

$$\begin{aligned}
O(r_{kj}(m), \lambda_j(m), \mu_{kj}(m)) = f\big(r_{kj}(m) \big) + \\
\sum_{j=1}^{N_t} \lambda_j(m) \left(\sum_{k=1}^{L} r_{kj}(m) - 1 \right) + \sum_{j=1}^{N_t} \sum_{k=1}^{L} \mu_{kj}(m) \big(-r_{kj}(m) \big),
\end{aligned} \tag{A10}$$

where $\lambda_j(m)$ and $\mu_{kj}(m)$ are the relevant restraint coefficients.

Generally, for convex optimization problem, the Karush–Kuhn–Tucker (KKT) condition is required and sufficient [42]. Thus, considering the KKT conditions of problem in Equation (A10), we have

$$\begin{cases}
-\dfrac{a_{kj}(m)}{r_{kj}(m)} + \lambda_j(m) - \mu_{kj}(m) = 0 \\
\lambda_j(m) \left(\sum\limits_{k=1}^{L} r_{kj}(m) - 1 \right) = 0 \\
\mu_{kj}(m) r_{kj}(m) = 0 \\
\lambda_j(m) > 0 \\
\mu_{kj}(m) \geq 0 \\
r_{kj}(m) \geq 0
\end{cases} \tag{A11}$$

When solving Equation (A11), we obtain

$$\begin{cases} r_{kj}(m) = \dfrac{a^n_{kj}(m)}{\lambda_j(m) - \mu_{kj}(m)} \\[2mm] \lambda_j(m) = \displaystyle\sum_{k=1}^{L} a^n_{kj}(m) \\[2mm] \mu_{kj}(m) = 0 \end{cases} \qquad (A12)$$

Comparing $\lambda_j(m)$ and $r_{kj}(m)$, we acquire the relationship between $r_{kj}(m)$ and $a^n_{kj}(m)$, i.e.,

$$r_{kj}(m) = \frac{a^n_{kj}(m)}{\displaystyle\sum_{k=1}^{L} a^n_{kj}(m)}. \qquad (A13)$$

Returning to our problem, the updating formula of $P(m)$ is given by

$$\begin{aligned} p(k|j,m) &= \frac{\displaystyle\sum_{n=1}^{\beta^1_m} g(x^n_j(m);k)}{\beta^1_m} \\[3mm] &= \frac{\displaystyle\sum_{n=1}^{c^1_m N} g(x^n_j(m);k)}{c^1_m N}, \end{aligned} \qquad (A14)$$

where $k \in \{1,\ldots,L\}$, $n \in \{1,\ldots,c^1_m N\}$, $c^1_m \in c_1$.

Therefore, we have successfully proven Theorem 2.

Appendix C

Calculating the updating formulas of N_m tasks continuously and correlatively is considered.

For the mth task, if $m = 1$, its optimal solution is taken from L schemes, and if $1 < m \le N_m$, its optimal scheme is only taken from the remaining $L - m + 1$ solutions since the $m - 1$ schemes selected before performing the mth task have been deleted.

Thus, referring to the proof process of Theorem 2, the updating formula of $Q(m)$ in problem $\mathcal{P}2$ is

$$\begin{aligned} q(k|j,m) &= \frac{\displaystyle\sum_{n=1}^{\beta^2_m} g(x^n_j(m);k)}{\beta^2_m} \\[3mm] &= \frac{\displaystyle\sum_{n=1}^{c^2_m N} g(x^n_j(m);k)}{c^2_m N}, \end{aligned} \qquad (A15)$$

where $k \in \{1,\ldots,L - m + 1\}$, $n \in \{1,\ldots,c^2_m N\}$, $c^2_m \in c_2$.

Hence, we have successfully proven Theorem 3.

References

1. Singh, H.; Sharma, M. Electronic Warfare System Using Anti-Radar UAV. In Proceedings of the 2021 8th International Conference on Signal Processing and Integrated Networks, Noida, India, 26–27 August 2021; pp. 102–107.
2. Deng, Z.; Gao, Y.; Hu, A.; Zhang, Y. A Mobile Phone Uplink CPDP-DTDOA Positioning Method Using UAVs for Search and Rescue. *IEEE Sens. J.* **2022**, *22*, 18170–18179. [CrossRef]
3. Fan, B.; Jiang, L.; Chen, Y.; Zhang, Y.; Wu, Y. UAV Assisted Traffic Offloading in Air Ground Integrated Networks With Mixed User Traffic. *IEEE T. Intell. Transp.* **2022**, *23*, 12601–12611. [CrossRef]
4. D'Arcy, S.; Gonzalez, F. Design and Flight Testing of a Rocket-Launched Folding UAV for Earth and Planetary Exploration Applications. In Proceedings of the 2022 IEEE Aerospace Conference, Big Sky, MT, USA, 5–12 March 2022; pp. 1–15.

5. Chen, X.; Liu, Y.; Yin, L.; Qi, Y. Cooperative Task Assignment and Track Planning For Multi-UAV Attack Mobile Targets. *J. Intell. Robot. Syst.* **2020**, *100*, 1383–1400.
6. Sabo, C.; Kingston, D.; Cohen, K. A Formulation and Heuristic Approach to Task Allocation and Routing of UAVs under Limited Communication. *Unmanned Syst.* **2014**, *2*, 1–17. [CrossRef]
7. Wang, J.; Zhang, Y.F.; Geng, L.; Fuh, J.Y.H.; Teo, S.H. A Heuristic Mission Planning Algorithm for Heterogeneous Tasks with Heterogeneous UAVs. *Unmanned Syst.* **2015**, *3*, 205–219. [CrossRef]
8. Gou, Q.; Li, Q. Task assignment based on PSO algorithm based on Logistic function inertia weight adaptive adjustment. In Proceedings of the 2020 3rd International Conference on Unmanned Systems, Harbin, China, 1–4 September 2020; pp. 825–829.
9. Li, Y.; Zhang, S.; Chen, J.; Jiang, T.; Ye, F. Multi-UAV Cooperative Mission Assignment Algorithm Based on ACO method. In Proceedings of the 2020 International Conference on Computing, Networking and Communications, Big Island, HI, USA, 17–20 February 2020; pp. 304–308.
10. Ma, Y.; Zhang, H.; Zhang, Y.; Gao, R.; Xu, Z.; Yang, J. Coordinated Optimization Algorithm Combining GA with Cluster for Multi-UAVs to Multi-tasks Task Assignment and Path Planning. In Proceedings of the 2019 IEEE 15th International Conference on Control and Automation, Edinburgh, UK, 22–26 August 2019; pp. 1026–1031.
11. Duan, X.; Liu, H.; Tang, H.; Cai, Q.; Zhang, F.; Han, X. A Novel Hybrid Auction Algorithm for Multi-UAVs Dynamic Task Assignment. *IEEE Access* **2020**, *8*, 86207–86222. [CrossRef]
12. Chen, J.; Wu, Q.; Xu, Y.; Qi, N.; Guan, X.; Zhang, Y.; Xue, Z. Joint Task Assignment and Spectrum Allocation in Heterogeneous UAV Communication Networks: A Coalition Formation Game-Theoretic Approach. *IEEE Trans. Wirel. Commun.* **2021**, *20*, 440–452. [CrossRef]
13. Qie, H.; Shi, D.; Shen, T.; Xu, X.; Li, Y.; Wang, L. Joint Optimization of Multi-UAV Target Assignment and Path Planning Based on Multi-Agent Reinforcement Learning. *IEEE Access* **2019**, *7*, 146264–146272. [CrossRef]
14. Tang, J.; Chen, X.; Zhu, X.; Zhu, F. Dynamic Reallocation Model of Multiple Unmanned Aerial Vehicle Tasks in Emergent Adjustment Scenarios. *IEEE Trans. Aerosp. Electron. Syst.* **2022**, 1–43. [CrossRef]
15. Qie, H.; Shi, D.; Shen, T.; Xu, X.; Li, Y.; Wang, L. Distributed Cooperative Search Algorithm With Task Assignment and Receding Horizon Predictive Control for Multiple Unmanned Aerial Vehicles. *IEEE Access* **2021**, *9*, 6122–6136.
16. Fu, X.; Feng, P.; Gao, X. Swarm UAVs Task and Resource Dynamic Assignment Algorithm Based on Task Sequence Mechanism. *IEEE Access* **2019**, *7*, 41090–41100. [CrossRef]
17. Chen, Y.; Yang, D.; Yu, J. Multi-UAV Task Assignment With Parameter and Time-Sensitive Uncertainties Using Modified Two-Part Wolf Pack Search Algorithm. *IEEE Trans. Aerosp. Electron. Syst.* **2018**, *54*, 2853–2872. [CrossRef]
18. Zhu, F.; Wu, F.; Chen, C.F.; Li, D.; Guo, Y.; Zhang, J.G.; Zhao, X. A coordinated assignment method for multi-UAV area search tasks. In Proceedings of the CSAA/IET International Conference on Aircraft Utility Systems, Nanchang, China, 17–20 August 2022; pp. 751–756.
19. Chen, Y.; Chen, J.; Du, C. Allocation of Multi-UAVs Timing-dependent Tasks based on Completion Time. In Proceedings of the 2022 WRC Symposium on Advanced Robotics and Automation, Beijing, China, 20 August 2022; pp. 71–76.
20. Yan, S.; Xu, J.; Song, L.; Pan, F. Heterogeneous UAV collaborative task assignment based on extended CBBA algorithm. In Proceedings of the 2022 7th International Conference on Computer and Communication Systems, Wuhan, China, 22–25 April 2022; pp. 825–829.
21. Yan, S.; Pan, F.; Zhang, D.; Xu, J. Research on Task Reassignment Method of Heterogeneous UAV in Dynamic Environment. In Proceedings of the 2022 6th International Conference on Robotics and Automation Sciences, Wuhan, China, 9–11 June 2022; pp. 57–61.
22. Liu, C.; Guo, Y.; Li, N.; Song, X. AoI-Minimal Task Assignment and Trajectory Optimization in Multi-UAV-Assisted IoT Networks. *IEEE Internet Things J.* **2022**, *9*, 21777–21791. [CrossRef]
23. Zhu, C.; Zhang, G.; Yang, K. Fairness-Aware Task Loss Rate Minimization for Multi-UAV Enabled Mobile Edge Computing. *IEEE Wirel. Commun. Lett.* **2023**, *12*, 94–98. [CrossRef]
24. Seid, A.M.; Lu, J.; Abishu, H.N.; Ayall, T.A. Blockchain-Enabled Task Offloading with Energy Harvesting in Multi-UAV-assisted IoT Networks: A Multi-agent DRL Approach. *IEEE J. Sel. Areas Commun.* **2022**, *40*, 3517–3532. [CrossRef]
25. Hu, N.; Qin, X.; Ma, N.; Liu, Y.; Yao, Y.; Zhang, P. Energy-efficient Caching and Task offloading for Timely Status Updates in UAV-assisted VANETs. In Proceedings of the 2022 IEEE/CIC International Conference on Communications in China, Sanshui, Foshan, China, 11–13 August 2022; pp. 1032–1037.
26. Gao, H.; Feng, J.; Xiao, Y.; Zhang, B.; Wang, W. A UAV-assisted Multi-task Allocation Method for Mobile Crowd Sensing. *IEEE Trans. Mob. Comput.* **2022**. [CrossRef]
27. Rubinstein, R.Y. Optimization of computer simulation models with rare events. *Eur. J. Oper. Res.* **1997**, *99*, 89–112. [CrossRef]
28. Rubinstein, R.Y. The cross-entropy method for combinatorial and continuous optimization. *Methodol. Comput. Appl. Probab.* **1999**, *1*, 127–190. [CrossRef]
29. Rubinstein, R.Y. Combinatorial optimization, cross-entropy, ants and rare events. In *Stochastic Optimization: Algorithms and Applications*; Springer: Boston, MA, USA, 2001; pp. 303–363.
30. De Boer, P.-T.; Kroese, D.P.; Mannor, S.; Rubinstein, R.Y. A tutorial on the cross-entropy method. *Ann. Oper. Res.* **2005**, *134*, 19–67. [CrossRef]

31. Chepuri, K.; Homem-de-Mello, T. Solving the vehicle routing problem with stochastic demands using the cross-entropy method. *Ann. Oper. Res.* **2005**, *134*, 153–181. [CrossRef]
32. Rubinstein, R.Y.; Kroses, D.P. The cross-entropy method: A unified approach to combinatorial optimization Monte-Carlo simulation and machine learning. *Technometrics* **2006**, *48*, 147–148.
33. Undurti, A.; How, J. A Cross-Entropy Based Approach for UAV Task Allocation with Nonlinear Reward. In Proceedings of the AIAA Guidance, Navigation, and Control Conference, Toronto, ON, Canada, 2–5 August 2010; pp. 1–16.
34. Le Thi, H.A.; Nguyen, D.M.; Dinh, T.P. Globally solving a nonlinear UAV task assignment problem by stochastic and deterministic optimization approaches. *Optim. Lett.* **2012**, *6*, 315–329. [CrossRef]
35. Huang, L.; Qu, H.;Zuo, L. Multi-Type UAVs Cooperative Task Allocation Under Resource Constraints. *IEEE Access* **2018**, *6*, 17841–17850. [CrossRef]
36. Cofta, P.; Ledziński, D.; Śmigiel, S.; Gackowska, M. Cross-Entropy as a Metric for the Robustness of Drone Swarms. *Entropy* **2020**, *22*, 597. [CrossRef] [PubMed]
37. Zhang, X.; Wang, K.; Dai, W. Multi-UAVs Task Assignment Based on Fully Adaptive Cross-Entropy Algorithm. In Proceedings of the 2021 11th International Conference on Information Science and Technology, Chengdu, China, 7–10 May 2021; pp. 286–291.
38. Wei, Y.; Wang, B.; Liu, W.; Zhang, L. Hierarchical Task Assignment of Multiple UAVs with Improved Firefly Algorithm Based on Simulated Annealing Mechanism. In Proceedings of the 2021 40th Chinese Control Conference, Shanghai, China, 26–28 July 2021; pp. 1943–1948.
39. Wang, Q.; Liu, L.; Tian, W. Cooperative Task Assignment of Multi-UAV in Road-network Reconnaissance Using Customized Genetic Algorithm. In Proceedings of the 2021 IEEE 4th Advanced Information Management, Communicates, Electronic and Automation Control Conference, Chongqing, China, 18–20 June 2021; pp. 803–809.
40. Costa, A.; Jones, O.D.; Kroese, D. Convergence properties of the cross-entropy method for discrete optimization. *Oper. Res. Lett.* **2007**, *35*, 573–580. [CrossRef]
41. Kennedy, J.; Eberhart, R. Particle swarm optimization. In Proceedings of the ICNN'95—International Conference on Neural Networks, Perth, WA, Australia, 27 November–1 December 1995; pp. 1942–1948.
42. Luo, Z.-Q.; Yu, W. An introduction to convex optimization for communications and signal processing. *IEEE J. Sel. Areas Commun.* **2006**, *24*, 1426–1438.

Article

A Distributed Collaborative Allocation Method of Reconnaissance and Strike Tasks for Heterogeneous UAVs

Hanqiang Deng, Jian Huang *, Quan Liu, Tuo Zhao, Cong Zhou and Jialong Gao

College of Intelligence Science and Technology, National University of Defense Technology, Changsha 410073, China
* Correspondence: nudtjHuang@hotmail.com

Abstract: Unmanned aerial vehicles (UAVs) are becoming more and more widely used in battlefield reconnaissance and target strikes because of their high cost-effectiveness, but task planning for large-scale UAV swarms is a problem that needs to be solved. To solve the high-risk problem caused by incomplete information for the combat area and the potential coordination between targets when a heterogeneous UAV swarm performs reconnaissance and strike missions, this paper proposes a distributed task-allocation algorithm. The method prioritizes tasks by evaluating the swarm's capability superiority to tasks to reduce the search space, uses the time coordination mechanism and deterrent maneuver strategy to reduce the risk of reconnaissance missions, and uses the distributed negotiation mechanism to allocate reconnaissance tasks and coordinated strike tasks. The simulation results under the distributed framework verify the effectiveness of the distributed negotiation mechanism, and the comparative experiments under different strategies show that the time coordination mechanism and the deterrent maneuver strategy can effectively reduce the mission risk when the target is unknown. The comparison with the centralized global optimization algorithm verifies the efficiency and effectiveness of the proposed method when applied to large-scale UAV swarms. Since the distributed negotiation task-allocation architecture avoids dependence on the highly reliable network and the central node, it can further improve the reliability and scalability of the swarm, and make it applicable to more complex combat environments.

Keywords: heterogeneous UAV swarm; reconnaissance and strike; distributed negotiate; time coordination; deterrent maneuver

Citation: Deng, H.; Huang, J.; Liu, Q.; Zhao, T.; Zhou, C.; Gao, J. A Distributed Collaborative Allocation Method of Reconnaissance and Strike Tasks for Heterogeneous UAVs. *Drones* **2023**, *7*, 138. https://doi.org/10.3390/drones7020138

Academic Editor: Oleg Yakimenko

Received: 13 January 2023
Revised: 10 February 2023
Accepted: 14 February 2023
Published: 15 February 2023

1. Introduction

The popularity of UAVs in civil aerial photography, agriculture, surveillance, and mapping [1] has made people see its application prospects in more fields. As a low-cost, low-risk, and cost-effective weapon or carrier, UAVs have frequently appeared on the battlefield. It has become the focus of researchers to endow decentralized, heterogeneous, and low-cost UAV swarms with autonomous coordination capabilities to complete more complex tasks, because this is an important way to improve the flexibility and reliability of small UAV swarms to perform combat tasks [2,3].

Since the battlefield is a highly confrontational environment, distributed collaboration architecture is an important way to achieve large-scale swarm collaboration [4]. The centralized architecture that has been widely researched and applied has the advantage of a simpler algorithm design, but it also has the problem of high requirements on the network and central computing nodes. In contrast, each UAV node in a distributed architecture communicates and cooperates with other UAVs as an independent entity. Since there are no critical nodes in the network, the architecture is highly scalable and reliable.

According to the analysis of relevant researchers, the commonly used methods of distributed task collaborative assignment can be divided into heuristic optimization algorithms [5,6], market-based methods [7–10] and alliance-based methods [2,11] and so on [12].

Heuristic optimization algorithms are widely used because they do not require gradient information and do not rely on problem models with good mathematical properties. For example, in the literature [13], an improved pigeon-inspired optimization algorithm is proposed to solve the optimization problem of cooperative target searches, while it adopts a centralized control architecture. For multi-UAV cooperative execution of reconnaissance missions, ref. [5] proposed an intelligent self-organized algorithm (ISOA) mission-planning method. UAVs exchange status and planning information with each other, and locally optimize route planning using the improved distributed ant colony algorithm to update route planning, and repeat the process until the task is completed. However, the article assumes that all UAVs are homogeneous and that the targets are find-and-destroy elements. Another paper [14] implements a distributed task assignment method for UAV swarm reconnaissance missions based on the wolf pack algorithm, including a cooperative search algorithm based on wolf reconnaissance behavior and a cooperative attack task assignment method after the target is discovered. The algorithm has good scalability, but it does not consider the risk of searching the unknown environment when optimizing the scheme.

The market-based method is one in which the bidders estimate the benefits of completing different tasks, broadcast the bids to each other, and win with the best one, and the bidders re-evaluate after the environment or allocation plan is updated until there is no conflict. Aiming at the task assignment problem of heterogeneous cooperative UAV, a paper [7] proposed a task assignment algorithm based on improved CBGA (improved consensus-based grouping algorithm, derived from CBBA [15]). The algorithm has a simple structure, but less consideration is given to factors such as the cooperative relationship between UAVs. To deal with real-time task allocation in resource-constrained wireless-sensor networks, the authors of [16] proposed a reverse auction-based scheme using an adaptive algorithm for each node (bidder) to locally calculate its best bid response with a non-smooth and concave payoff function.

The formation of the alliance divides the large-scale UAV swarm into several small UAV alliances through strategies such as cooperative games [11]. This architecture first distributes tasks among the alliances, and then redistributes the received tasks within the alliance to effectively reduce the dimension of the problem. Authors [2] use a layered extended contract network protocol to realize the collaborative control of UAV swarms, which has the advantage of solving speed when the swarm scale is large. However, this literature ignores the influence of the division method of UAV subsets on the effect of swarm behavior. For example, two UAVs that should have cooperated are divided into different alliances, resulting in a decrease in the quality of the solution.

Researchers have also tried to combine the advantages of different architectures. When the problem has complex constraints, it is difficult to converge to a good result by directly applying CBBA and other methods, and repeated negotiations will cause high communication costs. Therefore, some researches combine heuristic algorithms with market-based methods. For example, the paper [17] considers task-time constraints and obstacle constraints, uses intelligent optimization algorithms locally to optimize the scheme, and then negotiates with other UAVs. Similarly, ref. [18] regards the minimum distance sum and the minimum maximum completion time as the optimization goals, and first uses the genetic algorithm (GA) to locally optimize, and then uses the CBAA-derived algorithm to reach a consensus among nodes.

In terms of the factors concerned in the research of UAV swarm mission collaboration, the factors considered mainly include UAV maneuvering distance [8,19], area coverage [5,15], route planning [5,20], avoidance of no-fly zones [2], etc., while the threat of cooperation between enemy platforms is rarely considered.

Aiming at the scenario where there may be a potential cooperative relationship between enemy targets, this paper proposes a distributed collaborative optimization method for heterogeneous UAVs based on a negotiation mechanism and GA.

The main contributions of this paper include the following aspects:

- The priority of tasks is evaluated by the swarm's capability superiority over the tasks to reduce the search space. The capability superiority is represented by the spatial density and the capability availability of the tasks, and the attention mechanism is combined to suppress the distant tasks to evaluate the task priority;
- The time coordination mechanism and deterrent maneuver strategy is used to reduce the risk of reconnaissance missions. Due to the incomplete information of the task, multiple UAVs are used to reconnaissance the dense tasks synchronously, and the UAVs with strike capabilities are deployed with deterrent maneuver strategy to reduce the risk of reconnaissance missions;
- A distributed task-assignment negotiation mechanism is designed so that UAVs can run in a completely distributed manner. Compared with the centralized GA, the proposed method can reduce the problem search space, improve the optimization speed and the quality of the solution, and the distributed framework can also improve the scalability and reliability of the swarm.

The remainder of this paper is organized as follows: the problem is defined and described in Section 2. The distributed collaborative allocation method for heterogeneous UAVs is described in Section 3. In Section 4, a distributed simulation environment is built, and the proposed method is verified in this environment. Finally, we conclude the paper in Section 5.

2. Problem Description

Assuming that there are several suspicious areas on the battlefield, a heterogeneous UAV swarm with different reconnaissance and strike capabilities needs to be dispatched to perform the reconnaissance and strike mission, and the UAV nodes communicate with each other through a multi-hop ad hoc network. UAVs autonomously negotiate task-allocation schemes for reconnaissance and strike targets. Since there may be a synergistic relationship between enemy targets, the mission risk and mission completion time should be minimized during mission execution.

The problem can be formalized as the problem of N_U UAVs $U = \{u_i | i = 1, 2, \cdots, N_U\}$ completing N_T tasks $T = \{t_j | j = 1, 2, \cdots, N_T\}$.

The state of UAV u_i is denoted as $u_i = \left\langle p_i, v_i, a_i, T_i^p, T_i^b, \bar{T}_i^b, U_i \right\rangle$ where p_i is the current position; v_i is the maximum flight speed; a_i is the load capacity matrix of u_i, as shown in Table 1, the capacity between loads can be added but a single load cannot be split; T_i^p is the task set that is perceived but has not decided the assignment; T_i^b and $\bar{T}_i^b$ are the task queue that u_i will participate in and the task set that will not participate; U_i is the UAV swarm status perceived by the u_i, which can be updated through communication with other UAVs.

Table 1. Example of payload capacity of u_i.

Payload Type	Scout Speed	Penetration Ability	Damage Ability	Reusable
Scout payload	50	0	0	Y
Strike payload 1	0	40	60	N
Strike payload 2	0	80	40	N

The state of the task can be expressed as $t_j = \langle p_j, s_j, a_j \rangle$, where p_j is the position of the task, s_j is the area of the suspicious area where the task is located, and a_j is the strike capability required by the task. In the process of reconnaissance of suspicious areas by UAVs with reconnaissance capabilities, existing targets can be found and a_j can be obtained; but when there is no target in the area, this conclusion can only be drawn after the UAV has scouted the entire area, in this case $a_j = 0$.

The connection relationship between UAVs is expressed as an adjacency matrix $L = [l_{im}]_{i,m \in [1, N_U]}$, and there is $l_{im} = 1$ when distance $d_{im} \leq d_\delta$, otherwise $l_{im} = 0$, and d_δ is the maximum distance for single-hop communication. When the reconnaissance node completes the reconnaissance task, it broadcasts the reconnaissance result (that is, whether there is a target in the area and the required strike capability) to the swarm by using the ad hoc network. Each UAV utilizes the perceived task status and the status of other UAVs to optimize the distribution of reconnaissance and strike tasks by negotiating with neighboring UAVs.

3. The Proposed Method

To solve the above problems, this paper proposes a distributed collaborative allocation method of reconnaissance and strike tasks for heterogeneous UAVs, and its framework is shown in Figure 1.

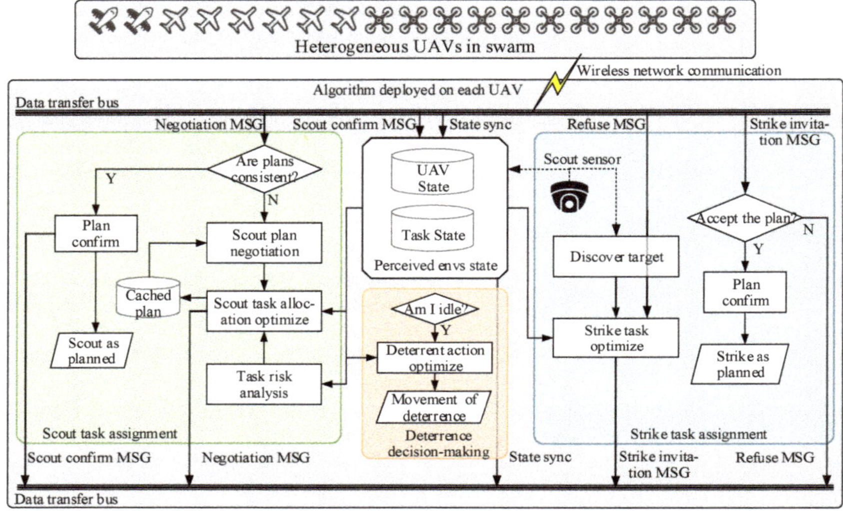

Figure 1. The framework of the collaborative allocation algorithm for reconnaissance and strike tasks.

This method consists of three main modules: negotiate for scout task assignment, strike task assignment, and deterrence decision-making. When negotiating scout tasks, this method first evaluates the tasks risk according to the degree of superiority of the UAVs over the enemy, and assigns tasks with the goal of minimizing the degree of task risk and the task completion time. Based on the perceived environmental information and the historical status information obtained by communicating with neighbor nodes, each UAV uses the GA to generate a local task-allocation and time-coordination plan after analyzing the priority of each task, and negotiates with neighbors to resolve conflicts. After the reconnaissance node discovers the enemy target, it will optimize the strike plan locally and request the relevant nodes to coordinate execution. If the request is rejected, it will re-optimize the strike plan until the strike mission is successfully assigned. When the nodes with strike capability are idle, they will fly to the reconnaissance nodes with weak strike capability to enhance their deterrence against the enemy and shorten the time from target discovery to striking.

3.1. Negotiate for Scout Task Assignment

When a large number of reconnaissance tasks need to be allocated, this paper first selects the reconnaissance tasks that should be completed first based on heuristic rules, and then uses the local optimization and distributed negotiation mechanism to allocate the tasks.

3.1.1. Heuristic Rules

To reduce the risk of UAVs executing reconnaissance missions, task allocation shall be based on the following rules:

- Give priority to the tasks that are isolated and in weak areas of the enemy;
- Give priority to the tasks where our strike capability is dominant;
- Give priority to nearby tasks.

Based on these rules, the priority evaluation method of tasks is defined as follows:

Definition 1. *S-Sig function. To make each UAV pay more attention to the local environment, referring to the sigmoid function, function $f_{ssig}(x)$ is defined as:*

$$f_{ssig}(x) = \frac{1}{1 + \exp(4x - 4)} \tag{1}$$

When $0 < x < 0.5$, $f_{ssig}(x)$ decays slowly. The decay speed increases with the increase of x and reaches the maximum at $x = 1$. When $x > 1$, its decay speed decreases and $\lim\limits_{x \to +\infty} f_{ssig}(x) = 0$. In task-priority evaluation, this function can be used to smoothly suppress the priority of tasks that are far away, while the priority of nodes that are close to the reference node is almost unaffected by distance.

Definition 2. *Spatial density of tasks. For the convenience of analysis, the typical influence radius of a single UAV is set to φ according to the cruising speed and combat radius of the UAV. Referring to the concept of kernel density estimation (KDE) in literature [21], we make the mutual influence between targets attenuate with the increase of distance, and assume that the probability of mutual cooperation between two targets within radius φ is large. Therefore, Equation (1) is used as the kernel function to calculate the task space density, and for any task $t_j \in T$, its space density ρ_j is defined as:*

$$\rho_j = \sum_{t_n \in T, j \neq k} f_{ssig}\left(d_{jn} / \varphi\right) \tag{2}$$

where d_{jn} is the Euclidean distance between task t_j and t_n. It can be inferred that ρ_j focuses on the radius within 2φ, because when $d_{jn} > 2\varphi$, $f_{ssig}\left(d_{jn} / \varphi\right) < 0.017$.

Definition 3. *Capability availability estimation of UAV. UAV u_i estimates the capability availability $f_{im}^{\mathbb{C}}(\tau)$ of neighboring UAV u_m according to its internal perception state at time τ, which can be expressed as:*

$$f_{im}^{\mathbb{C}}(\tau) = \prod_{c \in \mathbb{C}} f_{imc}(\tau)^{1/|\mathbb{C}|} \tag{3}$$

$$f_{imc}(\tau) = \sum_{u_k \in U_i} f_{ssig}(d_{mk} / \varphi) \lambda^{\tau - \tau_{ik}} \varsigma_{kc} \tag{4}$$

where $\mathbb{C}$ is the set of capability types involved in the problem; $f_{imc}(\tau)$ is the availability of capabilities of type c; U_i is the collection of UAVs perceived by u_i; d_{mk} is the distance between u_m and u_k; τ_{ik} is the timestamp when u_i receives the status of u_k; λ is the coefficient that the weight of the information from the neighboring UAV decays with time, that is, the longer the status of a UAV is updated, the lower the weight. ς_{kc} is the capability value of type c possessed by u_k.

Definition 4. *Capability availability estimation of task. Similar to Definition 3, UAV u_i estimates the capability availability $f_{ij}^{\mathbb{C}}(\tau)$ of task t_j according to its internal perception state at time τ, which can be expressed as:*

$$f_{ij}^{\mathbb{C}}(\tau) = \prod_{c \in \mathbb{C}} f_{ijc}(\tau)^{1/|\mathbb{C}|} \tag{5}$$

$$f_{ijc}(\tau) = \sum_{u_k \in \mathbf{U}_i} f_{ssig}\left(d_{jk}/\varphi\right) \lambda^{\tau - \tau_{ik}} \varsigma_{kc} \tag{6}$$

Definition 5. *Task prioritization assessment. Define* $\eta_{imj}(\tau)$ *as the priority of task* t_j *to* u_m *that is evaluated by* u_i*, and then* $\eta_{imj}(\tau)$ *can be expressed by the capability coverage of* u_m *at* t_j*, namely:*

$$\eta_{imj}(\tau) = \frac{1}{\rho_j} f_{im}^{\mathbb{C}}(\tau) \cdot f_{ssig}\left(\frac{d_{mj} + \alpha_1 \cdot \max\left(0, d_{ij} - \varphi\right)}{\varphi}\right) \tag{7}$$

the function $\max(\cdot)$ *means to take the maximum value, and* α_1 *is the weighting coefficient of the extra distance.* $d_{mj} + \alpha_1 \cdot \max\left(0, d_{ij} - \varphi\right)$ *indicates that the distance between* u_i *and* t_j *should also be considered when evaluating the capability coverage of* u_m *at* t_j*, and the farther the distance is, the greater the priority of task* t_j *is suppressed.*

It can be inferred from the definition of Equation (7) that the closer the task t_j is to u_i and u_m, the lower its spatial density, and the more sufficient the UAV capability that can cover it, the higher priority u_i thinks u_m will gives to t_j. This formula can be used for u_i to measure the superiority of our UAVs to different tasks, which is consistent with the heuristic rules.

3.1.2. Collaborative Optimization of Reconnaissance Tasks Assignment

When assigning the given reconnaissance and strike tasks, the algorithm should strive to maximize the proportion of task completion, minimize the total task completion time, and minimize the degree of task risk. Therefore, the objective function of reconnaissance task-allocation is defined as:

$$\max_{\beta^{sc}} J(\beta^{sc}) \tag{8}$$

where

$$J(\beta^{sc}) = \sum_{\left(u_i, t_j, \tau_j^s, g_j\right) \in \beta^{sc}} g_j(\beta^{sc}) \tag{9}$$

$$g_j(\beta^{sc}) = r_j(\beta^{sc}) \cdot e^{-\alpha_r\left(\tau_j^s - \tau\right)} \tag{10}$$

$$r_j(\beta^{sc}) = r_{\max} - \alpha_t \cdot \sum_{\left(u_m, t_n, \tau_n^s, g_n\right) \in \beta^{sc}} \left[\!\left[\tau_j^s < \tau_n^s \right]\!\right] \cdot f_{ssig}\left(d_{jn}/\varphi\right) \tag{11}$$

where the quaternion $\left(u_i, t_j, \tau_j^s, g_j\right)$ indicates that u_i will start reconnaissance tasks t_j at time τ_j^s, and the expected benefit is g_j; α_r is the coefficient of time for the discount of rewards; reconnaissance task plan β^{sc} is a collection of task assignment quaternions; τ_j^s and τ_n^s are the start execution times of tasks t_j and t_n, respectively; $r_j(\beta^{sc})$ is the expected reward for β^{sc} to complete t_j, which is composed of the maximum reward $r_{\max}$ and the estimated risk for completing the task, and α_t is the weight of risk; $[\![P]\!] = \begin{cases} 1 & \text{If } P \text{ is true} \\ 0 & \text{Otherwise} \end{cases}$ is an Iverson bracket, and if the start time τ_n^s is later than τ_j^s in the scheme, t_n will pose a threat to t_j.

It can be inferred from Equation (9) that if adjacent tasks can be scouted at the same time, the threat to each other can be reduced and the reward can be increased. However, starting reconnaissance at the same time also means that some tasks need to be deliberately postponed, leading to a decline in overall reward. Therefore, it is necessary to optimize the time synergy of each plan.

(1) Time-collaborative optimization of plan

The time-collaborative optimization of a plan is to optimize the specific start time of each assigned reconnaissance task. For any two assignments $\left(u_i, t_j, \tau_j^s, g_j\right)$ and $(u_m, t_n, \tau_n^s, g_n)$ in plan β^{sc}, if $\tau_j^s < \tau_n^s$, u_i can postpone its task start time to the same as u_m to improve its task reward $r_j(\beta^{sc})$. Express the updated plan as $\beta^{sc\prime}$, then there is $\tau_j^{s\prime} = \tau_n^s$. Let $\Delta \tau_{jn}^s = \tau_n^s - \tau_j^s$, then the gain of reward for time collaboration is:

$$
\begin{aligned}
\Delta g_{jn} &= g_j(\beta^{sc\prime}) - g_j(\beta^{sc}) = r_j(\beta^{sc\prime}) \cdot e^{-\alpha_r(\tau_n^s - \tau)} - r_j(\beta^{sc}) \cdot e^{-\alpha_r\left(\tau_j^s - \tau\right)} \\
&= e^{-\alpha_r\left(\tau_j^s - \tau\right)} \left[r_j(\beta^{sc\prime}) e^{-\alpha_r \cdot \Delta \tau_{jn}^s} - r_j(\beta^{sc}) \right] \\
&= e^{-\alpha_r\left(\tau_j^s - \tau\right)} \left[(r_j(\beta^{sc}) + \alpha_t \cdot f_{ssig}(d_{jn}/\varphi)) e^{-\alpha_r \cdot \Delta \tau_{jn}^s} - r_j(\beta^{sc}) \right] \\
&= e^{-\alpha_r\left(\tau_j^s - \tau\right)} \left[r_j(\beta^{sc}) \left(e^{-\alpha_r \cdot \Delta \tau_{jn}^s} - 1 \right) + \alpha_t \cdot f_{ssig}(d_{jn}/\varphi) e^{-\alpha_r \cdot \Delta \tau_{jn}^s} \right]
\end{aligned}
\tag{12}
$$

Since the time collaboration between any two assignments in a plan will depend on the recalculation of Equation (8), and the time collaborative optimization is an underlying algorithm that will be called repeatedly, we define Algorithm 1 based on Equation (12) to quickly optimize the time collaboration of the given plan.

Algorithm 1 Fast time-collaborative optimization

Input: The plan β^{sc} that needs to optimize its time collaboration
Output: The updated plan $\beta^{sc\prime}$.
1: **while** *True* **do**
2:    $\Delta g_{j^* n^*} \leftarrow \displaystyle\max_{\substack{\left(u_i, t_j, \tau_j^s, g_j\right) \in \beta^{sc} \\ (u_m, t_n, \tau_n^s, g_n) \in \beta^{sc} \\ \tau_j^s < \tau_n^s}} \left(\Delta g_{jn}\right)$ ▷ Find the best time collaboration pair using Equation (12)
3:    **if** $\Delta g_{j^* n^*} > \varepsilon_g$ **then**
4:       $\tau_{j^*}^s \leftarrow \tau_{n^*}^s$ ▷ Time collaboration when the gain of reward meets the threshold
5:    **else**
6:       **return** the updated β^{sc}
7:    **end if**
8: **end while**

---

(2) Optimization of task-assignment plan

The negotiation algorithm for reconnaissance tasks assignment consists of two parts: (i) optimizing the assignment scheme under specified conditions; (ii) negotiating with other nodes for conflict resolution.

The algorithm for optimizing the allocation plan under specified conditions is shown as Algorithm 2. The optional input $\overleftrightarrow{\beta^{sc}} = \left\{ \left(\overleftrightarrow{u}_i, \overleftrightarrow{t}_i \right) \mid i = 1, 2, \cdots, \overleftrightarrow{N^{sc}} \right\}$ is the specified partial of the task-allocation plan, where $\left(\overleftrightarrow{u}_i, \overleftrightarrow{t}_i \right)$ indicates that task $\overleftrightarrow{t}_i$ is assigned to $\overleftrightarrow{u}_i$, and $\overleftrightarrow{N^{sc}}$ is the number of assigned pairs. The optional input $\langle U^{sc}, T^{sc} \rangle$ is the set of scout UAVs U^{sc} and the set of tasks T^{sc} that need to be optimized for allocation.

If $\langle U^{sc}, T^{sc} \rangle$ is not given, the algorithm will automatically select the set of scout UAVs within n_c hops to u_i as U^{sc}, and select tasks according to the priority $\eta_{imj}(\tau)$ of each task t_j. This strategy meets the heuristic rules described in Section 3.1.1, and can reduce the optimization search space while preserving the high-quality solution space.

Algorithm 2 Scout plan optimization within UAV u_i

Input: $\overleftrightarrow{\beta^{sc}}$: Part of the task-assignment scheme that has been specified
$\quad\quad\quad \langle U^{sc}, T^{sc} \rangle$: The set of scout UAVs and tasks that need to be optimized
Output: The optimized plan β^{sc}.
1: **if** $\langle U^{sc}, T^{sc} \rangle$ is not given **then** $\quad\quad\quad\quad\quad$ ▷ Automatically select scout UAVs and tasks
2: $\quad\quad$ $U^{sc} \leftarrow$ the set of scout UAVs within n_c hops to u_i
3: $\quad\quad$ Init T_0 as an empty list
4: $\quad\quad$ **while** $|T_{sc}| < N^{sc\,\max}$ **do** ▷ Iteratively add tasks according to the evaluated priority
5: $\quad\quad\quad$ **for** u_m in U^{sc} **do**
6: $\quad\quad\quad\quad$ if the latest task added to T_0 for u_m is duplicated with existing tasks **then**
7: $\quad\quad\quad\quad\quad$ $T_0 \leftarrow$ add the next preferred task of u_m according to $\eta_{imj}(\tau)$ to T_0
8: $\quad\quad\quad\quad$ **end if**
9: $\quad\quad\quad$ **end for**
10: $\quad\quad\quad$ $T^{sc} \leftarrow \left\{ t_j | t_j \in T_0, (t_j, \cdot) \notin \overleftrightarrow{\beta^{sc}} \right\}$ $\quad\quad$ ▷ Remove duplicate tasks and tasks in $\overleftrightarrow{\beta^{sc}}$
11: $\quad\quad\quad$ **if** no task is added to T_0 **then**
12: $\quad\quad\quad\quad$ Break
13: $\quad\quad\quad$ **end if**
14: $\quad\quad$ **end while**
15: **end if**

16: **function** $f_\beta(\beta^{sc})$
17: $\quad\quad$ $\beta^{sc} \leftarrow \beta^{sc} \cup \overleftrightarrow{\beta^{sc}}$ $\quad\quad\quad\quad\quad\quad\quad\quad$ ▷ Merge the specified and generated plans
18: $\quad\quad$ Calculate τ_j^s of each assignment $\left(u_i, t_j, \tau_j^s, g_j \right) \in \beta^{sc}$ with the predicted previous
$\quad\quad\quad\quad\quad$ task finish time and the travel time between u_i and t_j
19: $\quad\quad$ Optimize the time collaborate of β^{sc} with Algorithm 1
20: $\quad\quad$ **return** $J(\beta^{sc})$ $\quad\quad\quad\quad\quad\quad$ ▷ Calculate the fitness of β^{sc} with Equation (9)
21: **end function**

22: $\beta^{sc^*} \leftarrow$ Using GA to optimize the assignment of tasks in T^{sc} to U^{sc} with the goal to
$\quad\quad$ maximize $f_\beta(\beta^{sc})$
23: **return** the best plan β^{sc^*}

After determining $\langle U^{sc}, T^{sc} \rangle$, the algorithm optimizes the assignment of the tasks based on GA, and its optimization goal is to maximize its fitness function $f_\beta(\beta^{sc})$. In $f_\beta(\beta^{sc})$, it first merges plan β^{sc} with the specified part of plan $\overleftrightarrow{\beta^{sc}}$; Then the start time of the task is estimated according to the predicted finish time of the preceding task of the UAV in each assignment and the travel time from the UAV to the corresponding task. Then, Algorithm 1 is used to quickly optimize the time collaboration between the assignments. Finally, the fitness is calculated according to the objective function defined by Equation (9).

When Algorithm 2 is used for conflict resolution optimization of multiple plans, the non conflict part of the plan can be regarded as $\overleftrightarrow{\beta^{sc}}$ and the conflict part as $\langle U^{sc}, T^{sc} \rangle$, which can reduce the search space of the optimization problem and improve the optimization speed.

(3) Negotiation-based conflict resolution

After each UAV has generated or updated the best plan β^{sc^*} for scout task allocation, it will send the plan to UAVs within n_c hops. Let $B_i^{sc} = \{ \beta_m^{sc} | hop(u_i, u_m) \leqslant n_c \}$ be the set of scout task-allocation plans received by u_i from other UAVs, then the scout task negotiation and conflict resolution algorithm can be expressed as Algorithm 3. The received UAV uses Algorithm 3 to resolve the conflict between its own plan and the received plan to update its plan until there is no conflict between the plans of neighboring nodes.

Algorithm 3 Scout plan conflict-resolving within UAV u_i

Input: β_i^{sc}: The latest local scout plan
$\qquad B_i^{sc}$: The set of received scout plans
Output: The updated local plan β_i^{sc}.

1: $\beta_i^{sc'} \leftarrow \beta_i^{sc}$
2: $U_i^{sc} \leftarrow \left\{ u_m | (u_m, t_n, \tau_n^s, g_n) \in \beta_i^{sc} \right\}$
3: **for** β_m^{sc} in B_i^{sc} **do**

4: $\qquad \overset{\leftrightarrow}{\beta_i^{sc}} \leftarrow$ consistent assignments between β_m^{sc} and $\beta_i^{sc'}$. $\qquad\qquad$ ▷ Lock the consistent part

5: $\qquad \beta_i^{conflict} \leftarrow \left(\beta_i^{sc'} \cup \beta_m^{sc} \right) - \overset{\leftrightarrow}{\beta_i^{sc}}$ $\qquad\qquad$ ▷ The part of conflict assignments

6: $\qquad U_i^{conflict} \leftarrow \left\{ u_m | (u_m, t_n, \tau_n^s, g_n) \in \beta_i^{conflict} \right\}$ $\qquad$ ▷ Extract the UAVs in conflict part

7: $\qquad T_i^{conflict} \leftarrow \left\{ t_n | (u_m, t_n, \tau_n^s, g_n) \in \beta_i^{conflict} \right\}$ $\qquad$ ▷ Extract the tasks in conflict part

8: $\qquad \beta_i^{sc^*} \leftarrow$ Re-optimize using Algorithm 2 with input $\left(\overset{\leftrightarrow}{\beta_i^{sc}}, \left\langle U_i^{conflict}, T_i^{conflict} \right\rangle \right)$

9: $\qquad \beta_i^{sc'} \leftarrow \left\{ (u_m, t_n, \tau_n^s, g_n) | (u_m, t_n, \tau_n^s, g_n) \in \beta_i^{sc^*}, u_m \in U_i^{sc} \right\}$
$\qquad\qquad\qquad$ ▷ Only the task assignments of the UAVs belonging to U_i^{sc} are retained
10: **end for**
11: **if** β_i^{sc} inconsistent with $\beta_i^{sc'}$ **then** $\qquad\qquad$ ▷ If the task assignment changes
12: $\qquad$ Broadcast $\beta_i^{sc'}$ $\qquad\qquad$ ▷ Broadcast the updated plan to neighbors within n_c hops
13: **end if**
14: $\beta_i^{sc} \leftarrow \beta_i^{sc'}$ $\qquad\qquad\qquad$ ▷ Replace the local plan with the new plan
15: $B_i^{sc} \leftarrow \varnothing$ $\qquad\qquad\qquad$ ▷ Clear the sets of received scout plan
16: **return** the updated β_i^{sc}

In Algorithm 3, the received plans are conflict resolved with the local plan one by one. For each task β_m^{sc}, the consistent part between it and $\beta_i^{sc'}$ is locked, the UAVs and tasks involved in the inconsistent part are extracted, and Algorithm 2 is used for re-optimization.

The reason for not merging all the plans at the same time is that the more plans received, the lower the probability of obtaining assignments containing consistent parts, which makes each iteration almost equal a full re-assignment, leading to low convergence efficiency.

3.2. Optimization of Strike Task Allocation

Let T_i^{st} represent the set of targets to strike in u_i, and the capability requirements a_j of each discovered target t_j is known; T_i^{sc} is the set of assigned reconnaissance tasks perceived by u_i; U_i^{st} is the idle UAVs with strike capability within n_c hops to u_i.

Since the targets that need to be struck are discovered dynamically, the scheduling of UAVs with strike capability is not only related to the currently discovered tasks, but also related to the tasks that may be discovered in the future. When optimizing strike capability scheduling, it is necessary to take the nearby reconnaissance tasks into consideration, that is, on the basis of ensuring that the capability requirements of discovered targets can be met, the capability of deterrent reconnaissance tasks should be enhanced as much as possible.

Therefore, the optimization objective of strike task allocation is defined as:

$$\min_{\beta^{st}} J\left(\beta^{st}\right) \tag{13}$$

where

$$J\left(\beta^{st}\right) = \sum_{\left(U_j, t_j, \tau_j^s\right) \in \beta^{st}} \left[\left[\!\left[t_j \in T_i^{st} \right]\!\right] \cdot \left(f^{us}\left(t_j, U_j\right) \cdot \Xi + \alpha_c \cdot \Delta_\varsigma\left(t_j, U_j\right) + \alpha_\tau \cdot \tau_{\max}\left(t_j, U_j\right) \right) \right]$$

$$- \alpha_{th} \cdot \min_{\substack{\left(U_j, t_j, \tau_j^s\right) \in \beta^{st} \\ t_j \in T_i^{sc}}} \left(\sum_{u_m \in U_j} \varsigma_m \right) \Bigg/ \left(\frac{1}{|U_j|} \sum_{u_m \in U_j} \frac{d_{mj}}{v_m} \right) \tag{14}$$

$$f^{us}\left(t_j, U_j\right) = \begin{cases} 0, & a_j \leqslant \sum_{u_m \in U_j} \varsigma m \\ 1, & else \end{cases} \tag{15}$$

$$\Delta\varsigma\left(t_j, U_j\right) = \sum_{c \in \mathbb{C}} \left(-a_{jc} + \sum_{u_m \in U_j} \varsigma mc\right) \tag{16}$$

$$\tau_{\max}\left(t_j, U_j\right) = \max_{u_m \in U_j}\left(d_{mj}/v_m\right) \tag{17}$$

β^{st} is the strike plan that composed of the strike capability assignment triplet $\left(U_j, t_j, \tau_j^{st}\right)$, and the triplet indicates that the set of UAVs U_j need to arrive and strike t_j at time τ_j^{st}. $f^{us}\left(t_j, U_j\right)$ is used to judge whether the capability requirements of task t_j can be met by the strike plan, and if not, a large constant Ξ will be added to the objective function to make the algorithm give priority to met the capability requirements of t_j. $\Delta\varsigma\left(t_j, U_j\right)$ represent the redundancy value of the strike capability assigned to t_j, and $\tau_{\max}\left(t_j, U_j\right)$ is the latest arrival time of the strike capability assigned to t_j, and these two are minimized by the algorithm on the basis of meeting the capability requirements. α_c and α_τ are weight coefficients. The part weighted by α_{th} is expected to maximize the minimum deterrent degree of reconnaissance tasks.

When a UAV discovers the target during reconnaissance, it triggers Algorithm 4 for strike task allocation, which uses GA to minimize the objective function Equation (13). After optimization, the number of strike loads required for each UAV is calculated in detail, and the invitations are send to the UAVs participating in the strike of the target that discovered by u_i in the plan.

Algorithm 4 Strike plan Optimization within UAV u_i

Input: n_{st}: Strike UAV invitation hops
$\quad\quad\quad U^{\neg st}$: The exclude set of strike UAVs
$\quad\quad\quad t_{ui}$: The target that discovered by u_i
$\quad\quad\quad T_i^{st}$: The set of strike targets discovered by other nodes and to be assigned
$\quad\quad\quad T_i^{sc}$: The set of assigned scout tasks perceived by u_i
Output: Strike plan or the result that failed
1: $\ U_i^{th} \leftarrow$ Idle strike UAVs within n_{st} to u_i, and not in $U^{\neg st}$
2: $\ T_i^{union} \leftarrow T_i^{st} \cup T_i^{sc} \cup \{t_{ui}\}$ $\quad\quad\quad\quad\quad\quad$ ▷ Taking strike and scout tasks into consideration
3: $\ \beta^{st^*} \leftarrow$ Using GA to optimize the assignment of U_i^{th} to T_i^{union} with the goal of Equation (13)
4: $\ \left(U_{ui}, t_{ui}, \tau_{ui}^{st}\right) \leftarrow \left(U_{ui}, t_{ui}, \tau_{ui}^{st}\right) \in \beta^{st^*}$ $\quad\quad\quad$ ▷ Extract the assignment for t_{ui} in β^{st^*}
5: **if** meets the capacity requirements of t_{ui} **then**
6: $\quad$ $U_{ui} \leftarrow$ Sort U_{ui} in ascending according to the distances between UAVs and t_{ui}
7: $\quad$ **for** u_m in U_{ui} **do** $\quad\quad\quad\quad$ ▷ Calculate the loads that each UAV will contribute in detail
8: $\quad\quad$ Occupy strike load of u_m one by one until t_{ui} is satisfied or all loads are occupied
9: $\quad$ **end for**
10: $\quad$ Recalculate the strike time τ_{ui}^{st} as the latest arrival time of the occupied UAVs
11: $\quad$ Send invitation to $u_m \in U_{ui}$ with the occupied loads and the strike time τ_{ui}^{st}
12: $\quad$ **return** $\left(U_{ui}, t_{ui}, \tau_{ui}^{st}\right)$
13: **else if** $n_{st} < n_{st\,\max}$ **then** $\quad\quad\quad$ ▷ Expand the request range for strike UAVs until $n_{st\,\max}$
14: $\quad$ $n_{st} \leftarrow n_{st} + 1$
15: $\quad$ Recursive optimize strike plan using Algorithm 4
16: **else**
17: $\quad$ **return** Failed $\quad\quad\quad\quad\quad$ ▷ There is not enough strike UAVs to execute this task
18: **end if**

However, if any UAV rejects the invitation, it will be excluded and the scheme will be optimized again until the strike task is successfully assigned.

3.3. Deterrence Maneuver Optimization

To enhance the capability deterrence against potential enemy targets in the reconnaissance area and shorten the time from target detection to strike execution, each idle UAV with strike capability tends to accompany other UAVs on reconnaissance tasks to provide potential capability deterrence.

In the deterrence maneuver optimization, each UAV u_i only considers the UAVs within n_c hops and the corresponding reconnaissance tasks of these UAVs. As there is no specific requirement on arrival time and capability for deterrence, each UAV takes maximizing the minimum task capability coverage as the optimization goal, and decides the destination according to the perceived situation without negotiating with other UAVs. Consistent with the part weighted by α_{th} in Equation (14), the objective function of deterrence maneuver optimization is as Equation (18) shows, and it periodically calls Algorithm 5 to update its deterrence maneuvers.

$$\max_{\beta^{th}} J\left(\beta^{th}\right) \tag{18}$$

where

$$J\left(\beta^{th}\right) = \min_{(U_j,t_j)\in\beta^{th}} \left(\sum_{u_m\in U_j} \varsigma_m\right) \Big/ \left(\frac{1}{|U_j|} \sum_{u_m\in U_j} \frac{d_{mj}}{v_m}\right) \tag{19}$$

Algorithm 5 Deterrence maneuver optimization within UAV u_i

1: $U_i^{th} \leftarrow$ Idle strike UAVs within n_c hops to u_i
2: $U_i^{sc} \leftarrow$ Scouting UAVs within n_c hop to u_i
3: $T_i^{sc} \leftarrow$ The tasks being scouted by U_i^{sc}
4: $\beta^{th*} \leftarrow$ Using GA to optimize the assignment of U_i^{th} to T_i^{sc} with the goal of Equation (18).
5: Extract the deterrence tasks of u_i from β^{th*} and maneuver to it.

---

4. Experiment and Result Analysis

4.1. Experiment Settings

In order to verify the distributed collaborative allocation method of reconnaissance and strike tasks for heterogeneous UAVs proposed in this paper, a simulation environment for heterogeneous UAV reconnaissance and strike tasks is built in Python 3.6 in this section, and its framework is shown in Figure 2. The simulation environment control module runs as an independent thread to support graphical user interface (GUI), scene generation, simulation progress control, UAV model scheduling, message exchange between UAVs, and interactive result determination between UAVs and tasks. The GUI is developed based on PyQt5 (5.15.4), and pyqtgraph (0.11.1) is used for real-time graphing of the status of UAVs and tasks. The GA used in Algorithms 2, 4 and 5 and the global optimization in Section 4.5 are from the package sko (0.6.6).

A screenshot of the GUI is shown in Figure 3, and the meanings of different elements are shown in the legend on the right. The display of elements such as text labels and topological connections can be controlled in this interface to better observe the experimental effect.

To verify the effectiveness of the method in the heterogeneous UAV swarm scenario, four kinds of UAVs are set in the simulation scene, including a mini scouter, mini striker, mini scout and strike UAV, and medium scout and strike UAV. The UAVs are different in flight velocity, scout speed, number of strike loads, and the capability of strike load. The UAV types and their parameter settings are shown as Table 2.

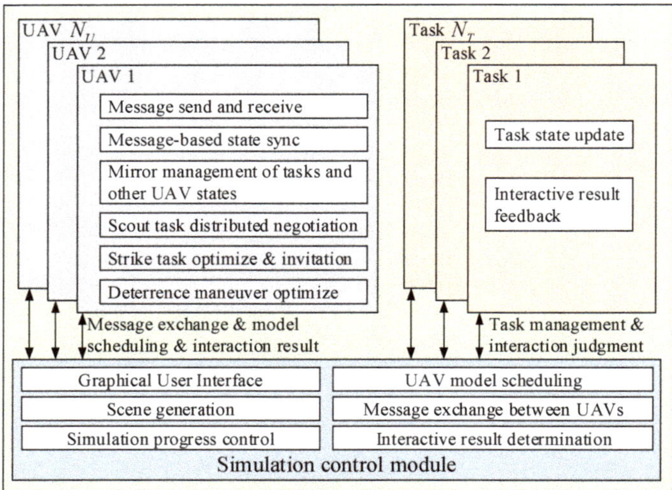

Figure 2. Distributed simulation environment for heterogeneous UAV reconnaissance and strike tasks.

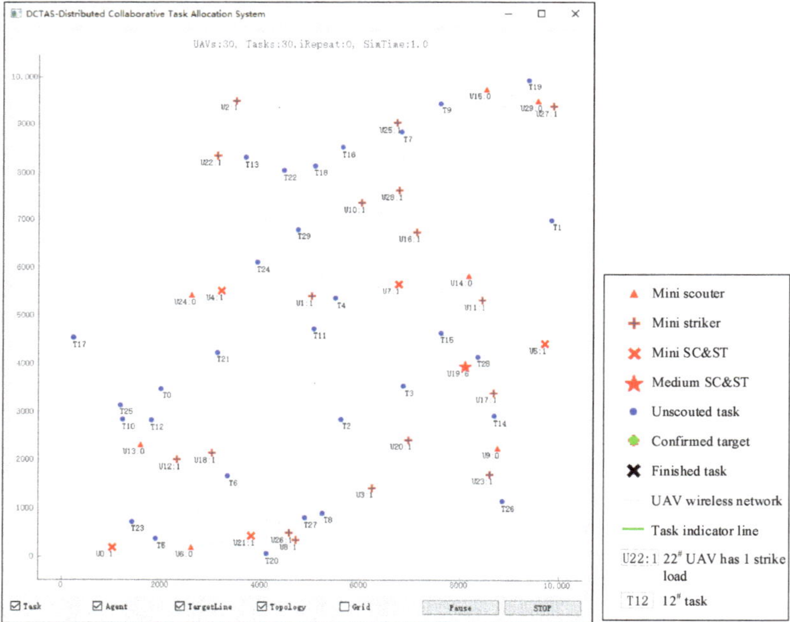

Figure 3. The generated initial scenario contains 30 UAVs and 30 tasks.

Table 2. UAV type and its parameter setting.

UAV Type	Velocity (m/s)	Scout Speed (m²/s)	Number of Strike Loads	Capability Vector of Strike Loads
Mini scouter	40	10,000	0	—
Mini striker	50	—	1	[40, 40]
Mini SC&ST	50	6000	1	[80, 80]
Medium SC&ST	80	15,000	6	[100, 100]

To simulate the war fog and the dynamic characteristics of the mission, five types of tasks are set up in the experiment, each of which has a differently sized suspicious area and required capability vector, as shown in Table 3. The fake target indicates that there is no actual target in the region, and before the completion of reconnaissance, the specific information of any target is unknown. Therefore, UAV reconnaissance and strike forces need to cooperate more flexibly to reduce mission risk and the time interval from discovery to strike.

Table 3. Task type and its parameter setting.

Task Type	Area Size (m^2)	Required Capability Vector
Fake target	1×10^6	—
Target type1	5×10^5	[25, 30]
Target type2	2×10^6	[100, 80]
Target type3	2×10^6	[40, 150]
Target type4	4×10^6	[200, 200]

The optimization result of GA is greatly affected by the population size π_n, iteration number π_i and mutation probability π_p. The larger the problem space, the larger π_n and π_i should be, so as to carry out a broader search. For the GA used in Algorithms 2 and 4, we let the number of permutations for assigning tasks to UAVs as n_{perm}, and the parameters of the GA are adaptively adjusted according to n_{perm} before each task assignment optimization.

4.2. Scene Generation

The experiments were set up in a rectangular area of 10×10 Km, without considering the height. In order to better present the cooperative effect of the UAVs and observe the operation effect, we set the total number of UAVs and tasks to 30, respectively. The proportions of four types of UAVs are 6:20:6:2, respectively, and the specific number is the total number of UAVs multiplied by the ratio and then rounded down, and the excess number is added to the mini scouter. Therefore, when the total number of UAVs is set to 30, the specific number of each UAV type is: 7, 17, 5, 1. Similarly, the ratio between the five types of tasks is set to 15:7:4:2:1, and when the total number of tasks is 30, the number of corresponding tasks is 16, 7, 4, 2, 1, respectively.

When generating a scene, the tasks are first randomly assigned to the experimental area with a uniform distribution, each task is randomly assigned a task type, and the quantity requirements of each task type is met. Similarly, initial positions and UAV types are randomly assigned to each UAV. Then, the generated UAVs and tasks are registered into the simulation control module of the experimental environment, and the information of all UAVs and tasks is broadcast to each UAV as the initial information for decision-making.

The maximum single-hop communication distance of radio between UAVs is set as $d_\delta = 3$ Km, and multi-hop transmission is supported. The synchronization of state information and the negotiation of task assignments can only take place between two UAVs when there is a communication link. Under the above constraints, the generated initial scenario contains 30 UAVs and 30 tasks, as Figure 3 shows, and the corresponding task types are shown in Table 4.

Table 4. Task IDs and corresponding task types in the scenario.

Task Type	Tasks
Fake target	T1, T5, T6, T7, T8, T12, T13, T18, T19, T20, T21, T22, T23, T24, T27, T28
Target type1	T9, T10, T11, T14, T16, T17, T29
Target type2	T3, T4, T15, T25
Target type3	T0, T26
Target type4	T2

4.3. Reconnaissance Task Priority Assessment Results

Before optimizing the reconnaissance task assignment, each evaluator, i.e., reconnaissance-capable UAV, evaluates the prior order of each task to each UAV using the evaluation method defined in Section 3.1.1. We set the number of negotiation hops to $n_c = 2$, and the UAVs will take other reconnaissance UAVs within 2 hops into consideration. For the scenario in Figure 3, the prior order of tasks to each UAV evaluated by different UAVs are shown in Table 5, and the tasks marked in bold are selected by each evaluator using Algorithm 2 to participate in this round of assignment.

Table 5. The prior order of tasks to each UAV evaluated by different UAVs at time=1.0. The bold numbers are the tasks selected to participate in this round of assignment.

Evaluator	Task Prior Order to Each UAV	Evaluator	Task Prior Order to Each UAV
U0	U0: (**5, 23, 6, 20, 12, 10**, 25, ...) U6: (**5, 23, 6, 20, 12, 10**, 25, ...) U13: (**23, 5, 6, 12, 10**, 25, 20, ...) U21: (**5, 6, 23, 20, 12**, 27, 10, ...)	U14	U14: (**15, 28, 3, 1, 14, 4, 7**, ...) U5: (**28, 15, 14, 3, 1, 4, 11**, ...) U7: (**15, 28, 3, 4, 14, 1, 11**, ...) U9: (**28, 15, 14, 3, 4, 1, 11**, ...) U15: (**1, 7**, 15, 28, 9, 4, 3, ...) U19: (**28, 15, 3, 14, 4, 1, 11**, ...)
U4	U4: (**24, 21, 29, 11**, 4, 0, 13, ...) U7: (**4**, 11, 29, 24, 21, 22, 13, ...) U24: (**24, 21, 29, 11**, 4, 0, 13, ...)	U15	U15: (**9, 19, 7, 16, 1, 18, 22**, ...) U7: (**7, 9, 16, 19, 1**, 18, 15, ...) U14: (**7, 9, 1, 19**, 16, 18, 15, ...) U29: (**19, 9, 7, 1, 16**, 18, 22, ...)
U5	U5: (**28, 14, 15, 3, 1, 26, 4**, ...) U7: (**15, 28, 14, 3, 1, 4**, 26, ...) U9: (**14, 28, 15, 3, 26, 1, 2**, ...) U14: (**28, 15, 14, 3, 1, 26, 4**, ...) U19: (**28, 14, 15, 3, 1, 26, 4**, ...)	U19	U19: (**14, 28, 15, 3, 2, 26, 4**, ...) U5: (**28, 14, 15, 3**, 26, 2, 4, ...) U7: (**15, 3, 28, 14**, 4, 11, 2, ...) U9: (**14, 28, 3, 15**, 26, 2, 4, ...) U14: (**15, 28, 3, 14, 4**, 2, 11, ...)
U6	U6: (**6, 20, 5, 23, 27, 8**, 12, ...) U0: (**5, 23, 6, 20**, 27, 12, 10, ...) U13: (**6, 23, 5, 20**, 12, 27, 10, ...) U21: (**6, 20, 5, 27, 23, 8**, 12, ...)	U21	U21: (**6, 20, 27, 8, 5, 23**, 2, ...) U0: (**5, 23, 6, 20, 27, 8**, 12, ...) U6: (**6, 20, 5, 27, 23, 8**, 12, ...) U9: (**8, 27, 2**, 20, 6, 3, 5, ...) U13: (**6, 5, 23, 20, 27, 8, 12**, ...)
U7	U7: (**15, 28, 3, 4, 11, 29, 24**, ...) U4: (**4, 11, 29, 24**, 3, 15, 28, ...) U5: (**28, 15, 3, 14, 4, 11, 1**, ...) U9: (**28, 3, 15, 14, 4, 11, 2**, ...) U14: (**15, 28, 3, 4, 11, 29, 14**, ...) U15: (**7**, 15, 28, 16, 4, 29, 9, ...) U19: (**15, 28, 3, 4, 14, 11, 2**, ...) U24: (**4, 11, 24**, 29, 3, 15, 28, ...)	U9	U9: (**14, 28, 26, 3, 15**, 2, 8, ...) U5: (**14, 28, 15, 3, 26, 2**, 8, ...) U7: (**28, 15, 3, 14, 26, 2, 4**, ...) U14: (**28, 15, 14, 3, 26, 2, 4**, ...) U19: (**14, 28, 3, 15, 26, 2**, 8, ...) U21: (**3, 14, 26, 2, 28**, 8, 15, ...)
U24	U24: (**21, 24, 0**, 29, 11, 4, 12, ...) U4: (**24, 21, 11, 29, 0, 4**, 13, ...) U7: (**24, 4**, 11, 29, 21, 13, 22, ...)	U29	U29: (**19, 9, 1**, 7, 16, 18, 15, ...) U15: (**19, 9, 7, 1**, 16, 18, 15, ...)
U13	U13: (**6, 23, 5, 12, 0**, 10, 25, ...) U0: (**23, 5, 6, 12**, 10, 25, 0, ...) U6: (**6, 5, 23, 12, 20**, 10, 0, ...) U21: (**6, 5, 23, 20**, 12, 0, 10, ...)	—	—

Taking the evaluation results of U7 as an example, the reconnaissance capable UAVs within 2 hops are U4, U5, U9, U14, U15, U19, U24, and U7 itself. The priority of task T15 ranks first for U7 since T15 is relatively close to U7, and the position of T15 can obtain more sufficient strike capability. Although T4 is the closest to U7, it gets a prior order of 4th for U7 because the unknown potential synergistic relationship between T11 and T4 increases the risk of T4, and the low coverage of UAVs at T4 further reduces its priority.

It can be found that there is a large gap between U4's task prior order as assessed by U7 and U4. T24 ranked after U4 because T24 has a great advantage in distance. However, the attention mechanism of U7 makes it pay more attention to the surrounding tasks, so the priority of T24 to U4 is suppressed in the evaluation of U7. This cognitive difference caused by inconsistent environmental cognition or subjective preferences can be corrected during the negotiation process with the other party.

Because each evaluator only considers several tasks with the highest priority in each round of allocation, the search space for task-allocation optimization can be effectively reduced.

4.4. Reconnaissance Task Assignment

After the reconnaissance UAVs generate a reconnaissance plan and share it with each other, each node uses Algorithm 3 to fuse the received plan with its own plan. The fusion process of U7 is as Figure 4 shows, and at time=1, U7 tends to give priority to nearby tasks due to its own attention mechanism, so its assignments include U4→T11, U14→T4, and U24→T29. However, this attention mechanism can cause inconsistencies among the generated plans when the distance between two UAVs is large. When merging the received plans with its own, U7 will take out the inconsistent part and redistribute it. At this time, its attention mechanism will be disabled, making the integrated plan more consistent with the views of each participant.

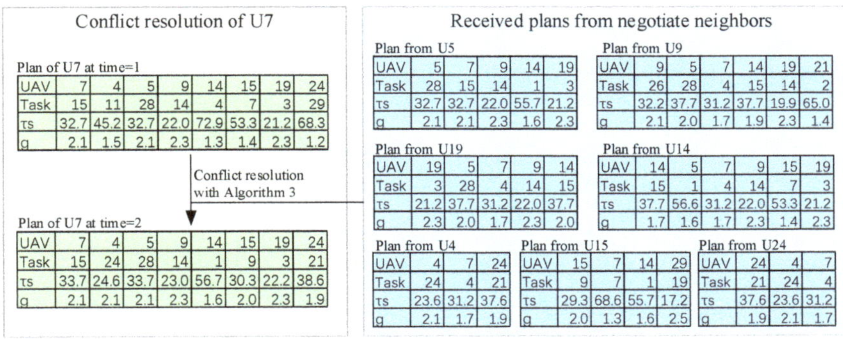

Conflict resolution of U7

Plan of U7 at time=1

UAV	7	4	5	9	14	15	19	24
Task	15	11	28	14	4	7	3	29
τ_s	32.7	45.2	32.7	22.0	72.9	53.3	21.2	68.3
g	2.1	1.5	2.1	2.3	1.3	1.4	2.3	1.2

Conflict resolution with Algorithm 3

Plan of U7 at time=2

UAV	7	4	5	9	14	15	19	24
Task	15	24	28	14	1	9	3	21
τ_s	33.7	24.6	33.7	23.0	56.7	30.3	22.2	38.6
g	2.1	2.1	2.1	2.3	1.6	2.0	2.3	1.9

Received plans from negotiate neighbors

Plan from U5

UAV	5	7	9	14	19
Task	28	15	14	1	3
τ_s	32.7	32.7	22.0	55.7	21.2
g	2.1	2.1	2.3	1.6	2.3

Plan from U9

UAV	9	5	7	14	19	21
Task	26	28	4	15	14	2
τ_s	32.2	37.7	31.2	37.7	19.9	65.0
g	2.1	2.0	1.7	1.9	2.3	1.4

Plan from U19

UAV	19	5	7	9	14
Task	3	28	4	14	15
τ_s	21.2	37.7	31.2	22.0	37.7
g	2.3	2.0	1.7	2.3	2.0

Plan from U14

UAV	14	5	7	9	15	19
Task	15	1	4	14	7	3
τ_s	37.7	56.6	31.2	22.0	53.3	21.2
g	1.7	1.6	1.7	2.3	1.4	2.3

Plan from U4

UAV	4	7	24
Task	24	4	21
τ_s	23.6	31.2	37.6
g	2.1	1.7	1.9

Plan from U15

UAV	15	7	14	29
Task	9	7	1	19
τ_s	29.3	68.6	55.7	17.2
g	2.0	1.3	1.6	2.5

Plan from U24

UAV	24	4	7
Task	21	24	4
τ_s	37.6	23.6	31.2
g	1.9	2.1	1.7

Figure 4. Negotiation process based on conflict resolution.

When the conflict is resolved, the assignment of U7 in Figure 4 is consistent with the updated assignments of other nodes, so this assignment is finally adopted and implemented. The allocation results of the first round of reconnaissance tasks are shown as the green target lines of each reconnaissance node in Figure 5a, and the specific allocation information is shown in Table 6. The topology after a period of execution is shown in Figure 5b.

From the allocation results, it can be found that the distributed allocation algorithm generally follows the principle of minimizing the completion time, but it also reflects the algorithm's expectation of enhancing the superiority over enemy and reducing mission risks. For example, U13 chose T6 instead of T12 as the first mission, because it is easier for the UAV swarms to form a superiority over enemy at T6. In contrast, the location of T12 is too dense and risky, and it should be executed after more UAVs are concentrated. The mission groups (T5, T23) and (T15, T28) are also relatively dense, but since the UAV swarm has a capability advantage here, the strategy of coordinating in time is adopted to reduce the mission risk. In the time-coordinated formation, the UAVs that could have arrived earlier choose to reduce the flight speed so that the formation can reach the targets at the same time, thereby avoiding the coordinated strike of the enemy due to individual exposure.

When each UAV adopts the distributed task assignment algorithm in this paper, the task scheduling Gantt chart is as Figure 6 shows. In it, the blue boxes represent the reconnaissance behavior, and the red boxes represent strike behavior. From the Gantt chart, it can be found that for UAVs with both reconnaissance and strike capabilities, such as U0 and U5, when their own capabilities can meet the target capability requirements, the strike can be carried out immediately. For example, U0's strike on T10 and U5's strike on T11 are instant. For targets with strong defense capabilities, the coordinated strike of multiple UAVs is required. For example, the strikes on T0 and T26 are all completed by the cooperation of four UAVs. Since the nodes with strike capability will maneuver to the reconnaissance nodes for deterrence when they are idle, it allows reconnaissance nodes that do not have strike capability can also strike quickly after discovering the target. For example, U25 can launch a strike on T9 (discovered by U15) within 8 s after confirming the strike mission.

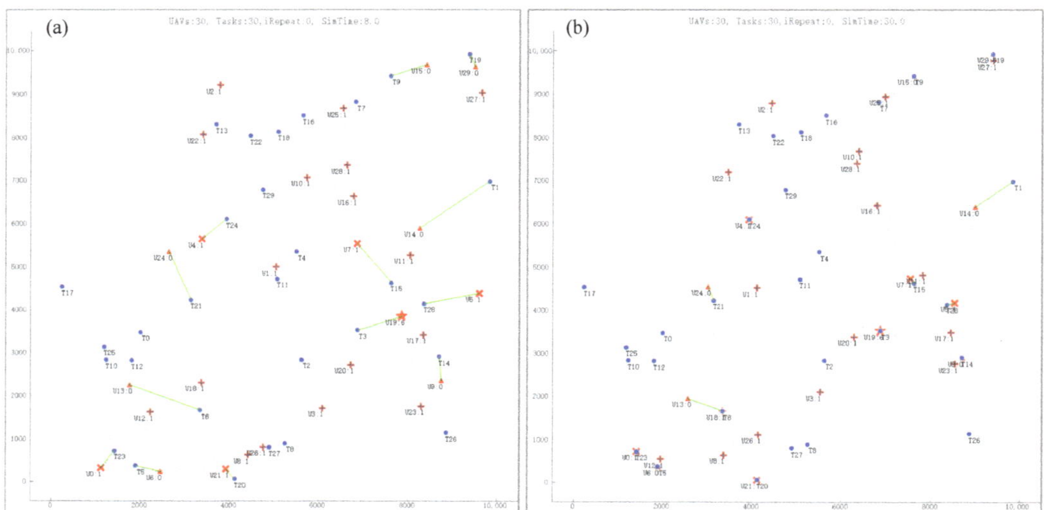

Figure 5. Topology diagram for different simulation times. (**a**) Each reconnaissance UAV has confirmed the reconnaissance task before time = 8. (**b**) Each reconnaissance node executes reconnaissance task, and the strike nodes perform deterrence maneuver at time = 30. The meanings of elements are consistent with those in Figure 3.

Table 6. Allocation information and time coordination relationship of the first round of reconnaissance tasks.

UAV ID	Current Pos	Task ID	Task Pos	τ_j^s (s)	Planned v_i (m/s)	Maximum v_i (m/s)	Collaborate UAVs
U0	(1029.8, 204.3)	T23	(1420.5, 729.8)	23.38	35.63	50	U6
U6	(2618.0, 202.9)	T5	(1904.0, 377.2)	23.38	40.00	40	U0
U5	(9722.0, 4410.0)	T28	(8364.7, 4143.4)	34.66	50.00	50	U7
U7	(6776.9, 5650.7)	T15	(7634.8, 4636.8)	33.66	48.01	50	U5
U4	(3223.9, 5507.1)	T24	(3931.8, 6108.0)	23.57	50.00	50	—
U9	(8777.8, 2235.3)	T14	(8712.1, 2910.3)	22.95	40.00	40	—
U13	(1596.7, 2313.2)	T6	(3348.8, 1669.4)	51.66	40.00	40	—
U14	(8179.1, 5827.7)	T1	(9846.0, 6986.1)	56.75	40.00	40	—
U15	(8543.8, 9725.0)	T9	(7617.9, 9429.2)	30.30	40.00	40	—
U19	(8114.8, 3937.6)	T3	(6881.7, 3528.7)	22.24	80.00	80	—
U21	(3826.9, 436.9)	T20	(4124.1, 67.6)	15.48	50.00	50	—
U24	(2615.7, 5425.9)	T21	(3144.3, 4235.8)	39.56	40.00	40	—
U29	(9578.5, 9477.7)	T19	(9387.9, 9928.5)	17.24	40.00	40	—

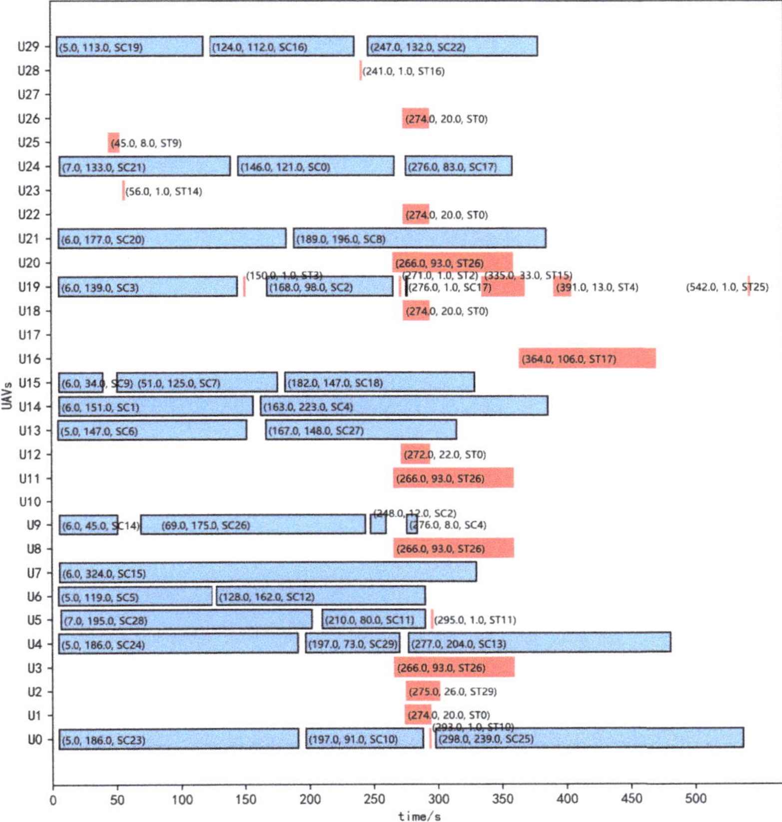

Figure 6. Gantt chart for reconnaissance and strike tasks. Blue boxes represent reconnaissance and red boxes represent strikes, and the triple elements represent mission confirmation time, duration of maneuver and task execution, and the concatenation of task type and task ID respectively.

4.5. Comparison with Centralized Global Optimization Based on GA

In the proposed distributed framework, each UAV only focuses on the nearby UAVs and tasks, and negotiates with the nearby UAVs to resolve the conflicts in the combat plans. This distributed solution not only realizes decoupling between UAVs, but also effectively reduces the search space of the task-allocation problem, and can improve the speed of solving the problem. Based on the scenario shown in Figure 3, we compare the proposed method with the centralized global optimization method based on GA in terms of solving speed and quality. The experimental platform is a MSI GS65 notebook installed with Windows system. Its CPU is i7-8750H, GPU is GTX1070 Max-Q, memory is 32 G, and the SSD is 512 G.

We have counted the decision-making time consumption of each reconnaissance UAV in steps 1-8 during the optimization of reconnaissance task allocation, because all UAVs have confirmed their tasks at the end of the 8th simulation step. The results are shown in Figure 7, and it can be found that the time consumed by each decision of each UAV is less than 0.25 s. Without considering the communication delay, if the maximum decision time of each step is taken as the cycle of this round of iteration, the total time of 8 simulation steps is 1.02 s. From the time-consumption distribution, it can also be found that the first step of decision-making only consumes a little time, while the second step of decision-making takes the most time, and then decreases gradually. This is because before the first decision, there is no communication between nodes, and each node makes decisions independently. The second decision is made after exchanging the results of the first round, and at this time, each node is performing conflict resolution on multiple collected plans, so it takes a lot of time. After the second step, the conflict between plans is gradually resolved, so the decision-making time is also shortened and the final task-allocation plan is formed. The optimization results are shown in Table 6, and from the global perspective, its fitness is 27.48 using Equation (9).

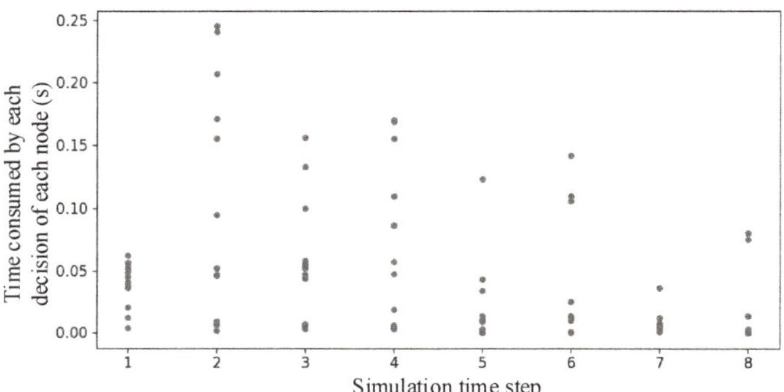

Figure 7. Time consumption of each UAV in the proposed method.

In the same scenario, we further use the centralized GA to optimize the reconnaissance task-allocation problem from a global perspective, and the fitness curve obtained is shown as the two CGA curves in Figure 8. In our proposed algorithm, the population number π_n and iteration number π_i of GA are automatically adjusted according to the number of permutations of the assignment problem. When this strategy is applied to the centralized method, the obtained parameters of GA are $\pi_n = 32$, $\pi_i = 30$ and $\pi_p = 0.1$. However, the search space for the global optimization problem of assigning 30 tasks to 13 UAVs is too large, so GA is difficult to converge to a good result under these parameters. In order to further expand the search of the global GA to obtain a better result, we adjust the parameters to $\pi_n = 400$, $\pi_i = 200$, $\pi_p = 0.3$. We can find that after 27 rounds of iteration, it has obtained a result with fitness close to that of the proposed paper, which consumes about 4.9 s.

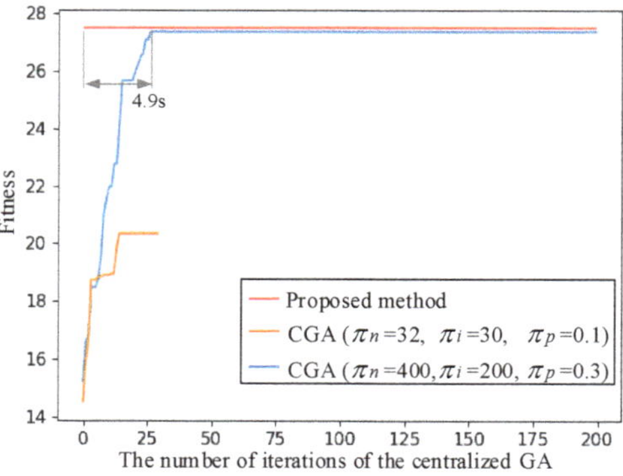

Figure 8. The fitness and time consumption of centralized GA.

Comparing the time consumption of the two methods, it can be found that the proposed method effectively reduces the computing load of each single UAV through the idea of divide and conquer, and can be applied to more types of small UAVs. By observing the change of fitness, we found that it is easy for centralized global optimization to fall into local optimal solution when the problem space is large. If the swarm size is further increased, it will be difficult for the centralized global optimization method to obtain good results, while distributed collaborative optimization has better scalability.

4.6. Comparison with No Time Coordination and Deterrence Maneuver

To verify the effect of the proposed method on reducing the risk of mission execution and improving the deterrence of strike capability, the spatial density of tasks represented by Equation (2) is taken as the scout risk; the strike capability availability of tasks represented by Equation (5) is taken as the capability coverage. In each simulation scenario, the average risk at the beginning of each scout task and the average strike capability coverage for scouting tasks during the whole simulation process are counted. The comparison results with the algorithms without time coordination or deterrence maneuver are shown in Table 7, and each result is the statistics of the mean and standard deviation of 10 simulation scenarios.

Table 7. The scout risk and strike capability coverage compared with no time coordination or deterrence maneuver. Each result is the mean and standard deviation of 10 simulation scenarios.

Deterrence Maneuver Type	Enable Time Collaboration		Disable Time Collaboration	
	Scout Risk	Cap. Coverage	Scout Risk	Cap. Coverage
Enable deterrence maneuver	0.238 ± 0.065	153.1 ± 61.3	0.311 ± 0.090	166.6 ± 63.9
Disable deterrence maneuver	0.234 ± 0.048	131.5 ± 33.7	0.311 ± 0.091	116.7 ± 31.5

The statistics of the results show that time coordination can reduce the scout risk by about 23%, and deterrence maneuver can improve the strike capability coverage by about 30%, which verifies the effectiveness of time coordination and maneuver deterrence.

4.7. Discussion

4.7.1. Computational Complexity Analysis

In the proposed method, the optimization of scout-task assignment needs to optimize task allocation and time coordination between UAVs, which is the part with high computational complexity of our proposed method. Therefore, analyzing the computational complexity of this part will aid further improvement.

In Algorithm 2, the most time-consuming process is to use line 24 to optimize the plan using GA. The computational complexity of GA can be expressed as $O(\pi_p \times \pi_i)$, where π_p is the population size and π_i is the number of iterations. For each plan generated by the GA, its fitness will be calculated through line 17-21 of Algorithm 2. Among them, line 20 uses Algorithm 1 for time collaborative optimization. Let the number of UAVs in the plan be N. The worst case of the while loop of Algorithm 1 will iterate $N - 1$ times, and each loop needs to calculate $N \times (N - 1)$ time alignment reward gains according to Equation (12), so the complexity of Algorithm 1 is about $O(N^3)$. Then line 21 of Algorithm 2 uses Equation (9) to calculate the fitness of the plan, in which the start time between any two UAVs needs to be compared to evaluate the task threat, and thus the complexity is about $O(N^2)$.

Therefore, the overall computational complexity of Algorithm 2 is about $O(\pi_p \times \pi_i \times (N^3 + N^2)) \approx O(\pi_p \times \pi_i \times N^3)$. Among them, the settings of π_p and π_i not only affect the calculation cost, but also affect the quality of the optimization results. In this paper, these two parameters are simply linearly mapped from the number of allocation combinations, and before deployment, it is necessary to further study the setting strategy of these parameters to compromise between the calculation cost and the optimization quality.

4.7.2. Method Characteristics under Different Network Connectivity

When the distributed UAV swarm is running, each UAV needs to perform regular state synchronization and event-triggered task assignment negotiation with UAVs within n_c hops, which also means that each UAV needs to process information from other UAVs within n_c hops. When the mission area is relatively scattered and the connectivity of the UAV network is low, the use of a limited range of autonomous collaboration can reduce the consumption of the network and reduce the computational power consumption of each UAV. While improving the reliability of the cluster, the distributed architecture can also enhance the scalability of the cluster.

However, when the UAVs are concentrated, such as when the network is fully connected in extreme cases, adopting a distributed architecture will reduce the operating efficiency of the system. The fully connected network requires each UAV to process the information of the entire battlefield, which indicates that each UAV should be equipped with high-performance computing resources. Moreover, the additional negotiation communication required by the distributed architecture will also increase the consumption of the network.

Therefore, the proposed distributed approach should be combined with centralized control when applied, and dynamically switch between the two according to the network connectivity status.

4.7.3. Influence of Network Instability on the Proposed Method

The mutual communication between UAVs is the basis for task-assignment negotiation and coordination during task execution. Due to the high-speed maneuvering of UAVs in 3D space, the topology of the flying ad hoc network (FANET) [22] changes rapidly, which may cause communication interruption, delay increase, and other problems. These problems may further make the negotiation period of task assignment longer, or even cause conflicts between the assigned plans due to the interruption of communication, and ultimately make the overall task assignment result worse. Therefore, the next step is to evaluate the impact of network instability on the method and study the corresponding strategies to improve its robustness, which can also avoid making this problem a vulnerable point to attack by the enemy.

Since the cooperation of UAVs in the proposed method mainly occurs between UAVs within 2 hops, the route-maintenance strategy of FANET can pay more attention to the optimization of short-distance routes to improve the QoS, which can not only improve the speed of distributed task allocation in this paper, but also help avoid collisions between UAVs and so on.

5. Conclusions

Due to the high risk of UAV clusters in executing reconnaissance and strike tasks under the condition of insufficient enemy information and potential synergy between targets, a distributed task-collaborative allocation method for heterogeneous UAV swarms is proposed. This method establishes a distributed task-allocation framework composed of a reconnaissance task-allocation method based on a negotiation mechanism and a strike task-allocation method based on an invitation mechanism. The reconnaissance task-allocation algorithm evaluates the task priority according to the superiority of the UAVs against the tasks to reduce the complexity of the optimization problem. Reconnaissance UAVs adopt a time-coordination strategy for reconnaissance, and UAVs with strike capabilities perform deterrent maneuvers when they are idle to reduce mission risks during mission execution. This method enables the UAV swarm to negotiate the allocation of tasks in a distributed framework, and at the same time, the evaluation of the capability advantage over the enemy, time coordination, and deterrence maneuver mechanism effectively reduce the risk of unknown targets to UAVs. The distributed framework not only improves the scalability of the swarm, but also enhances its reliability in the battlefield with a more complex electromagnetic environment.

Further research should include a more efficient algorithm that takes the negotiation mechanism and the network state of the swarm as prior information to replace the GA for the local optimization of UAVs, and should aim to obtain the best operating efficiency under different network connectivity. The proposed method should be further combined with a centralized or hierarchical task-allocation framework.

Author Contributions: Conceptualization, H.D. and J.H.; methodology and writing—original draft, H.D.; writing—review and editing, Q.L., C.Z., T.Z. and J.G. All authors have read and agreed to the published version of the manuscript.

Funding: This research received no external funding.

Institutional Review Board Statement: Not applicable.

Informed Consent Statement: Not applicable.

Data Availability Statement: Not applicable.

Conflicts of Interest: The authors declare no conflict of interest.

References

1. Shakhatreh, H.; Sawalmeh, A.H.; Al-Fuqaha, A.; Dou, Z.; Almaita, E.; Khalil, I.; Othman, N.S.; Khreishah, A.; Guizani, M. Unmanned Aerial Vehicles (UAVs): A Survey on Civil Applications and Key Research Challenges. *IEEE Access* **2019**, *7*, 48572–48634.
2. Qin, B.; Zhang, D.; Tang, S.; Wang, M. Distributed Grouping Cooperative Dynamic Task Assignment Method of UAV Swarm. *Appl. Sci.* **2022**, *12*, 2865. [CrossRef]
3. Zhang, J.; Xing, J. Cooperative task assignment of multi-UAV system. *Chin. J. Aeronaut.* **2020**, *33*, 2825–2827. [CrossRef]
4. Jiang, X.; Zeng, X.; Sun, J.; Chen, J. Research status and prospect of distributed optimization for multiple aircraft. *Acta Astronaut.* **2021**, *42*, 524551. (In Chinese) [CrossRef]
5. Zhen, Z.; Xing, D.; Gao, C. Cooperative search-attack mission planning for multi-UAV based on intelligent self-organized algorithm. *Aerosp. Sci. Technol.* **2018**, *76*, 402–411. [CrossRef]
6. Duan, H.; Zhao, J.; Deng, Y.; Shi, Y.; Ding, X. Dynamic Discrete Pigeon-Inspired Optimization for Multi-UAV Cooperative Search-Attack Mission Planning. *IEEE Trans. Aerosp. Electron. Syst.* **2021**, *57*, 706–720. [CrossRef]
7. Ma, Y.; Zhao, Y.; Bai, S.; Yang, J.; Zhang, Y. Collaborative task allocation of heterogeneous multi-UAV based on improved CBGA algorithm. In Proceedings of the 16th International Conference on Control, Automation, Robotics and Vision, Shenzhen, China, 13–15 December 2020; pp. 795–800.
8. Dai, W.; Lu, H.; Xiao, J.; Zeng, Z.; Zheng, Z. Multi-Robot Dynamic Task Allocation for Exploration and Destruction. *J. Intell. Robot. Syst.* **2019**, *98*, 455–479. [CrossRef]
9. Sheng, W.; Yang, Q.; Tan, J.; Xi, N. Distributed multi-robot coordination in area exploration. *Robot. Auton. Syst.* **2006**, *54*, 945–955. [CrossRef]
10. Ye, F.; Chen, J.; Sun, Q.; Tian, Y.; Jiang, T. Decentralized task allocation for heterogeneous multi-UAV system with task coupling constraints. *J. Supercomput.* **2020**, *77*, 111–132. [CrossRef]
11. Chen, J.; Wu, Q.; Xu, Y.; Qi, N.; Guan, X.; Zhang, Y.; Xue, Z. Joint Task Assignment and Spectrum Allocation in Heterogeneous UAV Communication Networks: A Coalition Formation Game-Theoretic Approach. *IEEE Trans. Wirel. Commun.* **2021**, *20*, 440–452. [CrossRef]
12. Jiang, Y. A Survey of Task Allocation and Load Balancing in Distributed Systems. *IEEE Trans. Parallel Distrib. Syst.* **2016**, *27*, 585–599. [CrossRef]
13. Li, L.; Xu, S.; Nie, H.; Mao, Y.; Yu, S. Collaborative Target Search Algorithm for UAV Based on Chaotic Disturbance Pigeon-Inspired Optimization. *Appl. Sci.* **2021**, *11*, 7358. [CrossRef]
14. Hu, J.; Wu, H.; Zhan, R.; Menassel, R.; Zhou, X. Self-organized search-attack mission planning for UAV swarm based on wolf pack hunting behavior. *J. Syst. Eng. Electron.* **2021**, *32*, 1463–1476.
15. Choi, H.; Brunet, L.; How, J.P. Consensus-Based Decentralized Auctions for Robust Task Allocation. *IEEE Trans. Robot.* **2009**, *25*, 912–926. [CrossRef]
16. Edalat, N.; Tham, C.; Xiao, W. An auction-based strategy for distributed task allocation in wireless sensor networks. *Comput. Commun.* **2012**, *35*, 916–928. [CrossRef]
17. Choi, H.; Kim, Y.; Kim, H.J. Genetic algorithm based decentralized task assignment for multiple unmanned aerial vehicles in dynamic environments. *Int. J. Aeronaut. Space Sci.* **2011**, 163–174. [CrossRef]
18. Patel, R.; Rudnick-Cohen, E.; Azarm, S.; Otte, M.; Xu, H.; Herrmann, J.W. Decentralized Task Allocation in Multi-Agent Systems Using a Decentralized Genetic Algorithm. In Proceedings of the IEEE International Conference on Robotics and Automation (ICRA), Paris, France, 31 May–31 August 2020; pp. 3770–3776.

19. Wu, H.; Li, H.; Xiao, R.; Liu, J. Modeling and simulation of dynamic ant colony's labor division for task allocation of UAV swarm. *Physica A* **2018**, *491*, 127–141. [CrossRef]
20. Cao, Y.; Wei, W.; Bai, Y.; Qiao, H. Multi-base multi-UAV cooperative reconnaissance path planning with genetic algorithm. *Clust. Comput.* **2019**, *22*, 5175–5184. [CrossRef]
21. Yu, W.; Ai, T.; Shao, S. The analysis and delimitation of Central Business District using network kernel density estimation. *J. Transp. Geogr.* **2015**, *45*, 32–47. [CrossRef]
22. Khan, M.A.; Safi, A.; Qureshi, I.M.; Khan, I.U. Flying ad-hoc networks (FANETs): A review of communication architectures, and routing protocols. In Proceedings of the 2017 First International Conference on Latest trends in Electrical Engineering and Computing Technologies, Karachi, Pakistan, 15–16 November 2017; pp. 1–9.

Article

Service Function Chain Scheduling in Heterogeneous Multi-UAV Edge Computing

Yangang Wang [1,2], Hai Wang [1,*], Xianglin Wei [2,*], Kuang Zhao [2], Jianhua Fan [2], Juan Chen [1], Yongyang Hu [2] and Runa Jia [1,2]

[1] College of Communication Engineering, Army Engineering University of PLA, Nanjing 210007, China
[2] The Sixty-Third Research Institute, National University of Defense Technology, Nanjing 210007, China
* Correspondence: hai_wang@aeu.edu.cn (H.W.); weixianglin@nudt.edu.cn (X.W.)

Abstract: Supporting Artificial Intelligence (AI)-enhanced intelligent applications on the resource-limited Unmanned Aerial Vehicle (UAV) platform is difficult due to the resource gap between the two. It is promising to partition an AI application into a service function (SF) chain and then dispatch the SFs onto multiple UAVs. However, it is still a challenging task to efficiently schedule the computation and communication resources of multiple UAVs to support a large number of SF chains (SFCs). Under the multi-UAV edge computing paradigm, this paper formulates the SFC scheduling problem as a 0–1 nonlinear integer programming problem. Then, a two-stage heuristic algorithm is put forward to solve this problem. At the first stage, if the resources are surplus, the SFCs are deployed to UAV edge servers in parallel based on our proposed pairing principle between SFCs and UAVs for minimizing the completion time sum of tasks. In contrast, a revenue maximization heuristic method is adopted to deploy the arrived SFCs in a serial service mode when the resource is insufficient. A series of experiments are conducted to evaluate the performance of our proposal. Results show that our algorithm outperforms other benchmark algorithms in the completion time sum of tasks, the overall revenue, and the task execution success ratio.

Keywords: edge computing; unmanned aerial vehicle; artificial intelligence; service function chain

Citation: Wang, Y.; Wang, H.; Wei, X.; Zhao, K.; Fan, J.; Chen, J.; Hu, Y.; Jia, R. Service Function Chain Scheduling in Heterogeneous Multi-UAV Edge Computing. *Drones* **2023**, *7*, 132. https://doi.org/10.3390/drones7020132

Academic Editor: Vishal Sharma

Received: 12 January 2023
Revised: 29 January 2023
Accepted: 9 February 2023
Published: 13 February 2023

1. Introduction

Artificial Intelligence (AI)—in particular Deep Learning (DL)—techniques have been widely adopted as a powerful tool in wireless communications and mobile computation areas due to their unique advantages such as automated feature extraction and high generalizability [1]. DL has been utilized from different perspectives for intelligent applications, e.g., wireless spectrum sensing, object tracking, and channel estimation [2–4]. Although the advantages are brought by DL techniques, fulfilling their intensive computation needs is a new challenge for resource-limited Internet of Things (IoT) devices. Mobile edge computing (MEC) is seen as a promising technology for solving this challenge [5–8]. It can make computation resources closer to IoT devices, so that the computation-intensive and delay-sensitive tasks can be offloaded to edge computing servers for executing.

For the IoT devices distributed in a rural area or even a hostile environment, less communications and computation infrastructures are available for processing the sensed big data. Due to the low operational cost, deployment flexibility, and high mobility, unmanned aerial vehicle (UAV) is considered to be the optimal temporary platform for emergency scenarios without infrastructure [9,10]. With this backdrop, introducing multi-UAV-empowered edge computing paradigm for processing edge-side big data is necessary [11,12]. Multiple UAVs can cooperate with each other in the sky to accept the data processing tasks from the IoT devices on the ground to conduct DL-involved tasks. Typically, the application of processing a DL-involved task can be treated as a service function chain (SFC) [13], in which several service functions (SFs) are connected in a sequential order. For instance, each SF could be a modular in the DL-involved application, such as data pre-processing functions, deep neural network components, and target tracking, etc. An SF is a static software template that can derive instances on demand based on the virtual machine (VM) or docker technology [13–15]. A corresponding SF instance (SI) has to be created whenever the UAV decides to process a task. Once all the SIs of SFs contained in an SFC are successfully created, a task can pass through each instance sequentially to obtain its required services. In addition, in order to be able to run SFCs with AI algorithms, it is necessary to equip UAVs with custom-made AI processors. Compared with graphics processing unit (GPU), field programmable gate array (FPGA) has obvious characteristics of low power consumption and small size [16,17], which has been treated as a promising solution on the UAV platforms without violating size, weight, and power constraints inherent to UAV design. Considering that the UAV has limited computation resources and storage resources, the SFs contained in one SFC and their corresponding SIs can be distributed on multiple UAVs. When a large number of tasks are offloaded onto the UAV network at the same time stage, massive SIs corresponding to the required SFs will have to be created. In order to make full use of the limited computation and communication resources of UAVs, an efficient SFC scheduling strategy is indispensable. However, it also faces many challenges: (1) There is a complex matching relationship between tasks, SIs, and SFCs due to the heterogeneity of UAVs and the resource requirements of the tasks. (2) There is a complex trade-off between the communication and computation resource scheduling. (3) It is hard to achieve a long-term multi-objective optimization for the scenario with continuous task arrival and unknown SFC requirements. Tremendous efforts have been made in designing task scheduling algorithms in multi-UAV edge computing paradigms [18–34]. However, they often assume that each task is served by only one UAV and less attention has been paid to SFC scheduling in a multi-UAV edge computing scenario. A detailed analysis of existing efforts is presented in Section 2. With this backdrop, this paper firstly formulates the SFC scheduling problem as a 0–1 nonlinear integer programming problem. Then, a two-stage heuristic algorithm is put forward to derive a sub-optimal solution of the problem. The main contributions of this paper are threefold:

(1) The SFC scheduling problem in heterogeneous "CPU + FPGA" computation architecture is formulated as a 0–1 nonlinear integer programming problem. The overall revenue of the system and the completion time sum of tasks are optimized with various resource constraints. To the best of our knowledge, this is the fist paper that has studied the SFC scheduling problem considering FPGA resources in the multi-UAV edge computing network;

(2) To solve the NP-hard problem with coupling variables, a two-stage heuristic algorithm called ToRu is put forward. At the first stage, i.e., when the resources are abundant, the SFCs of all tasks are deployed to UAV edge servers in parallel based on our proposed pairing principle between SFCs and UAVs for minimizing the sum of all tasks' completion time; at the second stage, i.e., when the resources are insufficient, a revenue maximization heuristic method is adopted to deploy the arrived SFCs in a serial service mode. In order to obtain the long-term optimization, a time-slot partitioning protocol is designed, based on which ToRu can operate repeatedly in each time-slot;

(3) A series of experiments are conducted to evaluate the performance of our proposal. Experimental results show that our proposed ToRu algorithm outperforms other benchmark algorithms in the the sum of all tasks' completion time, the overall revenue, and the task execution success ratio.

The remaining of this paper is organized as follows. Section 2 summarizes related work. The model and problem formulation of the proposed system are introduced in Section 3. Section 4 describes the details of our proposed ToRu algorithm. The experiments and analysis of the results are presented in Section 5. Finally, Section 6 concludes this paper.

2. Related Work

In the multi-UAV edge computing system, according to the granularity of services provided to users, the existing work can be roughly divided into two categories: task scheduling and SFC scheduling. The difference between the two is mainly reflected in the matching process between offloaded tasks and UAVs. The former only considers whether the UAV's hardware computation resources (such as CPU, RAM, etc.) meet the task requirements; the latter further considers whether the SFs deployed on UAVs match the offload tasks requirements and is closer to the real situation.

2.1. Task Scheduling

Considering the high complexity of the multi-UAV edge computing system, most task scheduling optimization problems need to be solved by jointly optimizing user association, computation resource allocation, trajectories of UAVs, the number of offloaded task bits, etc. Most corresponding optimization models are non-convex, and it is difficult to obtain the optimal solution in polynomial time. An alternating iteration method is currently widely used to solve such complex non-convex optimization problems. It decouples a complex non-convex problem into multiple simplified sub-problems that can be solved by typical convex optimization or heuristic algorithm. Based on the alternating iteration algorithm, Yu et al. [18] jointly optimized the number of local computing tasks and offloaded tasks, trajectories of UAVs, and offloading matching strategy between UAVs and user terminals for minimizing the energy consumption of user terminals; Zhang et al. [19] jointly optimized user association, allocation of CPU frequency, power and spectrum resources, as well as trajectory of UAVs based on the proposed double-loop structure with the aim of maximizing the computation efficiency maximization. Luo et al. [20] jointly optimized the task scheduling, bit allocation, and UAV trajectory for minimizing the energy consumption of ground users. Wang et al. [21] presented a two-layer optimization method for minimizing system energy consumption, through jointly optimizing the deployment of UAVs and offloading decision. Moreover, some scholars formulated the task scheduling problem as a Markov decision process (MDP). A series of methods based on deep reinforcement learning (DRL) are proposed: Chang et al. [22] proposed a reinforcement learning framework with applying synthetic considerations of the terminals' demand, risk, and geometric distance, so as to provide better Quality-of-Service and path planning; Ren et al. [23] proposed a real-time scalable scheduling approach in the dynamic edge computing environments based on a DRL method; Xue et al. [24] established the User, UAV cost, and UAV revenue model and then jointly optimized power control, resource allocation, and UE association for minimizing the system energy consumption by a multi-agent reinforcement deep learning algorithm (MADRL). Seid et al. [25] deployed a clustered multi-UAV to provide computing services to users' devices and proposed a MADRL-based approach for minimizing the overall network computation cost while ensuring the quality of service. Wu et al. [26] designed a pre-dispatch UAV-assisted vehicular edge computing networks to cope with the demand of vehicles in multiple traffic jams, and then proposed a DRL-based energy efficiency autonomous UAV deployment strategy. Furthermore, strategies based on game theory [27,28] are also adopted to solve optimization problems. Each UAV was considered as an individual player with private interests, the optimization problem was formulated as an offloading game with at least one Nash equilibrium. Asheralieva et al. [29] presented a novel game-theoretic and reinforcement learning framework for task offloading. A Stackelberg game approach is adopted in [30,31] for maximizing utilities.

2.2. SFC Scheduling

Qu et al. [32] studied the service provisioning in the UAV-enabled MEC networks, where the SFs placement, UAV trajectory, task scheduling, and computation resource allocation were jointly optimized, so as to minimize the overall energy consumption of all users. A sub-optimal solution was achieved based on a proposed two-stage alternating optimization algorithm. However, the application considered in [32] only contains one SF, so that SFC scheduling is not involved.

Wang et al. [33] proposed a reconfigurable service provisioning framework based on SFCs for Space–Air–Ground-Integrated Networks (SAGINs), where the computation and communications resource consumptions are balanced. Li et al. [34] investigated the online mapping and scheduling of dynamic virtual network functions (VNFs) in SAGINs, in which an Internet of Vehicles (IoV) service can be represented by a SFC formed with a set of chained VNFs. However, in the proposed SFC construction process, only the capacity constraints of CPU and buffer on NVF nodes are considered, and the channel resources constraints between NVF nodes are ignored. In fact, the channel resources between NFV nodes are very scarce in the wireless environment, and it often becomes the bottleneck factor affecting the SFC construction. In addition, the SFC scheduling mentioned in [33,34] aims to establish an end-to-end route for servicing data, where the data source and destination nodes are separately located in different geographical regions. However, in the edge computing scenario considered in this paper, the SFC scheduling aims to provide services for requesting users, where the input of raw data and the output of computing results are all located in the same access node. Moreover, dedicated hardware resources (such as GPU, FPGA, etc.) supporting AI algorithm operation are also not considered in [33,34].

In summary, it is difficult to solve the SFC scheduling problem involved in this paper with existing algorithms. To the best of our knowledge, compared with existing works, this is the first paper that considers the SFC scheduling problem in a multi-UAV edge computing network with FPGAs resources, which can provide services for complicated intelligent application-oriented tasks in the weak infrastructure areas.

3. System Model and Problem Formulation

3.1. Network Model

In order to achieve a long-term SFC scheduling optimization, the management and allocation of UAV network resources is particularly important. Inspired by [35], a two-layer UAV network architecture is designed, in which one UAV with relatively high computing and communication resource capability is used as the master, other UAVs as the slavers. As shown in Figure 1, the edge computing network consists of one master UAV and M slave UAVs with heterogeneous computation resources, which evenly hovers over the mission area in a fully interconnected manner for performing time-sensitive AI tasks. The master UAV controls all the computation and communication resources on the slave UAVs, and it is responsible for instantiating the required SFCs. The slave UAVs are mainly used to cover devices on the ground, receive the SFC requests and offloaded time-sensitive AI tasks. When a slave UAV receives the SFC requirement of a ground device, it immediately reports to the master UAV. Then, the master UAV creates an SFC instance for the requiring ground device based on the resource of one or multiple salve UAVs. Eventually, the task offloaded from the requiring ground device is processed on the created SFC instance. For simplicity of description, the "heterogeneous Multi-UAV edge computing network" will be referred to as "UAV network" for short in the rest of this paper. The slave UAV network can be denoted by a directed complete graph $\mathcal{G} = (\mathcal{U}, \mathcal{L})$, where $\mathcal{U}$ is the slave UAV set, denoted as $\mathcal{U} = \{U_1, U_2, \ldots, U_m, \ldots, U_M\}, 1 \leq m \leq M$. Several SFs are deployed on each slave UAV, which can be run by virtual machines (VMs) or docker technology [13–15]. $\mathcal{L}$ is the set of wireless links between slave UAVs, denoted as $\mathcal{L} = \{L_1, L_2, \ldots, L_m, \ldots, L_M\}, 1 \leq m \leq M$. $L_m = \{L_{m,1}, L_{m,2}, \ldots, L_{m,m'}, \ldots, L_{m,M}\}$ represents the wireless link sets of U_m transmitting data to other UAVs, in which $L_{m,m'} (1 \leq m' \leq M)$ indicates the wireless link of U_m to $U_{m'}$. We assume that UAVs use orthogonal frequency channels for communication without collisions (e.g., OFDMA [36]). Thus, $L_{m,m'}$ is represented as a two-tuple: $\{r_{m,m'}, N_m^c\}$. $r_{m,m'}$ is the data transmission rate on one sub-channel, N_m^c represents the current idle sub-channel number. We define N_{max}^c as the maximum number of sub-channels on each slave UAV. $N_m^c \leq N_{max}^c$ always holds.

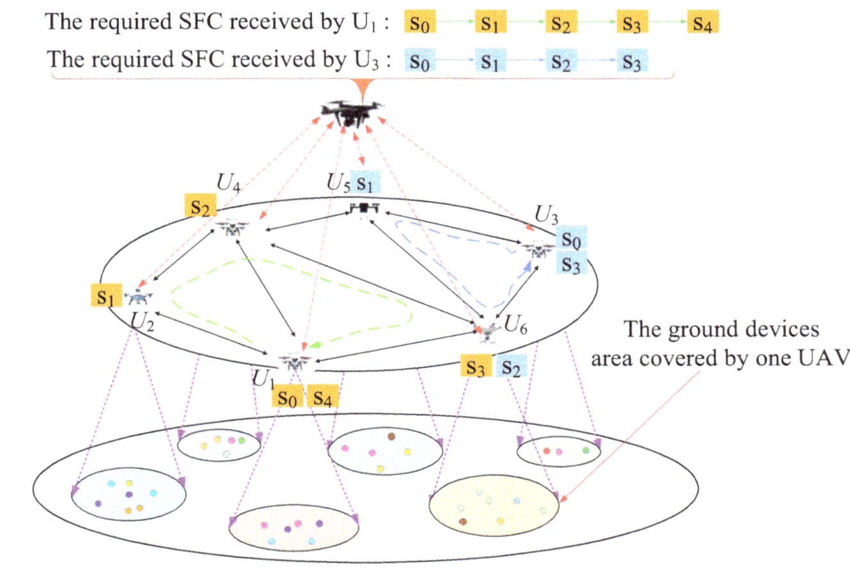

Figure 1. A heterogeneous Multi-UAV edge computing scenario. Each UAV deploys several SFs and can receive the SFC required from the devices on the ground. The master UAV controls the resources of the edge computing network, which can instantiate the required SFCs.

Considering that UAVs have the characteristics of line-of-sight (LoS) communication, a free space attenuation model is adopted [36]. Then, $r_{m,m'}$ can be expressed as:

$$r_{m,m'} = B \log_2\left(1 + \frac{p_m \beta_0}{N_0 B d_{m,m'}^2}\right), 1 \leq m, m' \leq M, m \neq m'. \tag{1}$$

where β_0 is the transmit power gain at a reference distance of one meter. p_m is the transmission power of U_m. N_0 is the noise power spectrum density. B is the bandwidth of a sub-channel. $d_{m,m'}$ is the distance between U_m and $U_{m'}$. Note that when $m = m'$, $r_{m,m'} = \infty$. In other words, this paper ignores the data exchanging time between two SF instances at the same UAV. For ease of description, the main notations used in this paper are listed in Table 1.

Table 1. Notations.

Notation	Definition
U_n	The n-th UAV in $\mathcal{U}$
SF	Service Function
SI	SF instances
SFC	Service Function Chain
SF_n	The n-th SF
$\widetilde{S_1}$	The set of FPGA-independent SFs
$\widetilde{S_2}$	The set of FPGA-dependent SFs
Q_1	The number of FPGA-independent SFs in $\widetilde{S_1}$
Q_2	The number of FPGA-dependent SFs in $\widetilde{S_2}$

Table 1. *Cont.*

Notation	Definition
$\mathcal{S}_m$	The set of SF deployed on U_m
$SF^*_{m,q}$	The q-th SF deployed on U_m
$\mathcal{T}$	The set of tasks currently requesting services
T_n	The n-th requesting task
N^{cpu}_m	The number of CPU cores on U_m
N^{fpga}_m	The number of FPGAs on U_m
$N^{cpu}_{m,a}$	The number of the current idle CPU cores on U_m
$N^{fpga}_{m,a}$	The number of the current idle FPGAs on U_m
$n^{si}_{m,q}$	The number of SIs currently created corresponding to $SF^*_{m,q}$
$f_{m,q}$	The frequency of CPU core on U_m
$a_{m,q}$	The processing speed of CP when running SI of $SF^*_{m,q}$
$L_{m,m'}$	The wireless link of U_m to $U_{m'}$
$r_{m,m'}$	The data transmission rate on one sub-channel of the link $L_{m,m'}$
N^c_m	The number of current idle sub-channels on the link $L_{m,m'}$
o_n	The source UAV that receives T_n
s_n	The required SFC of T_n
s^k_n	The k-th SF contained in the SFC required by T_n
l_n	The properties of T_n before its entering into an SFC
l^k_n	The properties of T_n when arriving at the instance of s^k_n in its SFC
$l^{k,0}_n$	The length of T_n when arriving at s^k_n
$l^{k,1}_n$	The number of CPU cycles required for processing one bit task on s^k_n
$l^{k,2}_n$	The number of AI accelerator operations required for processing one bit task on s^k_n
r_n	The minimum requirement set for the transmission rate between the instances of SFCs required by T_n
$r^{k,k+1}_n$	The minimum transmission rate requirement from the instance of s^k_n to s^{k+1}_n
c_n	The minimum computation resources requirement set of T_n for the instances of SFC.
$c^{k,0}_n$	The minimum CPU processing speed requirement of T_n on the instances of s^k_n.
$c^{k,1}_n$	The minimum AI accelerator processing speed requirement of T_n on the instances of s^k_n.
v_n	The revenue obtained through completing T_n
T^c_n	The completion time of T_n
$X^m_{n,k}$	The decision variable of creating the instance of s^k_n

3.2. Service Model

In this paper, SFs are divided into two categories based on their resource requirements: FPGA-independent SFs and FPGA-dependent SFs [15]. An SI corresponding to the FPGA-independent SF is created only based on CPU resource. In contrast, only the combination of CPU and FPGA resources can support the SI corresponding to the FPGA-dependent SF. The set of FPGA-independent SFs are denoted by $\widetilde{\mathcal{S}_1} = \{SF_1, SF_2, \ldots, SF_n, \ldots, SF_{Q_1}\}$. SF_n represents the n-th type of FPGA-independent SF, Q_1 represents the number of FPGA-independent SF types. Moreover, the set of FPGA-dependent SFs are denoted by $\widetilde{\mathcal{S}_2} = \{SF_{Q_1+1}, SF_{Q_1+2}, \ldots, SF_{n'}, \ldots, SF_{Q_1+Q_2}\}$. $SF_{n'}$ represents the n'-th type of FPGA-dependent SF, Q_2 is the number of FPGA-dependent SF types. Considering the heterogeneity of UAVs, the SF set deployed on each UAV may be different. Therefore, the set of SFs deployed on U_m are denoted by $\mathcal{S}_m = \{SF^*_{m,1}, SF^*_{m,2}, \ldots, SF^*_{m,q}, \ldots, SF^*_{m,Q_m}\}$, $Q_m \leq (Q_1 + Q_2)$ and $\mathcal{S}_m \subseteq (\widetilde{\mathcal{S}_1} \cup \widetilde{\mathcal{S}_2})$. $SF^*_{m,q}$ is the q-th SF deployed on U_m.

As shown in Figure 2, a general computation model for UAVs is considered in this paper, which includes two parts: CPU and FPGA resources. The former includes several CPU cores of the same type; the latter refers to some FPGAs, which can assist CPUs in performing AI workload, such as FPGA-based convolutional network accelerator (CNN) [17]. The connection between CPU cores and FPGAs can be established dynamically through internal bus (e.g., PCIe bus) [37]. The number of CPU cores and FPGAs on U_m are denoted by N^{cpu}_m and N^{fpga}_m, respectively. Accordingly, $N^{cpu}_{m,a}$ and $N^{fpga}_{m,a}$ separately represent the number of current idle CPU cores and FPGAs. As shown in Figure 2, when creating an SI of $SF^*_{1,1}$, U_1 firstly assigns idle CPU core #1 and FPGA #1 to it; then, the corresponding SF software is separately loaded onto them. In contrast, only idle CPU core #4 is assigned to $SF^*_{1,4}$ before loading its SF software. The processing capacity of an SI corresponding to $SF^*_{m,q}$

is expressed as a two-tuple: $\{f_{m,q}, a_{m,q}\}$. $f_{m,q}$ is the frequency of CPU core on U_m, measured in GHz; $a_{m,q}$ is the processing speed of FPGA when running the SI of $SF^*_{m,q}$, measured in GOP/s [38]. Note that $a_{m,q} = 0$ when $SF^*_{m,q}$ is an FPGA-independent SF.

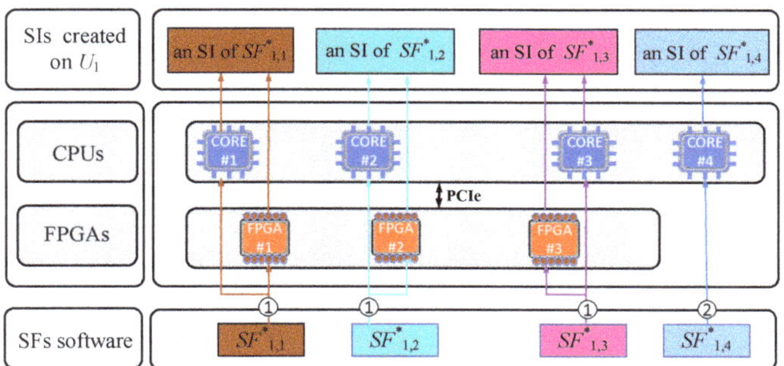

Figure 2. The creation process of SIs on U_1. The SI of FPGA-dependent SF (i.e., $SF^*_{1,1}$, $SF^*_{1,2}$, and $SF^*_{1,3}$) occupies a CPU core and an FPGA. Moreover, the SI of FPGA-independent SF (i.e., $SF^*_{1,4}$) only occupies a CPU core.

From the perspective of information security, we stipulate that an SI independently occupies a CPU core or a combination of 'CPU + FPGA'. Therefore, the maximum number of SIs that can be created simultaneously on one UAV is bounded. Define $n^{si}_{m,q}$ to represent the number of currently created SIs corresponding to $SF^*_{m,q}$ on U_m, and the following constraints must be satisfied [15].

$$n^{si}_{m,q} \leq N^{cpu}_m, SF^*_{m,q} \in \mathcal{S}_1, 1 \leq m \leq M. \tag{2}$$

$$n^{si}_{m,q} \leq N^{fpga}_m, SF^*_{m,q} \in \mathcal{S}_2, 1 \leq m \leq M. \tag{3}$$

$$\sum_{SF^*_{m,q} \in (\mathcal{S}_m \cap \widetilde{\mathcal{S}_2})} n^{si}_{m,q} \leq N^{fpga}_m, 1 \leq m \leq M. \tag{4}$$

$$\sum_{SF^*_{m,q} \in \mathcal{S}_m} n^{si}_{m,q} \leq N^{cpu}_m, 1 \leq m \leq M. \tag{5}$$

Equation (2) ensures that the number of currently created SIs corresponding to any FPGA-independent SF on U_m does not exceed the number of CPU cores on it. Equation (3) means that the number of currently created SIs corresponding to any FPGA-dependent SF on U_m does not exceed the number of FPGAs on it. Equation (4) guarantees that the total number of currently created SIs corresponding to all FPGA-dependent SFs on U_m does not exceed the number of FPGAs on it. Equation (5) restricts the total number of currently created SIs on U_m to not exceed the number of CPU cores on it.

3.3. Task Model

The tasks currently requesting service are denoted as a set $\mathcal{T} = \{T_1, T_2, \ldots, T_n, \ldots, T_{N_t}\}$. $N_t = |\mathcal{T}|$ denotes the total number of tasks. We define a six-tuple for T_n as follows: $T_n = \{o_n, s_n, l_n, r_n, c_n, v_n\}, 1 \leq n \leq N_t$. o_n shows the source UAV that receives T_n. s_n indicates its required SFC, denoted as $s_n = \{s^0_n, s^1_n, \ldots s^k_n, \ldots, s^{N^s_n}_n, s^{N^s_n+1}_n\}, 0 \leq k \leq N^s_n + 1$. s^0_n is the SF that receives the tasks from the ground and transmits it to the s^1_n, which has to be instantiated at the source UAV, as shown in Figure 1. $s^{N^s_n+1}_n$ is the SF that receives the computing result from $s_n^{N_n^s}$ and transmits it to the ground, which also has to be instantiated at the source UAV. N_n^s represents the total number of SFs contained in the required SFC. When $1 \leq k \leq N_n^s$, $s_n^k \in \mathcal{S}$ represents the k-th SF contained in the required SFC, which can be instantiated on any slave UAV that satisfies its resource requirements. l_n means the current properties of T_n when it arrives at each SF contained in s_n, denoted as $l_n = \{l_n^0, l_n^1, l_n^2, \ldots l_n^k, \ldots, l_n^{N_n^s}, l_n^{N_n^s+1}\}$, $0 \leq k \leq N_n^s + 1$. l_n^k is the property of T_n when it arrives at s_n^k, denoted as $l_n^k = \{l_n^{k,0}, l_n^{k,1}, l_n^{k,2}\}$. $l_n^{k,0}$ is the current length of T_n; $l_n^{k,1}$ is the number of CPU cycles required for processing one bit task; $l_n^{k,2}$ is the number of AI accelerator operations required for processing one bit task. The computation resources consumed by s_n^0 and $s_n^{N_n^s+1}$ are considered to be negligible in this paper, since their instances require far less computation resources than other AI instances. Therefore, $l_n^{0,0}$ and $l_n^{N_n^s+1,0}$ represents the initial length of T_n and its computing result, respectively; $l_n^{0,1}$, $l_n^{0,2}$, $l_n^{N_n^s+1,1}$ and $l_n^{N_n^s+1,2}$ are equal to 0. r_n represents the minimum transmission rate requirement between the SFs contained in the required SFC, denoted as $r_n = \{r_n^{0,1}, r_n^{1,2}, \ldots, r_n^{N_n^s-1,N_n^s}, r_n^{N_n^s,N_n^s+1}\}$. $r_n^{k,k+1}$ ($0 \leq k \leq N_n^s$) indicates the minimum transmission rate requirement from s_n^k to s_n^{k+1}. c_n represents the minimum computation resources requirement when creating the instances of SFs contained in the required SFC, denoted as $c_n = \{c_n^0, c_n^1, c_n^2, \ldots, c_n^k, \ldots, c_n^{N_n^s}, c_n^{N_n^s+1}\}$, $0 \leq k \leq N_n^s + 1$. c_n^k is the minimum computation resource requirement of initiating s_n^k, denoted as $c_n^k = \{c_n^{k,0}, c_n^{k,1}\}$. $c_n^{k,0}$ is the minimum processing speed requirement for CPU, measured in GHz; $c_n^{k,1}$ is the minimum processing speed requirement for AI accelerator, measured in GOP/s. For c_n^0 and $c_n^{N_n^s+1}$, their values are 0. v_n represents the revenue obtained through completing T_n.

As shown in Figure 1, an SFC is unidirectional, and the task goes through each SF sequentially. Furthermore, UAVs adopt the communication mode of orthogonal frequency division multiple access. Therefore, the communication resources consumed between adjacent SFs belong to the UAV instantiating the upstream SF. To sum up, for task T_n, it is more reasonable to create the SF instances in the reverse order of the required SFC, i.e., the instance of $s_n^{N_n^s+1}$ is created first, and the instance of s_n^0 is created last. Assume that the instance of s_n^{k+1} ($0 \leq k \leq N_n^s$) has been created on $U_{m'}$. If we want to continue creating the instance of s_n^k on U_m, the following conditions have to be satisfied at the same time:

$$s_n^k = SF_{m,q}^*, 1 \leq m \leq M, 1 \leq q \leq Q_m \tag{6}$$

$$c_n^{k,0} \leq f_{m,q}, 1 \leq m \leq M, 1 \leq q \leq Q_m \tag{7}$$

$$c_n^{k,1} \leq a_{m,q}, 1 \leq m \leq M, 1 \leq q \leq Q_m \tag{8}$$

$$\left\lceil \frac{r_n^{k,k+1}}{r_{m,m'}} \right\rceil \leq N_m^c, m \neq m'. \tag{9}$$

Equation (6) means that U_m has to deploy the SF matching s_n^k. Equations (7) and (8) ensure that the SI of s_n^k on U_m can satisfy the computing requirement of T_n. Equation (9) guarantees that U_m has enough idle sub-channels for transmitting T_n to $U_{m'}$ at a rate no smaller than the required. Moreover, define t_n^k as the stay time of T_n on the SI of $s_{n,k}$, which includes the executing time $t_n^{k,1}$ and transmitting time $t_n^{k,2}$. $t_n^{k,1}$ can be expressed as:

$$t_n^{k,1} = \begin{cases} \dfrac{l_n^{k,0} l_n^{k,1}}{f_{m,q}} + \dfrac{l_n^{k,0} l_n^{k,2}}{a_{m,q}}, & 1 \leq k \leq N_n^s. \\ 0, & k = 0, k = N_n^s + 1. \end{cases} \tag{10}$$

$$
t_n^{k,2} = \begin{cases} \dfrac{l_n^{k+1}}{\lceil \frac{r_n^{k,k+1}}{r_{m,m'}} \rceil r_{m,m'}}, & m \neq m', 0 \leq k \leq N_n^s \\[2ex] 0, & m = m', 0 \leq k \leq N_n^s. \end{cases}
\tag{11}
$$

We define the binary variable $X_{n,k}^m$ to represent the decision of creating the SI of s_n^k, which can be expressed as:

$$
X_{n,k}^m = \begin{cases} 1, & \text{if the SI of } s_n^k \text{ is created on } U_m \text{ with satisfying the constraints} \\ & \text{of Equations (6) to (9)} \\ 0, & \text{otherwise.} \end{cases}
\tag{12}
$$

Considering that the SI of each SF contained in the required SFC is created at most one time, the following constraints must be satisfied:

$$
\sum_{m=1}^{m=M} X_{n,k}^m \leq 1
\tag{13}
$$

Note that $X_{n,0}^{O_n} = 1$ and $X_{n,N_n^s+1}^{O_n} = 1$ always hold, since the SIs of s_n^0 and $s_n^{N_n^s+1}$ have to been created on the source UAV. T_n can be successfully executed only if the SIs of all SFs contained in its required SFC have been successfully created. At this time, the following formula holds.

$$
\prod_{k=0}^{k=N_n^s+1} \sum_{m=1}^{m=M} X_{n,k}^m = 1.
\tag{14}
$$

Lastly, we define T_n^c as the completion of T_n, which is expressed as follows:

$$
T_n^c = \sum_{m=1}^{m=M} \sum_{k=0}^{k=N_n^s+1} X_{n,k}^m (t_n^{k,1} + t_n^{k,2}), 1 \leq n \leq N
\tag{15}
$$

$$
s.t. : \text{Equations (1) to (14)}.
$$

3.4. Problem Formulation

For the required SFC, there may be multiple strategies of creating its SIs. For a task, it wants to be executed on the UAVs that can minimizes its completion time. On the other hand, the UAV network wants to process all received tasks by making full use of their limited computation resources and communication resources, so as to maximum the revenue. In this paper, we aim at minimizing the completion time sum of all task while maximizing the overall revenue of the UAV network by optimizing $X = \{X_{n,k}^m | 1 \leq n \leq N, 1 \leq m \leq M, 0 \leq k \leq N_n^s + 1\}$.

Therefore, the first optimization goal is formulated as:

$$
\mathcal{F}_1 = \sum_{n=1}^{n=N} T_n^c \prod_{k=0}^{k=N_n^s+1} \sum_{m=1}^{m=M} X_{n,k}^m
\tag{16}
$$

The second optimization goal is formulated as:

$$
\mathcal{F}_2 = \sum_{n=1}^{n=N} v_n \prod_{k=0}^{k=N_n^s+1} \sum_{m=1}^{m=M} X_{n,k}^m
\tag{17}
$$

According to Equations (16) and (17), a multi-objective optimization problem can be formulated as:

$$\mathbf{P_1} : \min_{\{X\}}(\mathcal{F}_1, -\mathcal{F}_2)$$

$$s.t. C_1 : 1 \leq n \leq N, 1 \leq m \leq M, 1 \leq q \leq Q_m, \, 0 \leq k \leq N_n^s + 1$$

$$C_2 : s_n^k = SF_{m,q}^*, \forall X_{n,k}^m = 1 \tag{18}$$

$$C_3 : X_{n,0}^{O_n} = 1, X_{n,N_n^s+1}^{O_n} = 1$$

$$C_4 : Equations \, (1) \sim (15)$$

Constraint (C_1) specifies the valid ranges of the involved variables in Constraint (C_2)~(C_4). Constraint (C_2) guarantees that the UAV of instantiating the required SF should deploy this SF in advance. Constraint (C_3) restricts that the SIs of the first SF and last SF of an SFC have to be created on the source UAV. Constraint (C_4) includes several constraints related to the resource requirements and the resource capacities, which are the described in detail after Equations (1)~(15).

4. Proposed Approach

In P_1, minimizing $\mathcal{F}_1$ is a 0–1 nonlinear integer programming problem. Minimizing $-\mathcal{F}_2$ is equivalent to maximizing $\mathcal{F}_2$, which is also a 0–1 nonlinear integer programming problem. At the same time, there is a close coupling relationship between $\mathcal{F}_1$ and $\mathcal{F}_2$. Furthermore, the task properties considered in this paper are not known in advance. Therefore, it is difficult to solve P_1 effectively with traditional optimization algorithms in an online manner. To tackle this problem in an online manner, we propose an efficient online two-stage heuristic algorithm named ToRu with much lower complexity. In ToRu, the minimization of the completion time sum of tasks is pursued in the case of abundant resources (i.e., the first stage); on the contrary, when the resources are insufficient (i.e., the second stage), the maximization of the UAV network revenue is pursued.

4.1. The ToRu Framework

The proposed ToRu is deployed on the master UAV. In order to providing a long-term service for unknown tasks in an online manner, a time-slot partition protocol is designed, as shown in Figure 3. The mission period is divided into time-slots with equal length. One time-slot is the basic unit for SFC scheduling. At the beginning of each time-slot, the master UAV starts ToRu with the requesting tasks arriving in the previous time-slot and the current idle resources on the UAV network as input parameters. ToRu completes SFC scheduling before the end of a time-slot and exits. According to the above process, ToRu is executed repeatedly in each time-slot, thus enabling the UAV network to provide the long-term service. Note that ToRu must complete SFC scheduling before the end of a time-slot, otherwise the time-slot length needs to be increased, which indicates that the number of task requesting service is too large. If ToRu completes SFC scheduling very early before the end of the time-slot, it means that the number of tasks requesting service is small and the time-slot length should be shortened, thus reducing the service waiting time of tasks. Therefore, the time-slot partition protocol proposed in this paper has good dynamic scalability.

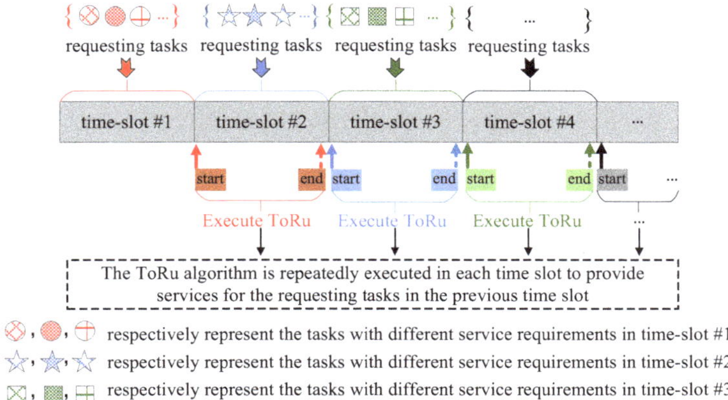

Figure 3. The time-slot partitioning protocol. The ToRu algorithm is started by the master UAV at the beginning of each time-slot and ends at the end of each time-slot, which is repeatedly executed in each time-slot to provide services for the requesting tasks in the previous time-slot. For example, ToRu is executed in time-slot #2 to only provide services for requesting tasks in time-slot #1.

The pseudocode of the ToRu algorithm is illustrated in Algorithm 1. Firstly, it calls Algorithm 2 with the task set $\mathcal{T}$, the idle communication resource set $\mathcal{L}$ and idle computation resource set $\mathcal{R}$ in the current time-slot as the input arguments. Note that $\mathcal{R}$ is expressed as $\mathcal{R} = \{R_1, R_2, \ldots, R_m, \ldots, R_M\}$. The $R_m = \{N_{m,a}^{cpu}, N_{m,a}^{fpga}, \mathcal{S}_m\}$ means the current idle computation resource and deployed SFs on U_m. Then, the result returned by Algorithm 2 is denoted as a four-tuple: $\{\widetilde{X}, \widetilde{S_u}, \widetilde{\mathcal{F}_1}, \widetilde{\mathcal{F}_2}\}$. $\widetilde{X}$ indicates a sub-optimal of problem P_1. $\widetilde{S_u}$ means the task execution success ratio based on $\widetilde{X}$, that is, the ratio of the number of successfully completed tasks to the total number. $\widetilde{\mathcal{F}_1}$ and $\widetilde{\mathcal{F}_2}$ shows the sum of the task completion time and the overall revenue based on $\widetilde{X}$, respectively. If $\widetilde{S_u} = 1$, the UAV network obtains the complete revenue from all tasks, and tasks also obtain the approximate minimum completion time sum; otherwise, it indicates that the UAV resources are insufficient, and Algorithm 3 is called for maximizing the overall revenue, regardless of the task completion time. This is reasonable, since when resources are insufficient, the number of the successfully completed tasks is more important than the task completion time. Algorithm 3 has the same type of the input parameters and output results with Algorithm 2. However, the $\widetilde{\mathcal{F}_1}$ returned by Algorithm 3 only represents the completion time sum of tasks that have been executed successfully.

Algorithm 1: General Framework of ToRu.

Input: the task set $\mathcal{T}$, the idle communication resource set $\mathcal{L}$, and the idle computation resource set $\mathcal{R}$.

Output: $\{X, S_u, \mathcal{F}_1, \mathcal{F}_2\}$.

1: Initialize: $X = 0$, $S_u = 0$, $\mathcal{F}_1 = 0$, $\mathcal{F}_2 = 0$; \\ S_u represents the task execution success ratio based on X.

2: Obtain $\{\widetilde{X}, \widetilde{S_u}, \widetilde{\mathcal{F}_1}, \widetilde{\mathcal{F}_2}\}$ by calling Algorithm 2 with $\mathcal{T}$, $\mathcal{L}$ and $\mathcal{R}$ as the input parameters.

3: **if** $S_u < 1$ **then**

4:     Obtain $\{\widetilde{X}, \widetilde{S_u}, \widetilde{\mathcal{F}_1}, \widetilde{\mathcal{F}_2}\}$ by calling Algorithm 3 with $\mathcal{T}$, $\mathcal{L}$ and $\mathcal{R}$ as the input parameters.

5: **end if**

6: $\{X, S_u, \mathcal{F}_1, \mathcal{F}_2\} = \{\widetilde{X}, \widetilde{S_u}, \widetilde{\mathcal{F}_1}, \widetilde{\mathcal{F}_2}\}$.

7: **return** $\{X, S_u, \mathcal{F}_1, \mathcal{F}_2\}$.

Algorithm 2: Minimizing the completion time sum.

Input: the task set $\mathcal{T}$, the idle communication resource set $\mathcal{L}$, the idle computation resource set $\mathcal{R}$.
Output: $\{X, S_u, \mathcal{F}_1, \mathcal{F}_2\}$.
1: Initialize: $X = \mathbf{0}$, $S_u = 0$, $\mathcal{F}_1 = 0$, $\mathcal{F}_2 = 0$, $L_{max} = 0$, $N_t = 0$, $sp = 0$, $n = 0$, $k = 0$.
2: $N_t = |\mathcal{T}|$.
3: L_{max} = the maximum length of SFCs in $\mathcal{T}$.
4: Instantiate the last SF of the SFC on its source UAV for each task.
5: $T_{temp} = Null.\backslash\backslash$ Define a temporary set.
6: **for** $p = 0$ to $p = L_{max} - 2$ **do**
7:    **for** $n = 1$ to N_t **do**
8:       $k = N_n^s - p$.
9:       **if** $k < 0$ **then**
10:          Continue.$\backslash\backslash$The SFC of T_n has been successfully instantiated.
11:       **else**
12:          Add the T_n to the set T_{temp}.
13:       **end if**
14:    **end for**
15:    **while** $T_{temp} \mathrel{!=} NULL$ **do**
16:       **for** $i = 1$ to $|T_{temp}|$ **do**
17:          Extract the i-th task T_{i^*} from the set T_{temp}, $T_{i^*} \in \mathcal{T}$.
18:          Compute the candidate UAVs of the SF $s_{i^*}^k$ for task T_{i^*} according to $C_2 \sim C_4$.
19:          **if** no the candidate UAVs **then**
20:             $S_u = 0$.
21:             **return** $\{X, S_u, \mathcal{F}_1, \mathcal{F}_2\}.\backslash\backslash$There are tasks that cannot be completed as required.
22:          **end if**
23:       **end for**
24:       Select one task T_{n^*} from the set T_{temp} and instantiate its SF $s_{n^*}^k$ on its optimal candidate UAV U_{m^*} based on our proposed principle in Section 4.2.
25:       The remaining resources of U_{m^*} are refreshed through subtracting the resources consumed by the instance of $s_{n^*}^k$.
26:       $X_{n^*,k}^{m^*} = 1$, and delete T_{n^*} from T_{temp}.
27:    **end while**
28: **end for**
29: $S_u = 1$.
30: Obtain the value of $\mathcal{F}_1$ and $\mathcal{F}_2$ according to Equation (16) and Equation (17).
31: **return** $\{X, S_u, \mathcal{F}_1, \mathcal{F}_2\}$.

Algorithm 3: Maximizing the overall revenue.

Input: the task set $\mathcal{T}$, the idle communication resource set $\mathcal{L}$, and the idle computation resource set $\mathcal{R}$.
Output: $\{X, S_u, \mathcal{F}_1, \mathcal{F}_2\}$.
1: Initialize: $X = \mathbf{0}$, $S_u = 0$, $\mathcal{F}_1 = 0$, $\mathcal{F}_2 = 0$, $N_t = 0$, $N_{fail} = 0$, $n = 0$, $k = 0$;
2: $N_t = |\mathcal{T}|$;
3: Sort the tasks in $\mathcal{T}$ according to the value of v_n / N_n^s. The larger the value, the higher the ranking.
4: **for** $n = 1$ to N_t **do**
5:    Instantiate the SF $s_n^{N_n^s+1}$ of the task T_n
6:    **for** $k = N_n^s$ to 0 **do**
7:       Compute the candidate UAVs of the SF s_n^k for the task T_n according to $C_2 \sim C_4$;
8:       **if** no the candidate UAVs **then**
9:          $N_{fail} + +$;
10:          Release the resources previously allocated to the task T_n, and clear the corresponding decision variable in X;
11:          Continue;
12:       **end if**
13:       Select an optimal candidate UAV U_m^* based on the principle proposed in Section 4.3 for instantiating the SF s_n^k;
14:       The remaining resources of U_{m^*} are refreshed through subtracting the resources consumed by the instance of $s_{n^*}^k$.
15:       $X_{n,k}^{m^*} = 1$;
16:    **end for**
17: **end for**
18: $S_u = (N_t - N_{fail}) / N_t$;
19: Obtain the value of $\mathcal{F}_1$ and $\mathcal{F}_2$ according to Equation (16) and Equation (17).
20: **return** $\{X, S_u, \mathcal{F}_1, \mathcal{F}_2\}$.

4.2. Suboptimal Solution to Minimize the Completion Time Sum

When instantiating an SFC for a requesting task, the UAV network not only needs to provide sufficient computation resources for each SF instance, but also to ensure that there are sufficient wireless link resources between adjacent SF instances. In order to improve the efficiency of SFC instantiation, this paper implements SF instantiation one-by-one according to the reverse order of SFC. As shown in Figure 4, the downstream SF is first instantiated on a UAV; then, the candidate UAVs that can instantiate the upstream SF are found according to $C_2 \sim C_4$; finally, in order to efficiently match SFCs and computing resources, so as to minimize the completion time sum of all tasks, this paper instantiates the SFCs of all tasks in parallel mode with sufficient UAV network resources. This parallel mode refers to the fact that only one SF in each SFC can be instantiated in a round of SF instantiation. After multiple rounds of SF instantiation, the instantiations of all SFCs can be completed. If all SFCs have the same length, the instantiations of all SFCs are completed at the same time.

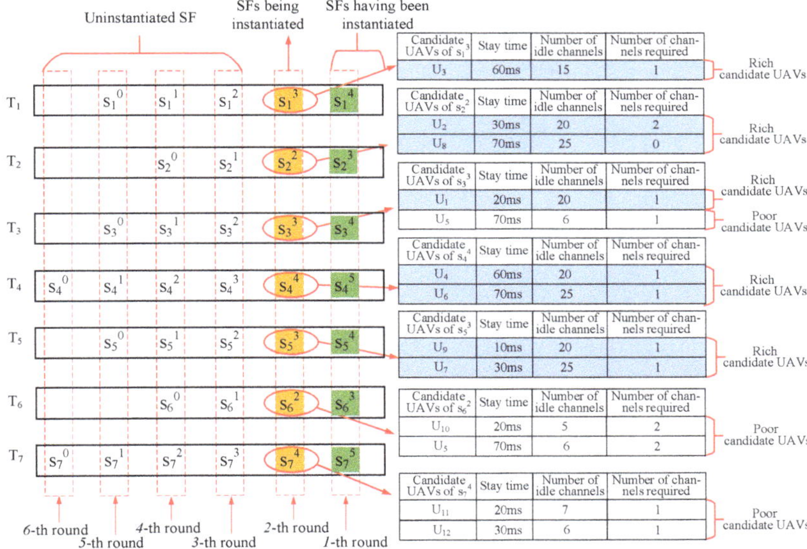

Figure 4. Workflow chart of Algorithm 2. The last SF of the SFCs of $T_1 \sim T_7$ is simultaneously instantiated on the source UAV in the 1-st round. In the 2-nd round, according to our proposed principle, the candidate UAVs for the upstream SFs (i.e., s_1^3, s_2^2, s_3^3, s_4^4, s_5^3, s_6^2 and s_7^4) is obtained, respectively; then, these upstream SFs is instantiated one by one; next, go into the next round. When the 6-th round is successfully completed, the SFCs of all tasks are successfully instantiated and Algorithm 2 exits.

A typical process of the SFCs instantiation in parallel mode is shown in Figure 4. Assume that the SFCs of task $T_1 \sim T_7$ need to be instantiated. The last SF of each SFC is first simultaneously instantiated on the source UAV (i.e., 1-st round). Then, according to $C_2 \sim C_4$, we separately identify candidate UAVs that can instantiate the upstream SF (i.e., 2-nd round). Furthermore, we calculate the corresponding task stay time and the number of sub-channels occupied when the upstream SF is instantiated on different candidate UAVs. Finally, we select one optimal UAV for each upstream SF from the candidates based on our proposed principle, which is described below:

1. The upstream SF with only one candidate UAV firstly are instantiated, which is beneficial to maximizing $\mathcal{F}_2$. As shown in Figure 4, s_1^3 contained in T_1 has one

candidate UAV, so it is firstly instantiated. When facing multiple such upstream SFs (like s_1^3), randomly select one of them for instantiation;

2. When all upstream SFs have multiple candidate UAVs, select the upstream SF with the candidate UAV that does not occupy the sub-channel, and instantiate it on this candidate UAV. As shown in Figure 4, s_2^2 contained in T_2 has multiple candidate UAVs, and if it is instantiated on the candidate U_8, no sub-channel is occupied. Therefore, s_2^2 should be first initialized in the current situation. When facing multiple such upstream SFs (like s_2^2), randomly select one of them for instantiation;

3. If there is no upstream SF that satisfies the above principle 1 or principle 2, the candidate UAVs of each upstream SF are divided into two categories based on the number of idle channels: the candidate UAVs with the number of idle channels greater than N_e are called "rich candidate UAVs", the remaining UAVs are called "poor candidate UAVs". Considering the shortage of UAV wireless link resources, the upstream SFs should be instantiated preferentially on candidate UAVs with abundant link resources, i.e.,"rich candidate UAVs", which is beneficial to maximizing $\mathcal{F}_2$. Therefore, we first select an optimal UAV for the upstream SF with "rich candidate UAVs", the specific principles are as follows:

   (a) The upstream SFs with only one "rich candidate UAV" are first instantiated. As shown in Figure 4, s_3^3 contained in T_3 has only one "rich candidate UAV", so it is instantiated first. When facing multiple such upstream SFs (e.g., s_3^3), randomly select one of them for instantiation;

   (b) When the remaining upstream SFs have multiple "rich candidate UAVs", we rank their candidate UAVs according to the stay time of a task executed on them, and the candidate UAV with short stay time is ranked higher. The upstream SF with the largest gap in the stay time between its first-ranked candidate UAV and its second-ranked candidate one will first be instantiated on the first candidate one, which is beneficial to minimizing $\mathcal{F}_1$. As shown in Figure 4, both s_4^4 contained in T_4 and s_5^3 contained in T_5 have two "rich candidate UAV". The gap in the stay time of s_4^4 (s_5^3) on its different candidate UAVs is 10 ms (20 ms), so s_5^3 is first instantiated on the candidate U_9. When the gap is the same, select one of them at random for instantiation

In addition, when all upstream SFs with "rich candidate UAVs" have been instantiated and there are still uninstantiated upstream SFs, i.e., the upstream SFs with only "poor candidate UAVs", we regard the "poor candidate UAVs" as "rich candidate UAVs" and select the optimal UAVs for the uninstantiated SFs according to the above principle (a) and (b). As shown in Figure 4, both s_6^2 contained in T_6 and s_7^4 contained in T_7 have only "poor candidate UAV", so they are lastly instantiated according to the above principle (a) and (b). Note that once an SF is successfully instantiated, all candidate UAVs belonging to uninstantiated SFs must be updated immediately before starting to select the optimal UAV for the next uninstantiated SF (i.e., step 16~25 in Algorithm 2). Repeat the above operations until the SFCs of all tasks are instantiated, whose pseudocode is as shown in Algorithm 2.

4.3. Suboptimal Solution to Maximize the Overall Revenue

When the UAV network resources are insufficient, in order to obtain more revenue, UAVs naturally give priority to serving tasks that can pay more on average for each SF instance, i.e., the task with the greatest value of v_n / N_n^s is given priority. Different from Algorithm 2, we adopt a serial service mode, that is, only after completing the SFC instantiation of one task, we start to instantiate the SFC of the next task. Thus, the principle of selecting one optimal UAV for each SF is also different from that in Section 4.2, which is described as follows:

1. The UAV network first instantiates the SFC for a task with the greatest value of v_n / N_n^s. When facing multiple tasks with the same payment, the UAV randomly selects one to serve;

2. An SF is preferentially instantiated on the candidate UAV that does not occupy the sub-channel;

3. If there is no task that satisfies the above principle 2, the candidate UAVs of each task are divided into "rich candidate UAVs" and "poor candidate UAVs" according to the principle in Section 4.2. Then, we do the following:

   (a) When the number of "rich candidate UAVs" is greater than 0, the candidate UAV with the lowest performance is selected, and the high-performance UAVs are left for subsequent tasks with higher computation requirement, which is beneficial to maximizing $\mathcal{F}_2$;

   (b) When the number of "rich candidate UAVs" is equal to 0, the above operations are performed among "poor candidate UAVs".

Repeat the above operations for the required SFCs of all tasks are instantiated, whose pseudocode is as shown in Algorithm 3.

4.4. Computational Complexity Analysis

In order to show the feasibility and efficiency of the proposed ToRu algorithm, we focus on its time computational complexity in this section. As shown in Algorithm 1, when the idle resources of the UAV network are rich and the number of tasks requesting service is small, ToRu only executes Algorithm 2; otherwise, it first executes Algorithm 2, and then executes Algorithm 3. Next, we analyze the time complexity of Algorithm 2 and Algorithm 3, respectively. As shown in Algorithm 2, the maximum value of variable $|T_{temp}|$ is equal to N_t, and its value decreases with the increase of the control variable p. Therefore, the worst time complexity of Algorithm 2 is $O(L_{max}.N_t)$, where L_{max} represents the maximum number of SFs contained in an SFC, N_t represents the number of tasks requesting service. Similarly, the maximum value of variable N_n^s in Algorithm 3 is equal to L_{max}, so that the worst time complexity of Algorithm 3 is also $O(L_{max}.N_t)$. To sum up, the worst time complexity of the proposed ToRu algorithm is $O(L_{max}.N_t)$, which can obtain the sub-optimal solution to the problem P_1 in polynomial time.

5. Simulation and Results Analysis

This section first presents the experimental settings, then analyzes the results.

5.1. Experimental Settings

Platform Settings. All experiments were conducted on a PC that runs Ubuntu 18.04 with 3.2 GHz CPU and 16 GB RAM. The proposed ToRu algorithm was designed and implemented using C++ language.

Parameter Settings. Consider a service scenario with 25 UAVs that are 500 m apart. The number of tasks input into the simulation environment varies from 10 to 190. The main parameter settings are included in Table 2.

Table 2. Parameter settings.

Parameter	Value	Parameter	Value
M	25	Q	30
$f_{m,q}$	$[1,10]$ GHz	$a_{m,q}$	$[2,20]$ GOPS
P_m	1 W	N_0	10^{-20} W/Hz
B	1 MHz	β_0	1.42×10^{-4}
$r_n^{k,k+1}$	10 Mbit	N_m^c	8
$l_n^{k,0}$	$[0.1,10]$ Mbit	$l_n^{k,1}$	$[100,1,000,000]$ Cycles
N_n^s	$[2,5]$	$l_n^{k,2}$	$[200,2,000,000]$ OPs
N_m^{cpu}	$[10,20]$	N_m^{fpga}	$[10,20]$
$d_{n,m}$	500 m	v_n	$[2,20]$

Experimental process. When the number of input tasks is given, the geographic location of each task is generated randomly, and then the attributes of each task (such as task length, required SFC, task complexity, etc.) are generated randomly. Finally, each task randomly requests service from a UAV that covers it. In addition, for a given number of input tasks, we simulate 100 times and take the average of the simulation results as the final value.

Comparison Benchmarks. To validate the necessity of each component of the comparison benchmarks design, we adopt a step-by-step evaluation philosophy in the experimental design. For each benchmark algorithm, there are two steps: the order in which these SFCs are instantiated, and the principle of instantiating SI contained in an SFC. For the first step, similar to the greedy algorithm [20], two sorting strategies are chosen as comparison benchmarks: (1) Revenue: the task with highest payment is firstly served; (2) Length: the task with the shortest SFC length is firstly served. For the second step, three strategies are chosen as comparison benchmarks: (1) Random: an SF is instantiated on a random candidate UAV, similar to the random algorithm [32]; (2) Greedy: an SF is instantiated on a candidate UAV with the best performance, similar to the greedy algorithm [20]; (3) Local: an SF is only instantiated on a local UAV, similar to the local algorithm [32]. Similar to these step-wise algorithms [20,32], we have 6 combination algorithms for comparison, marked as "Revenue + Random", "Revenue + Greedy", "Revenue + Local", "Length + Random", "Length + Greedy", and "Length + Local".

1. "Revenue + Random": it first selects the task with the highest payment and performs SFC scheduling for it; next, when instantiating one SF contained in an SFC, it always randomly selects one from the candidate UAVs to instantiate this SF.
2. "Revenue + Greedy": this algorithm first selects the task with the highest payment and performs SFC scheduling for it; next, when instantiating one SF contained in an SFC, it always selects the best performance one from the candidate UAVs to instantiate this SF.
3. "Revenue + Local": this algorithm first selects the task with the highest payment and performs SFC scheduling for it; next, when instantiating one SF contained in an SFC, it always selects the local one from the candidate UAVs to instantiate this SF.
4. "Length + Random": this algorithm first selects the task with the shortest SFC length and performs SFC scheduling for it; next, when instantiating one SF contained in an SFC, it always randomly selects one from the candidate UAVs to instantiate this SF.
5. "Length + Greedy": this algorithm first selects the task with the shortest SFC length and performs SFC scheduling for it; next, when instantiating one SF contained in an SFC, it always selects the best performance one from the candidate UAVs to instantiate this SF.
6. "Length + Local": this algorithm first selects the task with the shortest SFC length and performs SFC scheduling for it; next, when instantiating one SF contained in an SFC, it always selects the local one from the candidate UAVs to instantiate this SF.

5.2. Results and Analysis

5.2.1. The Completion Time Sum of Tasks

Figure 5 shows the completion time sum under the different tasks offloaded to the UAV network. We can see that our proposed ToRu algorithm significantly outperforms other algorithms in the completion time sum of tasks at the first stage, i.e., the stage before the number of tasks reaches critical point. Figure 6 shows the simulation results of the first stage in more detail. This is mainly because on the one hand, we instantiate each SF of the SFCs of all tasks in a parallel mode; on the other hand, the pairing rules between SFs and their candidate UAVs are designed from the global perspective, and each SF considers the impact on other SFs when selecting candidate UAVs. Then, the performance of "Revenue + Greedy" and "Length + Greedy" comes second. It is also reasonable, since both algorithms greedily choose the UAV with the highest processing performance to instantiate the SFC. Lastly, the Algorithm "Revenue + Random", "Revenue + Local", "Length + Random", and "Length + Local" have the worst performance because they do not consider the effect of UAVs on the completion time when making their choices. However, with the increase of the number of tasks, the completion time sum of tasks also increases. This is because the length and type of newly added tasks are random, so UAV computation resources are used more fully and more tasks are executed successfully. Furthermore, as the number of tasks continues to increase, the completion time sum of tasks in all algorithms no longer increases, it even starts to decrease (e.g., "Length + Greedy"). The reason is that tasks with less execution time are prioritized. To sum up, when the number of tasks exceeds the critical point (i.e., tasks can not be executed 100%), it is meaningless to evaluate the completion sum of tasks.

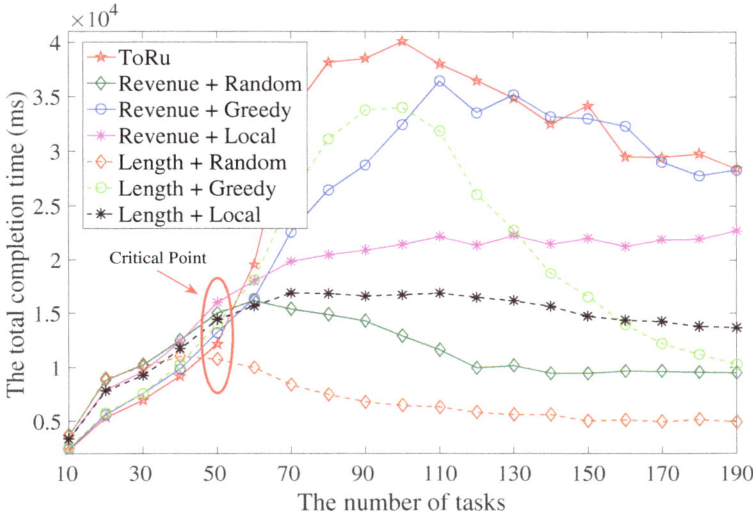

Figure 5. The completion time sum with different algorithms during the whole simulation phase.

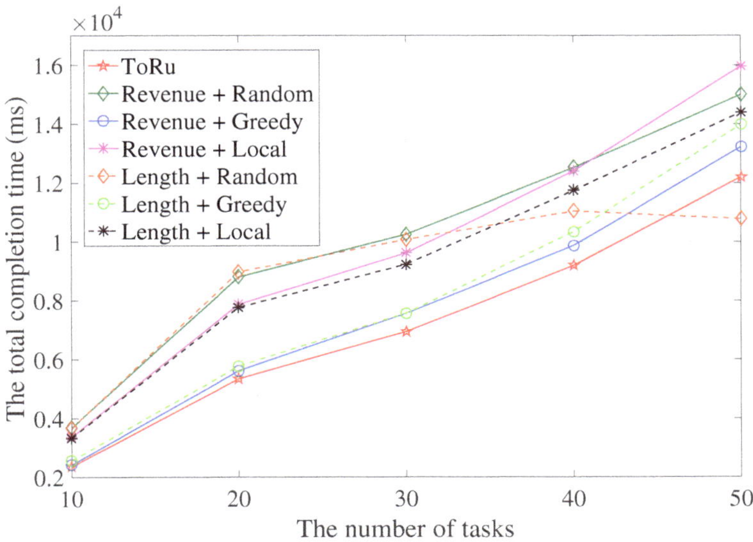

Figure 6. The completion time sum with different algorithms in the stage of abundant resources.

5.2.2. The Task Execution Success Ratio

Figure 7 shows the average execution success ratio of algorithms under different tasks. The result shows that algorithm ToRu has the highest success ratio, especially with the gradual increase of tasks, it can still maintain a high success ratio. This is mainly because ToRu only selects candidate UAVs with low performance to meet the requirements when instantiating SF, in other words, the improvement in the task execution success ratio comes at the expense of individual task execution performance. The algorithms "Length + Local" and "Revenue + Local" show the worst performance before the critical point, because they do not fully utilize the resources of the UAV network. With the increase of the number of tasks, "Length + Greedy" has the highest task execution success ratio. This is reasonable since it serves tasks with the shortest SFCs first, which consume less computation and communication resources. Finally, the algorithms "Length + Random" and "Revenue + Random" shows the poor performance when the network load is heavy.

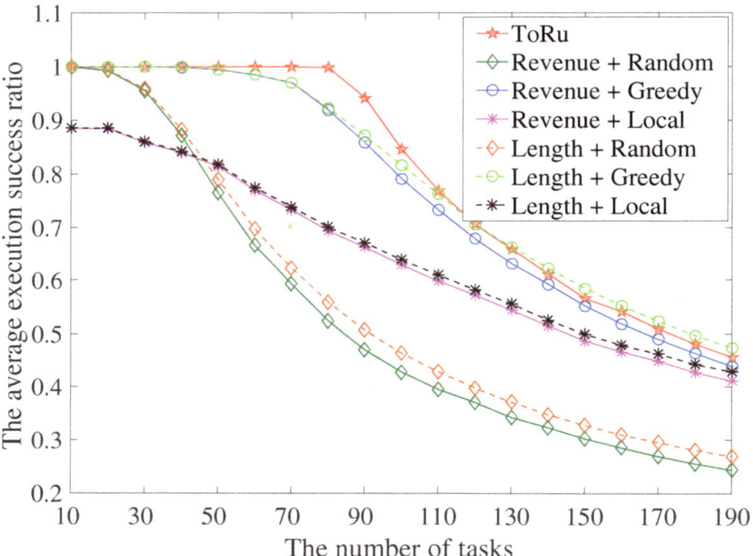

Figure 7. The task execution success ratio with different algorithms.

5.2.3. The Overall Revenue

Figure 8 presents the overall revenue under different tasks. As can be seen in Figure 7, when the number of tasks is small, the task execution success ratios of the algorithms except "Length + Local" and "Revenue + Local" are all 100%. Therefore, they have equal revenue. With the increase of tasks, the overall revenue of each algorithm increases rapidly before the critical point, then, the growth becomes slow. Moreover, the overall revenue of algorithm ToRu is always the largest among the seven algorithms because it not only ensures that tasks with higher payments are executed, but also ensures a higher task execution success rate. The algorithm "Revenue + Greedy" performs better, which is mainly because it first completes tasks that pay more. In addition, we can see that the performance of algorithms based on revenue sorting principle is better than that of algorithms based on length sorting, which is reasonable.

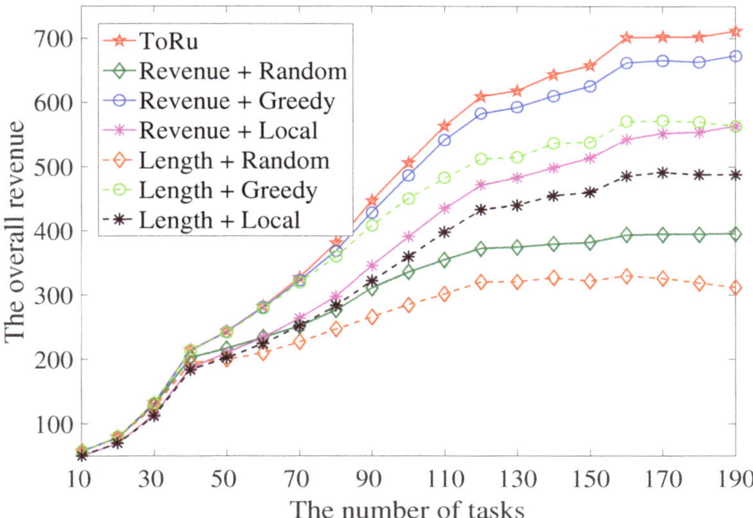

Figure 8. The overall revenue with different algorithms.

5.2.4. The Resources Utilization

Figures 9 and 10 show the utilization of channel and computation resources of different algorithms, respectively. We can find that the channel resources consumed by the algorithm ToRu increase with the number of tasks before the critical point. This is because the resources are sufficient before the critical point, and the algorithm ToRu seeks to minimize the completion time sum of tasks without caring about resource consumption. Due to the limited communication resources between UAVs, it often becomes the bottleneck of the task execution success ratio. When the number of tasks requesting service is small, Algorithm 2 is executed. It instantiates each requested SFC in parallel mode, that is, it selects the most preferred UAV for each SF contained in different SFCs at the same time, which will lead to the premature consumption of the most preferred UAVs early. Therefore, the probability that all SFs contained in one SFC are deployed on the same UAV will be reduced. Different SFs contained in one SFC have to interact with each other, resulting in the high channel utilization. On the contrary, when the number of tasks requesting service is large, Algorithm 3 is executed. It instantiates each requested SFC in serial mode, that is, it starts to instantiate the next SFC after having instantiated all SFs contained in the previous SFC, so that the probability that all SFs contained in one SFC are instantiated on the same UAV will be greatly increased. The interaction between different SFs on the same UAV will no longer consume channel resources, thus, the channel utilization will be rapidly reduced. As shown in Figure 9, after the critical point, the channel utilization in the algorithm ToRu decreases greatly and remains stable at a low value. This inevitably leads to the increase of the completion time sum of tasks (as shown in Figure 5), but it can improve the task execution success ratio (as shown in Figure 7). The algorithms "Revenue + Local" and "Length + Local" do not consume channels because they only execute tasks on the local UAV. The channel utilization of other algorithms increases sharply with the increase of the number of tasks, so that their computation resources cannot be fully utilized, as shown in Figure 10. This will reduce the task execution success ratio and the overall revenue, as shown in Figures 7 and 8. It is obvious that the computation utilization of the algorithm ToRu has reached 100%, which is the reason why its task execution success ratio and overall revenue are the highest.

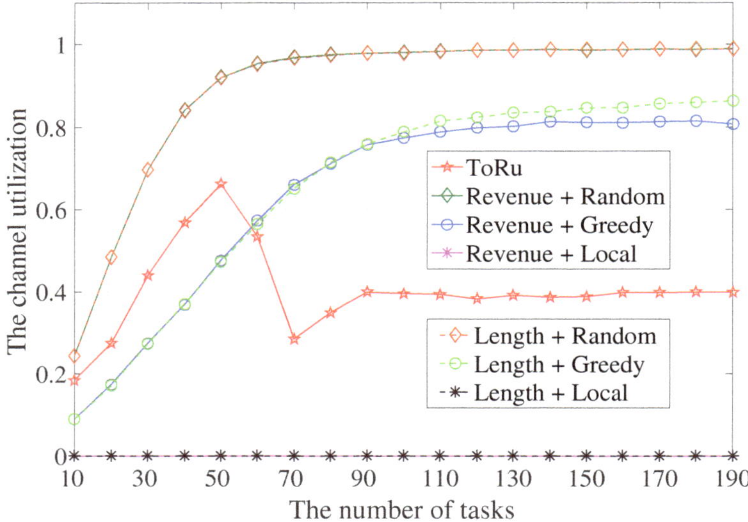

Figure 9. The channel utilization with different algorithms.

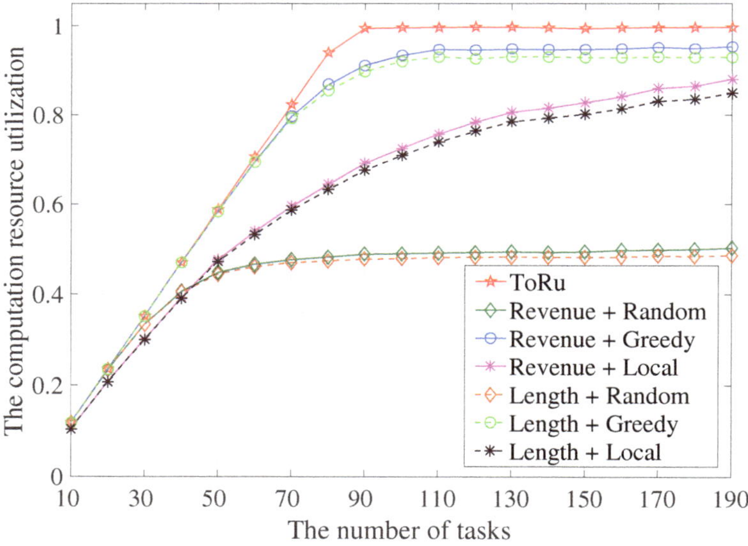

Figure 10. The computation resource utilization with different algorithms.

5.2.5. The Operation Time

Table 3 provides the operation time results of our proposed ToRu algorithm on average under different number of tasks requesting service. We can see that the operation time of ToRu is relatively small in the first stage (that is, before the number of tasks requesting service reaches "critical point", as shown in Figure 5). This is because as there are sufficient resources at this stage, ToRu can exit after executing Algorithm 2. When the number of tasks requesting service exceeds the "critical point" (i.e., 50), ToRu can judge that the resources are insufficient after executing Algorithm 2, and it then continues to execute Algorithm 3 for rescheduling SFCs. This undoubtedly increases the operation time, as shown in Table 3. However, as the number of tasks requesting service continues to increase, the shortage of resources will become more obvious, which can be easily judged by several loop operations of Algorithm 2. Therefore, the operation time of Algorithm 2 can be ignored, and the operation time of ToRu only includes the one of Algorithm 3. As shown in Table 3, the operation time of the ToRu algorithm decreased sharply after the number of tasks requesting service exceeds 110, and then maintained a stable small increase. This is consistent with our complexity analysis results in Section 4.4, indicating that our proposed algorithm has good execution efficiency and scalability.

Table 3. Operation time of ToRu algorithm.

Number of Tasks	Operation Time	Number of Tasks	Operation Time
$N_t = 20$	~ 2 ms	$N_t = 110$	~ 34 ms
$N_t = 30$	~ 6 ms	$N_t = 120$	~ 36 ms
$N_t = 40$	~ 14 ms	$N_t = 130$	~ 38 ms
$N_t = 50$	~ 32 ms	$N_t = 140$	~ 40 ms
$N_t = 60$	~ 50 ms	$N_t = 150$	~ 46 ms
$N_t = 70$	~ 90 ms	$N_t = 160$	~ 50 ms
$N_t = 80$	~ 95 ms	$N_t = 170$	~ 56 ms
$N_t = 90$	~ 110 ms	$N_t = 180$	~ 62 ms
$N_t = 100$	~ 126 ms	$N_t = 190$	~ 74 ms

6. Conclusions

This paper formulates the SFC scheduling problem as a 0–1 nonlinear integer programming problem in the multi-UAV edge computing network with CPU + FPGA computation architecture. A two-stage heuristic algorithm named ToRu is put forward to derive a sub-optimal solution of the problem. At the first stage, the SFCs of all tasks are scheduled to UAV edge servers in parallel based on the our proposed pairing principle between SFCs and UAVs for minimizing the completion time sum of tasks; at the second stage, a revenue maximization heuristic is adopted to schedule the arrived SFCs in a serial service method. A series of experiments were conducted to evaluate the performance of our proposal. The results show that our algorithm outperforms other benchmark algorithms in the completion time sum of tasks, the overall revenue, and the task execution success ratio.

The main limitation of ToRu algorithm lies in the fact that it is designed to realize the online long-term SFC scheduling based on the stable UAV network topology. In other words, it cannot be applied directly in the scenario where UAVs frequently join and exit. In our future work, we plan to design a supplemental algorithm with network topology prediction capability, which can help ToRu adapt to the dynamic scenario.

Author Contributions: Conceptualization, Y.W. and H.W.; methodology, Y.W. and X.W.; software, K.Z.; validation, J.F. and J.C.; formal analysis, Y.H.; investigation, R.J.; resources, K.Z.; data curation, K.Z.; writing—original draft preparation, Y.W.; writing—review and editing, X.W.; visualization, J.C.; supervision, H.W.; project administration, Y.H.. All authors have read and agreed to the published version of the manuscript.

Funding: This research was supported by the Natural Science Foundation of China under Grant No. 62171465.

Institutional Review Board Statement: Not applicable.

Informed Consent Statement: Not applicable.

Data Availability Statement: Not applicable.

Acknowledgments: The authors would like to thank all coordinators and supervisors involved and the anonymous reviewers for their detailed comments that helped to improve the quality of this article.

Conflicts of Interest: The authors declare no conflict of interest.

References

1. Chen, W.; He, R.; Wang, G.; Zhang, J.; Wang, F.; Xiong, K.; Ai, B.; Zhong, Z. Ai assisted phy in future wireless systems: Recent developments and challenges. *China Commun.* **2021**, *18*, 285–297. [CrossRef]
2. Sarikhani, R.; Keynia, F. Cooperative spectrum sensing meets machine learning: Deep reinforcement learning approach. *IEEE Commun. Lett.* **2020**, *24*, 1459–1462. [CrossRef]
3. Zheng, S.; Chen, S.; Qi, P.; Zhou, H.; Yang, X. Spectrum sensing based on deep learning classification for cognitive radios. *China Commun.* **2020**, *17*, 138–148. [CrossRef]
4. Xie, J.; Liu, C.; Liang, Y.-C.; Fang, J. Activity pattern aware spectrum sensing: A cnn-based deep learning approach. *IEEE Commun. Lett.* **2019**, *23*, 1025–1028. [CrossRef]
5. Deng, S.; Zhao, H.; Fang, W.; Yin, J.; Dustdar, S.; Zomaya, A.Y. Edge intelligence: The confluence of edge computing and artificial intelligence. *IEEE Internet Things J.* **2020**, *7*, 7457–7469. [CrossRef]
6. Wang, J.; Wei, X.; Fan, J.; Duan, Q.; Liu, J.; Wang, Y. Request pattern change-based cache pollution attack detection and defense in edge computing. *Digit. Commun. Netw.* **2022**. [CrossRef]
7. Liu, Z.; Cao, Y.; Gao, P.; Hua, X.; Zhang, D.; Jiang, T. Multi-uav network assisted intelligent edge computing: Challenges and opportunities. *China Commun.* **2022**, *19*, 258–278. [CrossRef]
8. Wu, W.; Zhou, F.; Wang, B.; Wu, Q.; Dong, C.; Hu, R.Q. Unmanned Aerial Vehicle Swarm-Enabled Edge Computing: Potentials, Promising Technologies, and Challenges. *IEEE Wirel. Commun.* **2022**, *29*, 78–85. [CrossRef]
9. Zhao, N.; Lu, W.; Sheng, M.; Chen, Y.; Tang, J.; Yu, F.R.; Wong, K.K. Uav-assisted emergency networks in disasters. *IEEE Wirel. Commun.* **2019**, *26*, 45–51. [CrossRef]
10. Cao, B.; Li, M.; Liu, X.; Zhao, J.; Cao, W.; Lv, Z. Many-Objective Deployment Optimization for a Drone-Assisted Camera Network. *IEEE Trans. Netw. Sci. Eng.* **2021**, *8*, 2756–2764. [CrossRef]
11. Wang, X.; Han, Y.; Leung, V.C.M.; Niyato, D.; Yan, X.; Chen, X. Convergence of edge computing and deep learning: A comprehensive survey. *IEEE Commun. Surv. Tutor.* **2020**, *22*, 869–904. [CrossRef]
12. Dong, C.; Shen, Y.; Qu, Y.; Wang, K.; Zheng, J.; Wu, Q.; Wu, F. Uavs as an intelligent service: Boosting edge intelligence for air-ground integrated networks. *IEEE Netw.* **2021**, *35*, 167–175. [CrossRef]
13. Behravesh, R.; Harutyunyan, D.; Coronado, E.; Riggio, R. Time-sensitive mobile user association and sfc placement in mec-enabled 5g networks. *IEEE Trans. Netw. Serv. Manag.* **2021**, *18*, 3006–3020. [CrossRef]
14. Xu, Z.; Gong, W.; Xia, Q.; Liang, W.; Rana, O.F.; Wu, G. Nfv-enabled iot service provisioning in mobile edge clouds. *IEEE Trans. Mob. Comput.* **2021**, *20*, 1892–1906. [CrossRef]
15. Wang, Y.; Wei, X.; Wang, H.; Fan, J.; Chen, J.; Zhao, K.; Hu, Y. Joint UAV deployment, SF placement, and collaborative task scheduling in heterogeneous multi-UAV-empowered edge intelligence. *IET Commun. Early Access Artic.* **2023**. [CrossRef]
16. Xu, C.; Jiang, S.; Luo, G.; Sun, G.; An, N.; Huang, G.; Liu, X. The case for fpga-based edge computing. *IEEE Trans. Mob. Comput.* **2022**, *21*, 2610–2619. [CrossRef]
17. Li, J.; Un, K.-F.; Yu, W.-H.; Mak, P.-I.; Martins, R.P. An fpga-based energy-efficient reconfigurable convolutional neural network accelerator for object recognition applications. *IEEE Trans. Circuits Syst. Express Briefs* **2021**, *68*, 3143–3147. [CrossRef]
18. Yu, X.; Niu, W.; Zhu, Y.; Zhu, H. UAV-assisted cooperative offloading energy efficiency system for mobile edge computing. *Digit. Commun. Netw.* **2022**. [CrossRef]
19. Zhang, J.; Zhou, L.; Zhou, F.; Seet, B.-C.; Zhang, H.; Cai, Z.; Wei, J. Computation-efficient offloading and trajectory scheduling for multi-uav assisted mobile edge computing. *IEEE Trans. Veh. Technol.* **2020**, *69*, 2114–2125. [CrossRef]
20. Luo, Y.; Ding, W.; Zhang, B. Optimization of task scheduling and dynamic service strategy for multi-uav-enabled mobile-edge computing system. *IEEE Trans. Cogn. Commun. Netw.* **2021**, *7*, 970–984. [CrossRef]
21. Wang, Y.; Ru, Z.-Y.; Wang, K.; Huang, P.-Q. Joint deployment and task scheduling optimization for large-scale mobile users in multi-uav-enabled mobile edge computing. *IEEE Trans. Cybern.* **2020**, *50*, 3984–3997. [CrossRef]
22. Chang, H.; Chen, Y.; Zhang, B.; Doermann, D. Multi-uav mobile edge computing and path planning platform based on reinforcement learning. *IEEE Trans. Emerg. Top. Comput. Intell.* **2022**, *6*, 489–498. [CrossRef]
23. Ren, T.; Niu, J.; Dai, B.; Liu, X.; Hu, Z.; Xu, M.; Guizani, M. Enabling efficient scheduling in large-scale uav-assisted mobile-edge computing via hierarchical reinforcement learning. *IEEE Internet Things J.* **2022**, *9*, 7095–7109. [CrossRef]
24. Xue, J.; Wu, Q.; Zhang, H. Cost optimization of UAV-MEC network calculation offloading: A multi-agent reinforcement learning method. *Ad Hoc Netw.* **2022**, *136*, 102981. [CrossRef]
25. Seid, A.M.; Boateng, G.O.; Mareri, B.; Sun, G.; Jiang, W. Multi-Agent DRL for Task Offloading and Resource Allocation in Multi-UAV Enabled IoT Edge Network. *IEEE Trans. Netw. Serv. Manag.* **2021**, *18*, 4531–4547. [CrossRef]
26. Wu, Z.; Yang, Z.; Yang, C.; Lin, J.; Liu, Y.; Chen, X. Joint deployment and trajectory optimization in UAV-assisted vehicular edge computing networks. *J. Commun. Netw.* **2022**, *24*, 47–58. [CrossRef]
27. Moura, J.; Hutchison, D. Game Theory for Multi-Access Edge Computing: Survey, Use Cases, and Future Trends. *IEEE Commun. Surv. Tutor.* **2019**, *2*, 260–288. [CrossRef]
28. Liu, L.; Zhang, S.; Zhang, L.; Pan, G.; Yu, J. Multi-UUV Maneuvering Counter-Game for Dynamic Target Scenario Based on Fractional-Order Recurrent Neural Network. *IEEE Trans. Cybern. Access Artic.* **2022**. [CrossRef]

29. Asheralieva, A.; Niyato, D. Hierarchical game-theoretic and reinforcement learning framework for computational offloading in uav-enabled mobile edge computing networks with multiple service providers. *IEEE Internet Things J.* **2019**, *6*, 8753–8769. [CrossRef]

30. Wu, Q.; Chen, J.; Xu, Y.; Qi, N.; Fang, T.; Sun, Y.; Jia, L. Joint Computation Offloading, Role, and Location Selection in Hierarchical Multicoalition UAV MEC Networks: A Stackelberg Game Learning Approach. *IEEE Internet Things J.* **2022**, *9*, 18293–18304. [CrossRef]

31. Zhou, H.; Wang, Z.; Min, G.; Zhang, H. UAV-aided Computation Offloading in Mobile Edge Computing Networks: A Stackelberg Game Approach. *IEEE Internet Things J. Early Access Artic.* **2022**. [CrossRef]

32. Qu, Y.; Dai, H.; Wang, H.; Dong, C.; Wu, F.; Guo, S.; Wu, Q. Service provisioning for uav-enabled mobile edge computing. *IEEE J. Sel. Areas Commun.* **2021**, *39*, 3287–3305. [CrossRef]

33. Wang, G.; Zhou, S.; Zhang, S.; Niu, Z.; Shen, X. SFC-Based Service Provisioning for Reconfigurable Space-Air-Ground Integrated Networks. *IEEE J. Sel. Areas Commun.* **2020**, *38*, 1478–1489. [CrossRef]

34. Li, J.; Shi, W.; Wu, H.; Zhang, S.; Shen, X. Cost-Aware Dynamic SFC Mapping and Scheduling in SDN/NFV-Enabled Space–Air–Ground-Integrated Networks for Internet of Vehicles. *IEEE Internet Things J.* **2022**, *9*, 5824–5838. [CrossRef]

35. Xia, J.; Wang, P.; Li, B.; Fei, Z. Intelligent task offloading and collaborative computation in multi-UAV-enabled mobile edge computing. *China Commun.* **2022**, *19*, 244–256. [CrossRef]

36. Liu, S.; Yang, T. Delay aware scheduling in uav-enabled ofdma mobile edge computing system. *IET Commun.* **2020**, *14*, 3203–3211. [CrossRef]

37. Tan, G.; Shui, C.; Wang, Y.; Yu, X.; Yan, Y. Optimizing the linpack algorithm for large-scale pcie-based cpu-gpu heterogeneous systems. *IEEE Trans. Parallel Distrib. Syst.* **2021**, *32*, 2367–2380. [CrossRef]

38. Kowsalya, T. Area and power efficient pipelined hybrid merged adders for customized deep learning framework for FPGA implementation. *Microprocess. Microsyst.* **2020**, *72*, 102906. [CrossRef]

MDPI

St. Alban-Anlage 66

4052 Basel

Switzerland

www.mdpi.com

Drones Editorial Office

E-mail: drones@mdpi.com

www.mdpi.com/journal/drones